Walter Greiner

CLASSICAL ELECTRODYNAMICS

Foreword by D. Allan Bromley

With 284 Figures

Springer

Walter Greiner
Institut für Theoretische Physik
Johann Wolfgang Goethe-Universität
Robert Mayer Strasse 10
Postfach 11 19 32
D-60054 Frankfurt am Main
Germany

Library of Congress Cataloging-in-Publication Data
Greiner, Walter, 1935–
    [Klassische Elektrodynamik. English]
    Classical electrodynamics/Walter Greiner.
        p.   cm.—(Classical theoretical physics)
    Includes bibliographical references and index.
    ISBN 0-387-94799-X (softcover: alk. paper)
    1. Electrodynamics.   I. Title.   II. Series.
    QC631.G79513   1996
    537.6—dc20   96-15530

Printed on acid-free paper.

Production managed by Francine McNeill; manufacturing supervised by Joe Quatela.
Photocomposed copy prepared from the author's LaTeX files.
Printed and bound by R.R. Donnelley and Sons, Harrisonburg, VA.
Printed in the United States of America.

9 8 7 6 5 4 3 2 1

ISBN 0-387-94799-X Springer-Verlag New York Berlin Heidelberg   SPIN 10539352

Foreword

More than a generation of German-speaking students around the world have worked their way to an understanding and appreciation of the power and beauty of modern theoretical physics—with mathematics, the most fundamental of sciences—using Walter Greiner's textbooks as their guide.

The idea of developing a coherent, complete presentation of an entire field of science in a series of closely related textbooks is not a new one. Many older physicians remember with real pleasure their sense of adventure and discovery as they worked their ways through the classic series by Sommerfeld, by Planck, and by Landau and Lifshitz. From the students' viewpoint, there are a great many obvious advantages to be gained through the use of consistent notation, logical ordering of topics, and coherence of presentation; beyond this, the complete coverage of the science provides a unique opportunity for the author to convey his personal enthusiasm and love for his subject.

These volumes on classical physics, finally available in English, complement Greiner's texts on quantum physics, most of which have been available to English-speaking audiences for some time. The complete set of books will thus provide a coherent view of physics that includes, in classical physics, thermodynamics and statistical mechanics, classical dynamics, electromagnetism, and general relativity; and in quantum physics, quantum mechanics, symmetries, relativistic quantum mechanics, quantum electro- and chromodynamics, and the gauge theory of weak interactions.

What makes Greiner's volumes of particular value to the student and professor alike is their completeness. Greiner avoids the all too common "it follows that ...," which conceals several pages of mathematical manipulation and confounds the student. He does not hesitate to include experimental data to illuminate or illustrate a theoretical point, and these data, like the theoretical content, have been kept up to date and topical through frequent revision and expansion of the lecture notes upon which these volumes are based.

Moreover, Greiner greatly increases the value of his presentation by including something like one hundred completely worked examples in each volume. Nothing is of greater importance to the student than seeing, in detail, how the theoretical concepts and tools

v

under study are applied to actual problems of interest to working physicists. And, finally, Greiner adds brief biographical sketches to each chapter covering the people responsible for the development of the theoretical ideas and/or the experimental data presented. It was Auguste Comte (1789–1857) in his *Positive Philosophy* who noted, "To understand a science it is necessary to know its history." This is all too often forgotten in modern physics teaching, and the bridges that Greiner builds to the pioneering figures of our science upon whose work we build are welcome ones.

Greiner's lectures, which underlie these volumes, are internationally noted for their clarity, for their completeness, and for the effort that he has devoted to making physics an integral whole. His enthusiasm for his sciences is contagious and shines through almost every page.

These volumes represent only a part of a unique and Herculean effort to make all of theoretical physics accessible to the interested student. Beyond that, they are of enormous value to the professional physicist and to all others working with quantum phenomena. Again and again, the reader will find that, after dipping into a particular volume to review a specific topic, he or she will end up browsing, caught up by often fascinating new insights and developments with which he or she had not previously been familiar.

Having used a number of Greiner's volumes in their original German in my teaching and research at Yale, I welcome these new and revised English translations and would recommend them enthusiastically to anyone searching for a coherent overview of physics.

D. Allan Bromley
Henry Ford II Professor of Physics
Yale University
New Haven, Connecticut, USA

Preface

Classical Electrodynamics contains the lectures that form part of the course of study in theoretical physics at the Johann Wolfgang Goethe University in Frankfurt am Main. There they are given for students in physics and mathematics in their third semester and are preceded by Theoretical Mechanics I (first semester) and Theoretical Mechanics II (second semester). Quantum Mechanics I—An Introduction then completes the first part of the lectures. Graduate course work continues with Quantum Mechanics II—Symmetries and Relativistic Quantum Mechanics (fifth semester), Thermodynamics and Statistics, Quantum Electrodynamics, Field Quantization, Gauge Theory of Weak Interaction, Quantum Chromodynamics, General Relativity and Cosmology, Nuclear and Solid State Theory, and other, more specialized courses in Many Particle Theory, etc.

As in all other fields mentioned, we present classical electrodynamics according to the inductive method that comes closest to the methodology of the research physicist. Starting from some key experimental observations, the framework of the theory is developed step by step, and after the basic equations, that is, the Maxwell equations, have been obtained, new phenomena are investigated from thereon.

This leads to electrostatics and magnetostatics and their application to macroscopic problems and further to the theory of electromagnetic waves in vacuum, which are among the most fascinating consequences of Maxwell's equations. We follow Maxwell theory with respect to basic, field theoretical questions (energy, momentum of the field) and its application to establish optics (laws of reflection and refraction, frequency dependency and conductivity, polarization and index of refraction), as well as in the sector of practical application (propagation of waves, wave guides, resonance cavity, etc.). Also, the covariant formulation of electrodynamics in the framework of the theory of special relativity is presented, ending with the relativistically covariant Lagrange formalism. Many worked-out examples and exercises illustrate the general theory and its applications.

Finally, biographical and historical footnotes as well as an extra section on the history of electrodynamics anchor the scientific development within the general context of scientific progress and evolution. In this context, I thank the publishers Harri Deutsch and

F.A. Brockhaus (*Brockhaus Enzyklopädie*, F.A. Brockhaus, Wiesbaden—marked by BR) for giving permission to extract the biographical data of physicists and mathematicians from their publications.

The lectures are now up for their fifth German edition. Over the years, many students and collaborators have helped to work out exercises and illustrative examples. For the first English edition, I enjoyed the help of Ulrich Eichmann, Nils Hammon, Oliver Martin, and Panajotis Papazoglou. The coordinatory help of Sven Soff is particularly appreciated.

Finally, I am pleased to acknowledge the agreeable collaboration with Dr. Thomas von Foerster and his team at Springer-Verlag New York, Inc. The English manuscript was copyedited by Margaret Marynowski, and the production of the book was supervised by Francine McNeill.

Walter Greiner
Johann Wolfgang Goethe-Universität
Frankfurt am Main

Contents

Foreword   v

Preface   vii

I Electrostatics   1

1 Introduction and Fundamental Concepts   3

2 Green's Theorems   45

3 Orthogonal Functions and Multipole Expansion:
Mathematical Supplement   70

4 Elementary Considerations on Function Theory:
Mathematical Supplement   100

II Macroscopic Electrostatics   121

5 The Field Equations for Space Filled with Matter   123

6 Simple Dielectrics and the Susceptibility   132

7 Electrostatic Energy and Forces in a Dielectric   156

III Magnetostatics

179

8 Foundations of Magnetostatics 181

9 The Vector Potential 205

10 Magnetic Moment 213

11 The Magnetic Field in Matter 222

IV Electrodynamics

235

12 Faraday's Law of Induction 237

13 Maxwell's Equations 250

14 Quasi-Stationary Currents and Current Circuits 276

15 Electromagnetic Waves in Vacuum 302

16 Electromagnetic Waves in Matter 316

17 Index of Reflection and Refraction 333

18 Wave Guides and Resonant Cavities 354

19 Light Waves 379

20 Moving Charges in Vacuum 409

21 The Hertzian Dipole 429

22 Covariant Formulation of Electrodynamics 458

23 Relativistic-Covariant Lagrangian Formalism 478

24 Systems of Units in Electrodynamics: Supplement 494

25 About the History of Electrodynamics 499

Index 551

PART I

ELECTROSTATICS

1 Introduction and Fundamental Concepts

In the investigation of the properties of charged bodies at rest the following results have been obtained experimentally: charged bodies (charges) exert a force on each other. There are two kinds of charges, positive and negative ones. Unlike charges attract each other, like charges repel each other. The force between two charges q_1 and q_2 is proportional to their product:

$$F_{12} \sim q_1 q_2$$

The force decreases with the square of the mutual distance, that is,

$$F_{12} \sim \frac{1}{|\mathbf{r}_1 - \mathbf{r}_2|^2}$$

The electrostatic forces are central forces. Thus, for the force exerted by the charge 2 on the charge 1 we can write

$$\mathbf{F}_{12} = k q_1 q_2 \frac{\mathbf{r}_1 - \mathbf{r}_2}{|\mathbf{r}_1 - \mathbf{r}_2|^3} \tag{1.1}$$

k is a constant of proportionality still to be fixed. See Figure 1.1. This equation for the

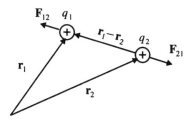

Figure 1.1. On Coulomb's law: Like charges repel each other.

force acting between two charges is called *Coulomb's law*. Furthermore, the principle of *superposition* is valid: The electric forces exerted by several charges q_2, q_3, ... on a test charge q_1 superpose each other undisturbed without the force between q_1 and a certain charge (e.g., q_2) being changed due to the presence of other charges. In particular, this implies that the forces between charges can be merely *two-body forces*; many-body forces do not occur. For *many-body forces* the force between two bodies 1 and 2 depends also on the positions of the other bodies r_3, r_4, ... For example, a three-body force would be

$$\mathbf{F}_{12} = kq_1q_2 \frac{\mathbf{r}_1 - \mathbf{r}_2}{\left| (\mathbf{r}_1 - \mathbf{r}_2)\left(1 + \frac{q_3^2}{q_1q_2}\frac{|\mathbf{r}_1 - \mathbf{r}_2|}{|\mathbf{r}_s - \mathbf{r}_3|^3}\right)\right|^3} \tag{1.2}$$

Here, $\mathbf{r}_s$ is the center of gravity between q_1 and q_2. See Figure 1.2. This three-body force would tend to a two-body force (1.1) as $r_3 \to \infty$, as should be. Microscopically, one can imagine that the force (i.e., a force field) originates from the virtual exchange of particles. These are thrown back and forth between the centers, like tennis balls, and in this way they bind the centers to each other. For two-body forces, this exchange proceeds between two centers only; for three- (many-)body forces, a detour via the third center (or several centers) occurs. (See Figure 1.3.) In the Coulomb interaction photons are exchanged; in the weak interaction, Z- and W-bosons; in the gravitational interaction, gravitons; and in the strong (nuclear) interaction, π-mesons (or, on a deeper level, gluons). The photons and gravitons have a rest mass equal to zero. Therefore, these forces are of infinite range. On the other hand, the short range of the strong interaction ($\sim 2\text{fm} = 2 \cdot 10^{13}\text{cm}$) is based on the finite rest mass of the π-meson. Nowadays, we know that pions and nucleons are built up out of quarks. The quarks interact by the exchange of heavy photons (interacting intensively among each other and coupling to the so-called glue-balls). These heavy photons are called

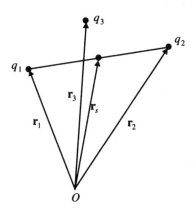

Figure 1.2. For the explanation of a three-body force: The charges q_i are placed at the position vectors r_i. The vector to the center of gravity of the charges q_1 and q_2 is r_s.

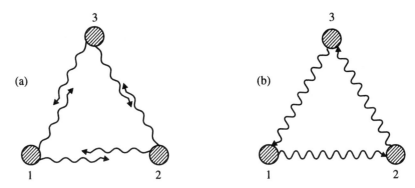

Figure 1.3. (a) Exchange of particles in a two-body interaction. (b) Exchange of particles in a three-body force.

gluons. For the ordinary Coulomb force in the presence of further charges q_i the force exerted on charge q_1 reads

$$\mathbf{F} = kq_1 \sum_{i=2}^{N} q_i \frac{\mathbf{r}_1 - \mathbf{r}_i}{|\mathbf{r}_1 - \mathbf{r}_i|^3} \tag{1.3}$$

In this form, Coulomb's law is valid exactly only for point charges and for uniformly charged spherical bodies. For charges of arbitrary shape, deviations appear which will be discussed later on. Nevertheless, one should wonder about the $1/r^2$-dependence of the Coulomb force. This particular force law is related to the fact that the rest mass of the photons exchanged by the charges is zero. According to Heisenberg's uncertainty relation they can then be produced virtually with a long range R. The uncertainty relation states

$$\Delta E \Delta t \sim \hbar, \quad \Delta E = \text{uncertainty of energy} \approx \mu c^2$$
$$\Delta t \sim \frac{\hbar}{\Delta E}, \quad \Delta t = \text{uncertainty of time} \approx \frac{\hbar}{\mu c^2} \tag{1.4}$$

The latter expression gives the lifetime of a virtual particle having the rest mass $\Delta E \approx \mu c^2$, and thus the range is

$$R = c\Delta t \sim \frac{\hbar c}{\Delta E}$$

This explains the long range of the Coulomb force. If the photon had the rest mass μ, then the Coulomb potential (compare the following pages) would have to be of the Yukawa type, namely,

$$V(r) \sim \frac{e^{-r/\lambda}}{r} \tag{1.5}$$

Here, $\lambda = h/\mu c$ would be the so-called Compton wavelength of a photon having the rest mass μ. For $\mu = 0$ one obtains the Coulomb potential of a point charge. Nowadays, the best precision measurements of the photon mass yield a value of $\mu c^2 \leq 5 \cdot 10^{-16} \text{eV}$.[1] The charge represents a new physical property of bodies. Now, we may introduce a new dimension for the charge, or express it in terms of the dimensions mass, length, and time used in mechanics. The product kq_1q_2 is fixed in equation (1.1). Subject to this condition the dimension of the single factors, charge and constant of proportionality, can still be chosen freely. *Depending on the choice of k we get different systems of units.* In textbooks, mainly two different systems of units are still used nowadays, the Gaussian and the rationalized system of units. In the *Gaussian system of units* the constant of proportionality k takes the numerical value 1 and remains nondimensional. Then, the charge is no longer an independent unit. In the CGS-system, one obtains from equation (1.1) the unit $1 \text{cm}^{3/2} \text{g}^{1/2} \text{s}^{-1}$ for the charge, which is also denoted the electrostatic unit (esu) or statCoulomb. This explicit tracing back of electromagnetic quantities to mechanical units can be found virtually only in older textbooks; in more recent textbooks on atomic and nuclear physics or quantum mechanics using the Gaussian system of units the charge is handled like an independent unit; thereby the physical interrelations often become clearer. Setting $|\mathbf{r}_1 - \mathbf{r}_2| = r$, equation (1.1) takes the simple form:

$$F = \frac{q_1 q_2}{r^2}$$

$$(1.6)$$

The opposite line is taken in the so-called rationalized system of units. Here, the unit is fixed by the charge. Its value is determined by measuring the force exerted by two current-carrying conductors on each other. According to the definition, when a current of one Ampere is flowing through two parallel, rectilinear, infinitely long conductors placed at a distance of one meter from each other, a force of $2 \cdot 10^{-7}$ newton per meter of their length acts between them. The product of current and time gives the quantity of charge:

   1 Coulomb (C) = 1 Ampere second (As)

By this (arbitrary) definition, the constant of proportionality k takes a dimension as well as a fixed numerical value; one sets

$$k = \frac{1}{4\pi \epsilon_0}$$

$$(1.7)$$

The constant ϵ_0 is called the *permittivity of vacuum*; it has the value

$$\epsilon_0 = 8.854 \cdot 10^{-12} \left[\frac{\text{As}}{\text{Vm}} \right] \approx \frac{1}{4\pi \cdot 9 \cdot 10^9} \left[\frac{\text{As}}{\text{Vm}} \right]$$

$$(1.8)$$

---

[1] We refer to the seminar held by W. Martienssen and his graduate students P. Kurowski and J. Wagner; prepr. Physikalisches Institut, Universität Frankfurt/M. (1974). A lab test of Coulomb's law is described by E.R. Williams, J.E. Faller, and H.A. Hill in "New Experimental Test of Coulomb's Law: A Laboratory Upper Limit on the Photon Rest Mass," *Phys. Rev. Letters* **26** (1971) 721.

(For the moment, we consider the unit volt (V) as an abbreviation of $1V = 1A^{-1}m^2kgs^{-3} = 1NmC^{-1}$.) In the framework of this system of units, Coulomb's law reads

$$F = \frac{1}{4\pi\epsilon_0} \cdot \frac{q_1 q_2}{r^2} \tag{1.9}$$

A comparison of (1.9) and (1.6) then yields the relation between the charges in the Gaussian system of units (q) and in the rationalized system of units (q^*) (also called the mksA— (meter kilogram second Ampere)—system). It reads

$$q = \frac{1}{\sqrt{4\pi\epsilon_0}}q^* \tag{1.10}$$

The unit of q^* is 1 Coulomb = 1 Ampere second. In the Gaussian system, this corresponds to

$$1\,\text{Coulomb} \cdot \frac{1}{\sqrt{4\pi\epsilon_0}} \approx 1\text{As}\frac{1}{\sqrt{4\pi\frac{1}{4\pi 9\cdot10^9}\frac{\text{As}}{\text{Vm}}}}$$

$$= 1\text{As}\sqrt{9\cdot10^9\frac{\text{Vm}}{\text{As}}} = \sqrt{9\cdot10^9\,\text{Vm As}}$$

$$= \sqrt{9\cdot10^9\frac{\text{m}^3\text{kg}}{\text{s}^2}} = \sqrt{9\cdot10^{18}\frac{\text{cm}^3\,\text{g}}{\text{s}^2}}$$

$$= 3\cdot10^9\sqrt{\text{erg cm}} = 3\cdot10^9\text{cgs charge units}$$

$$\equiv 3\cdot10^9\,\text{statCoulomb} \tag{1.11}$$

In macroscopic physics and experimental physics, the rationalized system of units is used predominantly. In atomic physics, nuclear physics, and many textbooks in theoretical physics, the Gaussian system of units is used mostly. *Here, we will use exclusively the Gaussian system of units.*

The electric field intensity

To explain the notion of the electric field intensity, we start from the force **F** exerted by a charge q_1 on a test charge q that is as small as possible. The field intensity caused by q_1 at the position **r** of the charge q is defined by the quotient:

$$\mathbf{E}(\mathbf{r}) = \frac{\mathbf{F}}{q} \tag{1.12}$$

Since, in general, the electric field is altered by the test charge q, we take the limit of an infinitely small charge:

$$\mathbf{E} = \lim_{\Delta q\to 0}\frac{\Delta\mathbf{F}}{\Delta q} = \frac{d\mathbf{F}}{dq} \tag{1.13}$$

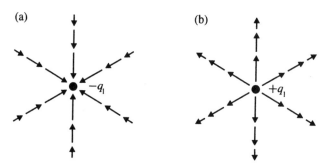

Figure 1.4. (a) **E**-field of a negative point charge. (b) **E**-field of a positive point charge.

With Coulomb's law (1.1) the electric field of a point charge q_1 is

$$\mathbf{E}(\mathbf{r}) = \frac{q_1(\mathbf{r} - \mathbf{r}_1)}{|\mathbf{r} - \mathbf{r}_1|^3} \tag{1.14}$$

The electric field vector $\mathbf{E}(\mathbf{r})$ of a positive point charge is directed radially outward; that of a negative point charge is directed radially inward, as depicted in Figure 1.4. According to the *principle of superposition* (1.3), for a system of point charges we have

$$\mathbf{E}(\mathbf{r}) = \sum_i \frac{q_i(\mathbf{r} - \mathbf{r}_i)}{|\mathbf{r} - \mathbf{r}_i|^3} = \sum_i \mathbf{E}_i \tag{1.15}$$

This is shown in Figure 1.5. In the presence of a continuous charge distribution, we have to go from a summation over the point charges to an integration over the spatial distribution (see Figure 1.6). Instead of the point charge q_i we have to insert the charge element $\rho(\mathbf{r}') \, dV'$. Here, ρ and dV are the charge density and the volume element, respectively:

$$\mathbf{E}(\mathbf{r}) = \int \rho(\mathbf{r}') \frac{\mathbf{r} - \mathbf{r}'}{|\mathbf{r} - \mathbf{r}'|^3} dV' \tag{1.16}$$

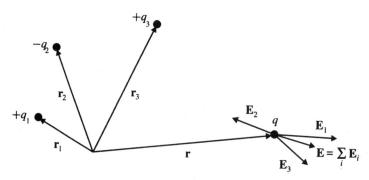

Figure 1.5. The electric field intensity at the position **r** for a sum of point charges q_i.

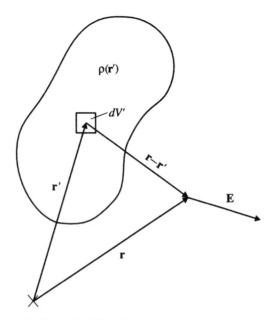

Figure 1.6. The electric field intensity **E(r)** at the position **r** for a continuous charge distribution $\rho(\mathbf{r}')$.

Tacitly, we have assumed the principle of superposition to be valid. But its validity is not self-evident. As we will see later, it is identical to the assumption that the fundamental electromagnetic equations are linear equations.

Gauss' law

We can determine the *flux of the field intensity produced by a point charge through a surface s* enclosing this charge.

Suppose that the point charge q is placed at the origin of the coordinate system; then the flux through the surface element da is given by

$$\mathbf{E} \cdot \mathbf{n}\, da = \frac{q}{r^2} \frac{\mathbf{r}}{r} \cdot \mathbf{n}\, da \tag{1.17}$$

where **n** is the normal vector to the surface (see Figure 1.7). But, the field intensity crossing the area $\mathbf{n}\, dF$ is equal to that crossing the area $\cos\theta\, da$. Expressed in terms of the solid angle, this implies (Figures 1.8 and 1.9):

$$\cos\theta\, da = r^2\, d\Omega \tag{1.18}$$

Therefore,

$$\mathbf{E} \cdot \mathbf{n}\, da = \frac{q}{r^2} \cos\theta\, da = \frac{q}{r^2} r^2\, d\Omega = q\, d\Omega \tag{1.19}$$

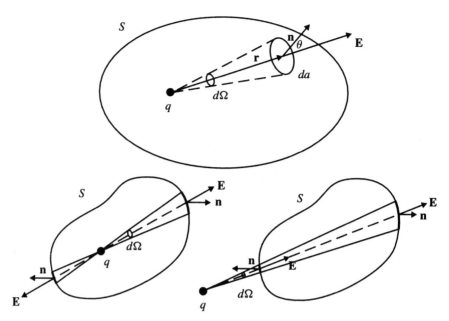

Figure 1.7. On Gauss' law: The normal component of the electric field is integrated over the surface s. If the charge q is located inside (outside) s, the total solid angle about the charge is equal to $4\pi(0)$.

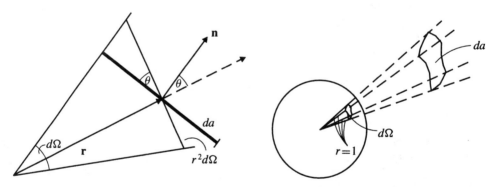

Figure 1.8. Projection of the area element da.   **Figure 1.9.** The solid angle $d\Omega$ is the central projection of the area da onto the unit sphere.

The flux crossing the surface is obtained by integration

$$\oint_s \mathbf{E} \cdot \mathbf{n}\, da = q \oint_s d\Omega = 4\pi q \tag{1.20}$$

because the integration over the solid angle yields 4π. So, we obtain the result

$$\oint_s \mathbf{E} \cdot \mathbf{n}\, da = \begin{cases} 4\pi q & \text{for } q \text{ inside the closed surface} \\ 0 & \text{for } q \text{ outside the closed surface} \end{cases} \tag{1.21}$$

The relation derived here for a point charge is called *Gauss' law*. If the surface encloses several charges, then according to the principle of superposition,

$$\oint_s \mathbf{E} \cdot \mathbf{n}\, da = \oint_s \sum_i \mathbf{E}_i \cdot \mathbf{n}\, da = \sum_i \oint_s \mathbf{E}_i \cdot \mathbf{n}\, da = 4\pi \sum_i q_i \tag{1.22}$$

or for a continuous charge distribution

$$\oint_s \mathbf{E} \cdot \mathbf{n}\, da = 4\pi \int_V \rho(\mathbf{r})\, dV \tag{1.23}$$

The surface integral on the left-hand side is transformed into a volume integral with the aid of the *Gauss theorem*:

$$\oint_s \mathbf{E} \cdot \mathbf{n}\, da = \int_V \operatorname{div} \mathbf{E}\, dV \tag{1.24}$$

Hence,

$$\int_V \operatorname{div} \mathbf{E}\, dV = 4\pi \int_V \rho(\mathbf{r})\, dV \tag{1.25}$$

or

$$\int_V (\operatorname{div} \mathbf{E} - 4\pi\rho(\mathbf{r}))\, dV = 0 \tag{1.26}$$

Since this is valid for an arbitrary volume, the integrand has to be zero, and we obtain the relation

$$\operatorname{div} \mathbf{E}(\mathbf{r}) = 4\pi\rho(\mathbf{r}) \tag{1.27}$$

between the field intensity and the charge distribution producing it. So, in space the charges are the sources (positive charges) and sinks (negative charges) of the electric field.

The electric potential

Now we demonstrate that the electric field can be written as the gradient of a potential (see Figure 1.10). We have

$$\mathbf{E}(\mathbf{r}) = \int_V \rho(\mathbf{r}') \frac{(\mathbf{r} - \mathbf{r}')}{|\mathbf{r} - \mathbf{r}'|^3}\, dV' \tag{1.28}$$

Differentiating the expression $1/|\mathbf{r} - \mathbf{r}'|$ with respect to the unprimed coordinate $\mathbf{r}$, then we see that

$$\frac{\partial}{\partial x} \frac{1}{\sqrt{(x - x')^2 + (y - y')^2 + (z - z')^2}} = -\frac{(x - x')}{(\sqrt{(x - x')^2 + (y - y')^2 + (z - z')^2})^3}$$

thus

$$\nabla \frac{1}{|\mathbf{r} - \mathbf{r}'|} \equiv \operatorname{grad} \frac{1}{|\mathbf{r} - \mathbf{r}'|} = -\frac{(\mathbf{r} - \mathbf{r}')}{|\mathbf{r} - \mathbf{r}'|^3} \tag{1.29}$$

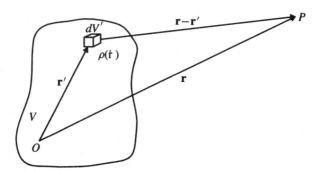

Figure 1.10. On the calculation of the potential of a charge cloud.

From this relation, one obtains for the field intensity

$$\mathbf{E}(\mathbf{r}) = - \int \rho(\mathbf{r}')\nabla \frac{1}{|\mathbf{r} - \mathbf{r}'|} \, dV' = -\nabla \int \frac{\rho(\mathbf{r}')}{|\mathbf{r} - \mathbf{r}'|} \, dV' \tag{1.30}$$

Thus, the field intensity can be derived as the gradient of a potential. The potential $\Phi(\mathbf{r})$ is obtained as the integral over the entire charge distribution:

$$\Phi(\mathbf{r}) = \int \frac{\rho(\mathbf{r}')}{|\mathbf{r} - \mathbf{r}'|} \, dV' \tag{1.31}$$

With this definition, we can write for the field intensity

$$\mathbf{E}(\mathbf{r}) = - \operatorname{grad} \Phi(\mathbf{r}) \equiv -\nabla\Phi(\mathbf{r}) \tag{1.32}$$

Since the curl of a gradient always vanishes ($\nabla \times \nabla = 0$), we obtain

$$\operatorname{curl} \mathbf{E} = 0 \tag{1.33}$$

So, we have shown that the electrostatic field can be described by the two differential equations

$$\nabla \cdot \mathbf{E} = 4\pi\rho \tag{1.34}$$

$$\nabla \times \mathbf{E} = 0 \tag{1.35}$$

The second equation implies that electrostatic forces are conservative forces. In other words, the electrostatic field is irrotational. Taking (1.32) into account, we obtain from the divergence of equation (1.34)

$$\nabla \cdot \mathbf{E}(\mathbf{r}) = -\nabla^2\Phi(\mathbf{r}) = 4\pi\rho(\mathbf{r}) \tag{1.36}$$

or

$$\Delta\Phi(\mathbf{r}) = -4\pi\rho(\mathbf{r}) \tag{1.37}$$

This equation is called *Poisson's equation*; for a charge-free region $\rho = 0$, and Poisson's equation reduces to the *Laplace equation*

$$\Delta\Phi(\mathbf{r}) = 0 \tag{1.38}$$

Now we will show that a potential of the form

$$\Phi(\mathbf{r}) = \frac{1}{|\mathbf{r} - \mathbf{r}'|} \tag{1.39}$$

(a point charge $q = 1$ has been chosen) satisfies the Poisson equation for a point charge. For that, we place the charge at the origin of the coordinate system ($\Phi(\mathbf{r}) = 1/r$) and apply the Laplace operator to it. For a point charge the problem is spherically symmetric, and we have to consider the r-coordinate only. For $r \neq 0$ we obtain by simple arithmetic

$$\Delta \frac{1}{r} = \frac{1}{r} \frac{\partial^2}{\partial r^2} \left(r \frac{1}{r} \right) = \frac{1}{r} \frac{\partial^2}{\partial r^2} (1) = 0 \tag{1.40}$$

But for $r = 0$ the expression is not defined. Therefore, we have to perform a limiting procedure: we integrate $\Delta \phi$ in a neighborhood of $r = 0$ and transform this volume integral by means of Gauss' theorem into a surface integral independent of r

$$\int_V \Delta \left(\frac{1}{r} \right) dV = \int_V \text{div} \left(\text{grad} \left(\frac{1}{r} \right) \right) dV = \oint_s \text{grad} \left(\frac{1}{r} \right) \cdot \mathbf{n} \, da = \oint_s \frac{\partial}{\partial r} \left(\frac{1}{r} \right) r^2 \, d\Omega$$

$$= -4\pi \tag{1.41}$$

So, we have shown that

$$\Delta \left(\frac{1}{r} \right) = 0, \quad \text{for } r \neq 0 \tag{1.42}$$

and that for the volume integral the following relation is valid:

$$\int_V \Delta \frac{1}{r} \, dV = -4\pi \tag{1.43}$$

Mathematical supplement: The δ-function

At this point, it is useful to introduce Dirac's δ-function. Dirac introduced the δ-function in analogy to Kronecker's δ_{ik}-symbol, as a generalization for continuous indices

$$f(a) = \int_{-\infty}^{\infty} f(x)\delta(x - a) \, dx \tag{1.44}$$

Hence, by the δ-function $\delta(x - a)$, the function value at the point $x = a$ is assigned to the function $f(x)$; by the expression given above it is defined only as a functional. Quantities like $\delta(x)$ cannot be regarded as functions in the usual sense. They are not integrable in the framework of Riemann's notion of the integral. The δ-function is treated in a mathematically exact manner within the *theory of distributions*.[2] Here, we restrict

---

[2] *Distribution* means a generalization of the notion of function in functional analysis; linear functional on certain abstract spaces. The Dirac δ-function for example, although very important in theoretical physics, is a distribution, not a function. The theory of distribution was developed by L. Schwartz between 1945 and 1950.

ourselves to giving some properties of the δ-function with a heuristic proof. The quantity $\delta(x)$ can be viewed in terms of the limit of a function having the property that it vanishes everywhere and becomes singular at the point $x = 0$ in such a way that

$$\int_{0-\epsilon}^{0+\epsilon} \delta(x)\,dx = 1, \qquad \epsilon > 0 \tag{1.45}$$

The δ-function is then defined by the following properties:

$$\delta(x - a) = 0 \qquad \text{when } x \neq a \tag{1.46}$$

and

$$\int_{b_1}^{b_2} \delta(x - a)\,dx = \begin{cases} 1 & \text{for } b_1 \leq a \leq b_2 \\ 0 & \text{otherwise} \end{cases} \tag{1.47}$$

For the product of an arbitrary function and the derivative of the δ-function one obtains

$$\int_{-\infty}^{\infty} f(x)\delta'(x - a)\,dx = [f(x)\delta(x - a)]_{-\infty}^{\infty} - \int_{-\infty}^{\infty} f'(x)\delta(x - a)\,dx$$

$$= -f'(a) \tag{1.48}$$

Just the same easy way one finds $\delta(ax) = (1/|a|)\delta(x)$, since

$$\int_{-\epsilon}^{\epsilon} \delta(ax)\,dx = \begin{cases} \int_{-\epsilon a}^{\epsilon a} \dfrac{1}{a}\delta(z)\,dz & \text{for } a > 0 \\[3mm] \int_{-\epsilon a}^{\epsilon a} -\dfrac{1}{|a|}\delta(z)\,dz = \int_{-\epsilon|a|}^{\epsilon|a|} \dfrac{1}{|a|}\delta(z)\,dz & \text{for } a < 0 \end{cases} \tag{1.49}$$

Therefore, in general, $\delta(ax) = \delta(x)/|a|$.

If the δ-function contains a function $f(x)$ of the independent variable x in the argument then one can transform in the following way:

$$\delta(f(x)) = \sum_i \frac{1}{\left|\dfrac{df}{dx}(x_i)\right|} \delta(x - x_i) \tag{1.50}$$

if $f(x)$ has simple zeros at $x = x_i$ (see Figure 1.11). In the vicinity of a zero x_i, the same arguments apply. The factor a is replaced by $df/dx(x_i)$ because in the nearest neighborhood of x_i the function $f(x)$ can be approximated by $f(x) = (df/dx)_i (x - x_i) = a(x - x_i)$. This can be proved mathematically in more detail in the following way.

Let the function $f(x)$ have N simple zeros x_i ($i = 1, \ldots N$), that is, $f(x_i) = 0$, $df/dx|_{x_i} \neq 0$. The δ-function contributes to the integral $I = \int_{-\infty}^{\infty} g(x)\delta(f(x))\,dx$ only if its argument vanishes. Hence it is sufficient to consider only those contributions to the

---

Since then, it has been made use of in many fields of analysis, e.g., in the theory of differential equations, and in modern physics. An especially clear explanation of distributions has been given by two mathematicians, Laugwitz and Schmieden.

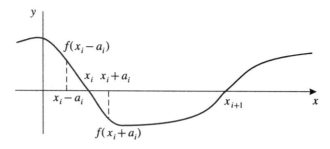

Figure 1.11.

integral given above resulting from (arbitrarily small) neighborhoods of the individual zeros x_i of f:

$$I = \sum_{i=1}^{N} \int_{x_i-a_i}^{x_i+a_i} \delta[f(x)]g(x)\,dx \tag{1.51}$$

If $f'(x_i) \neq 0$, it is also nonzero in an interval about x_i (let f' be continuous). But, let a be chosen so small that $f'(x) \neq 0$ for $x \epsilon (x_i - a, x_i + a), i = 1, \ldots n$. Then, the function is invertible in any neighborhood $(x_i - a, x_i + a)$; let the inverse function be $f_i^{-1}(y)$. With the help of the substitution formula we then obtain

$$I = \sum_{i=1}^{N} \int_{f(x_i-a)}^{f(x_i+a)} \delta(y)g[f_i^{-1}(y)]\frac{dy}{f'[f_i^{-1}(y)]} = \sum_{i=1}^{N} g[f_i^{-1}(0)]\frac{1}{|f'(f_i^{-1}(0))|}$$

$$= \sum_{i=1}^{N} g(x_i)\frac{1}{|f'(x_i)|} \tag{1.52}$$

since $f_i^{-1}(0) = x_i$. The absolute value appears since $f'(x_i) < 0, f(x_i - a) > f(x_i + a)$; that is, the limits of integration have to be interchanged in the integral. But this expression can be written as

$$I = \int_{-\infty}^{\infty} \sum_{i=1}^{N} \frac{1}{|f'(x_i)|}\delta(x - x_i)g(x)\,dx \tag{1.53}$$

Distributions can be understood also as a *limit of functions*. For example, we consider the family of functions

$$f(x, \sigma) = \frac{1}{\sqrt{\pi}\sigma}e^{-x^2/\sigma^2} \tag{1.54}$$

where σ is a parameter. The integral of this function is

$$\frac{1}{\sqrt{\pi}\sigma} \int_{-\infty}^{\infty} e^{-x^2/\sigma^2}\,dx = \frac{1}{\sqrt{\pi}} \int_{-\infty}^{\infty} e^{-y^2}\,dy = 1 \tag{1.55}$$

that is, it is independent of σ. Now for any point $x \neq 0$ the limit of these functions approaches zero as $\sigma \to 0$. Hence,

$$\lim_{\sigma \to 0} f(x, \sigma) = \lim_{\sigma \to 0} \frac{1}{\sqrt{\pi}\sigma} e^{-x^2/\sigma^2} = \delta(x) \tag{1.56}$$

Such representations of distributions in terms of limits of functions are often very useful to identify the properties of distributions. Another approach is offered by the notion of *Weyl's eigendifferentials*, with which we will be concerned in more detail in the lectures on quantum mechanics. For spatial and multidimensional problems we write the δ-function with a vectorial argument by forming the δ-function of the individual Cartesian components:

$$\delta(\mathbf{r} - \mathbf{a}) = \delta(x - a_x)\delta(y - a_y)\delta(z - a_z) \tag{1.57}$$

is a functional vanishing everywhere except $\mathbf{r} = \mathbf{a}$. For example, by means of the δ-function we can describe a sequence of point charges, like a continuous charge distribution. A charge density of the form

$$\rho(\mathbf{r}) = \sum_{i=1}^{N} q_i \delta(\mathbf{r} - \mathbf{a}_i) \tag{1.58}$$

describes N point charges of the magnitude q_i sitting at the points $\mathbf{a}_i$. In the same way, with the help of the δ-function we can write the two relations (1.42) and (1.43) in a closed form:

$$\Delta \frac{1}{r} = -4\pi \delta(\mathbf{r}) \tag{1.59}$$

If the charge is not placed at the origin, one has correspondingly

$$\Delta \frac{1}{|\mathbf{r} - \mathbf{r}'|} = -4\pi \delta(\mathbf{r} - \mathbf{r}') \tag{1.60}$$

If the field is produced by a continuous charge distribution $\rho(\mathbf{r})$, then

$$\Delta \phi(\mathbf{r}) = \int \rho(\mathbf{r}') \Delta \frac{1}{|\mathbf{r} - \mathbf{r}'|} \, dV'$$
$$= \int \rho(\mathbf{r}')(-4\pi \delta(\mathbf{r} - \mathbf{r}')) \, dV' = -4\pi \rho(\mathbf{r}) \tag{1.61}$$

corresponding exactly to the Poisson equation (1.37).

The potential energy of a charge in an electric field

When a charge q is moved in an electric field of intensity $\mathbf{E}(\mathbf{r})$ from A to B, the force $\mathbf{F}(\mathbf{r}) = q\mathbf{E}(\mathbf{r})$ acts on it.

The work done during this motion is given by the path integral over the force (Figure 1.12)

$$W = -\int_A^B \mathbf{F}(\mathbf{r}') \cdot d\mathbf{r}' = -q \int_A^B \mathbf{E}(\mathbf{r}') \cdot d\mathbf{r}' \tag{1.62}$$

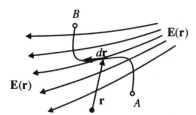

Figure 1.12. Work has to be done to move a charged particle in an electric field.

The minus sign appears because work done against the field is cosidered to be positive. If in this equation the field intensity is expressed by the potential, then

$$W = q \int_A^B \nabla\phi \cdot dr' = q \int_A^B \left(\frac{\partial\phi}{\partial x}dx + \frac{\partial\phi}{\partial y}dy + \frac{\partial\phi}{\partial z}dz \right) = q \int_A^B d\phi$$

$$W = q \int_A^B d\phi = q(\phi(B) - \phi(A)) \tag{1.63}$$

that is, the work done along the path from A to B corresponds exactly to the difference in the potential energy at these two points. Thus, the work is independent of the path.

The field intensity across charged surfaces

We now consider the behavior of the components of the electric field intensity across a charged surface. The surface charge is described by the charge density σ, having the dimension of charge per area:

$$\sigma(\mathbf{r}) = \lim_{\Delta a \to 0} \frac{\Delta q(\mathbf{r})}{\Delta a} \tag{1.64}$$

An area element Δa carries the charge $q = \int_{\Delta a} \sigma(\mathbf{r})da$. The area element Δa should lie inside the volume element ΔV. Applying Gauss' law, we obtain

$$\int_{\Delta V} \text{div } \mathbf{E} \, dV = \oint_{\Delta s(\Delta V)} \mathbf{E} \cdot \mathbf{n} \, da = 4\pi q = 4\pi \int_{\Delta a} \sigma(\mathbf{r}) \, da \tag{1.65}$$

where the integral extends over the surface Δs of the volume ΔV enclosing the area element Δa.

Now, we choose the volume element to be a cuboid of negligible thickness; then the flux of the E-field through the lateral surfaces can be neglected, and we obtain the flux through the area by integration. For bookkeeping purposes, we will label the field intensity before and after crossing the surface by the indices 1 and 2, respectively. Hence, one obtains (Figure 1.13)

$$(\mathbf{E}_2 \cdot \mathbf{n} - \mathbf{E}_1 \cdot \mathbf{n})\Delta a = 4\pi \sigma(\mathbf{r})\Delta a$$

(a)

(b)

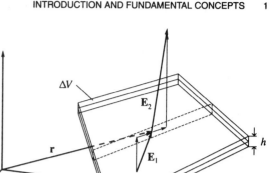

Figure 1.13. The electric field intensity across a charge surface: (a) In a cut. (b) In perspective.

By cancellation and factoring out, we further obtain

$$(\mathbf{E}_2 - \mathbf{E}_1) \cdot \mathbf{n} = 4\pi\sigma(\mathbf{r}) \tag{1.66}$$

that is, *the normal component of the* **E**-*field changes by the amount* $4\pi\sigma$ *in crossing a charged surface.* Since the field intensity can be derived from the potential, we know that curl $\mathbf{E} = -\,\mathrm{curl\,grad}\,\phi = 0$, and the circulation integral vanishes due to the Stokes theorem:

$$\oint \mathbf{E} \cdot d\mathbf{r} = \int_s \mathrm{curl}\,\mathbf{E} \cdot \mathbf{n}\,da = 0$$

We integrate now along the closed path parallel to the components of the **E**-field tangential to the surface Δs of the volume element ΔV where we assume that the thickness of the volume element can be neglected (see Figure 1.14). Then the contributions from the front surfaces can be also neglected, and we obtain

$$\oint \mathbf{E} \cdot d\mathbf{r} = (\mathbf{E}_{2t} - \mathbf{E}_{1t}) \cdot \Delta\mathbf{r} = 0 \tag{1.67}$$

Hence, $\mathbf{E}_{2t} = \mathbf{E}_{1t}$; that is, *the tangential component of the* **E**-*field does not change across a charged surface.* The refraction of the electric field crossing a (positively) charged layer is illustrated in Figure 1.13.

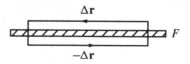

Figure 1.14. Illustration of the integration path used in equation (1.67).

Example 1.1:   The parallel-plate capacitor

As an example for the application of Gauss' law, we will calculate the field intensity in a parallel-plate capacitor. This device consists of two conducting plates arranged parallel to each other at a separation d (see Figures 1.15 and 1.16). The charge density on the plates is constant: $\sigma = q/a$ or $-q/a$, where q represents the entire charge on one plate.

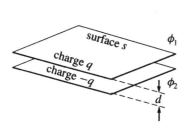

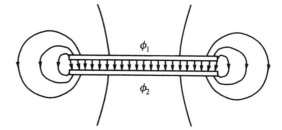

Figure 1.15. A parallel-plate capacitor.

Figure 1.16. A cut through a parallel-plate capacitor showing field lines. The electric field points from the positively charged plate to the negatively charged plate.

To simplify the problem, we will neglect the stray fields at the edge of the capacitor and assume that there is no field outside the capacitor; then, we can calculate the field intensity $\mathbf{E}$ from equation (1.66) by applying Gauss' law for one plate:

$$(\mathbf{E} - \mathbf{0}) \cdot \mathbf{n} = E = 4\pi\sigma = 4\pi\frac{q}{a}$$

$\mathbf{n}$ is a normal vector to the plates and points from the positively charged plate to the negatively charged plate; as does the electric field $\mathbf{E}$. The field intensity inside the capacitor is constant. For the potential difference between the plates we obtain

$$\phi_2 - \phi_1 = -\int_0^d \mathbf{E} \cdot d\mathbf{r} = -E\int_0^d dx = -Ed$$

Inserting the relation found for E yields the voltage V:

$$V = -(\phi_2 - \phi_1) = \phi_1 - \phi_2 = 4\pi\frac{q}{a}d$$

The voltage is the difference of the potentials at the positively charged plate (ϕ_1) and the negatively charged plate (ϕ_2). Now, we define the capacitance C of a capacitor by the equation

$$C = \frac{q}{V}$$

This tells us the charge q the capacitor can carry for a given voltage V. Thus, the capacitance of a parallel-plate capacitor is

$$C = \frac{a}{4\pi d} \tag{1.68}$$

Example 1.2: Exercise: The spherical capacitor

Calculate the capacitance of a capacitor consisting of two concentric spherical shells (see Figure 1.17). Each sphere carries a uniformly distributed charge. The radii of the spheres are r_1 and r_2. The outer sphere has the negative charge $-q$, and the inner sphere has the positive charge q.

Solution Because of the spherically symmetric arrangement, the field distribution is also spherically symmetric.

Using spherical coordinates, we need to consider only the r-component; the field has no other components.

The interior of a charged spherical shell is field-free. This follows immediately from Gauss' law, since no charge is present inside the sphere, and always

$$\oint_s E_n \, da = 4\pi r^2 E_n = 0$$

where r gives the radius of a spherical surface lying inside the charge distribution.

For the spherical capacitor, this means that the electric field **E** originates only from the inner sphere. Placing a spherical surface of radius r between the spherical shells, then according to Gauss' law:

$$\oint_s E_n \, da = 4\pi r^2 E_n(r) = 4\pi q$$

Hence, the field intensity depends quadratically on the radius:

$$E_n(r) = \frac{q}{r^2} \tag{1.69}$$

The potential difference between the inner and the outer spheres is

$$V = \int_{r_1}^{r_2} E_n \, dr = q \int_{r_1}^{r_2} \frac{dr}{r^2} = q \left(\frac{1}{r_1} - \frac{1}{r_2} \right)$$

So, we obtain for the capacitance of a spherical capacitor

$$C = \frac{q}{V} = \frac{r_1 r_2}{r_2 - r_1} \tag{1.70}$$

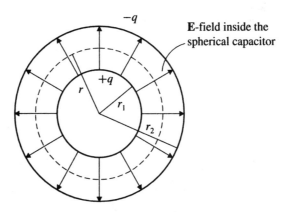

Figure 1.17. Spherical capacitor.

Concerning the result for the field intensity, we note that the field intensity is the same as that of a point charge of equal strength placed at the center of the sphere. This result is valid for all spherically symmetric charge distributions.

Example 1.3: Exercise: The cylindrical capacitor

Calculate the capacitance of a capacitor consisting of two coaxial cylinders of height h and radii r_1 and r_2 (see Figure 1.18). Neglect the stray fields at the edges.

Solution   Again, we place a corresponding cylindrical surface of radius ρ between both cylinders and apply Gauss' law:

$$\oint E_\rho da = E_\rho \cdot 2\pi\rho h = 4\pi q \tag{1.71}$$

Here, q is the charge on the inner cylinder. We have used the fact that the field is axially symmetric, and so stray fields should vanish. The voltage is obtained by integration over the field intensity (1.71).

$$V = \int_{r_1}^{r_2} E_\rho \, d\rho = \frac{2q}{h} \int_{r_1}^{r_2} \frac{d\rho}{\rho} = \frac{2q}{h} \ln \frac{r_2}{r_1}$$

Hence, the capacitance is

$$C = \frac{h}{2 \ln \dfrac{r_2}{r_1}}$$

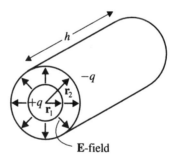

Figure 1.18. Cylindrical capacitor.

Dipole layer–dipole moment

Let two parallel surfaces at a separation $d(\mathbf{r}')$ carry the charge densities $\sigma(\mathbf{r}')$ and $-\sigma(\mathbf{r}')$, respectively.

Diminishing the distance between the surfaces with the potential difference between them remaining constant, the charge density tends to infinity. The limit taken by the product of charge density and local distance is denoted as the *dipole density* $\mathbf{D}(\mathbf{r})$:

$$\mathbf{D}(\mathbf{r}') = D(\mathbf{r}')\mathbf{n} = \lim_{d \to 0} \sigma(\mathbf{r}') \, d(\mathbf{r}')\mathbf{n}(\mathbf{r}') \tag{1.72}$$

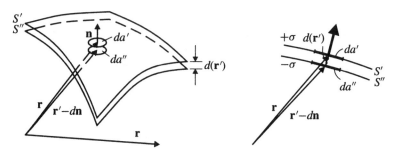

Figure 1.19. Dipole layer.

We define the *magnitude of the dipole moment* by the integral $p = \int D\,da$; its direction is per definition perpendicular to the surface a' and *points from the negative charge to the positive charge*. The normal vector $\mathbf{n}$ points also in the direction from the negative charge to the positive one (see Figure 1.19). The potential of the dipole layer is composed additively of the potentials of the single layers:

$$\phi(\mathbf{r}) = \int \frac{\sigma(\mathbf{r}')}{|\mathbf{r} - \mathbf{r}'|}\,da' - \int \frac{\sigma(\mathbf{r}')}{|\mathbf{r} - \mathbf{r}' + d(\mathbf{r}')\mathbf{n}|}\,da' \tag{1.73}$$

At a large distance from the dipole layer, one has $|\mathbf{r} - \mathbf{r}'| \gg d$, and the denominator of the second integrand can be expanded in the following way:

$$\frac{1}{|\mathbf{r} - \mathbf{r}' + d(\mathbf{r}')\mathbf{n}|} = \frac{1}{\sqrt{(\mathbf{r} - \mathbf{r}')^2 + 2(\mathbf{r} - \mathbf{r}') \cdot d\mathbf{n} + (d\mathbf{n})^2}}$$

$$= \frac{1}{|\mathbf{r} - \mathbf{r}'|} \cdot \frac{1}{\sqrt{1 + \left(\dfrac{d\mathbf{n}}{\mathbf{r} - \mathbf{r}'}\right)^2 + \dfrac{2(\mathbf{r} - \mathbf{r}') \cdot d\mathbf{n}}{(\mathbf{r} - \mathbf{r}')^2}}}$$

$$\approx \frac{1}{|\mathbf{r} - \mathbf{r}'|} \cdot \left(1 - \frac{(\mathbf{r} - \mathbf{r}') \cdot d\mathbf{n}}{(\mathbf{r} - \mathbf{r}')^2 + \cdots}\right) \tag{1.74}$$

where we have neglected $(d(\mathbf{r}')\mathbf{n})^2/(\mathbf{r} - \mathbf{r}')^2$.

Substituting this into (1.73), we obtain for the potential

$$\phi(\mathbf{r}) = \int \frac{\sigma(\mathbf{r}')}{|\mathbf{r} - \mathbf{r}'|}\,da' - \int \frac{\sigma(\mathbf{r}')}{|\mathbf{r} - \mathbf{r}'|} \left(1 - \frac{(\mathbf{r} - \mathbf{r}') \cdot d(\mathbf{r}')\mathbf{n}}{|\mathbf{r} - \mathbf{r}'|^2}\right) da'$$

$$= \int \sigma(\mathbf{r}') \frac{(\mathbf{r} - \mathbf{r}') \cdot d(\mathbf{r}')\mathbf{n}}{|\mathbf{r} - \mathbf{r}'|^3}\,da'$$

$$\phi(\mathbf{r}) = \int D(\mathbf{r}')\mathbf{n} \cdot \frac{\mathbf{r} - \mathbf{r}'}{|\mathbf{r} - \mathbf{r}'|^3}\,da' \tag{1.75}$$

To simplify the integrand, we introduce the solid angle $d\Omega$ subtended at the point of observation $\mathbf{r}$ by the dipole layer.

From Figures 1.20 and 1.21, we obtain the following expression:

$$|\mathbf{r'} - \mathbf{r}|^2 \, d\Omega = da'\mathbf{n} \cdot \frac{\mathbf{r'} - \mathbf{r}}{|\mathbf{r'} - \mathbf{r}|} = da' \cos\theta \tag{1.76}$$

Hence, the potential of the dipole layer is

$$\phi(\mathbf{r}) = -\int D(\mathbf{r'}) \, d\Omega \tag{1.77}$$

If the dipole density of the layer is constant, then the potential of the layer at the point $\mathbf{r}$ is determined by the solid angle subtended at $\mathbf{r}$ by the layer. An *infinitesimal dipole layer* is denoted as a *point dipole*.

The potential of a point dipole is

$$\phi(\mathbf{r}) = -\int D(\mathbf{r'}) \, d\Omega$$

$$= -\int_{\Delta a} \sigma(\mathbf{r'}) \, d(\mathbf{r'})\mathbf{n} \frac{\mathbf{r'} - \mathbf{r}}{|\mathbf{r'} - \mathbf{r}|^3} \, da'$$

$$= -\frac{\mathbf{r'} - \mathbf{r}}{|\mathbf{r'} - \mathbf{r}|^3} \cdot \mathbf{n} \, d(\mathbf{r}) \int_{\Delta a} \sigma(\mathbf{r'}) \, da'$$

$$= -\frac{\mathbf{r'} - \mathbf{r}}{|\mathbf{r'} - \mathbf{r}|^3} \cdot \mathbf{n}(\mathbf{r}) \, d(\mathbf{r}) q \tag{1.78}$$

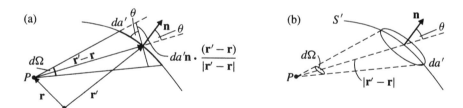

Figure 1.20. The potential at P originating from the dipole density D in the area element da' is the negative product of D and the solid angle $d\Omega$ subtended at P by da'. (a) In a cut. (b) In perspective.

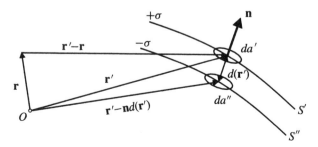

Figure 1.21. On the geometry of a dipole layer.

The definition of the *dipole vector (dipole moment)* **p** is

$$\mathbf{p} = \mathbf{n}\,dq \tag{1.79}$$

that is, the magnitude of the charge times the vector pointing from the negative charge to the positive one. We obtain for the potential

$$\phi(\mathbf{r}) = -\mathbf{p} \cdot \frac{(\mathbf{r}' - \mathbf{r})}{|\mathbf{r}' - \mathbf{r}|^3} = -\frac{|\mathbf{p}|}{|\mathbf{r}' - \mathbf{r}|^2}\cos\theta \tag{1.80}$$

where θ is the angle between $\mathbf{p(n)}$ and $\mathbf{r}' - \mathbf{r}$. As pointed out above, **p** points from the negative charge to the positive one.

The potential across a dipole layer

Now we will investigate the behavior of the potential across a dipole layer.

For that, the point of observation **r** is placed directly in front of a dipole layer (Figure 1.22); then (Figure 1.23)

$$\int d\Omega = 2\pi$$

and for the potentials ϕ_1 and ϕ_2 on both sides of the dipole layer we obtain from (1.77):

$$\phi_1 = -D \int d\Omega = -2\pi D$$

$$\phi_2 = D \int d\Omega = 2\pi D \tag{1.81}$$

Comparing this expression with (1.80), we see that the change of the sign results from the change of the sign of $\mathbf{r}' - \mathbf{r}$. The difference of both potentials is

$$\phi_2 - \phi_1 = 4\pi D \tag{1.82}$$

that is, in crossing the dipole layer the potential changes by the amount $4\pi D$. In other words, the potential loss inside the dipole layer is $4\pi D$.

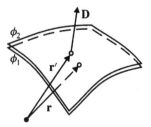

Figure 1.22. Discontinuity in the potential at a dipole layer.

Figure 1.23. Illustration of the solid angle element $d\Omega$.

Example 1.4: Exercise: Dipole moment

Calculate the potential of a dipole consisting of two charges e and $-e$ at the distance a (Figure 1.24).

Solution   The potential at the point P is given by

$$\phi(x, y, z) = \frac{e}{r_1} - \frac{e}{r_2} \qquad (1.83)$$

With $r^2 = x^2 + y^2 + z^2$, the distances satisfy

$$r_1 = \sqrt{\left(x - \frac{a}{2}\right)^2 + y^2 + z^2} = r\sqrt{1 - \frac{ax}{r^2} + \frac{a^2}{4r^2}}$$

We neglect the quadratic term and expand the root ($r \gg a$):

$$r_1 \approx r\left(1 - \frac{1}{2}\frac{ax}{r^2}\right)$$

While for r_2, we have

$$r_2 \approx r\left(1 + \frac{1}{2}\frac{ax}{r^2}\right)$$

Substituting both expansions into (1.83), we have

$$\phi(x, y, z) = \frac{e}{r}\left(\frac{1}{1 - \frac{1}{2}\frac{ax}{r^2}} - \frac{1}{1 + \frac{1}{2}\frac{ax}{r^2}}\right)$$

Transforming the bracket to the least common denominator $1 - \frac{1}{4}\left(ax/r^2\right)^2$, we can neglect the quadratic term; then

$$\phi(x, y, z) = \frac{eax}{r^3}$$

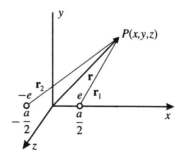

Figure 1.24. A point dipole: The dipole moment $\mathbf{p} = (ea, 0, 0)$ points from $-e$ to e, and it has the magnitude charge (e) times distance (a).

Now we introduce the dipole moment $\mathbf{p}$, which points from $-e$ to $+e$ and has the magnitude $|\mathbf{p}| = ea$. Thus, we obtain for the potential

$$\phi(\mathbf{r}) = \frac{\mathbf{p} \cdot \mathbf{r}}{r^3} \qquad (1.84)$$

The same result is obtained from equation (1.80) for $\mathbf{r}' = 0$.

With the unit radial vector $\mathbf{e}_r$, the dipole potential can be written as

$$\phi(r, \theta) = \frac{\mathbf{p} \cdot \mathbf{e}_r}{r^2} = \frac{p \cos \theta}{r^2} \qquad (1.85)$$

The electric field follows as (Figure 1.25)

$$\begin{aligned}
\mathbf{E} &= -\operatorname{grad} \phi \\
&= -\frac{\partial \phi}{\partial r} \mathbf{e}_r - \frac{1}{r} \frac{\partial \phi}{\partial \theta} \mathbf{e}_\theta - \frac{1}{r} \frac{1}{\sin \varphi} \frac{\partial \phi}{\partial \varphi} \mathbf{e}_\varphi \\
&= \frac{2p \cos \theta}{r^3} \mathbf{e}_r + \frac{p \sin \theta}{r^3} \mathbf{e}_\theta = \frac{3\mathbf{e}_r (\mathbf{p} \cdot \mathbf{e}_r) - \mathbf{p}}{r^3}
\end{aligned} \qquad (1.86)$$

But this expression is not complete. As we see at once, the singular point $r = 0$ needs particular consideration. Both the dipole potential and the corresponding $\mathbf{E}$-field are represented in Figure 1.25. Equation (1.86) describes the dipole field for $r \neq 0$. Since the integral $\int_V \mathbf{E} \, dV$ over a small sphere about the dipole lying in the center can be calculated easily, we have

$$\int_V \mathbf{E} \, dV = \int_V -\nabla \frac{(\mathbf{p} \cdot \mathbf{r})}{r^3} r^2 \, dr \, d\Omega = -\int_F \frac{\mathbf{p} \cdot \mathbf{e}_r}{r^2} \mathbf{e}_r r^2 \, d\Omega = -\frac{4\pi}{3} \mathbf{p}$$

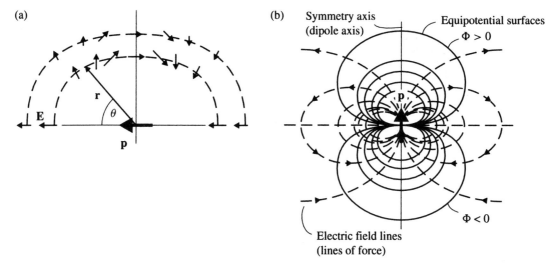

Figure 1.25. (a) The field of an electric dipole. At some points the field intensities are shown as arrows the magnitude of which decreases proportional to $1/r^3$. (b) The field lines and the equipotential surfaces. The potential surfaces of the dipole are rotationally symmetric about the $\mathbf{p}$-axis. The potential becomes $+\infty$ at the position of the positive charge and $-\infty$ at the position of the negative charge.

but, on the other hand, according to (1.21) $\int_V \mathbf{E}\,dV = 0$, and a contradiction appears between the equations (1.84) and (1.86). It is removed by adding the term $(4\pi/3)\mathbf{p}\delta(\mathbf{r})$ to (1.86). This consideration resembles the discussion of Gauss' law in the context of equation (1.43). It makes clear to us that divergences (like, e.g., the dipole potential at $r = 0$) demand special consideration.

Remark   In the transformation of the volume integral to a surface integral we have used Gauss' theorem in a generalized form:

$$\int_V \nabla \circ \hat{A}\,dV = \int_a \mathbf{n} \circ \hat{a}\,da \tag{1.87}$$

Here, the operation $\circ$ may be a scalar product $(\cdot)$ with a vector (then $\hat{A}$ is a vector so that the whole makes sense) the gradient of a scalar (then $\hat{A}$ is a scalar), or the cross product $(\times)$ with a vector (then $\hat{A}$ is a vector). In our example the operation has been the gradient formation and $\hat{A} = \mathbf{p} \cdot \mathbf{e}_r/r^2$; the vector normal to the spherical surface is $\mathbf{n} = \mathbf{e}_r$. (Compare this to the following exercise.)

Example 1.5:   Exercise: General integral theorems

The particular importance of the integral theorems of Gauss and Stokes arise from the transition from a volume integral to a surface integral and from a surface integral to a curvilinear integral, respectively. These integral theorems are special forms of more general theorems. Show that more general integral theorems can be derived from the theorems of Gauss and Stokes by an appropriate choice of $\mathbf{A}$, especially

$$\int_V dV\,\nabla\circ = \int_{a(V)} d\mathbf{a}\circ \quad \text{and} \quad \int_a (d\mathbf{a}\circ\nabla)\times = \oint_{C_a} d\mathbf{r}\circ \tag{1.88}$$

Solution   At first, we start from *Gauss' theorem*:

$$\int_V \operatorname{div}\mathbf{A}\,dV = \int_V \nabla\cdot\mathbf{A}\,dV = \int_a \mathbf{A}\,da \tag{1.89}$$

where $\mathbf{A}$ is an arbitrary vector field. The generalization of the Gauss theorem is achieved by the following ansatz for $\mathbf{A}$

$$\mathbf{A} = \rho(\mathbf{r}) \cdot \mathbf{c} \tag{1.90}$$

and

$$\mathbf{A} = \mathbf{c} \times \mathbf{B}(\mathbf{r}) \tag{1.91}$$

Here, $\mathbf{c}$ denotes an arbitrary but constant vector. From (1.89) we find for the choice (1.90) and (1.91):

$$\mathbf{c} \cdot \int_V \operatorname{grad}\rho\,dV = \int_V dV\,\nabla\cdot(\rho\mathbf{c}) = \int_a d\mathbf{a}\cdot\mathbf{c}\rho = \mathbf{c}\int_a d\mathbf{a}\rho \tag{1.92}$$

and

$$\begin{aligned}
-\mathbf{c}\cdot\int_V dV\operatorname{curl}\mathbf{B} &= \int_V dV\,(\mathbf{B}\cdot\operatorname{curl}\mathbf{c} - \mathbf{c}\cdot\operatorname{curl}\mathbf{B}) \\
&= \int_V dV\operatorname{div}(\mathbf{c}\times\mathbf{B}) = \int_a d\mathbf{a}\cdot(\mathbf{c}\times\mathbf{B}) \\
&= -\mathbf{c}\cdot\int_a d\mathbf{a}\times\mathbf{B}
\end{aligned} \tag{1.93}$$

Equations (1.92) and (1.93) are valid for any **c** . Hence,

$$\int_V dV \nabla \rho = \int_a d\mathbf{a}\rho \tag{1.94}$$

$$\int_V dV (\nabla \times \mathbf{B}) = \int_a d\mathbf{a} \times \mathbf{B} \tag{1.95}$$

Comparing the results (1.94) and (1.95), we find at once the rule

$$\int_V dV \nabla \circ = \int_{a(V)} d\mathbf{a} \circ$$

where ∘ replaces the ordinary, scalar, or cross products. With the same ansatz for **A** we find from the *Stokes theorem*

$$\int_a d\mathbf{a} \cdot \text{curl} \, \mathbf{A} = \int_a d\mathbf{a} \cdot (\nabla \times \mathbf{A}) = \oint_{C_a} d\mathbf{r} \cdot \mathbf{A}$$

the general rule

$$\int_a (d\mathbf{a} \circ \nabla) \times = \oint_{C_a} d\mathbf{r} \circ$$

The energy of a charge distribution

The potential energy of a point charge q in an electrostatic field **E** is the product of both: $W = q\mathbf{E}$. Now we calculate the potential energy of a number of point charges. We proceed in the following way: We imagine that the charges q_i are infinitely far from each other and calculate the work required to bring them from infinity to a certain separation $\mathbf{r}_i$. The work to be spent corresponds then to the potential energy of the charge distribution. If the charges are infinitely far from each other then the space is field-free, and we set the potential equal to zero: $\phi_0 = 0$. (We could give also a value different from zero to the potential ϕ_0: the potential energy is determined only up to a constant.) Now, the charge q_1 is shifted from infinity to $\mathbf{r}_1$. For this no work has to be done since the space is still field-free. But the charge q_1 causes a potential

$$\phi_1(\mathbf{r}) = \frac{q_1}{|\mathbf{r} - \mathbf{r}_1|} \tag{1.96}$$

against which one has to do work when the charge q_2 is brought to $\mathbf{r}_2$. This work is

$$W_2 = q_2 \phi_1(\mathbf{r}_2) \tag{1.97}$$

For the charge q_3 we have now to expend work against the potentials ϕ_1 and ϕ_2:

$$W_3 = q_3(\phi_1(\mathbf{r}_3) + \phi_2(\mathbf{r}_3)) \tag{1.98}$$

where ϕ_2 is given by

$$\phi_2(\mathbf{r}) = \frac{q_2}{|\mathbf{r} - \mathbf{r}_2|} \tag{1.99}$$

For other charges everything proceeds correspondingly. For the transport of the charge q_n, we have to spend the work

$$W_n = q_n [\phi_1(\mathbf{r}_n) + \phi_2(\mathbf{r}_n) + \cdots + \phi_{n-1}(\mathbf{r}_n)] = q_n \sum_{k=1}^{n-1} \phi_k(\mathbf{r}_n) \tag{1.100}$$

The total potential energy is given by the sum of all W_i:

$$W = \sum_{i=2}^{N} W_i = \sum_{i=2}^{N} q_i \sum_{k=1}^{i-1} \phi_k(\mathbf{r}_i) \tag{1.101}$$

Now we replace the potentials by

$$\phi_k(\mathbf{r}_i) = \frac{q_k}{|\mathbf{r}_i - \mathbf{r}_k|} \tag{1.102}$$

and such the energy can be expressed by the charge distribution:

$$W = \sum_{i=2}^{N} \sum_{k=1}^{i-1} \frac{q_i q_k}{|\mathbf{r}_i - \mathbf{r}_k|} = \frac{1}{2} \sum_{i}^{N} \sum_{k \neq i}^{N} \frac{q_i q_k}{|\mathbf{r}_i - \mathbf{r}_k|} \tag{1.103}$$

The last sum runs over the indices i as well as k, where the terms $i = k$ are omitted. Now we proceed to a continuous charge distribution. The point charge q_i is replaced by the charge element $\rho(\mathbf{r})dV$, and the summation becomes an integration

$$W = \frac{1}{2} \iint \frac{\rho(\mathbf{r}')\rho(\mathbf{r})}{|\mathbf{r} - \mathbf{r}'|} dV \, dV' \tag{1.104}$$

Translating the sum (1.103) to an integral (1.104), we have not taken into account the condition $i \neq k$ appearing in (1.103) (see Figure 1.26). Really, in (1.104) the integration extends over the point $\mathbf{r} = \mathbf{r}'$, so that (1.104) contains automatically *self-energy parts* which become infinitely large for point charges. In the following example, we will discuss this in more detail. At first, we continue our considerations starting from the expression (1.104). Since the potential of a charge distribution is given by

$$\phi(\mathbf{r}) = \int \frac{\rho(\mathbf{r}')}{|\mathbf{r} - \mathbf{r}'|} dV' \tag{1.105}$$

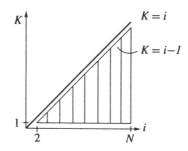

Figure 1.26.

we obtain equation (1.104) in the form

$$W = \frac{1}{2} \int \rho(\mathbf{r})\phi(\mathbf{r})\, dV \tag{1.106}$$

We want to express now the energy in terms of an integral over the field intensity. The charge density satisfies the Poisson equation

$$\Delta\phi(\mathbf{r}) = -4\pi\rho(\mathbf{r}) \tag{1.107}$$

In equation (1.106), we substitute the charge density according to this equation

$$W = -\frac{1}{8\pi} \int \phi\nabla^2\phi\, dV \tag{1.108}$$

Furthermore, the relation $\phi\nabla \cdot \nabla\phi = \nabla \cdot (\phi\nabla\phi) - (\nabla\phi)^2$ is valid. So the integrand can be rewritten

$$W = \frac{1}{8\pi} \int (\nabla\phi)^2\, dV - \frac{1}{8\pi} \int \nabla \cdot (\phi\nabla\phi)\, dV \tag{1.109}$$

The second volume integral can be converted to a surface integral

$$\int_V \nabla \cdot (\phi\nabla\phi)\, dV = \int_a \phi\nabla\phi \cdot \mathbf{n}\, da \tag{1.110}$$

This integral vanishes since we can displace the area a to infinity; we have $\phi \sim 1/r$, $\nabla\phi \sim 1/r^2$, so that the integrand with $1/r^3$ tends more rapidly to zero than the area element ($\sim r^2$) tends to infinity. If we further introduce the field intensity $\mathbf{E} = -\nabla\phi$, for the total energy we obtain the result

$$W = \frac{1}{8\pi} \int \mathbf{E}^2\, dV \tag{1.111}$$

The integrand represents the *energy density of the electric field:*

$$w = \frac{1}{8\pi}\mathbf{E}^2 \tag{1.112}$$

Example 1.6:   Interaction energy of two point charges

As an illustration, we want to derive the interaction energy of two point charges $W = q_1 q_2 / |\mathbf{r}_1 - \mathbf{r}_2|$ with the aid of the relation (1.111). The fields of the individual point charges are

$$\mathbf{E}_1(\mathbf{r}) = \frac{q_1(\mathbf{r} - \mathbf{r}_1)}{|\mathbf{r} - \mathbf{r}_1|^3}$$

and

$$\mathbf{E}_2(\mathbf{r}) = \frac{q_2(\mathbf{r} - \mathbf{r}_2)}{|\mathbf{r} - \mathbf{r}_2|^3}$$

The entire field is $\mathbf{E} = \mathbf{E}_1 + \mathbf{E}_2$. Substituting into (1.111) yields

$$W = \frac{1}{8\pi} \int (\mathbf{E}_1^2 + \mathbf{E}_2^2 + 2\mathbf{E}_1 \cdot \mathbf{E}_2)\, dV \tag{1.113}$$

Inserting the fields, then we have

$$W = \frac{1}{8\pi} \int \frac{q_1^2}{|\mathbf{r} - \mathbf{r}_1|^4}\, dV + \frac{1}{8\pi} \int \frac{q_2^2}{|\mathbf{r} - \mathbf{r}_2|^4}\, dV + \frac{1}{4\pi} \int \frac{q_1 q_2 (\mathbf{r} - \mathbf{r}_1) \cdot (\mathbf{r} - \mathbf{r}_2)}{|\mathbf{r} - \mathbf{r}_1|^3 |\mathbf{r} - \mathbf{r}_2|^3}\, dV$$
$$= W_1 + W_2 + W_{12} \tag{1.114}$$

In the equations (1.113) and (1.114), the first two terms W_1 and W_2 represent the *self-energy of the charges*. They do not change as the two charges approach each other. The self-energy of a charge corresponds to the electric part of the work to be done if the charge q would be condensed from an infinitely distributed charge cloud. The self-energy of the charge is obtained from equation (1.111) since in the transition from point charges (equation (1.103)) to continuous charge distributions (equation (1.104)) we have to integrate over the entire space, though we had excluded identical indices before. The occurrence of (infinitely large) self-energies is exhibited also in the fact that equation (1.111) is positively definite, while the interaction energy can be positive or negative depending on the sign of the charge. Considering realistic charged spheres of finite radius R instead of point charges, the self-energies are very large but finite, namely,

$$W = \frac{3}{5} \frac{q^2}{R}$$

In this case, we can simply calculate them and subtract them from the total energy to obtain the interaction energy. (See also the discussion of mirror nuclei on the following pages.) To solve the integral of the interaction energy, the origin of the coordinate system is positioned at r_1, that is, at the charge q_1. Then

$$W_{12} = \frac{q_1 q_2}{4\pi} \int \frac{\mathbf{r} \cdot (\mathbf{r} - \mathbf{r}_2)}{r^3 |\mathbf{r} - \mathbf{r}_2|^3}\, dV$$

We have $|\mathbf{r} - \mathbf{r}_2|^2 = r^2 + r_2^2 - 2rr_2 \cos\theta$. Using spherical coordinates so that θ is the angle between $\mathbf{r}$ and $\mathbf{r}_2$ (Figure 1.27), then

$$W_{12} = \frac{q_1 q_2}{4\pi} \int_{-\infty}^{\infty} \int_0^{\pi} \int_0^{2\pi} \frac{r^2 - rr_2 \cos\theta}{r^3 \left(\sqrt{r^2 + r_2^2 - 2rr_2 \cos\theta}\right)^3} r^2\, dr\ \sin\theta\, d\theta\, d\varphi$$

$$= \frac{q_1 q_2}{2} \int_{r=0}^{\infty} \int_{-1}^{1} \frac{r - r_2 \cos\theta}{\left(\sqrt{r^2 + r_2^2 - 2rr_2 \cos\theta}\right)^3}\, dr\, d(\cos\theta)$$

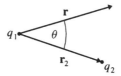

Figure 1.27. Illustration of the angle θ.

With the substitution $u = \sqrt{r^2 + r_2^2 - 2rr_2\cos\theta}$, the integral yields $W_{12} = q_1q_2/r_2$. Due to the complicated integrals appearing here, relation (1.104) can be applied to simple charge distributions only. For example, the exact calculation of the Coulomb energy in ternary or multiple fission already (fragmentation of a nucleus into three or more, mostly unlike parts) leads already to considerable numerical difficulties.

Example 1.7:   Energy density in a parallel-plate capacitor

The electric field intensity in the parallel-plate capacitor has been computed already; in Exercise 1.1, we obtained

$$E = 4\pi\sigma$$

For the energy density, we obtain from that expression

$$w = \frac{1}{8\pi}E^2 = 2\pi\sigma^2$$

Hence, a parallel-plate capacitor of the area a and plate separation x posseses the energy content

$$W = 2\pi\sigma^2 ax$$

From this expression, we can calculate easily the force acting between the plates of the capacitor. The force is obtained as the derivative of the potential energy with respect to the path

$$F = -\frac{dW}{dx} = -2\pi\sigma^2 a$$

Potential and charge distribution of an atomic nucleus

The atomic nucleus is regarded to be a uniformly charged sphere of radius R. Let the charge of the Z protons be Ze. Then the charge density is constant and is given by

$$\rho(\mathbf{r}) = \rho_0 = \frac{Ze}{\frac{4}{3}\pi R^3} \tag{1.115}$$

The whole problem is spherically symmetric; therefore, using spherical coordinates, in Poisson's equation

$$\Delta\phi(\mathbf{r}) = -4\pi\rho(\mathbf{r}) \tag{1.116}$$

we need to consider only the r-dependent part, and we obtain the equation

$$\frac{1}{r^2}\frac{d}{dr}\left(r^2\frac{d}{dr}\phi(r)\right) = -4\pi\rho(r) \tag{1.117}$$

This equation is solved for the regions $r \leq R$ and $r \geq R$. For $r \geq R$, the charge density vanishes (see Figure 1.28). Then, from Poisson's equation

$$\frac{d}{dr}\left(r^2\frac{d}{dr}\phi_a(r)\right) = 0 \tag{1.118}$$

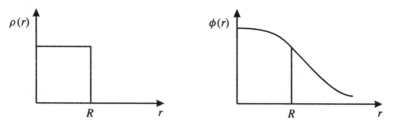

Figure 1.28. Charge distribution as well as Coulomb potential of an atomic nucleus of radius R.

or integrated once,

$$\frac{d}{dr}\phi_a(r) = \frac{C_1}{r^2} \tag{1.119}$$

The constant C_1 is fixed with the help of Gauss' law; it reads

$$E(R') \cdot 4\pi R'^2 = 4\pi Ze \tag{1.120}$$

if the surface of a sphere of radius $R' \geq R$ is taken as the Gaussian surface. With $\frac{d}{dr}\phi_a = -E = C_1/r^2$, we get $C_1 = -Ze$. The further integration yields

$$\phi_a = \frac{Ze}{r} + C_2 \tag{1.121}$$

The constant C_2 is set equal to zero because the potential has to vanish as $r \to \infty$. In the interior of the nucleus we have

$$\frac{d}{dr}\left(r^2\frac{d}{dr}\phi_i(r)\right) = -4\pi\rho_0 r^2 \tag{1.122}$$

A first integration yields

$$r^2\frac{d}{dr}\phi_i = -\frac{4\pi}{3}\rho_0 r^3 + k_1 \tag{1.123}$$

The second integration yields

$$\phi_i = -\frac{4\pi}{6}\rho_0 r^2 - \frac{k_1}{r} + k_2 \tag{1.124}$$

Since the potential has to remain finite for $r = 0$, one has $k_1 = 0$. We introduce the total charge $Ze = \frac{4}{3}\pi R^3\rho_0$:

$$\phi_i(r) = -\frac{Ze}{2}\frac{r^2}{R^3} + k_2 \tag{1.125}$$

The potentials in the different regions have to be equal for $r = R$, $\phi_i(R) = \phi_a(R)$; thus,

$$k_2 = \frac{3}{2}\frac{Ze}{R} \tag{1.126}$$

Then the potential distribution is

$$
\phi(r) = \begin{cases} \dfrac{Ze}{R}\left(\dfrac{3}{2} - \dfrac{1}{2}\dfrac{r^2}{R^2}\right) & \text{for } r \le R \\[2ex] \dfrac{Ze}{r} & \text{for } r \ge R \end{cases} \tag{1.127}
$$

In the sketch, the potential distribution and the charge distribution are given once more.

Potential distribution and charge distribution in an atom

When a potential distribution is given, the charge distribution can be determined in a similar manner. The field of a point nucleus is screened by the inner electrons so that an outer electron (valence electron) moves in a smaller field (see Figure 1.29).

Such a Coulomb potential is given by

$$
\phi(r) = \frac{Ze}{r}e^{-\alpha r}(1 + \alpha r) \tag{1.128}
$$

Obviously, this potential decreases faster than the bare Coulomb potential. Now, we start from Poisson's equation again and determine the charge density $\rho(r)$. Due to the spherical symmetry we have

$$
\Delta\phi(r) = \frac{1}{r^2}\frac{d}{dr}\left(r^2\frac{d}{dr}\phi\right) = -4\pi\rho \tag{1.129}
$$

Introducing the potential, we cannot use

$$
\Delta = \frac{1}{r^2}\frac{d}{dr}r^2\frac{d}{dr} \tag{1.130}
$$

since $\phi(r)$ is divergent for $r = 0$; that is, it is not differentiable. Therefore, we split off the singularity:

$$
\phi(r) = Ze\left[\frac{1}{r} + \frac{e^{-\alpha r} - 1}{r} + \alpha e^{-\alpha r}\right] \tag{1.131}
$$

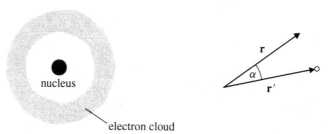

Figure 1.29. Nucleus with electron cloud and valence electron.

Now,

$$\Delta\left(\frac{1}{r}\right) = -4\pi\,\delta(\mathbf{r}) \tag{1.132}$$

$$\Delta\left(\frac{e^{-\alpha r} - 1}{r}\right) = \frac{1}{r^2}\frac{d}{dr}\left(r^2\frac{d}{dr}\frac{e^{-\alpha r} - 1}{r}\right)$$

$$= \frac{1}{r^2}\frac{d}{dr}\left(r^2\frac{-\alpha e^{-\alpha r}r - e^{-\alpha r} + 1}{r^2}\right)$$

$$= -\frac{1}{r^2}\frac{d}{dr}\left[e^{-\alpha r}(1 + \alpha r)\right]$$

$$= -\frac{1}{r^2}\left[-\alpha e^{-\alpha r}(1 + \alpha r) + \alpha e^{-\alpha r}\right]$$

$$= \frac{\alpha^2}{r}e^{-\alpha r} \tag{1.133}$$

$$\Delta\left(\alpha e^{-\alpha r}\right) = \frac{\alpha}{r^2}\frac{d}{dr}\left(r^2\frac{d}{dr}e^{-\alpha r}\right) = \frac{-\alpha^2}{r^2}\frac{d}{dr}r^2 e^{-\alpha r}$$

$$= \frac{-\alpha^2}{r^2}\left[2re^{-\alpha r} - \alpha r^2 e^{-\alpha r}\right]$$

$$= \frac{-\alpha^2}{r}e^{-\alpha r}(2 - \alpha r) \tag{1.134}$$

Altogether,

$$\rho(\mathbf{r}) = \frac{\Delta\phi(\mathbf{r})}{-4\pi} = Ze\left[\delta(\mathbf{r}) + \frac{\alpha^2}{4\pi r}e^{-\alpha r}(1 - \alpha r)\right]$$

Thus, we have a point charge at the origin surrounded by a space charge. The space charge is negative for $r > 1/\alpha$, as to be expected for an electron cloud. That it is positive for $r < 1/\alpha$ makes less sense and must be viewed as a shortcoming of our simple model for the screened potential $\phi(\mathbf{r})$. Now, we still calculate the total charge $Q(r)$ inside a sphere of radius r:

$$Q(r) = \int_{\text{sphere}} \rho(r')\,dV'$$

$$= Ze + Ze\alpha^2\int_0^r dr' r'^2 \cdot \frac{e^{-\alpha r'}}{r'}(1 - \alpha r')$$

$$= Ze\left(1 + \alpha^2\int_0^r dr' r' e^{-\alpha r'} - \alpha^3\int_0^r dr' r'^2 e^{-\alpha r'}\right)$$

$$= Ze\left(1 + \alpha^2 \cdot \frac{e^{-\alpha r'}}{\alpha^2}\left(-\alpha r' - 1\right)\Big|_0^r - \alpha^3 \cdot e^{-\alpha r'}\left(-\frac{r'^2}{\alpha} - \frac{2r'}{\alpha^2} - \frac{2}{\alpha^3}\right)\Big|_0^r\right)$$

$$= Ze \left[1 + e^{-\alpha r} \left(-\alpha r - 1 + \alpha^2 r^2 + 2\alpha r + 2 \right) - (-1 + 2) \right]$$
$$= Ze\, e^{-\alpha r} \left(\alpha^2 r^2 + \alpha r + 1 \right) \tag{1.135}$$

The total charge is

$$Q = \lim_{r \to \infty} Q(r) = 0 \tag{1.136}$$

So altogether, the atom is neutral.

For small distances ($r \to 0$) the charge density is

$$\rho(\mathbf{r}) = Ze \left[\delta(\mathbf{r}) + \frac{\alpha^2}{4\pi r} \right] \tag{1.137}$$

so that the space charge has the same sign as the point charge ($r \to 0$).

If we start from the potential distribution

$$\phi(r) = \frac{Ze}{r} e^{\alpha r} \tag{1.138}$$

then

$$\rho(r) = Ze \left[\delta(\mathbf{r}) - \frac{\alpha^2}{4\pi r} e^{-\alpha r} \right] \tag{1.139}$$

and

$$Q(r) = Ze \left[1 + (1 + \alpha r)e^{-\alpha r} - 1 \right] = Ze\, e^{-\alpha r}(1 + \alpha r) \tag{1.140}$$

as well as

$$Q = \lim_{r \to \infty} Q(r) = 0 \tag{1.141}$$

Methods for the determination of nuclear radii

(a) **With the help of μ-mesonic atoms:** By the capture of a μ-meson (200 times heavier than an electron) a μ-mesonic atom is created with a life time of 10^{-6} s. After that time the μ-meson is captured, usually by the nucleus; that is, it interacts with the proton according to the reaction

$$\mu^- + p \to n + \nu_\mu \tag{1.142}$$

Here, ν_μ is the *muonic neutrino*.

The μ-meson observes a different potential distribution than the electron in the ordinary atom since it is $\sim$200 times closer to the nucleus than the corresponding electron. This is caused by its larger mass. Namely, the Bohr radius of the atom with an electron is

$$a_e = \frac{\hbar^2}{m_e e^2} = 0.529\,\text{Å} = 0.529 \cdot 10^{-8}\text{cm} \tag{1.143}$$

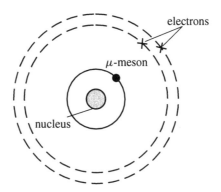

Figure 1.30. A μ-mesonic atom.

and that of the atom with the muon is

$$a_\mu = \frac{\hbar^2}{m_\mu e^2} = 256\,\text{fm} = 2,56 \cdot 10^{-11}\text{cm} \tag{1.144}$$

Being close to the nucleus the muon feels the deviations of the electric potential from the pure $1/r$-Coulomb potential (see Figure 1.30). From that the nuclear radii may be determined. The dimension of a nucleus is $R \approx 1.2A^{1/3}10^{-13}$ cm, where A is the number of nucleons (protons and neutrons).

(b) By mirror nuclei: The energy content of a nucleus amounts to

$$E_{\text{Coul}} = \frac{1}{2} \int \rho_0\, \phi(r)\, dV$$

where

$$\rho_0 = \frac{Ze}{\frac{4\pi}{3} R^3}$$

and according to (1.128),

$$\phi(r) = \frac{Ze}{R} \left[\frac{3}{2} - \frac{1}{2} \left(\frac{r^2}{R^2} \right) \right] \qquad \text{for } r < R$$

Hence,

$$E_{\text{Coul}} = \frac{1}{2}\rho_0 \int_0^R \int_0^{4\pi} \frac{Ze}{R} \left[\frac{3}{2} - \frac{1}{2} \left(\frac{r}{R} \right)^2 \right] r^2\, dr\, d\Omega = \frac{3}{5} \frac{(Ze)^2}{R} \tag{1.145}$$

This formula for the energy contains also the self-energies of the protons. Since the self-energy of a proton assumed to be smeared out throughout the nucleus follows from equation (1.145) with $Z = 1$, as $E_{\text{self}} = \frac{3}{5}e^2/R$, the energy content of the nucleus is

$$E_{\text{Coul}}(Z) = \frac{3}{5} \frac{(Ze)^2}{R} - Z\frac{3}{5}\frac{e^2}{R} = \frac{3}{5}\frac{Z(Z-1)}{R}e^2 \tag{1.146}$$

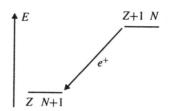

Figure 1.31. Energy relation in the β^+ decay of an atomic nucleus.

Mirror nuclei are nuclei having the same nucleon number, where one nucleus has $Z+1$ protons and Z neutrons and the other one has Z protons and $Z+1$ neutrons. The nuclei differ in their Coulomb energies by the value

$$\Delta E = E(Z+1) - E(Z)$$
$$= \frac{3}{5}\frac{e^2}{R}[(Z+1)Z - Z(Z-1)] = \frac{6}{5}\frac{e^2}{R}Z \tag{1.147}$$

The nuclear radius can be determined by measuring the energy difference of the mirror nuclei. This energy is measured by the energy released in the β^+ decay. In the β^+ decay, a neutron is created from the proton along with a positron e^+ and a neutrino ν_e (Figure 1.31)

$$p \rightarrow n + e^+ + \nu_e \tag{1.148}$$

The classical electron radius

If the electron is regarded as a classical uniformly charged sphere, then its energy is $E_{\text{Coul}} = \frac{3}{5}e^2/R$ (R = electron radius). Following Einstein, we have $E = mc^2$; that is, the mass of a particle is a measure of the self-energy. But the electrostatic energy is not the only factor since the magnetic and the gravitational interactions contribute also to the total energy of the particle. The latter one is negative (binding). Therefore, $E_{\text{Coul}} \geq mc^2$; thus $\frac{3}{5}e^2/R \geq mc^2$. Then, $R \leq \frac{3}{5}e^2/(mc^2)$. The factor $e^2/(mc^2)$ is called the *classical electron radius*. It is an approximate measure for the dimension of the electron, $R_{\text{el}} \sim 2,81 \cdot 10^{-13}$cm.

Example 1.8:   Exercise: The potential of an uniformly charged rod

The charge q is distributed uniformly on a straight line of the length $2c$, as in Figure 1.32. The potential distribution is required. For the discussion of the equipotential surfaces elliptic coordinates have to be used $u = \frac{1}{2}(l_1 + l_2)$ and $v = \frac{1}{2}(l_1 - l_2)$.

Solution   The potential is given by

$$\phi(r,z) = \frac{q}{2c}\int_{-c}^{c}\frac{dz'}{\sqrt{r^2 + (z-z')^2}}$$

if the notations of Figure 1.32 are used.

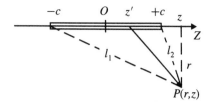

Figure 1.32. Uniformly charged rod.

The charge density is $q/(2c)$. The integration yields

$$\phi(r, z) = \frac{q}{2c} \left\{ \ln \left[2\sqrt{r^2 + (z - z')^2} + 2(z' - z) \right] \right\} \Big|_{-c}^{c}$$

$$= \frac{q}{2c} \ln \left[\frac{\sqrt{r^2 + (z - c)^2} - (z - c)}{\sqrt{r^2 + (z + c)^2} - (z + c)} \right]$$

The roots of this expression give the distances of the end points of the charge distribution from P. With the notations

$$l_1 = \sqrt{r^2 + (z + c)^2}, \qquad l_2 = \sqrt{r^2 + (z - c)^2}$$

we have

$$\phi(r, z) = \frac{q}{2c} \ln \left[\frac{l_2 - (z - c)}{l_1 - (z + c)} \right]$$

Introducing now the definition of the elliptic coordinates with $l_1 = u + v$, $l_2 = u - v$, and $cz = uv$, the argument is

$$\frac{l_2 + c - z}{l_1 - c - z} = \frac{u - v + c - z}{u + v - c - z} = \frac{c(u - v + c) - uv}{c(u + v - c) - uv}$$

$$= \frac{(u + c)(cu - cv + c^2 - uv)(u - c)}{(u + c)(cu + cv - c^2 - uv)(u - c)}$$

$$= \frac{(u + c)(cu^2 - cuv + c^2u - u^2v - c^2u + c^2v + cuv - c^3)}{(u - c)(cu^2 + cuv - c^2u - u^2v + c^2u + c^2v - cuv - c^3)}$$

$$= \frac{(u + c)(cu^2 - u^2v + c^2v - c^3)}{(u - c)(cu^2 - u^2v + c^2v - c^3)} = \frac{u + c}{u - c}$$

and hence,

$$\phi = \frac{q}{2c} \ln \frac{u + c}{u - c} \tag{1.149}$$

This is the exact solution valid in the neighborhood of the rod (near field) as well as at large distances (distant field). The area u = constant is thus an equipotential surface. It is a surface on which the sum of the distances remains constant: $l_1 + l_2 =$ constant. This surface is an ellipsoid of revolution about the z-axis with the foci $\pm c$. For large distances $u \gg c$, we have

$$\frac{u+c}{u-c} = \frac{1 + \dfrac{c}{u}}{1 - \dfrac{c}{u}} \sim 1 + 2\frac{c}{u}$$

For the logarithm, we have $\ln(1 + x) = x/1 - x^2/2 + x^3/3 - x^4/4 + \cdots$; therefore,

$$\ln\left(1 + 2\frac{c}{u}\right) \sim 2\frac{c}{u}$$

In this approximation (distant field), u is equal to the distance from the origin, and the potential has been transformed into that of a point charge at the origin.

Example 1.9:   Exercise: The capacitance of a conducting ellipsoid of revolution

Calculate the capacitance of a conducting ellipsoid of revolution (Figure 1.33) with the help of Exercise 1.8.

Solution   The conducting ellipsoid is placed about the charged segment $2c$ such that the ellipsoid coincides with an equipotential surface. The electrons in the conducting surface displace themselves such that the potential outside the ellipsoid does not change. The main semi-axis a is equal to the value of u at the surface of the ellipsoid. Hence, the potential of the ellipsoid is

$$\phi_1 = \frac{q}{2c} \ln \frac{a+c}{a-c}$$

Outside the ellipsoid, the potential is equal to that of the rod in the preceding exercise. The charge of the rod is determined in such a way that the potential takes the given value ϕ_1. At infinity the potential vanishes. As a point of comparison for the calculation of the capacitance we introduce an infinitely distant sphere with the potential $\phi_2 = 0$, enclosing the ellipsoid. Then the capacitance C is

$$\frac{1}{C} = \frac{\phi_1 - \phi_2}{q} = \frac{1}{2c} \ln \frac{a+c}{a-c}$$

For the minor axis of the ellipsoid b, we have $b^2 = a^2 - c^2$. For the capacitance of the conducting ellipsoid we get

$$C = \frac{2\sqrt{a^2 - b^2}}{\ln\left(\dfrac{a + \sqrt{a^2 - b^2}}{a - \sqrt{a^2 - b^2}}\right)} = \frac{2\sqrt{a^2 - b^2}}{\ln \dfrac{\left(a + \sqrt{a^2 - b^2}\right)\left(a + \sqrt{a^2 - b^2}\right)}{\left(a - \sqrt{a^2 - b^2}\right)\left(a + \sqrt{a^2 - b^2}\right)}}$$

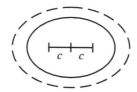

Figure 1.33. Conducting ellipsoid of revolution.

$$= \frac{2\sqrt{a^2 - b^2}}{\ln \dfrac{\left(a + \sqrt{a^2 - b^2}\right)^2}{a^2 - a^2 + b^2}} = \frac{\sqrt{a^2 - b^2}}{\ln \left[\dfrac{(a + \sqrt{a^2 - b^2})^2}{b^2}\right]^{1/2}} = \frac{\sqrt{a^2 - b^2}}{\ln \dfrac{a + \sqrt{a^2 - b^2}}{b}}$$

For stretched ellipsoids, that is, for small quotients b/a one obtains

$$C = \frac{a}{\ln \dfrac{2a}{b}}$$

This is approximately the capacitance of a wire of the length $2c$ and of the diameter $2b$.

Example 1.10: Exercise: Capacitances and self-induction coefficients

Consider n conductors. The ith conductor carries the charge q_i and is at potential V_i. There is a linear relation between the charges and the potentials

$$q_i = \sum_{j=1}^{n} C_{ij} V_j$$

with constant quantities C_{ij}. The coefficients C_{ii} are called capacitances, and the $C_{ij} (i \neq j)$ are the self-induction coefficients.

(a) Explain the relation given above.

(b) Calculate the C_{ij} for two concentric spherical shells of radii r_i and r_a. What relation exists between the C_{ij} and the "capacitance" of the spherical capacitor?

Solution **(a)** Let the potential at infinity be equal to zero. If all of the conductors are grounded ($V_i = 0$), then also all charges are 0. If the first conductor receives the potential V_1 while the other conductors remain grounded (see Figure 1.34), then the charge $q_i^{(1)}$, flowing to the ith conductor depends linearly on V_1: $q_i^{(1)} = C_{i1} V_1$. If now the second conductor is brought to potential V_2, an additional charge $q_i^{(2)} = C_{i2} V_2$ flows to the ith conductor, etc. Altogether, $q_i = \sum_j q_i^{(j)} = \sum_j C_{ij} V_j$.

The C_{ij} depend on the dimension, the shape, and the arrangement of the conductors. The potential energy of the system can be expressed easily by these quantities

$$W = \frac{1}{2} \sum_i q_i V_i = \frac{1}{2} \sum_{ij} C_{ij} V_i V_j$$

$q_1,\ V_1 \neq 0$

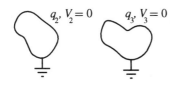
$q_2,\ V_2 = 0$ $q_3,\ V_3 = 0$

r_a q_a r_i q_i

Figure 1.34. Conductors at potentials $V_i \neq 0$ and $V_j = 0$ (grounded).

Figure 1.35. Spherical capacitor: The inner and outer shells carry charges q_i and q_o, respectively.

(b) Let the charges q_i and q_a reside on the inner and outer spherical shells, respectively (see Figure 1.35). The field intensity produced by these charges is

$$\mathbf{E}(\mathbf{r}) = q_a \frac{\mathbf{e}_r}{r^2}\theta(r - r_a) + q_i \frac{\mathbf{e}_r}{r^2}\theta(r - r_i)$$

Hence,

$$V_a = \int_{r_a}^{\infty} \mathbf{E}(\mathbf{r}) \cdot d\mathbf{r} = \int_{r_a}^{\infty} \frac{q_a + q_i}{r^2} dr = (q_a + q_i)\frac{1}{r_a}$$

$$V_i = \int_{r_i}^{\infty} \mathbf{E}(\mathbf{r}) \, d\mathbf{r} = \int_{r_i}^{\infty} \frac{q_i}{r^2} dr + \int_{r_a}^{\infty} \frac{q_a}{r^2} dr = \frac{q_a}{r_a} + \frac{q_i}{r_i}$$

If this system of equations is solved for q_i and q_a, then

$$\begin{pmatrix} q_i \\ q_a \end{pmatrix} = \begin{pmatrix} C_{ii} & C_{ia} \\ C_{ai} & C_{aa} \end{pmatrix}\begin{pmatrix} V_i \\ V_a \end{pmatrix} = \frac{r_a}{r_i - r_a}\begin{pmatrix} r_i & -r_i \\ -r_i & r_i \end{pmatrix}\begin{pmatrix} V_i \\ V_a \end{pmatrix}$$

The "capacitance" of a system of two conductors is defined by

$$C = \frac{q}{\Delta V}$$

where $q_1 = q$ and $q_2 = -q$ are the charges on the first and second conductor, respectively, and $\Delta V = V_1 - V_2$ represents the potential difference between them. Now,

$$\begin{pmatrix} q_1 \\ q_2 \end{pmatrix} = \begin{pmatrix} C_{11} & C_{12} \\ C_{21} & C_{22} \end{pmatrix}\begin{pmatrix} V_1 \\ V_2 \end{pmatrix}$$

and therefore,

$$\begin{pmatrix} V_1 \\ V_2 \end{pmatrix} = \frac{1}{\det(C_{ij})}\begin{pmatrix} C_{22} & -C_{12} \\ -C_{21} & C_{11} \end{pmatrix}\begin{pmatrix} q_1 \\ q_2 \end{pmatrix}$$

For the capacitance one obtains

$$C = \frac{q}{V_1 - V_2} = q \cdot \frac{\det(C_{ij})}{q(C_{22} + C_{21} + C_{12} + C_{11})} = \frac{C_{11}C_{22} - C_{12}C_{21}}{C_{11} + C_{12} + C_{21} + C_{22}}$$

For the spherical capacitor, we get

$$C = \frac{\left(\dfrac{r_a}{r_a - r_i}\right)^2 \cdot (r_i r_a - r_i^2)}{\left(\dfrac{r_a}{r_a - r_i}\right) \cdot (r_a - r_i)} = \frac{r_i r_a}{r_a - r_i}$$

This agrees with our previous result (Exercise 1.2).

Biographical notes

Charles Auguste Coulomb, b. June 14, 1736, Angoulème–d. Aug, 23, 1806, Paris, French physicist. A corresponding member of the Académie des Sciences from 1774, Coulomb became a full member of its successor organization (Institut National) in 1795. Until 1776 he held the position of Lieutenant-Colonel du génie in Martinique, and after that he was Inspecteur général de l'Université in Paris. (see also p. 512) [BR]

Carl Friedrich Gauß, b. Apr. 30, 1777, Braunschweig–d. Feb. 23, 1855, Göttingen. Gauss' father was a day laborer. Gauss' extraordinary talent for mathematics was realized very early. In 1791 he became a protégé of the Duke of Braunschweig, who paid for his education. From 1795 to 1798, Gauss studied in Göttingen, and in 1799 he did a doctorate in Helmstedt. In 1807 he became head of the observatory and professor at Göttingen university. He turned down all offers for better jobs. Gauss started working practically in 1791 with tests of the geometrical-arithmetical mean, of distribution of prime numbers, and in 1792 with "Grundlagen der Geometrie." Just two years later, he discovered the "method of smallest squares." From 1795 he dealt with the theory of numbers, e.g., the squared reciprocity theorem. In 1796 Gauss published his first work, which proved that it was possible to construct a regular polygon of n angles with a pair of compasses and a ruler, as long as n stands for Fermat's prime numbers. This holds especially for $n = 17$. Gauss' thesis (1799) succeeded in proving the fundamental theorem of algebra. Several other proofs were to follow. It is known that at that time Gauss already had a basic knowledge of a theory of elliptical and modular functions. The first extensive treatise written by him was published in 1801. "Disquistiones Arithmeticae" marked the beginning of a new theory of numbers. Among other topics it contains a theory of divided circles. From about 1801 Gauss was interested in astronomy, too. In 1801 he managed to calculate the trajectory of the planetoid Ceres; in 1809 and 1818 he experimented with the attraction of general ellipsoids. A treatise on the hypergeometric series was published in 1812. It includes the first correct and systematic tests with convergence. In 1820 Gauss turned to geodesy. His most important theoretical achievement was the "surface theory" (1827) with the "Theorema egregium." He worked in practical geodesy, too. Between 1821 and 1825 he executed extensive measurings. In spite of such time-consuming work, he published two works on "biquadratic rests" in 1825 and 1831, the latter including a presentation of complex numbers in a plane, and a new theory of prime numbers. During the last years of his life, Gauss was interested in problems of physics. Important achievements in this field are the invention of an electric telegraph (1833–1834, together with W. Weber), and potential theory (1839–1840). Several important findings are only known from his diaries and letters, e.g., the non-euclidian geometry (1816). The reason for his being so secretive was the strict standard he applied to the publication of his thoughts, as well as his attempt to avoid unnecessary arguments. (see also p. 520)

Siméon Denis Poisson, b. June 21, 1781, Pithiviers–d. Apr. 25, 1840, Paris. Poisson was a student at the Ecole Polytechnique. After having finished his studies, he worked there. In 1802, he got a chair. Poisson was a member of the Bureau of Longitudes and of the Académie des Sciences. From 1787 he was Pair of France. Poisson worked in several fields, e.g., general mechanics, heat conduction, potential theory, differential equations, and probability calculus. (see also p. 525)

Pierre Simon Laplace, b. March 28, 1749, Beaumont-en-Age–d. March 5, 1827, Paris. After having finished school, Laplace became teacher in Beaumont and—with the help of d'Alembert—professor at the Ecole Militaire in Paris. Since Laplace tended to change his political opinion quite often, he was honoured by Napoleon as well as by Louis XVIII. His most important work is "Analytical Theory of Probability" (1799–1825). Probability calculus includes the method of generating functions, Laplace transformations, and the ultimate formulation of mechanical materialism. Celestial mechanics includes Laplace's cosmological hypothesis, theories on the shape of Earth and of Moon's motion, perturbation theory of planets, and the potential theory of the Laplace equation. (see also p. 522)

Paul Maurice Dirac, b. Aug. 8, 1902, Bristol–d. Oct. 20, 1984, Tallahassee, Florida, USA. Dirac studied in Bristol and Cambridge, and at several foreign universities. In 1932 he got a chair for mathematics. Dirac is said to be one of the founders of quantum mechanics. The mathematical equivalent developed by him mainly consists of a noncommutative algebra to calculate the properties of atoms. Dirac developed a relativistic theory of electrons, predicted the discovery of the positron in 1928, and made important contributions to quantum field theory. He received the Nobel Prize in 1933.

2 Green's Theorems

For the determination of the electrostatic potential we encountered the relation

$$\phi(\mathbf{r}) = \int \frac{\rho(\mathbf{r}')}{|\mathbf{r} - \mathbf{r}'|} dV' \tag{2.1}$$

This relation can be employed if the charge distribution $\rho(\mathbf{r}')$ is known. But frequently *in electrostatics the problem is a different one*. Besides charge distributions $\rho(\mathbf{r})$ also the potential distributions $\phi(\mathbf{r})$ or distributions of the electric field $\mathbf{E}(\mathbf{r}) = -\nabla\phi$ on certain surfaces and bodies are given (see Figure 1.1). The charge distributions producing them are mostly not known. These *boundary conditions* fix the potential at all points in space. For the solution of this boundary-value problem the so-called Green's theorems are required, which will be derived in the following.

Let $\mathbf{A}(\mathbf{r})$ denote a vector field, and let $\varphi(\mathbf{r})$ and $\psi(\mathbf{r})$ be two scalar fields. Let the vector field $\mathbf{A}$ be constructed out of the φ and ψ-fields in the following way:

$$\mathbf{A} = \varphi\nabla\psi \tag{2.2}$$

Thus, the product rule yields

$$\nabla \cdot \mathbf{A} = \nabla \cdot (\varphi\nabla\psi) = \varphi\nabla^2\psi + (\nabla\varphi) \cdot (\nabla\psi) \tag{2.3}$$

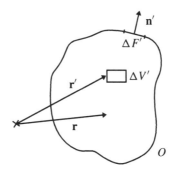

Figure 2.1. Illustation of the geometrical relations in the integrations for Green's theorems.

We start from Gauss' theorem

$$\int_V \nabla' \cdot \mathbf{A} \, dV' = \oint_S \mathbf{A} \cdot \mathbf{n}' \, da' \tag{2.4}$$

and replace $\nabla' \cdot \mathbf{A}$ by equation (2.3). Furthermore, we write $\mathbf{A} \cdot \mathbf{n}' = \varphi(\nabla'\psi) \cdot \mathbf{n}' = \varphi \, \partial\psi/\partial n'$, the derivative of the scalar field ψ in the direction of the normal vector to the surface. The primes mean that the derivatives are taken with respect to the coordinates which are also integrated. With this replacement the *first Green's theorem* follows from Gauss' theorem:

$$\int_V \left[\varphi \, \nabla'^2 \psi + (\nabla'\varphi) \cdot (\nabla'\psi) \right] dV' = \oint_S \varphi \frac{\partial\psi}{\partial n'} \, da' \tag{2.5}$$

In general, the charge density $\rho(\mathbf{r})$ and the potential distribution $\phi(\mathbf{r})$, or the electric field $\mathbf{E}(\mathbf{r})$ on a surface S (position vector $\mathbf{r}$) are given. The *second Green's theorem* is derived by interchanging the arbitrarily chosen scalar fields in (2.5) and subsequent subtraction of the equation obtained from equation (2.3). The second Green's theorem then reads (with $\nabla^2 = \Delta$):

$$\int_V \left[\varphi(\mathbf{r}')\Delta'\psi(\mathbf{r}') - \psi(\mathbf{r}')\Delta'\varphi(\mathbf{r}') \right] dV' = \oint_S \left[\varphi(\mathbf{r}')\frac{\partial\psi(\mathbf{r}')}{\partial n'} - \psi(\mathbf{r}')\frac{\partial\varphi(\mathbf{r}')}{\partial n'} \right] da' \tag{2.6}$$

From Green's theorems and employing Poisson's differential equation we want to find now an integral representation of the potential more general than the relation (2.1). For the arbitrarily chosen scalar fields we set

$$\psi(\mathbf{r}') = \frac{1}{|\mathbf{r} - \mathbf{r}'|} \quad \text{and} \quad \varphi(\mathbf{r}') = \phi(\mathbf{r}') \text{ (potential)} \tag{2.7}$$

Then

$$\Delta'\psi = \Delta' \frac{1}{|\mathbf{r} - \mathbf{r}'|} = -4\pi \, \delta(\mathbf{r} - \mathbf{r}') \tag{2.8}$$

and

$$\Delta'\varphi = \Delta'\phi(\mathbf{r}') = -4\pi\rho(\mathbf{r}') \tag{2.9}$$

With these relations we get, according to equation (2.6)

$$-4\pi \int_V \phi(\mathbf{r}') \, \delta(\mathbf{r} - \mathbf{r}') \, dV' + 4\pi \int_V \frac{\rho(\mathbf{r}')}{|\mathbf{r} - \mathbf{r}'|} \, dV'$$
$$= \oint_S \left[\phi(\mathbf{r}')\frac{\partial}{\partial n'}\frac{1}{|\mathbf{r} - \mathbf{r}'|} - \frac{1}{|\mathbf{r} - \mathbf{r}'|}\frac{\partial}{\partial n'}\phi(\mathbf{r}') \right] da' \tag{2.10}$$

If the point of observation $\mathbf{r}$ lies inside the integration volume, then the first term on the left-hand side yields the potential at $\mathbf{r}$:

$$\phi(\mathbf{r}) = \int_V \frac{\rho(\mathbf{r}')}{|\mathbf{r} - \mathbf{r}'|} \, dV' + \frac{1}{4\pi} \oint_S \left[\frac{1}{|\mathbf{r} - \mathbf{r}'|}\frac{\partial}{\partial n'}\phi(\mathbf{r}') - \phi(\mathbf{r}')\frac{\partial}{\partial n'}\frac{1}{|\mathbf{r} - \mathbf{r}'|} \right] da' \tag{2.11}$$

Now we treat two special cases of this equation. At first, we shift the integration surface S to infinity. Then, the integral goes faster to zero, e.g., for a point charge:

$$\frac{1}{|\mathbf{r} - \mathbf{r}'|} \frac{\partial}{\partial n'} \phi(\mathbf{r}') \sim \frac{1}{r'^3}$$

than the surface element tends to infinity. Consequently, the surface integral vanishes, and the known form (2.1) remains:

$$\phi(\mathbf{r}) = \int_V \frac{\rho(\mathbf{r}')}{|\mathbf{r} - \mathbf{r}'|} dV'$$

On the other hand, if the integration volume is free of charges, then the first term of equation (2.11) becomes zero, and the potential is determined only by the values of the potential and the values of its derivatives at the boundary of the integration region (the surface S).

One should note that the integration volume always has to contain the point $\mathbf{r}$ in order to get a solution for the potential.

According to equation (2.11) the potential $\phi(\mathbf{r})$ is defined by the potential and its normal derivatives on the boundary surface. But, by giving both values $\phi(S)$ and $\partial\phi(S)/\partial n$ the problem is overdetermined: as we will see soon, it is enough to give one condition to fix the potential. The boundary conditions are called *Dirichlet boundary conditions* if ϕ is given at the boundary, and *Neumann boundary conditions* if the normal derivative $\partial\phi/\partial n$ (that is the normal component of the electric field intensity) is given there.

Uniqueness of the solutions

We will now show that the potential is determined uniquely by giving Dirichlet *or* Neumann boundary conditions.

The two solutions $\phi_1(\mathbf{r})$ and $\phi_2(\mathbf{r})$, assumed to be different from each other, obey the Poisson equation (or the Laplace equation):

$$\Delta\phi_{1,2}(\mathbf{r}) = -4\rho(\mathbf{r}), \qquad \text{that is,} \qquad \Delta(\phi_1 - \phi_2) = 0 \tag{2.12}$$

At the boundary both functions satisfy the same conditions

$$\phi_1(S) = \phi_2(S) \qquad \text{or} \qquad \frac{\partial\phi_1}{\partial n}(S) = \frac{\partial\phi_2}{\partial n}(S) \tag{2.13}$$

Now we set $u = \phi_1 - \phi_2$ and use the first Green's theorem with $\varphi = \psi = u$, so that

$$\int_V [u(\mathbf{r}')\,\Delta'u(\mathbf{r}') + (\nabla'u(\mathbf{r}'))^2]\,dV' = \oint_S u(\mathbf{r}')\frac{\partial u(\mathbf{r}')}{\partial n'}\,da' \tag{2.14}$$

Due to the condition (2.12) the first term of the volume integral vanishes. The surface integral becomes zero since either u (Dirichlet) or $\partial u/\partial n$ (Neumann) vanish on the surface due to the boundary conditions. This means:

$$\int_V (\nabla' u(\mathbf{r}'))^2 \, dV' = 0 \qquad \text{or} \qquad \nabla' u(\mathbf{r}') = 0 \tag{2.15}$$

Thus, u is constant throughout the volume V. As for Dirichlet boundary conditions, u is zero at the boundary, and we have $u = 0$ everywhere, that is, $\phi_1 = \phi_2$, so that the uniqueness is demonstrated. For Neumann boundary conditions, both solutions for ϕ differ at most in an insignificant additive constant. With the proved uniqueness of the solution of the potential problem it becomes clear that the potential is fixed uniquely by either Dirichlet or Neumann boundary conditions. If $\phi(S)$ and $\partial\phi(S)/\partial n$ are given as required in equation (2.11) the problem is overdetermined. In other words, $\phi(S)$ and $\partial\phi(S)/\partial n$ depend on each other.

Green function

With the help of the second Green's theorem we can calculate now the solution of the Poisson equation or the Laplace equation within a certain bounded volume with known Dirichlet or Neumann boundary conditions. To get the equations (2.11) and (2.12), we choose $\psi = 1/|\mathbf{r} - \mathbf{r}'|$ corresponding to the potential of a unit charge for which holds

$$\nabla^2 \frac{1}{|\mathbf{r} - \mathbf{r}'|} = -4\pi \delta(\mathbf{r} - \mathbf{r}') \tag{2.16}$$

This function ψ is one of the class of functions G depending on $\mathbf{r}$ and $\mathbf{r}'$ and for which we have

$$\nabla^2 G(\mathbf{r}, \mathbf{r}') = -4\pi \delta(\mathbf{r} - \mathbf{r}') \qquad \text{with} \qquad G(\mathbf{r}, \mathbf{r}') = \frac{1}{|\mathbf{r} - \mathbf{r}'|} + F(\mathbf{r}, \mathbf{r}') \tag{2.17}$$

Here, F has to fulfil the Laplace equation $\Delta F = 0$. G is called the *Green function*, and it is symmetric in $\mathbf{r}$ and $\mathbf{r}'$: $G(\mathbf{r}, \mathbf{r}') = G(\mathbf{r}', \mathbf{r})$.

It should be stated again that the particular solution $\psi(\mathbf{r}, \mathbf{r}') = 1/|\mathbf{r} - \mathbf{r}'|$ obeys Poisson's equation (2.16), but it does not satisfy Dirichlet or Neumann boundary conditions, except if the surface lies at infinity. For the Green function $G(\mathbf{r}, \mathbf{r}')$ the boundary conditions can be taken into account via the functions $F(\mathbf{r}, \mathbf{r}')$ in equation (2.17). We will continue to follow this idea.

Let us consider equation (2.11), which is a conditional equation for $\phi(\mathbf{r})$. But it still contains both conditions, $\phi(\mathbf{r}')|_S$ as well as $\partial\phi(\mathbf{r}')/\partial n|_S$. Due to the Green function and the freedom contained in it, it is possible to choose a function $F(\mathbf{r}, \mathbf{r}')$ in such a way that one of the surface integrals vanishes and, thus, an equation with either Dirichlet or Neumann boundary condition is obtained. Setting $\varphi = \phi$ and $\psi = G(\mathbf{r}, \mathbf{r}')$ in (2.6), and taking into account the relations found above for the Green function, we find in analogy to equation (2.11)

$$\phi(\mathbf{r}) = \int_V \rho(\mathbf{r}')G(\mathbf{r}, \mathbf{r}')\, dV' + \frac{1}{4\pi} \oint_S \left[G(\mathbf{r}, \mathbf{r}')\frac{\partial \phi(\mathbf{r}')}{\partial n'} - \phi(\mathbf{r}')\frac{\partial G(\mathbf{r}, \mathbf{r}')}{\partial n'} \right] da' \quad \textbf{(2.18)}$$

By an appropriate choice of $G(\mathbf{r}, \mathbf{r}')$ one or the other of the surface integrals can be eliminated in the following way: for Dirichlet boundary conditions, we set $G_D(\mathbf{r}, \mathbf{r}') = 0$, if $\mathbf{r}'$ lies on the surface S, that is, if $\mathbf{r}' = \mathbf{r}'(S)$. Then, the problem is formulated only for Dirichlet boundary conditions since the equation given above becomes

$$\phi(\mathbf{r}) = \int_V \rho(\mathbf{r}')G_D(\mathbf{r}, \mathbf{r}')\, dV' - \frac{1}{4\pi} \oint_S \phi(\mathbf{r}')\frac{\partial G_D(\mathbf{r}, \mathbf{r}')}{\partial n'}\, da' \quad \textbf{(2.19)}$$

For Neumann boundary conditions the obvious ansatz $\partial/\partial G_D(\mathbf{r}, \mathbf{r}')n' = 0$ leads to a wrong result since it does not fulfil the requirement of Gauss' law (for a unit charge):

$$\oint_S \frac{\partial G_N(\mathbf{r}, \mathbf{r}')}{\partial n'}\, da' = -4\pi \quad \textbf{(2.20)}$$

Therefore, the simplest ansatz is

$$\frac{\partial G_N(\mathbf{r}, \mathbf{r}')}{\partial n'} = -\frac{4\pi}{S} \quad \textbf{(2.21)}$$

if S is the entire surface and $\mathbf{r}'$ lies on the surface. Then we obtain

$$\phi(\mathbf{r}) = \langle \phi \rangle_S + \int_V \rho(\mathbf{r}')G_N(\mathbf{r}, \mathbf{r}')\, dV' + \frac{1}{4\pi} \oint_S \frac{\partial \phi(\mathbf{r}')}{\partial n'}G_N(\mathbf{r}, \mathbf{r}')\, da' \quad \textbf{(2.22)}$$

Here, $\langle \phi \rangle_S = \frac{1}{S} \oint_S \phi(\mathbf{r}')da'$ is the average value of the potential at the surface. This average value can be absorbed always into the additive constant in which the potential is arbitrary.

If the entire surface S tends to infinity, then it vanishes if $\phi(\mathbf{r})$ decreases faster than $1/r$, considering $r \to \infty$. Here, the physical meaning of $F(\mathbf{r}, \mathbf{r}')$ should be mentioned. $F(\mathbf{r}, \mathbf{r}')$

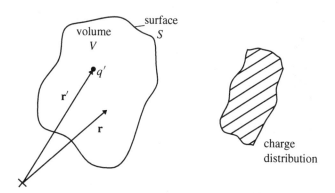

Figure 2.2. The hatched region shows the charge distribution outside the volume V generating the "image potential" $F(\mathbf{r}, \mathbf{r}')$. Depending on where q' lies in V, that is, depending on $\mathbf{r}'$, the image charge $F(\mathbf{r}, \mathbf{r}')$ will be different.

solves the Laplace equation inside V. Hence, the function $F(\mathbf{r}, \mathbf{r}')$ denotes the potential of a charge distribution outside the volume V so that, together with the potential $1/|\mathbf{r}-\mathbf{r}'|$ of the point charge at the point $\mathbf{r}'$, the Green function can just reach the values $G_D(\mathbf{r}, \mathbf{r}') = 0$ or $\partial/\partial G_D(\mathbf{r}, \mathbf{r}')n' = -4\pi/S$ at the boundary surface ($\mathbf{r}' = \mathbf{r}'(S)$). It is clear that the external charge distribution depends on the point charge in the volume whose potential or the normal derivative of whose potential it has to compensate for $\mathbf{r}' = \mathbf{r}'(S)$. This means that $F(\mathbf{r}, \mathbf{r}')$ depends on the parameter $\mathbf{r}'$, which gives the position of the point charge distribution in the volume. The *method of images* (Figure 2.2) that is used in the following examples is based on this knowledge.

Example 2.1: The grounded conducting sphere in the field of a charge q

Consider a conducting sphere of radius a held at ground potential ($\phi = 0$). *Grounded* means that the potential is the same as that of the surface of the earth, and therefore, the same as very far from the sphere, at infinity.

By $\mathbf{y}$ we denote the vector from the center of the sphere to the position of the charge q, and by $\mathbf{r}$ the point of observation (see Figure 2.3). The boundary conditions are $\phi(a) = \phi(|\mathbf{r}| = \infty) = 0$. This is a *Dirichlet* problem. First, we determine the Green function G_D obeying the boundary condition, using the *method of electric images*. That is, we try to achieve the boundary condition by placing a second charge of magnitude q' at an appropriate position (e.g., at $\mathbf{y}'$ in the figure) such that by superposition with the first charge q the boundary condition is satisfied.

As will be shown by the calculation, it is sufficient to use one image charge q' possessing, for symmetry reasons, a position vector $\mathbf{y}'$ parallel to $\mathbf{y}$. We choose the center of the sphere as the origin. To determine the magnitude and position, we set up the common potential of both charges and try to fulfil the boundary conditions. The potential of the two charges is

$$\phi(\mathbf{r}, \mathbf{y}) = \frac{q}{|\mathbf{r} - \mathbf{y}|} + \frac{q'}{|\mathbf{r} - \mathbf{y}'|}$$

If $\mathbf{n} = \mathbf{r}/|\mathbf{r}|$ and $\mathbf{n}' = \mathbf{y}/|\mathbf{y}|$ are the unit vectors in $\mathbf{r}$- and $\mathbf{y}$-direction, respectively, then

$$\phi(\mathbf{r}, \mathbf{y}) = \frac{q}{|r\mathbf{n} - y\mathbf{n}'|} + \frac{q'}{|r\mathbf{n} - y'\mathbf{n}'|}$$

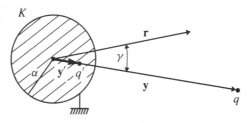

Figure 2.3. Grounded conducting sphere in the field of a charge q.

From this expression, by factoring out r or y' at the point $r = a$ we obtain the following relation:

$$\phi(|\mathbf{r}| = a, \mathbf{y}) = \frac{q}{a\left|\mathbf{n} - \dfrac{y}{a}\mathbf{n}'\right|} + \frac{q'}{y'\left|\mathbf{n}' - \dfrac{a}{y'}\mathbf{n}\right|}$$

$$= \frac{q}{a\sqrt{1 + \dfrac{y^2}{a^2} - 2\dfrac{y}{a}\cos\gamma}} + \frac{q'}{y'\sqrt{1 + \dfrac{a^2}{y'^2} - 2\dfrac{a}{y'}\cos\gamma}} = 0$$

according to the assumption (Dirichlet boundary condition for the Green function). This equation is correct for all $\mathbf{n} \cdot \mathbf{n}' = \cos\gamma$ only if

$$\frac{q}{a} = -\frac{q'}{y'} \quad \text{and} \quad \frac{y}{a} = +\frac{a}{y'}$$

Thus, the charge and the dimension of the image are[1]

$$q' = -\frac{a}{y}q; \qquad y' = \frac{a^2}{y}$$

We can see that q' and y' approach zero if y becomes large, that is, the charge vanishes at infinity. Furthermore, q and q' become opposite and equal when $|y' - y|$ tends to zero. With these values for q' and y', the potential of the charge q of a grounded sphere is

$$\phi(\mathbf{r}, \mathbf{y}) = q\left(\frac{1}{|\mathbf{r} - \mathbf{y}|} - \frac{a}{y\left|\dfrac{a^2}{y^2}\mathbf{y} - \mathbf{r}\right|}\right)$$

This holds for the region external to the sphere, i.e., for $r \geq a$. Namely, this region is the charged volume V of equation (2.19). The surface of the sphere is the bounding surface S. Since according to our general considerations the solution of the boundary value problem is unique, the solution found for the problem is the only one. The Green function for this potential distribution is

$$G_D(\mathbf{r}, \mathbf{y}) = \frac{1}{|\mathbf{r} - \mathbf{y}|} - \frac{a}{y\left|\dfrac{a^2}{y^2}\mathbf{y} - \mathbf{r}\right|}$$

$$= \frac{1}{\sqrt{r^2 + y^2 - 2ry\cos\gamma}} - \frac{1}{\sqrt{a^2 + \dfrac{r^2y^2}{a^2} - 2yr\cos\gamma}}$$

It may be checked easily that $G_D(\mathbf{a}, \mathbf{y})$ as well as $G_D(\mathbf{r}, \mathbf{a})$ vanish as required. Furthermore, we have $G_D(\mathbf{r}, \mathbf{y}) = G_D(\mathbf{y}, \mathbf{r})$, as always. To determine the charge induced by q at the sphere, we start from the statement that the jump of the field intensity across charged surfaces is equal to the charge density at the surface

$$(\mathbf{E}_2 - \mathbf{E}_1) \cdot \mathbf{n} = \mathbf{E}_2 \cdot \mathbf{n} = -\frac{\partial\phi}{\partial n} = 4\pi\sigma(\mathbf{r})$$

---

[1] There is another, more trivial, solution $q = -q'$, $y = y' = a$, but here the charge would vanish.

because $\mathbf{E}_1 = 0$ since the fields inside a homogeneous conducting sphere break down. The electrons of the conductor shift until no further forces are acting, that is, $\mathbf{E}(\mathbf{r}) = 0$ for $r < a$. This follows from the fact that the whole sphere, including its interior region, has the potential $\phi = 0$. Hence, the system is spherically symmetric

$$\sigma(r = a) = -\frac{1}{4\pi}\frac{\partial\phi}{\partial n}(r = a) = -\frac{1}{4\pi}\frac{\partial\phi}{\partial r}(r = a)$$

$$= -\frac{q}{4\pi}\left[-\frac{r - y\cos\gamma}{\left(\sqrt{r^2 + y^2 - 2ry\cos\gamma}\right)^3} + \frac{ry^2/a^2 - y\cos\gamma}{\left(\sqrt{a^2 + r^2y^2/a^2 - 2yr\cos\gamma}\right)^3}\right]_{r=a}$$

$$= -\frac{q}{4\pi ay}\cdot\frac{\left(1 - \dfrac{a^2}{y^2}\right)}{\left(1 + \dfrac{a^2}{y^2} - \dfrac{2a}{y}\cos\gamma\right)^{3/2}}$$

where γ is the angle between $\mathbf{r}$ and $\mathbf{y}$. Drawing the density of the surface charge against the angle γ for two ratios y/a, one obtains Figure 2.4.

The total induced charge Q is

$$Q = \int_S \sigma(y, \gamma)a^2 d\Omega = -\frac{a}{y}q = q'$$

just equal to the image charge, as it must be according to Gauss' law. The next question to be answered concerns the strength of the attractive force between the charge q and the charge induced by it at the surface of the sphere. We could perform this calculation by integration using the charge density of the

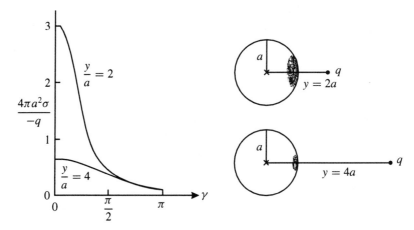

Figure 2.4. Induced charge, which decreases with increasing distance y of the charge q. Note, that the induced charge is plotted in units of $4\pi a^2\sigma/-q$.

surface found above. But the induced charge, as we have seen, behaves just like a charge q' sitting at the position $\mathbf{y}'$, so it is easier to compute the force between the original charge and the image charge:

$$F = \frac{q\,q'}{|\mathbf{y} - \mathbf{y}'|^2} = -\frac{q^2\dfrac{a}{y}}{y^2\left(1 - \dfrac{a^2}{y^2}\right)^2}$$

Hence, the two charges always attract each other. For smaller distances we have $y \cong a$ and therefore

$$F \approx -q^2 \cdot a^2 \cdot \frac{1}{\left(y^2 - a^2\right)^2} = -q^2 \cdot a^2 \cdot \frac{1}{(y+a)^2(y-a)^2} \approx -\frac{q^2}{4}\frac{1}{(y-a)^2} = -\frac{q^2}{4}\frac{1}{d^2}$$

The quantity $d = y - a$ denotes the distance between the charge q and the spherical surface. In the vicinity of the surface of the sphere the force decreases as $1/d^2$, but far outside it varies as $1/y^3$, similar to the force in a dipole field. This is the typical behavior of the charge and the induced charge.

Obviously, if a charge leaves the surface of the sphere and further moves to infinity work has to be done against the attractive force of the induced charge. This is the *work function*. It is the reason that, in general, the charges remain on a conductor (the sphere in our case) and do not leave it to move to infinity. Due to the repulsion of like charges one would not expect this, at first. But, when a charge moves off the surface the corresponding image charge appears immediately, attracting the original charge (to get it back). ·

Example 2.2:   A conducting ungrounded sphere in the field of a charge

Let a conducting but ungrounded sphere be in the vicinity of a point charge q. If this sphere, in the charge-free space, has a total charge Q distributed uniformly over its surface, then this charge is conserved if q is located in the exterior region because the charge cannot drain off. Of course, a charge q' will be distributed over the surface in such a way that the surface-charge distribution calculated in Example 2.1 will be observed. The residual charge $(Q - q')$ will be distributed uniformly over the sphere because the forces of q are balanced already by $q' = -aq/y$ at the position $\mathbf{y}' = (a^2/y)\mathbf{y}/y$. Therefore, the charge $Q - q'$ will act like it is concentrated at the center. Hence, in the exterior region the potential found in the Example 2.1 will get an additional term of the form $(Q - q')/r$:

$$\phi(\mathbf{r}, \mathbf{y}) = \frac{q}{|\mathbf{r} - \mathbf{y}|} - \frac{\dfrac{aq}{y}}{\left|\mathbf{r} - \dfrac{a^2}{y^2}\mathbf{y}\right|} + \frac{Q + \dfrac{aq}{y}}{r}$$

Similarly, the attractive force between q and the sphere can be determined again

$$\mathbf{F}(\mathbf{y}) = q\mathbf{E}(\mathbf{y}) = q\left[-\operatorname{grad}\phi(\mathbf{r}, \mathbf{y})|_{\mathbf{r}=\mathbf{y}}\right]$$

$$= q\left[-\frac{\dfrac{aq}{y}\left(\mathbf{r} - \dfrac{a^2}{y^2}\mathbf{y}\right)}{\left|\mathbf{r} - \dfrac{a^2}{y^2}\mathbf{y}\right|^3} + \frac{Q + \dfrac{aq}{y}}{r^3}\mathbf{r}\right]_{\mathbf{r}=\mathbf{y}} \tag{2.23}$$

$$= \frac{q}{y^2}\left[Q - \frac{qa^3(2y^2 - a^2)}{y(y^2 - a^2)^2}\right]\frac{\mathbf{y}}{y} \tag{2.24}$$

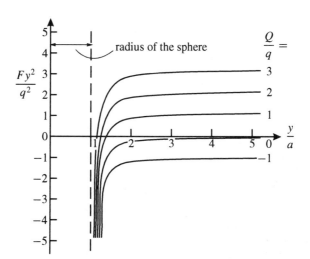

Figure 2.5. The force between a charge and its image charge for a conducting ungrounded sphere.

The divergent self-interaction of the charge q represented for $\mathbf{r} = \mathbf{y}$ by the first term in the potential $\phi(\mathbf{r}, \mathbf{y})$ has been omitted here.

While the force between the charge and the corresponding image charge is always attractive, for small distances always a strong attractive force appears in this case, but for $Q/q > 0$ the sphere and the charge repel each other at larger distances (see Figure 2.5). The strong attractive force occurring in close vicinity of the surface and the *work of escape* to be done to overwhelm it are responsible for the fact that like charges remain on the surface of the sphere in spite of the electrostatic repulsion. The work function is independent of the signs of Q and q. If these signs are equal, work can be gained beyond a certain distance. But, if the sphere is grounded, work has to be done up to infinity.

Example 2.3:   A conducting sphere kept at a potential V in the field of a charge

Let a point charge q be in the vicinity of a conducting sphere whose surface is kept at a constant potential V.

This problem can be reduced to the former example if we bear in mind that a charge of magnitude $V \cdot a$ sitting in the center of the sphere produces the required potential V at the surface. Hence, the required equations are obtained by replacing $(Q - q')$ by Va in the expressions of the previous example:

$$\phi(\mathbf{r}, \mathbf{y}) = \frac{q}{|\mathbf{r} - \mathbf{y}|} - \frac{aq}{y\left|\mathbf{r} - \dfrac{a^2}{y^2}\mathbf{y}\right|} + \frac{Va}{r}$$

and

$$F(\mathbf{y}) = \frac{q}{y^2}\left[Va - \frac{qay^3}{\left(y^2 - a^2\right)^2}\right]\frac{\mathbf{y}}{y}$$

These equations describe the potential and the forces outside the sphere, that is, for $r > a$. The potential Va/r satisfies $\Delta(Va/r) = 0$ outside the sphere, so it is a particular solution $F(\mathbf{r}, \mathbf{r'})$ of the homogeneous Poisson equation, that is, of the Laplace equation. It is chosen in such a way that the boundary condition is fulfilled at the spherical surface. Since $\phi(\mathbf{r}, \mathbf{y})$ obeys Poisson's equation $\Delta\phi(\mathbf{r}, \mathbf{y}) = -q\,\delta(\mathbf{r} - \mathbf{y})$ as well as the boundary condition, the potential constructed in this manner is unique (compare equations (2.12)-(2.15)). Hence, the solution $\phi(\mathbf{r}, \mathbf{y})$ found is the only one. As can be seen, attractive forces occur in the vicinity of the surface of the sphere in this case, too.

Example 2.4: A conducting sphere in an uniform electric field

Let a conducting sphere be placed in an uniform electric field as in Figure 2.6. As we will see later, it is not necessary to distinguish between a grounded sphere and an ungrounded sphere in this case.

This case is reduced to our first example by assuming that the electric field is generated by two equal and opposite charges q and $-q$ located at equal distances on opposite sides of the sphere. These two charges generate a field of magnitude $2q/(R^2)$ at the origin.

To get a uniform field, we assume that q and R tend to infinity in such a way that the magnitude of the field intensity $E_0 = 2q/R^2$ is always conserved.

The total potential is the superposition of the potentials of the two charges generating the field and their image charges. The magnitude and the position of the image charges can be taken from Example 2.1, and thus we obtain

$$\phi(\mathbf{r}, \mathbf{y}) = -\frac{q}{|\mathbf{r} - \mathbf{y}|} + \frac{q}{|\mathbf{r} + \mathbf{y}|} + \frac{q\dfrac{a}{y}}{|\mathbf{r} - \mathbf{y}'|} - \frac{q\dfrac{a}{y}}{|\mathbf{r} + \mathbf{y}'|}$$

Setting $y = R$, then due to

$$\mathbf{y}' = \frac{a^2}{y}\frac{\mathbf{y}}{y} = \frac{a^2}{R}\frac{\mathbf{y}}{y}$$

the potential is

$$\phi(r, R, \gamma) = \frac{-q}{\sqrt{R^2 + r^2 - 2rR\cos\gamma}} + \frac{q}{\sqrt{R^2 + r^2 + 2rR\cos\gamma}}$$
$$+ \frac{qa}{R\sqrt{r^2 + \dfrac{a^4}{R^2} - 2r\dfrac{a^2}{R}\cos\gamma}} - \frac{qa}{R\sqrt{r^2 + \dfrac{a^4}{R^2} + 2r\dfrac{a^2}{R}\cos\gamma}}$$

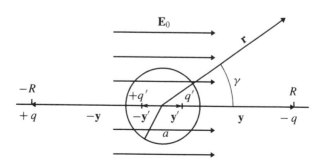

Figure 2.6. Conducting sphere in an uniform electric field.

$$= -\frac{q}{R\sqrt{1 + \left(\dfrac{r}{R}\right)^2 - 2\dfrac{r}{R}\cos\gamma}} + \frac{q}{R\sqrt{1 + \left(\dfrac{r}{R}\right)^2 + 2\dfrac{r}{R}\cos\gamma}}$$

$$+ \frac{qa}{Rr\sqrt{1 + \dfrac{a^2}{r^2}\dfrac{a^2}{R^2} - 2\dfrac{a}{r}\dfrac{a}{R}\cos\gamma}} - \frac{qa}{Rr\sqrt{1 + \dfrac{a^2}{r^2}\dfrac{a^2}{R^2} + 2\dfrac{a}{r}\dfrac{a}{R}\cos\gamma}}$$

If $r/R \ll 1$ the root can be expanded, and taking into account the term linear in r/R we obtain

$$\phi(r,\gamma) = -\frac{2q}{R^2}r\cos\gamma + \frac{2q}{R^2}\frac{a^3}{r^2}\cos\gamma, = -E_0\left(r - \frac{a^3}{r^2}\right)\cos\gamma$$

The expression $-\mathbf{E} \cdot \mathbf{r} = -E_0 r\cos\gamma$ is the potential in the uniform field, while $E_0(a^3/r^2)\cos\gamma$ obviously comes from the image charges. The two image charges q' and $-q'$ form a dipole with the dipole moment

$$p = 2q'y' = \frac{qa}{R}\frac{2a^2}{R} = a^3 E_0$$

The second term in the expression for the potential calculated above is nothing but the potential of the dipole.

Just as in the first example here, the induced surface-charge density can be calculated, too:

$$\sigma(\gamma) = -\frac{1}{4\pi}\frac{\partial\phi}{\partial r}(r = a) = \frac{3}{4\pi}E_0\cos\gamma$$

Then, the total induced charge is

$$Q = \int_{S_{\text{sphere}}} \sigma a^2 \sin\gamma\, d\gamma\, d\phi = 0$$

Since no additional charge is induced, there is no need to distinguish between a grounded and an ungrounded sphere in this example.

Example 2.5: The inversion of a potential with respect to a sphere

If two point charges q_i sitting at the positions $(r_i, \vartheta_i, \varphi_i)$ create the potential $\phi(r, \vartheta, \varphi)$ at the point (r, ϑ, φ), then it is claimed that the potential originating from the charges $q_i' = (a/r_i)q_i$ inverted

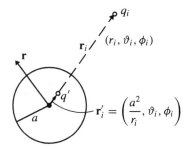

Figure 2.7. Illustration of the image charge.

with respect to a sphere of radius a and thus sitting at the positions $(a^2/r_i, \vartheta_i, \varphi_i)$ is given by $\phi'(r, \vartheta, \varphi) = (a/r)\phi(a^2/r, \vartheta, \varphi)$.

If the inverted charges, defined in this way, are negative, i.e., $q_i' = -(a/r_i)q_i$, then they are identical with the image charges generated at a conducting grounded sphere of radius a (see Figure 2.7). If γ_i is the angle between $\mathbf{r}$ and $\mathbf{r}_i$, then the potential of the charges lying at the positions $\mathbf{r}_i$ is

$$\phi(r, \vartheta, \varphi) = \sum_i \frac{q_i}{\sqrt{r^2 + r_i^2 - 2rr_i \cos \gamma_i}}$$

and because of $\gamma_i = \gamma_i'$ we obtain for the potential of the inverted charges $q_i' = -(a/r_i)q_i$ at the positions $\mathbf{r}_i' = (a^2/r_i, \vartheta_i, \varphi_i)$ the following expression:

$$\phi'(r, \vartheta, \varphi) = \sum_i \frac{q_i'}{\sqrt{r^2 + r_i'^2 - 2rr_i' \cos \gamma_i}}$$

$$= \sum_i \frac{\dfrac{aq_i}{r_i}}{\sqrt{r^2 + \left(\dfrac{a^2}{r_i}\right)^2 - 2r\dfrac{a^2}{r_i} \cos \gamma_i}}$$

$$= \frac{a}{r} \sum_i \frac{q_i}{\sqrt{\left(\dfrac{a^2}{r}\right)^2 + r_i^2 - 2r_i\dfrac{a^2}{r} \cos \gamma_i}} = \frac{a}{r}\phi\left(\frac{a^2}{r}, \vartheta, \varphi\right)$$

By means of this relation we can calculate easily the potential of the image charges when the potential of the initial charge is known. This method is called *inversion with respect to a sphere*; the center of the sphere is called the *center of inversion*. This method has been presented for point charges, but it can be applied easily to extended charge distributions.

Example 2.6: Exercise: Point charges in front of a conducting plane

Let a point charge q be at a distance a in front of an infinitely extending conducting wall. What charge density will be induced at the wall? What is the magnitude of the total charge of the plane? Treat the problem by introducing an image charge. (See Figure 2.8.)

Solution

Since q is a point charge, in the absence of the conducting plane its electric field would have to be derived from the potential

$$\tilde{\phi} = \frac{q}{r}$$

But, by no means does this potential fulfill the condition to be constant at the conducting plane. In a strict mathematical manner, a certain solution $\phi_H(\mathbf{r})$ of the Laplace equation $\Delta\phi_H(\mathbf{r}) = 0$ has to be superposed to the potential of the point charge to fulfill this condition. We obtain a field for which the plane is an equipotential surface, and hence fulfills the boundary condition if the opposite charge $-q$, arranged mirror-symmetric at the distance $-a$ from the plane, is associated with the charge q. If r' is the distance of the point of observation from the image charge, then

$$\phi(\mathbf{r}) = \frac{q}{r} - \frac{q}{r'} = \tilde{\phi}(\mathbf{r}) + \phi_H(\mathbf{r})$$

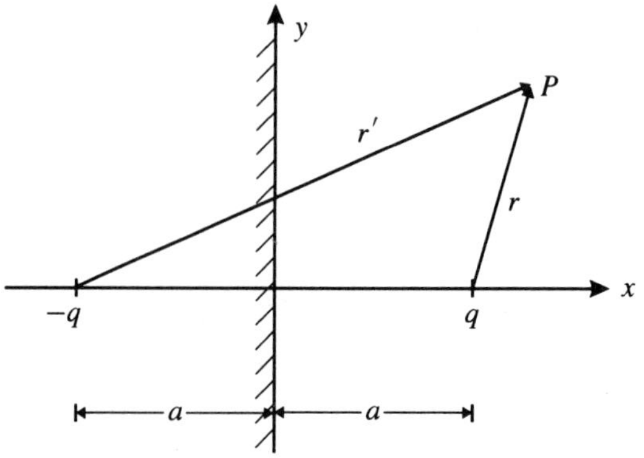

Figure 2.8. Point charge in front of a conducting wall.

represents the potential of the total field. It is zero at the boundary plane since here $r = r'$. This $\phi(\mathbf{r})$ satisfies the Poisson equation $\Delta\phi(\mathbf{r}) = -4\pi q\delta(\mathbf{r} - a\mathbf{e_x})$ and the boundary condition. Due to the uniqueness of the solution (see equations (2.12) and (2.15)) this is the only solution. Then the component of the field intensity normal to the boundary plane is

$$E_n = -\frac{\partial\phi}{\partial x} = -q\left(\frac{\partial\frac{1}{r}}{\partial x} - \frac{\partial\frac{1}{r'}}{\partial x}\right)\Bigg|_{x=0} \tag{2.25}$$

From the Pythagorean theorem, $r^2 = (a - x)^2 + y^2$ and $r'^2 = (a + x)^2 + y^2$, and we get

$$\frac{\partial\frac{1}{r}}{\partial x} = +\frac{2(a - x)}{2\left((a - x)^2 + y^2\right)^{3/2}} = \frac{a - x}{r^3}$$

Analogously, $(\partial 1/r')/\partial x = -(a + x)/r'^3$.

Substituting these values into (2.25) and taking into account that $r = r'$ at the point $x = 0$, we obtain

$$E_n = -q\left(\frac{\partial\frac{1}{r}}{\partial x} - \frac{\partial\frac{1}{r'}}{\partial x}\right)\Bigg|_{x=0}$$

$$= -q\left(\frac{a - x}{r^3} + \frac{a + x}{r'^3}\right)\Bigg|_{x=0} = \frac{-2qa}{r^3}$$

Since according to Gauss' law (for flux through the surface) $E_n = 4\pi\sigma$, the induced surface density is obtained by solving for σ:

$$\sigma = \frac{E_n}{4\pi} = -\frac{2qa}{4\pi r^3} = -\frac{qa}{2\pi r^3}$$

Here, r lies on the plane.

The charge is distributed over the plane surface of the conducting wall in such a way that the surface density is inversely proportional to the third power of the distance from the electric point charge. To obtain the total charge of the plane we introduce polar coordinates (ρ, φ). The surface element is $ds = \rho \, d\rho \, d\varphi$, and $r^2 = \rho^2 + a^2$. The total charge is obtained by the following integration:

$$\int \sigma \, ds = -\frac{qa}{2\pi} \int_0^{2\pi} d\varphi \int_0^{\infty} \frac{\rho \, d\rho}{(\rho^2 + a^2)^{3/2}}$$

The solution of the integral is

$$\int \sigma ds = -qa \left. \frac{-1}{(\rho^2 + a^2)^{1/2}} \right|_{\rho=0}^{\infty} = -q$$

So, we obtain the result that in the plane an equal but opposite charge is induced. This is not surprising, keeping in mind our knowledge about the charges induced at a conducting sphere (Example 2.1).

Example 2.7: The Green function for a sphere: General solution of the special potential problem

According to our general considerations in this chapter, the solution of the potential of a point charge near a grounded sphere ($\Phi = 0$ at the surface) obtained in Example 2.1 is just the Green function of the Poisson equation for Dirichlet boundary conditions,

$$G_D(\mathbf{r}, \mathbf{y}) = \frac{1}{|\mathbf{r} - \mathbf{y}|} - \frac{a}{y \left| \frac{a^2}{y^2} \mathbf{y} - \mathbf{r} \right|}$$

Here, the exterior region of the sphere and the spherical surface whose normal vector $\mathbf{n}'$ points to the center of the sphere (see Figure 2.9) are the volume V and the surface S, respectively, in the general considerations.

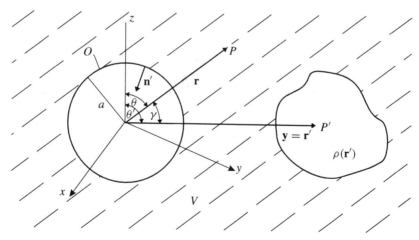

Figure 2.9. The volume V (exterior region of the sphere) and the surface S whose normal vector $\mathbf{n}'$ points to the center of the sphere. For convenience the origin of the coordinate system has been placed at the center of the sphere.

Setting $\mathbf{y} = \mathbf{r}'$ and introducing spherical coordinates, we obtain

$$G_D(\mathbf{r}, \mathbf{r}') = \frac{1}{\sqrt{r^2 + r'^2 - 2rr' \cos \gamma}} - \frac{1}{\sqrt{\dfrac{r^2 r'^2}{a^2} + a^2 - 2rr' \cos \gamma}}$$

This representation exhibits explicitly the symmetry of the Green function $G(\mathbf{r}, \mathbf{r}') = G(\mathbf{r}', \mathbf{r})$, following from its definition. The δ-function is symmetric under $\mathbf{r} \leftrightarrow \mathbf{r}'$; $\delta(\mathbf{r} - \mathbf{r}') = \delta(\mathbf{r}' - \mathbf{r})$.

For the general solution of the Dirichlet problem for the grounded sphere, we need further

$$\frac{\partial G_D(\mathbf{r}, \mathbf{r}')}{\partial n'}\bigg|_{r'=a} = - \frac{r^2 - a^2}{a(r^2 + a^2 - 2ar \cos \gamma)^{3/2}}$$

Here, we have taken into account that $\mathbf{n}'$ points in the negative $\mathbf{r}'$-direction, that is, $\partial/\partial n' = -\partial/\partial r'$. Now, if a charge distribution $\rho(\mathbf{r}')$ is given outside the sphere the general solution reads

$$\phi(\mathbf{r}) = \int_V G_D(\mathbf{r}, \mathbf{r}') \rho(\mathbf{r}') \, dV' - \frac{1}{4\pi} \oint_S \frac{\partial G_D(\mathbf{r}, \mathbf{r}')}{\partial n'} \phi(\mathbf{r}') \, da'$$

$$= \int_V \frac{1}{(r^2 + r'^2 - 2rr' \cos \gamma)^{1/2}} \rho(\mathbf{r}') r'^2 \, dr' \, d\Omega'$$

$$- \int_V \frac{1}{\left(\dfrac{r^2 r'^2}{a^2} + a^2 - 2rr' \cos \gamma \right)^{1/2}} \rho(\mathbf{r}') r'^2 \, dr' \, d\Omega'$$

$$+ \frac{1}{4\pi} \oint_S \phi(a, \theta', \varphi') \frac{(r^2 - a^2)}{a(r^2 + a^2 - 2ar \cos \gamma)^{3/2}} a^2 \, d\Omega'$$

In the integrals one has to note that the angle γ depends on the integration variable $d\Omega'$ because

$$\cos \gamma = \frac{\mathbf{r}}{r} \cdot \frac{\mathbf{r}'}{r'} = \{\sin \theta \cos \varphi, \sin \theta \sin \varphi, \cos \theta\} \cdot \{\sin \theta' \cos \varphi', \sin \theta' \sin \varphi', \cos \theta'\}$$

$$= \sin \theta \sin \theta' (\cos \varphi \cos \varphi' + \sin \varphi \sin \varphi') + \cos \theta \cos \theta'$$

$$= \cos \theta \cos \theta' + \sin \theta \sin \theta' \cos(\varphi - \varphi')$$

If the potential problem is in particular such that no external charges occur ($\rho(\mathbf{r}') \equiv 0$) but only the potential distribution $\phi(a, \theta', \varphi')$ at the spherical surface (relative to the earth), then the potential in the region $V (r \geq a)$ is given by

$$\phi(\mathbf{r}) = \frac{1}{4\pi} \oint_S \phi(a, \theta', \varphi') \frac{a(r^2 - a^2)}{(r^2 + a^2 - 2ar \cos \gamma)^{3/2}} d\Omega'$$

Example 2.8: Exercise: Conducting hemispheres at distinct potentials

As an example for the potential distribution outside the sphere with a given potential at the surface we consider a conducting sphere of radius a consisting of two hemispheres separated by an insulating disk (see Figure 2.10). The hemispheres are kept at distinct potentials. It is sufficient to consider the potentials $\pm V$ because different potentials at the hemispheres can be obtained as a superposition of this solution (refering to $\pm V$) with that belonging to the potential $\overline{V}$ at the entire sphere.

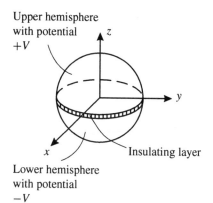

Upper hemisphere
with potential
$+V$

Insulating layer

Lower hemisphere
with potential
$-V$

Figure 2.10. A conducting sphere, separated by an insulating disk.

As derived in Example 2.7, the potential outside a sphere of radius a is

$$\phi(\mathbf{r}) = \frac{1}{4\pi} \int_\Omega \phi(a, \theta', \varphi') \frac{a(r^2 - a^2)}{(r^2 + a^2 - 2ar \cos \gamma)^{3/2}} \, d\Omega$$

where $\phi(a, \theta', \varphi')$ denotes the potential at the spherical surface, and γ is the angle between the vectors of the source point and the point of observation. Determine the potential between the two hemispheres separated by an insulating disk perpendicular to the z-axis and possessing the potentials $\pm V$.

Solution We have

$$\phi(r, \theta, \phi) = \frac{1}{4\pi} \int_\Omega \phi(a, \theta', \varphi') \frac{a(r^2 - a^2)}{(r^2 + a^2 - 2ar \cos \gamma)^{3/2}} \, d\Omega'$$

with

$$\phi(a, \theta', \varphi') = \begin{cases} +V & 0° \leq \theta' < 90° \\ -V & 90° < \theta' \leq 180° \end{cases}$$

and

$$d\Omega' = \sin \theta' \, d\theta' \, d\varphi' = -d(\cos \theta') \, d\varphi'$$

Substitution yields

$$\phi(r, \theta, \varphi) = \frac{V}{4\pi} \int_0^{2\pi} d\varphi' \left[\int_0^1 \frac{a(r^2 - a^2)}{(r^2 + a^2 - 2ar \cos \gamma)^{3/2}} \, d(\cos \theta') \right.$$

$$\left. - \int_{-1}^0 \frac{a(r^2 - a^2)}{(r^2 + a^2 - 2ar \cos \gamma)^{3/2}} \, d(\cos \theta') \right]$$

With the transformation $\theta' \to \pi - \theta'$ and $\varphi' \to \varphi' + \pi$ the second integral can be combined with the first one:

$$\phi(r, \theta, \varphi) = \frac{Va(r^2 - a^2)}{4\pi} \int_0^{2\pi} d\varphi' \int_0^1 d(\cos \theta')$$
$$\times \left[(a^2 + r^2 - 2ar \cos \gamma)^{-3/2} - (a^2 + r^2 + 2ar \cos \gamma)^{-3/2} \right] \qquad (2.26)$$

Since the dependence of $\cos \gamma$ on θ' and φ' is rather complicated, these integrals cannot be solved in a closed form. Only the potential along the positive z-axis can be given in a closed form since there $\cos \gamma = \cos \theta'$ because $\theta = 0$.

Thus, the integral can be simplified

$$\phi(z) = \phi(r, \theta = 0, \varphi)$$
$$= \frac{Va(r^2 - a^2)2\pi}{4\pi} \int_0^1 d(\cos \theta') \left[(a^2 + r^2 - 2ar \cos \theta')^{-3/2} - (a^2 + r^2 + 2ar \cos \theta')^{-3/2} \right]$$

These integrals can be executed in an elementary manner and one obtains

$$\Phi(z) = V \left[1 - \frac{z^2 - a^2}{z\sqrt{z^2 + a^2}} \right] \qquad (2.27)$$

Especially, at the point $z = a$ we have $\Phi(z = a) = V$, as required.

In order to go further in the general case the integrand in (2.26) can be expanded in a power series and integrated term by term. Factoring $(a^2 + r^2)$ out of the root expression we obtain

$$\Phi(r, \theta, \phi) =$$
$$\frac{Va(r^2 - a^2)}{4\pi(r^2 + a^2)^{3/2}} \int_0^{2\pi} d\varphi' \int_0^1 d(\cos \theta') \left[(1 - 2\alpha \cos \gamma)^{-3/2} - (1 + 2\alpha \cos \gamma)^{-3/2} \right]$$

where $\alpha = ar/(a^2 + r^2)$. Since the even powers of $2\alpha \cos \gamma$ cancel we obtain

$$\left[(1 - 2\alpha \cos \gamma)^{-3/2} - (1 + 2\alpha \cos \gamma)^{-3/2} \right] = 6\alpha \cos \gamma + 35\alpha^3 \cos^3 \gamma + \cdots$$

with only odd powers of $2\alpha \cos \gamma$. Due to

$$\int_0^{2\pi} d\varphi' \int_0^1 d(\cos \theta') \cos \gamma = \pi \cos \theta$$
$$\int_0^{2\pi} d\varphi' \int_0^1 d(\cos \theta') \cos^3 \gamma = \frac{\pi}{4} \cos \theta (3 - \cos^2 \theta)$$

the potential becomes finally

$$\Phi(r, \theta, \phi) = \frac{3Va^2}{2r^2} \left[\frac{r^3(r^2 - a^2)}{(r^2 + a^2)^{5/2}} \right] \cos \theta \left[1 + \frac{35}{24} \frac{a^2 r^2}{(a^2 + r^2)^2} (3 - \cos^2 \theta) + \cdots \right]$$

Obviously, only odd powers of $\cos \theta$ occur, which is required by the symmetry of the problem. For $r \gg a$ (distant zone) the expansion parameter is

$$\alpha^2 = \frac{a^2 r^2}{(r^2 + a^2)^2} \to \frac{a^2}{r^2}$$

and the potential simplifies to the form

$$\Phi(r, \theta, \phi) = \frac{3Va^2}{2r^2} \left[\cos\theta - \frac{7a^2}{12r^2} \left(\frac{5}{2}\cos^3\theta - \frac{3}{2}\cos\theta \right) + \cdots \right]$$

This series expansion converges very fast: For $r/a = 5$ the second term in the square brackets is only $\sim 2/100$. Furthermore, this solution tends to (2.27) as $\theta = 0$. Obviously, the first term corresponds to a dipole lined up with the z-axis at the origin with a dipole moment $P = \frac{3}{2}Va^2$.

Example 2.9: Exercise: Point-like charge between grounded conducting planes

A point charge q is located in between two grounded conducting planes which intersect at the angles $\alpha = 90^0, 60^0, 45^0$ (see Figure 2.11). Find the electric field in each case by the method of images. Furthermore, calculate the potential $\phi(\rho, z)$, the components of the electric field $E_R(\rho, z)$ and $E_\phi(\rho, z)$, as well as the surface-charge density $\sigma(\rho)$ in the neighborhood of $\rho = 0$.

Note: Use the Laplace equation in two dimensions.

Solution

By a repeated reflection of the charges in the planes one obtains the displayed arrangement in the three cases. One also realizes immediately that the boundary conditions are fulfilled because the potential of each charge is compensated (at the planes) by the corresponding opposite charge. Let two planes intersect at the angle ψ, in general, in between which a charge q is located with the coordinates (ρ, ϕ) (the intersection line of the planes is the z-axis). A reflection in the planes ① or ② maintains ρ as constant and changes only φ; i.e., all mirror charges lie on a circle. If a charge is lying at the position (ρ, φ), then for a reflection and in the plane ① $q(\rho, \varphi') \rightarrow -q(\rho, 2\psi - \varphi')$ in the plane ② $q(\rho, \varphi') \rightarrow -q(\rho, -\varphi')$. In our case

$\underline{90°}: q(\rho, \varphi) \rightarrow -q(\rho, 180° - \varphi) \rightarrow q(\rho, -180° + \varphi) \rightarrow -q(\rho, 360° - \varphi) \rightarrow$
$\quad q(\rho, \varphi)$

$\underline{60°}: q(\rho, \varphi) \rightarrow -q(\rho, 120° - \varphi) \rightarrow q(\rho, -120° + \varphi) \rightarrow -q(\rho, 240° - \varphi) \rightarrow$
$\quad q(\rho, -240° + \varphi) \rightarrow -q(\rho, 360° - \varphi) \rightarrow q(\rho, \varphi)$

$\underline{45°}: q(\rho, \varphi) \rightarrow -q(\rho, 90° - \varphi) \rightarrow q(\rho, -90° + \varphi) \rightarrow -q(\rho, 180° - \varphi) \rightarrow$
$\quad q(\rho, -180° + \varphi) \rightarrow -q(\rho, 270° - \varphi) \rightarrow q(\rho, -270° + \varphi) \rightarrow$
$\quad -q(\rho, 360° - \varphi) \rightarrow q(\rho, \varphi)$

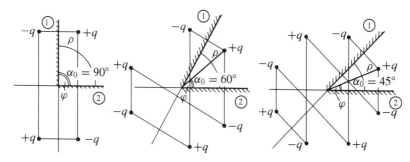

Figure 2.11. On the construction of image charges.

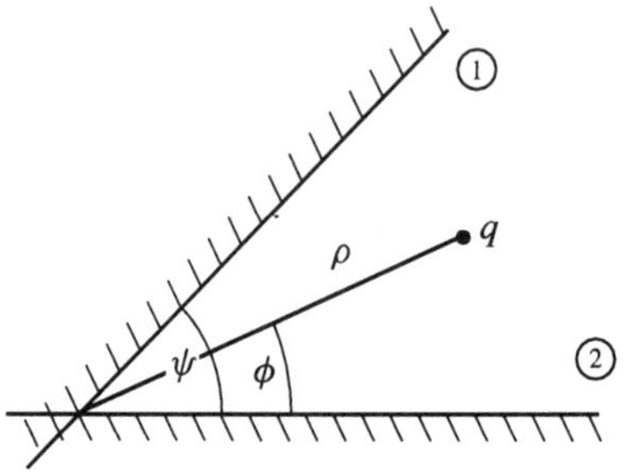

Figure 2.12.

In general, this method can be applied only for $\psi = 180°/n$, if n is an even number, and $2n - 1$ image charges are needed (see Figure 2.12). The required potential is

$$\phi(x, y, z) = \sum_{i=1}^{2n} \frac{Q_i}{\left[(\rho \cos \varphi_i - x)^2 + (\rho \sin \varphi_i - y)^2 + z^2\right]^{1/2}}$$

with $|Q_i| = q$. The sign of Q_i and the angles φ_i are obtained by the construction. It is $|Q_i| = q(-1)^{i+1}$. Thus, we obtain for the required **E**-Field

$$\mathbf{E}(\mathbf{r}) = q \sum_{i=1}^{2n} \frac{(-1)^{i+1}}{\left[(\rho \cos \varphi_i - x)^2 + (\rho \sin \varphi_i - y)^2 + z^2\right]^{3/2}} \begin{pmatrix} x - \rho \cos \varphi_i \\ y - \rho \sin \varphi_i \\ z \end{pmatrix}$$

We assume that the two conducting planes are kept at the potential $\phi = V$ and form an angle β between each other. As we are interested in the behavior $\rho \approx 0$ we neglect further external charge distributions defining the boundary conditions of the system uniquely. Due to the symmetry of the arrangement, we use plane polar coordinates in which the Laplace equation reads

$$\frac{1}{\rho} \frac{\partial}{\partial \rho} \left(\rho \frac{\partial \phi}{\partial \rho} \right) + \frac{1}{\rho^2} \frac{\partial^2 \phi}{\partial \varphi^2} = 0 \tag{2.28}$$

For the solution we make the separation ansatz:

$$\phi(\rho, \varphi) = R(\rho) F(\varphi) \tag{2.29}$$

Multiplying (2.28) by ρ^2/ϕ, we obtain

$$\frac{\rho^2}{RF} \cdot \left| \frac{F}{\rho} \frac{d}{d\rho} \left(\rho \frac{dR}{d\rho} \right) + \frac{R}{\rho^2} \frac{d^2 F}{d\varphi^2} = 0 \right. \tag{2.30}$$

$$\Longleftrightarrow \frac{\rho}{R} \frac{d}{d\rho} \left(\rho \frac{dR}{d\rho} \right) + \frac{1}{F} \frac{d^2 F}{d\varphi^2} = 0 \tag{2.31}$$

If equation (2.31) has to be valid for all ρ, φ, then any term has to be a constant

$$\frac{\rho}{R} \frac{d}{d\rho} \left(\rho \frac{dR}{d\rho} \right) = v^2 = -\frac{1}{F} \frac{d^2 F}{d\varphi^2} \tag{2.32}$$

Since an elementary integration of (2.32) is not successful, we try an ansatz in terms of a power series for the differential equation in ρ:

$$\text{Ansatz:} \quad R(\rho) = \sum_{n=0}^{\infty} c_n \rho^n \tag{2.33}$$

$$\frac{dR}{d\rho} = \sum_{n=1}^{\infty} n c_n \rho^{n-1}, \qquad \frac{d^2R}{d\rho^2} = \sum_{n=2}^{\infty} n(n-1) c_n \rho^{n-2}$$

Substitution into (2.32) yields

$$\rho^2 \frac{d^2R}{d\rho^2} + \rho \frac{dR}{d\rho} = R\nu^2 \tag{2.34}$$

$$\Longleftrightarrow \sum_{n=0}^{\infty} n(n-1) c_n \rho^n + \sum_{n=0}^{\infty} n c_n \rho^n = \sum_{n=0}^{\infty} \nu^2 c_n \rho^n \tag{2.35}$$

Equation (2.35) can be valid only if

$$n(n-1) + n = \nu^2 \Longleftrightarrow n = \pm\nu, \qquad \text{i.e.,} \quad R(\rho) = c_\nu \rho^\nu + c_{-\nu} \rho^{-\nu} \tag{2.36}$$

For $\nu = 0$, (2.32) can be integrated easily:

$$\rho \frac{dR}{d\rho} = b_0, \qquad\qquad \frac{da}{d\varphi} = B_0 \tag{2.37}$$

$$\Longrightarrow R(\rho) = a_0 + b_0 \ln(\rho), \qquad F(\varphi) = A_0 \varphi + B_0 \tag{2.38}$$

If $\nu \neq 0$ again, then we know the solution of $d^2F/d\varphi^2 + \nu^2 F = 0$ (harmonic oscillator in φ):

$$F(\varphi) = A\cos(\nu\varphi) + B\sin(\nu\varphi) \tag{2.39}$$

Rewriting the coefficients in equation (2.36), $c_\nu \equiv a_\nu$, $c_{-\nu} \equiv b_\nu$, the solutions can be combined:

$$\left.\begin{array}{l} R(\rho) = a_\nu \rho^\nu + b_\nu \rho^{-\nu} \\[4pt] F(\varphi) = A\cos(\nu\varphi) + B\sin(\nu\varphi) \end{array}\right\} \quad \nu \neq 0 \tag{2.40}$$

$$\left.\begin{array}{l} R(\rho) = a_0 + b_0 \ln(\rho) \\[4pt] F(\varphi) = A_0 + B_0 \varphi \end{array}\right\} \quad \nu = 0 \tag{2.41}$$

If we want to construct the general solution of the two-dimensional potential problem then we have to note that the potential ϕ has to be unique. From that it follows, for example, that ν has to be a positive or negative integer or $\nu = 0$. Furthermore, B_0 has to vanish since φ is a cyclic coordinate. Thus, the general solution of the problem reads

$$\phi(\rho, \varphi) = a_0 + b_0 \ln(\rho) + \sum_{n=1}^{\infty} a_n \rho^n \sin(n\varphi + \alpha_n) + \sum_{n=1}^{\infty} b_n \rho^{-n} \sin(n\varphi + \beta_n) \tag{2.42}$$

With $0 \leq \varphi \leq \beta$, in our example the boundary conditions are $\phi = V$ for all $\rho \geq 0$, if $\varphi = 0$ or $\varphi = \beta$.

Equation (2.41) yields $b_0 = B_0 = 0$ ($A_0 \cdot a_0 = V$). Furthermore, the b_ν from (2.40) have to vanish, as well as A. (That the b_ν have to vanish is obvious due to the intended investigation $\rho \approx 0$.)

$$\Rightarrow \phi(\rho, 0) = \phi(\rho, \beta) \Leftrightarrow a_\nu \rho^\nu \cdot 0 = a_\nu \rho^\nu B \sin(\nu\beta) \tag{2.43}$$

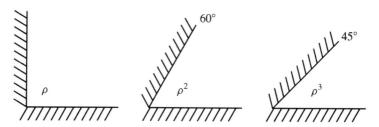

Figure 2.13. Radial dependence of the E-field and of the charge density σ close to the edge formed by two conducting grounded planes.

(2.43) can be fulfilled only if

$$v\beta = m\pi \qquad \text{i.e.,} \quad v = \frac{m\pi}{\beta}, \; m = 1, 2, \ldots \tag{2.44}$$

Thus, we have the solution:

$$\phi(\rho, \varphi) = V + \sum_{m=1}^{\infty} a_m \rho^{m\pi/\beta} \sin(m\pi\varphi/\beta) \tag{2.45}$$

The still undetermined a_m are fixed by the potential distribution distant from $\rho \approx 0$. We start from $a_1 \neq 0$. For $a_1 = 0$ a separate consideration would be necessary. From (2.45) we have for $\rho \approx 0$

$$\phi(\rho, \varphi) \approx V + a_1 \rho^{\pi/\beta} \sin(\pi\varphi/\beta) \tag{2.46}$$

For the electric field $\left(\nabla\psi = (\partial\psi/\partial\rho)\mathbf{e}_\rho + (1/\rho)(\partial\psi/\partial\varphi)\mathbf{e}_\varphi \right)$, we obtain

$$\mathbf{E}(\rho, \varphi) = -\frac{\pi a_1}{\beta} \rho^{(\pi/\beta)-1} \begin{pmatrix} \sin(\pi\varphi/\beta) \\ \cos(\pi\varphi/\beta) \end{pmatrix} \tag{2.47}$$

The surface-charge densities at $\varphi = 0$ and $\varphi = \beta$ are equal to each other for $\rho \approx 0$, namely,

$$\sigma(\rho) = \frac{E_\varphi(\rho, 0)}{4\pi} \simeq -\frac{a_1}{4\beta} \rho^{(\pi/\beta)-1} \tag{2.48}$$

From (2.47) and (2.48), we can see that the **E**-field as well as $\sigma(\rho)$ depend on $\rho^{(\pi/\beta)-1}$ as ρ, ρ^2, ρ^3 for the cases $\beta = \pi/2, \pi/3, \pi/4$. The region becomes more field-free, as the angle becomes more acute. Within a very acute angle no charge accumulates.

The result is shown in a more pictorial form in Figure 2.13: for $\beta = \pi$, **E** and σ become independent of ρ.

**Example 2.10:   Exercise: Green function of the Poisson equation
in various dimensions**

Calculate the Green function for the Poisson equation in the dimensions $d = 1, 2, 3$. Use the spherical coordinates:

$$d = 1: \; \Delta = \frac{d^2}{dx^2} = \frac{d^2}{dr^2}, r = |x|$$

$$d = 2: \quad \Delta = \frac{\partial^2}{\partial x^2} + \frac{\partial^2}{\partial y^2} = \frac{1}{r}\frac{\partial}{\partial r}\left(r\frac{\partial}{\partial r}\right) + \frac{1}{r^2}\frac{\partial^2}{\partial \varphi^2}$$

$$d = 3: \quad \Delta = \frac{\partial^2}{\partial x^2} + \frac{\partial^2}{\partial y^2} + \frac{\partial^2}{\partial z^2}$$

$$= \frac{1}{r^2}\frac{\partial}{\partial r}\left(r^2\frac{\partial}{\partial r}\right) + \frac{1}{r^2\sin\vartheta}\frac{\partial}{\partial\vartheta}\left(\sin\vartheta\frac{\partial}{\partial\vartheta}\right) + \frac{1}{r^2\sin^2\vartheta}\frac{\partial^2}{\partial\varphi^2}$$

Solution The Green function has to obey the following equation in d dimensions:[2]

$$\Delta_x G(\mathbf{x}, \mathbf{x}') = \delta^{(d)}(\mathbf{x} - \mathbf{x}')$$

where $\mathbf{x}$ is a d-dimensional vector. At first, we look for a special solution of this equation. Since the Laplace operator is invariant under translations and rotations, we expect the existence of a translational-invariant and rotational-invariant solution. Hence, we make the ansatz

$$G_s(\mathbf{x}, \mathbf{x}') = G_s(\mathbf{x} - \mathbf{x}') = G_s(\mathbf{r}) = G_s(r)$$

With the previous results one obtains

$$d = 1 \quad : \quad \frac{d^2}{dr^2}G_s(r) = \delta(x) = \frac{\delta(r)}{2}$$

$$\Rightarrow \frac{d}{dr}G_s(r) = \frac{1}{2} \Rightarrow G_s(r) = \frac{1}{2}r \Rightarrow G_s(x) = \frac{1}{2}|x|$$

With

$$\text{sgn } x = \begin{cases} +1 \text{ for } x > 0 \\ -1 \text{ for } x < 0 \end{cases} \quad \text{and} \quad \frac{d}{dx}\text{sgn } x = 2\delta(x)$$

the check yields

$$\frac{d^2}{dx^2}G_s(x) = \frac{d^2}{dx^2}\left(\frac{1}{2}x \cdot \text{sgn } x\right)$$

$$= \frac{d}{dx}\left(\frac{1}{2}\text{sgn } x + \frac{1}{2}x2\delta(x)\right) = \frac{d}{dx}\frac{1}{2}\text{sgn } x = \delta(x)$$

$$d = 2 \quad : \quad \frac{1}{r}\frac{\partial}{\partial r}\left(r\frac{\partial}{\partial r}\right)G_s(r) = \delta^{(2)}(\mathbf{r}) = \frac{\delta(r)}{2\pi r}$$

$$\Rightarrow \frac{\partial}{\partial r}\left(r\frac{\partial}{\partial r}\right)G_s(r) = \frac{\delta(r)}{2\pi}$$

$$\Rightarrow r\frac{\partial}{\partial r}G_s(r) = \frac{1}{2\pi} \Rightarrow \frac{\partial}{\partial r}G_s(r) = \frac{1}{2\pi r}$$

$$\Rightarrow G_s(r) = \frac{1}{2\pi}\ln r \Rightarrow G_s(\mathbf{r}) = \frac{1}{2\pi}\ln|\mathbf{r}|$$

---

[2]Here, the Green function is defined without the factor -4π in front of the δ-function.

$$d = 3 \;:\; \frac{1}{r^2}\frac{\partial}{\partial r}r^2\frac{\partial}{\partial r}G_s(r) = \delta^{(3)}(\mathbf{r}) = \frac{\delta(r)}{4\pi r^2}$$

$$\Rightarrow \frac{\partial}{\partial r}r^2\frac{\partial}{\partial r}G_s(r) = \frac{\delta(r)}{4\pi} \Rightarrow r^2\frac{\partial}{\partial r}G_s(r) = \frac{1}{4\pi}$$

$$\Rightarrow \frac{\partial}{\partial r}G_s(r) = \frac{1}{4\pi r^2}$$

$$\Rightarrow G_s(r) = -\frac{1}{4\pi r} \Rightarrow G_s(\mathbf{r}) = -\frac{1}{4\pi |\mathbf{r}|}$$

The proof, that for $d = 3$ $G_s(\mathbf{r})$ satisfies Poisson's equation for a point charge, has been given above (compare equation (1.60)). For $d = 2$ the proof proceeds analogously. The general solution of the Poisson equation is

$$G(\mathbf{x}, \mathbf{x}') = G_s(\mathbf{x} - \mathbf{x}') + F(\mathbf{x}, \mathbf{x}')$$

with $\Delta F(\mathbf{x}, \mathbf{x}') = 0$.

The choice of $F(\mathbf{x}, \mathbf{x}')$ is fixed by the boundary conditions of the problem.

Example 2.11:   Exercise: Symmetry of the Green function for the Dirichlet problem

Prove that the Green function of the Dirichlet problem is symmetric, i.e.,

$$G_D(\mathbf{x}, \mathbf{x}') = G_D(\mathbf{x}', \mathbf{x})$$

Solution   We start from the second Green's theorem

$$\int_V [\varphi(\mathbf{y})\Delta_y\psi(\mathbf{y}) - \psi(\mathbf{y})\Delta_y\varphi(\mathbf{y})]d^3y = \int_{\partial V}\left[\varphi(\mathbf{y})\frac{\partial\psi(\mathbf{y})}{\partial n_y} - \psi(\mathbf{y})\frac{\partial\varphi(\mathbf{y})}{\partial n_y}\right]d\sigma(y)$$

and set $\varphi(\mathbf{y}) = G_D(\mathbf{y}, \mathbf{x})$ and $\psi(\mathbf{y}) = G_D(\mathbf{y}, \mathbf{x}')$. Since $G_D(\mathbf{y}, \mathbf{x})$ is the Green function of Δ,

$$\Delta_y G_D(\mathbf{y}, \mathbf{x}) = -4\pi\delta(\mathbf{y} - \mathbf{x})$$

Furthermore, it satisfies Dirichlet's boundary condition $G_D(\mathbf{y}, \mathbf{x}) = 0$ for $y\epsilon\{$all boundary points of $V\}$.

We obtain

$$\int_V [G_D(\mathbf{y}, \mathbf{x})(-4\pi)\delta(\mathbf{y} - \mathbf{x}') - G_D(\mathbf{y}, \mathbf{x}')(-4\pi)\delta(\mathbf{y} - \mathbf{x})]d^3y = 0$$

or

$$G_D(\mathbf{x}', \mathbf{x}) - G_D(\mathbf{x}, \mathbf{x}') = 0$$

Biographical notes

George Green, b. July 14, 1793, Nottingham–d. March 31, 1841, Sneinton (n. Nottingham). Green not only succeeded his father as a baker and milliner, he attentively followed all discoveries in electrophysics and studied Laplace's works, too. After further studies in Cambridge he worked there at the Cajus-College. His main work "Essays on the Application of Mathematical Analysis to Theories of Electricity and Magnetism" (1828) presented the first attempt to describe phenomena of electricity mathematically. Together with some of Gauß works it marked the beginning of potential theory. (see also p. 526)

Peter Gustav Lejeune-Dirichlet, b. Feb. 13, 1805, Düren–d. May 5, 1859, Göttingen. Dirichlet lived in Paris from 1822 to 1827, then went to Breslau (Wroclaw) as a lecturer. From 1832 he worked in Berlin. In 1855 he was appointed to Gauß former chair in Göttingen. Dirichlet made important contributions to the theory of Fourier series, potential theory, variational calculus, and especially to the theory of numbers. He was the first to extensively use analytical functions to solve arithmetical problems, and thus founded the analytical theory of numbers.

Carl Gottfried Neumann, b. May 7, 1832, Königsberg (Kaliningrad)–d. March 27, 1925, Leipzig, son of F.E. Neumann (see p. 526). Neumann was a student from 1850 to 1855. In 1856 he did a doctorate in Königsberg. Neumann had chairs in Halle, Basel, Tübingen, and Leipzig (since 1868). Neumann made important contributions to *potential theory*, e.g., Neumann's potential function as solution for boundary value problems. The Neumann series was named after him.

3 Orthogonal Functions and Multipole Expansion

MATHEMATICAL SUPPLEMENT

Expansion of arbitrary functions in terms of a complete set of functions

In mathematical physics the expansion of an arbitrary function in terms of an orthogonal system of functions plays an important role. The Fourier series encountered in the problem of the vibrating string is an example of such an expansion.

Let us consider now a finite or infinite system of real or complex functions

$$U_1(x), U_2(x), \ldots, U_n(x)$$

in the interval $[a, b]$. This system of functions is called to be orthogonal if the functions satisfy the orthogonality relation:

$$\int_a^b U_n(x)U_m^*(x)dx = s_n\delta_{nm} \tag{3.1}$$

where U_m^* is the function complex conjugate to U_m. If $s_n = 1$ for $n = 1, 2, \ldots$, then the system is said to be *normalized* (orthonormal system). If $U(x)$ is not identical with the null function, that is,

$$\int_a^b U(x)U^*(x)dx > 0 \tag{3.2}$$

then this integral is denoted the norm of the function $U(x)$. So, a function is said to be normalized if

$$\int_a^b U(x)U^*(x)dx = 1 \tag{3.3}$$

Any function different from the null function can be normalized by multiplying with an appropriate constant so that

$$U'(x) = \frac{U(x)}{\sqrt{\int_a^b U(x')U^*(x')dx'}} \tag{3.4}$$

is the normalized function corresponding to $U(x)$. The analogy to the orthogonal vectors $\mathbf{e}_\nu$ with $\mathbf{e}_\nu \cdot \mathbf{e}_\mu = \delta_{\nu\mu}$ in R^n is obvious. The relation (3.4) corresponds to the normalization of a vector $\mathbf{e} = \mathbf{a}/\sqrt{\mathbf{a} \cdot \mathbf{a}}$; the left-hand side of equation (3.1) corresponds to the scalar product of vectors. Now, we want to expand an arbitrary function $f(x)$ in $[a, b]$ in terms of an orthogonal system of functions $U_n(x)$. This expansion is given by

$$f(x) \Rightarrow \sum_{n=1}^{N} a_n U_n(x) \tag{3.5}$$

This corresponds to the expansion of a vector $\mathbf{a}$ in the basis $\mathbf{e}_\nu$

$$\mathbf{a} = \sum_\nu a_\nu \mathbf{e}_\nu = \sum_\nu (\mathbf{a} \cdot \mathbf{e}_\nu)\mathbf{e}_\nu$$

Analogous to the corresponding notion in the theory of Fourier series, the numbers a_n are often denoted as generalized Fourier coefficients. Now it has to be investigated whether this series (3.5) converges to $f(x)$ as $N \to \infty$ that is, whether the equation

$$f(x) = \sum_{n=1}^{\infty} a_n U_n(x)$$

is valid, and what the coefficients a_n look like.

Let the function $f(x)$ be described by (3.5). We take the square error M_N

$$M_N(a_1, \ldots, a_N) = \int_a^b \left| \sum_{n=1}^{N} a_n U_n(x) - f(x) \right|^2 dx \tag{3.6}$$

and determine the coefficients a_n in such a way that we minimize M_N.

Squaring the integrand we obtain

$$M_N = \sum_{j,k=1}^{N} \delta_{jk} a_j a_k^* - \sum_{j=1}^{N} l_j a_j^* - \sum_{j=1}^{N} l_j^* a_j + d$$

with

$$\delta_{jk} = \int_a^b U_j(x)U_k^*(x)dx, \qquad l_j = \int_a^b U_j^*(x)f(x)dx, \qquad l_j^* = \int_a^b U_j(x)f^*(x)dx$$

and $d = \int_a^b f(x)f^*(x)\,dx$ as the norm of $f(x)$.

M_N becomes minimal for those values of a_j for which

$$\left.\begin{array}{l} \dfrac{\partial M_N}{\partial a_j} = 0 = a_j^* - l_j^* \\[2ex] \dfrac{\partial M_N}{\partial a_j^*} = 0 = a_j - l_j \end{array}\right\} \qquad \text{i.e.,} \quad a_j = \int_a^b U_j^*(x) f(x) dx \qquad \textbf{(3.7a)}$$

is valid.[1] This minimal value is

$$M_{N,\min} = d - \sum_{j=1}^N a_j a_j^* \qquad \textbf{(3.7b)}$$

and equation (3.16) follows immediately. In the sense of the least squares method, the series formed with the a_k

$$F_N(x) = \sum_{k=1}^N a_k U_k(x) \qquad \textbf{(3.8)}$$

is the best approximation to $f(x)$ which can be achieved by N functions of an orthonormal system. But, the system of functions is a suitable approximation to $f(x)$ only if

$$\lim_{N \to \infty} M_{N,\min} = \lim_{N \to \infty} \int_a^b |F_N(x) - f(x)|^2 dx = 0 \qquad \textbf{(3.9)}$$

This is the condition for the so-called *convergence in square mean*. In general, this notation is transferred to any function sequence in the function space, i.e.,

$$\lim_{N \to \infty} F_N(x) = f(x) \qquad \textbf{(3.10)}$$

in $[a, b]$, if

$$\lim_{N \to \infty} \int_a^b |F_N(x) - f(x)|^2 dx = 0 \qquad \textbf{(3.11)}$$

This yields immediately the statement

$$f(x) = \sum_{n=1}^\infty a_n U_n(x) \qquad \textbf{(3.12)}$$

Forming the product of this relation and $U_m^*(x)$, due to the orthonormality relation, one obtains the expression

$$a_m = \int_a^b f(x) U_m^*(x) dx \qquad \textbf{(3.13)}$$

---

[1] M_N is a function of a_i and a_j^* $(i, j = 1, 2 \ldots N)$. The variables a_i, a_j^* can be seen as independent variables in differentiation; being complex numbers, they are indeed independent; i.e., $a_i \neq a_j^*$.

for the generalized Fourier coefficients. This is again equation (3.7a). Now, we still want to consider Bessel's inequality

$$\sum_{n=1}^{\infty} a_n^2 \leq \int_a^b |f(x)|^2 dx \tag{3.14}$$

known from analysis. Truncating the expansion of the function $f(x)$ after N terms, in any case we have

$$\int_a^b \left| f(x) - \sum_n^N a_n U_n(x) \right|^2 dx \geq 0 \tag{3.15}$$

Taking into account (3.13) and the possibility to interchange integration and summation, Bessel's inequality is valid directly. Since the right-hand side of

$$\sum_{n=1}^{N} a_n^2 \leq \int_a^b |f(x)|^2 dx \tag{3.16}$$

is an expression independent of n, on the left-hand side N can be chosen arbitrarily (also $N \to \infty$); so (3.14) is demonstrated. This can be concluded also directly from (3.7b) taking into account the fact $M_N \geq 0$. If in this relation the equality sign is valid, then the system belonging to the a_n describes the function $f(x)$ completely. For that reason this (Parseval's) equation is also called the *completeness relation*.

Analogous to (3.12) we obtain an expansion of a function of two independent variables $f(x, y)$ in the intervals: $x\epsilon[a, b]$ and $y\epsilon[c, d]$.

$$f(x, y) = \sum_{m=1}^{\infty} \sum_{n=1}^{\infty} a_{mn} U_m(x) V_n(y) \tag{3.17}$$

with

$$a_{mn} = \int_a^b dx \int_c^d f(x, y) U_m^*(x) V_n^*(y) dy \tag{3.18}$$

Here, $U_m(x)$ and $V_n(y)$ are orthonormal. A similar generalization holds in higher dimensional spaces.

Examples of systems of orthonormal functions

Fourier series

Since for positive integers n, m the relation

$$\int_{-\pi}^{\pi} \sin(nx') \cdot \sin(mx') \, dx' = \pi \delta_{nm}$$

is valid, the sequence of functions

$$\overline{U}_n(x') = \frac{1}{\sqrt{\pi}} \sin(nx'), \qquad (n = 1, 2, \ldots)$$

represents an orthonormal system in the interval $-\pi \leq x' \leq \pi$. However, as this sequence consists of odd functions only ($f(x') = -f(x')$) it can be used to approximate only odd functions. If one takes, in addition, the orthonormal system of even functions ($f(-x') = f(x')$):

$$\overline{V}_n(x') = \frac{1}{\sqrt{\pi}} \cos(nx'), \qquad (n = 1, 2, \ldots)$$

$$\overline{V}_0(x') = \frac{1}{\sqrt{2\pi}}, \qquad (n = 0)$$

then arbitrary functions can be represented by

$$f(x') = \frac{a'_0}{2} + \sum_{n=1} a'_n \cos(nx') + b'_n \sin(nx') \qquad \textbf{(3.19)}$$

For even functions $f(x)$ b'_n vanishes; for odd functions a'_n vanishes. The Fourier coefficients are, from (3.13),

$$a'_n = \frac{1}{\pi} \int_{-\pi}^{\pi} f(x') \cos(nx') dx'$$

$$b'_n = \frac{1}{\pi} \int_{-\pi}^{\pi} f(x') \sin(nx') dx'$$

From the functions $\overline{U}_n, \overline{V}_n$, the functions being orthonormal in the interval $-d/2 \leq x \leq +d/2$

$$u_n(x) = \sqrt{\frac{2}{d}} \sin \frac{2\pi nx}{d}$$

$$v_n(x) = \sqrt{\frac{2}{d}} \cos \frac{2\pi nx}{d}$$

may be constructed by the coordinate transformation

$$x' = \frac{2\pi x}{d}$$

For these functions we have

$$\int_{-d/2}^{d/2} u_n(x) u_{n'}(x) \, dx = \delta_{nn'}$$

$$\int_{-d/2}^{d/2} v_n(x)v_{n'}(x)\,dx = \delta_{nn'}$$

$$\int_{-d/2}^{d/2} u_n(x)v_n(x)\,dx = 0$$

and for the expansion

$$f(x) = a_0 + \sum_{n=1}^{\infty} a_n v_n(x) + b_n u_n(x)$$

one obtains the coefficients

$$a_n = \int_{-d/2}^{d/2} v_n(x)f(x)\,dx$$

$$b_n = \int_{-d/2}^{d/2} u_n(x)f(x)\,dx$$

Fourier integrals

Any function $f(x)$ can be represented also in the form (3.19). The following elementary relations hold:

$$v_n(x) = \sqrt{\frac{2}{d}} \cos \frac{(2\pi nx)}{d} = \sqrt{\frac{2}{d}} \left(\frac{e^{i2\pi nx/d} + e^{-i2\pi nx/d}}{2} \right) = \frac{g_n(x) + g_{-n}(x)}{\sqrt{2}} \qquad \textbf{(3.20)}$$

and

$$u_n(x) = \sqrt{\frac{2}{d}} \sin \frac{(2\pi nx)}{d} = \sqrt{\frac{2}{d}} \left(\frac{e^{i2\pi nx/d} - e^{-i2\pi nx/d}}{2i} \right) = \frac{g_n(x) - g_{-n}(x)}{\sqrt{2}i} \qquad \textbf{(3.21)}$$

By rewriting we find that the functions

$$g_n(x) = \sqrt{\frac{1}{d}} e^{i2\pi nx/d} \qquad (n = 0, \pm 1, \pm 2 \ldots) \qquad \textbf{(3.22)}$$

also form an orthogonal system for which

$$\int_{-d/2}^{d/2} g_n^*(x)g_{n'}(x)\,dx = \delta_{nn'} \qquad \textbf{(3.23a)}$$

Namely,

$$g_n(x) = \frac{v_n(x) + i u_n(x)}{\sqrt{2}}$$

$$g_{-n}(x) = \frac{v_n(x) - iu_n(x)}{\sqrt{2}} \qquad (n \geq 0) \tag{3.23b}$$

and therefore, e.g.,

$$\int_{-d/2}^{d/2} g_n^*(x)g_{-n}(x)\,dx$$

$$= \frac{1}{2} \int_{-d/2}^{d/2} [v_n(x)v_n(x) - u_n(x)u_n(x) - i(v_n(x)u_n(x) + u_n(x)v_n(x))]\,dx$$

$$= 0$$

and

$$\int_{-d/2}^{d/2} g_n^*(x)g_{n'}(x)\,dx$$

$$= \frac{1}{2} \int_{-d/2}^{d/2} [v_n(x)v_n'(x) + u_n(x)u_n'(x) + i(v_n(x)u_n'(x) - u_n(x)v_n'(x))]\,dx$$

$$= \frac{1}{2}(\delta_{nn'} + \delta_{nn'}) = \delta_{nn'}$$

The fact that the factor $1/\sqrt{2}$ appears in (3.20) and (3.21), and not $1/2$, guarantees the normalization of the functions $g_n(x)$ (3.23a). For any function in the interval $-d/2 \leq x \leq d/2$ that can be considered periodic, we have

$$f(x) = \sum_{n=-\infty}^{\infty} a_n g_n(x) = \sum_{n=-\infty}^{\infty} a_n \frac{e^{i2\pi nx/d}}{\sqrt{d}} \tag{3.24}$$

with

$$a_n = \int_{-d/2}^{d/2} g_n^*(x)f(x)dx = \int_{-d/2}^{d/2} \left(\frac{e^{-i2\pi nx/d}}{\sqrt{d}} f(x)dx \right)$$

In particular, for the $\delta(x - x')$-function with $d/2 \leq x' \leq d/2$ we have

$$\delta(x - x') = \sum_{n=-\infty}^{\infty} g_n(x)g_n^*(x') = \sum_{n=-\infty}^{\infty} \frac{e^{i(2\pi n/d)(x-x')}}{d} \tag{3.25a}$$

Let us consider now the limit $d \to \infty$. The periodicity interval of the Fourier series of $-d/2$ to $+d/2$ is then extended to the whole region $-\infty \leq x \leq +\infty$. Obviously,

$$f(x) = \sum_{n=-\infty}^{\infty} a_n \frac{e^{i2\pi nx/d}}{\sqrt{d}} = \sum_{n=-\infty}^{\infty} a_n \frac{e^{i2\pi nx/d}}{\sqrt{d}} \Delta n \tag{3.25b}$$

where $\Delta n = 1$, and for the $\delta(x - x')$-function

$$\delta(x - x') = \sum_{n=-\infty}^{\infty} \frac{e^{i(2\pi n/d)(x-x')}}{d} \Delta n = \sum_{n=-\infty}^{\infty} \frac{e^{ik_n(x-x')}}{d} \frac{d}{2\pi} \Delta k_n$$

$$\Rightarrow \lim_{d \to \infty} \int_{-\infty}^{+\infty} \frac{e^{ik(x-x')}}{2\pi} dk = \int_{-\infty}^{+\infty} \left(\frac{e^{ikx}}{\sqrt{2\pi}}\right)\left(\frac{e^{ikx'}}{\sqrt{2\pi}}\right)^* dk \qquad \textbf{(3.25c)}$$

Here, the variable $k_n = 2\pi n/d$ has been introduced, which in the limit $d \to \infty$ converts to the continuous variable k. Obviously, the functions

$$g(k, x) = \frac{e^{ikx}}{\sqrt{2\pi}} \qquad \textbf{(3.26)}$$

form an orthonormal complete basis depending on the continuous parameter k. Due to the symmetry of k and x in $e^{ikx}/\sqrt{2\pi}$, analogous to equation (3.25c) we obtain

$$\delta(k - k') = \int_{-\infty}^{+\infty} \left(\frac{e^{ikx}}{\sqrt{2\pi}}\right)\left(\frac{e^{ik'x}}{\sqrt{2\pi}}\right)^* dx \qquad \textbf{(3.27)}$$

For any function $f(x)$,

$$f(x) = \int_{-\infty}^{+\infty} A(k)g(k, x)\, dk = \int_{-\infty}^{+\infty} A(k)\frac{e^{ikx}}{\sqrt{2\pi}} dk$$

with

$$A(k) = \int_{-\infty}^{+\infty} f(x)g^*(k, x)\, dx = \int_{-\infty}^{+\infty} f(x)\frac{e^{-ikx}}{\sqrt{2\pi}} dx \qquad \textbf{(3.28)}$$

since

$$rl \int_{-\infty}^{+\infty} f(x)\frac{e^{-ikx}}{\sqrt{2\pi}} dx = \int_{-\infty}^{\infty}\int_{-\infty}^{+\infty} A(k')\frac{e^{ik'x}}{\sqrt{2\pi}} \frac{e^{-ikx}}{\sqrt{2\pi}} dx\, dk'$$

$$= \int_{-\infty}^{\infty} A(k')\delta(k - k')\, dk'$$

$$= A(k)$$

The symmetry between $f(x)$ and $A(x)$ is a remarkable fact.

$A(k)$ is called the *Fourier transform* or the *spectral function* of $f(x)$. The expansion of a periodic function (with the period d) in a Fourier series is a representation in terms of a sum of harmonic oscillators with the frequencies $k_n = 2\pi n/d$, while the Fourier integral expresses the function $f(x)$ as a sum of an infinite number of oscillators with different frequencies. One says that the fourier integral yields an expansion of the function in terms of a continuous spectrum, where the distribution of the continuous frequencies k corresponds to the spectral density $A(k)$.

To represent a function $f(x)$ as a Fourier integral one has to assume the convergence of the following integral:

$$\int_{-\infty}^{+\infty} |f(x)| \, dx$$

thus,

$$\lim_{x \to \infty} f(x) = \lim_{x \to -\infty} f(x) = 0$$

Furthermore, the function has also to obey Dirichlet boundary conditions. It should be mentioned that the Fourier integral simplifies when $f(x)$ is an even or odd function. (Then one obtains cosine or sine terms only.)

Furthermore, we note Parseval's identity: let

$$f(x) = \int_{-\infty}^{+\infty} F(k) \frac{e^{ikx}}{\sqrt{2\pi}} \, dk$$

$$g(x) = \int_{-\infty}^{+\infty} G(k') \frac{e^{ik'x}}{\sqrt{2\pi}} \, dk'$$

Then

$$\int_{-\infty}^{+\infty} f(x)g^*(x) \, dx = \int_{-\infty}^{+\infty} F(k) \int_{-\infty}^{+\infty} G^*(k') \int_{-\infty}^{+\infty} \frac{e^{ikx}}{\sqrt{2\pi}} \frac{e^{-ik'x}}{\sqrt{2\pi}} \, dx \, dk \, dk'$$

$$= \int_{-\infty}^{+\infty} F(k) \int_{-\infty}^{\infty} G^*(k') \delta(k - k') \, dk \, dk'$$

$$= \int_{-\infty}^{\infty} F(k) G^*(k) \, dk$$

and similarly also

$$\int_{-\infty}^{+\infty} f(u)g^*(u - x) \, du = \int_{-\infty}^{+\infty} F(k)G^*(k) \frac{e^{ikx}}{\sqrt{2\pi}} \, dk$$

The last relation is known as the *convolution theorem*.

Example 3.1:   Exercise: Orthonormalization of the polynomials x^n

Show that

(a) In the space $L^2(-1, 1)$ of the functions quadratic integrable in the interval $(-1, 1)$ the set of all polynomials $x^n (n = 0, 1, \dots)$ is linearly independent with the scalar product

$$(f, g) = \int_{-1}^{1} f^*(x)g(x) \, dx$$

Apply Schmidt's orthogonalization procedure to this set.

(b) The functions $e^{im\phi}$ $(m \in \mathbb{Z})$ form an orthogonal system in the space $L^2(0, 2\pi)$. Are they normalized?

Solution

(a) **Linear independence:** A set of functions is linearly independent if any finite subset is linearly independent. Thus, we consider the set $M = \{x^{n_1}, \ldots, x^{n_N}\}$ and the linear combination

$$a_1 x^{n_1} + \cdots + a_n x^{n_N} = 0 \qquad (n_1, \ldots, n_N \quad \text{pairwise distinct})$$

However, such a polynomial vanishes only if $a_1 = \cdots = a_N = 0$ (identity theorem for power series) implying the linear independence of M. Since M is an arbitrary finite subset of $\{x^n : n = 0, 1, \ldots\}$, This set is also linearly independent.

Orthonormalization: The Schmidt orthonormalization procedure changes the linearly independent vectors $a_1, a_2, \ldots$ of a vector space with the scalar product $(\cdot, \cdot)$ to an orthonormal system

$$\hat{a}_1 = \frac{a_1}{\|a_1\|}$$

$$\hat{a}_2 = \frac{a_2 - (\hat{a}_1, a_2)\hat{a}_1}{\|a_2 - (\hat{a}_1, a_2)\hat{a}_1\|}$$

etc. In our case,

$$n = 0: \quad \int_{-1}^{1} x^0 x^0 \, dx = \int_{-1}^{1} dx = 2 \qquad \qquad \text{hence } P_0'(x) = \frac{1}{\sqrt{2}}$$

$$n = 1: \quad \int_{-1}^{1} P_0'(x) x^1 \, dx = \frac{1}{\sqrt{2}} \int_{-1}^{1} x \, dx = 0$$

$$\int_{-1}^{1} x^1 x^1 \, dx = \int_{-1}^{1} x^2 \, dx = \frac{2}{3} \qquad \qquad \text{hence } P_1'(x) = \sqrt{\frac{3}{2}} x$$

$$n = 2: \quad \int_{-1}^{1} P_0'(x) x^2 \, dx = \frac{1}{\sqrt{2}} \int_{-1}^{1} x^2 \, dx = \frac{\sqrt{2}}{3}$$

$$\int_{-1}^{1} P_1'(x) x^2 \, dx = \frac{3}{\sqrt{2}} \int_{-1}^{1} x^3 \, dx = 0$$

$$\int_{-1}^{1} \left(x^2 - \frac{1}{3}\right)\left(x^2 - \frac{1}{3}\right) dx = \frac{8}{45} \qquad \text{hence } P_2'(x) = \sqrt{\frac{45}{8}} \left(x^2 - \frac{1}{3}\right)$$

etc. Although the polynomials $P_n'(x)(n = 0, 1, 2, \ldots)$ form an orthonormal system they are not of use in physics. Instead, the so-called Legendre polynomials are used, which are defined by

$$P_n(x) = \frac{P_n'(x)}{P_n'(1)}$$

which yields

$$P_0(x) = 1, \quad P_1(x) = x, \quad P_2(x) = \frac{1}{2}\left(3x^2 - 1\right), \ldots \tag{3.29}$$

Remark　　The polynomials $P_n'(x)$, as well as the Legendre polynomials, form even a complete orthonormal system or orthogonal system in $L^2(-1, 1)$, that means, any function of this space can be represented as

$$f(x) = \sum_{n=0}^{\infty} a_n P_n(x) = \sum_{n=0}^{\infty} a_n' P_n'(x) = \sum_{n=0}^{\infty} b_n x^n$$

Here, the convergence is defined by the scalar product; in general, it is not pointwise. The proof of that is not trivial. The x^n, $P_n'(x)$, or $P_n(x)$ do not form a basis of $L^2(-1, 1)$, since a basis $\{b_c\}$, $c \in I$ of a vector space has to have the property that any vector f of this vector space may be represented as a finite linear combination

$$f = \sum_{n=1}^{N} c_n b_{c_n}$$

This is not guaranteed for the vector systems mentioned because not all functions of $L^2(-1, 1)$ are polynomials of degree n (e.g., e^x, $\sin x$, ...).

But frequently the notion *vector orthonormal system* is used synonymously with *basis*.

(b) We have

$$(e^{im\phi}, e^{in\phi}) = \int_0^{2\pi} e^{-im\phi} e^{in\phi} \, d\phi = \int_0^{2\pi} e^{-i(m-n)\phi} \, d\phi$$

$$= \left\{ \begin{array}{ll} 2\pi & \text{if } m = n \\ \dfrac{i}{m-n} \left[e^{-2\pi i(m-n)} - 1 \right] & \text{if } m \neq n \end{array} \right\} = 2\pi \delta_{mn}$$

Therefore, the functions $e^{im\phi}$ are orthogonal; their norm is $\sqrt{2\pi}$.

Spherical harmonics

We want to consider three further orthogonal systems of functions of particular importance in mathematical physics. In the solution of potential problems, one encounters always the function (Figure 3.1)

$$\frac{1}{|\mathbf{r} - \mathbf{r}'|} \tag{3.30}$$

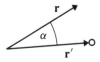

Figure 3.1. Illustration of the angle α.

which for a point charge $\mathbf{r}'$ takes the form

$$\frac{1}{\sqrt{r^2 - 2rr'\cos\alpha + r'^2}} \tag{3.31}$$

The root has to be expanded in a power series in the ratio of the distances r and r'. We denote by $r_<$ and $r_>$ the smaller and the greater value, respectively, of r and r'. Then, $r_</r_> < 1$, and we obtain the following expansion:

$$\frac{1}{r_>\sqrt{1 - 2\left(\dfrac{r_<}{r_>}\right)\cos\alpha + \left(\dfrac{r_<}{r_>}\right)^2}}$$

$$= \frac{1}{r_>}\left\{1 + (\cos\alpha)\left(\frac{r_<}{r_>}\right) + \frac{1}{2}\left(3\cos^2\alpha - 1\right)\left(\frac{r_<}{r_>}\right)^2 + \cdots\right\}$$

The coefficients occurring are the Legendre functions

$$\frac{1}{r_>\sqrt{1 - 2\left(\dfrac{r_<}{r_>}\right)\cos\alpha + \left(\dfrac{r_<}{r_>}\right)^2}}$$

$$= \frac{1}{r_>}\left\{P_0 + P_1(\cos\alpha)\left(\frac{r_<}{r_>}\right) + P_2(\cos\alpha)\left(\frac{r_<}{r_>}\right)^2 + P_3(\cos\alpha)\left(\frac{r_<}{r_>}\right)^3 + \cdots\right\}$$

$$= \sum_{l=0}^{\infty} \frac{r_<^l}{r_>^{l+1}} P_l(\cos\alpha) \tag{3.32}$$

Setting $x = \cos\alpha$, the first few Legendre polynomials are:

$$\begin{aligned}
P_0(x) &= 1 \\
P_1(x) &= x \\
P_2(x) &= \tfrac{1}{2}\left(3x^2 - 1\right) \\
P_3(x) &= \tfrac{1}{2}\left(5x^3 - 3x\right)
\end{aligned} \tag{3.33}$$

The Legendre polynomials with even indices are even functions, those with odd indices are odd functions. The Legendre polynomials may be calculated also by the relation (Rodrigues formula)

$$P_l(x) = \frac{1}{2^l l!}\frac{d^l}{dx^l}(x^2 - 1)^l \tag{3.34}$$

They form a complete orthogonal system of functions in the interval $x \in [-1, 1]$, ($\alpha \in [0, 2\pi]$), and

$$\int_{-1}^{1} P_l(x)P_{l'}(x)\,dx = \frac{2}{2l + 1}\delta_{ll'} \tag{3.35}$$

If we want to expand the potential of arbitrary extended charge distributions, we have to perform two further steps given here without a detailed proof.

A complete orthonormal system of functions defined on the unit sphere is represented by the spherical harmonics. Here, we point out briefly their relationship with the Legendre polynomials,

$$Y_{lm}(\theta, \varphi) = \sqrt{\frac{2l+1}{4\pi}\frac{(l-m)!}{(l+m)!}} P_l^m(\cos\theta)e^{im\varphi} \tag{3.36}$$

The index m is an integer in the bounds $-l \le m \le l$.

We give the explicit form of the first few spherical harmonics:

$$Y_{00} = \frac{1}{\sqrt{4\pi}}$$

$$Y_{22} = \frac{1}{4}\sqrt{\frac{15}{2\pi}}\sin^2\theta e^{i2\varphi}$$

$$Y_{11} = -\sqrt{\frac{3}{8\pi}}\sin\theta e^{i\varphi}$$

$$Y_{21} = -\sqrt{\frac{15}{8\pi}}\sin\theta\cos\theta e^{i\varphi}$$

$$Y_{10} = \sqrt{\frac{3}{4\pi}}\cos\theta$$

$$Y_{20} = \sqrt{\frac{5}{4\pi}}\left(\frac{3}{2}\cos^2\theta - \frac{1}{2}\right)$$

$$Y_{33} = -\frac{1}{4}\sqrt{\frac{35}{4\pi}}\sin^3\theta e^{i3\varphi} \tag{3.37}$$

$$Y_{32} = \frac{1}{4}\sqrt{\frac{105}{2\pi}}\sin^2\theta\cos\theta e^{i2\varphi}$$

$$Y_{31} = -\frac{1}{4}\sqrt{\frac{21}{4\pi}}\sin\theta\left(5\cos^2\theta - 1\right)e^{i\varphi}$$

$$Y_{30} = \sqrt{\frac{7}{4\pi}}\left(\frac{5}{2}\cos^3\theta - \frac{3}{2}\cos\theta\right)$$

The Y_{lm} with negative index m result from these functions according to equation (3.42) below.

The functions $P_l^m(\cos\theta)$ are the associated Legendre polynomials. They arise from the Legendre polynomials:

$$P_l^m(x) = (-1)^m(1-x^2)^{m/2}\frac{d^m}{dx^m}P_l(x) \tag{3.38}$$

Here, the index m is positive. The associated Legendre polynomials with negative index m differ from those with positive m only by a pre-factor.

$$P_l^{-m}(x) = (-1)^m\frac{(l-m)!}{(l+m)!}P_l^m(x) \tag{3.39}$$

Substituting relation (3.38) into (3.34),

$$P_l^m(x) = \frac{(-1)^m}{2^l l!}(1-x^2)^{m/2}\frac{d^{l+m}}{dx^{l+m}}(x^2-1)^l \tag{3.40}$$

For a fixed m the associated Legendre polynomials form a set of orthogonal functions:

$$\int_{-1}^{1} P_l^m(x) P_{l'}^m(x)\,dx = \frac{2}{2l+1}\frac{(l+m)!}{(l-m)!}\delta_{ll'} \tag{3.41}$$

The functions $e^{im\varphi}$ form a set of orthogonal functions as we have seen above. The product of the associated Legendre polynomials P_l^m and the function $e^{im\varphi}$ form, together with the normalization factor, the spherical harmonics Y_{lm}.

The spherical harmonics satisfy the relation

$$Y_{l,m}^*(\theta,\varphi) = (-)^m Y_{l,-m}(\theta,\varphi) \tag{3.42}$$

These functions are orthonormal

$$\int_0^{2\pi} d\varphi \int_0^{\pi} Y_{l'm'}^*(\theta,\varphi) Y_{lm}(\theta,\varphi)\sin\theta\,d\theta = \delta_{ll'}\delta_{mm'} \tag{3.43}$$

The term $\sin\theta$ occurs since the area element on the unit sphere is $\sin\theta\,d\theta\,d\varphi$. The completeness of the set of spherical harmonics is expressed by the relation

$$\sum_{l=0}^{\infty}\sum_{m=-l}^{l} Y_{lm}^*(\theta,\varphi) Y_{lm}(\theta',\varphi') = \delta(\varphi-\varphi')\,\delta(\cos\theta - \cos\theta') \tag{3.44}$$

The spherical harmonics Y_{lm} are related to Legendre polynomials P_l by a relation known as the addition theorem (see Figure 3.2):

$$P_l(\cos\alpha) = \frac{4\pi}{2l+1}\sum_{m=-1}^{l} Y_{lm}^*(\theta',\varphi') Y_{lm}(\theta,\varphi)$$

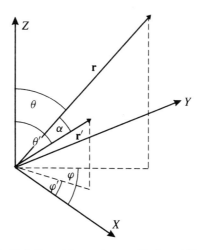

Figure 3.2. Definition of the angles used in the addition theorem.

The angle α is the angle between the vectors $\mathbf{r}$ and $\mathbf{r}'$, $\mathbf{r} \cdot \mathbf{r}' = r \cdot r' \cos \alpha$. Expressing both of the position vectors in spherical coordinates we obtain for α

$$\mathbf{r} \cdot \mathbf{r}' = r \cdot r'(\sin \theta \sin \theta' \cos(\varphi - \varphi') + \cos \theta \cos \theta') = r \cdot r' \cos \alpha \qquad (3.45)$$

The addition theorem offers one the possibility to extend the expansion valid for a point charge (axially symmetric distribution) to an arbitrary charge distribution. In this case, $\cos \alpha$ is replaced in equation (3.32) according to (3.45), and with the addition theorem we obtain

$$\frac{1}{|\mathbf{r} - \mathbf{r}'|} = \sum_{l=0}^{\infty} \sum_{m=-l}^{l} \frac{4\pi}{2l + 1} \frac{r_<^l}{r_>^{l+1}} Y_{lm}^*(\theta', \varphi') Y_{lm}(\theta, \varphi) \qquad (3.46)$$

The spherical harmonics are also obtained in solving the Laplace equation in spherical coordinates. In the separation of the variables one obtains the numbers l and m as separation constants. For the functions Θ depending on the angle θ, the generalized Legendre differential equation is valid ($x = \cos \theta$):

$$(1 - x^2)\frac{d^2\Theta}{dx^2} - 2x\frac{d\Theta}{dx} + \left[l(l + 1) - \frac{m^2}{1 - x^2} \right] \Theta = 0$$

whose solutions are the associated Legendre polynomials P_l^m. For $m = 0$ the equation becomes the Legendre differential equation possessing the Legendre polynomials P_l as solutions.

In this section we have gathered the most important properties of the spherical harmonics mostly without proof. In volume 4 of this series, the lectures on quantum mechanics, the theory of spherical harmonics will be outlined in detail.

Multipole expansions

If we have a bounded charge distribution vanishing outside a sphere of radius a (see Figure 3.3) about the origin, then the potential in the external region (without boundary surfaces with boundary conditions in finiteness, the Dirichlet surface term vanishes), for $r > a$, is given by

$$\phi(\mathbf{r}) = \int \frac{\rho(\mathbf{r}')}{|\mathbf{r} - \mathbf{r}'|} dV' \qquad (3.47)$$

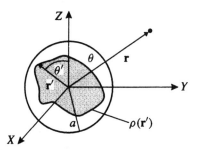

Figure 3.3. The charge distribution has the dimension a.

Since the charge distribution is bounded and the potential in the exterior region is considered, with $r > r'$ we can substitute the expansion according to equation (3.46):

$$\phi(\mathbf{r}) = \sum_{l=0}^{\infty} \sum_{m=-l}^{l} \frac{4\pi}{2l+1} \frac{1}{r^{l+1}} Y_{lm}(\theta, \varphi) \int \rho(\mathbf{r}') r'^l Y_{lm}^*(\theta', \varphi') \, dV' \tag{3.48}$$

The integrals occurring in this sum depend only on the particular charge distribution. They describe the outward action of the charge distribution completely. The expressions

$$q_{lm}^* = \int Y_{lm}^*(\theta', \varphi') r'^l \rho(\mathbf{r}') \, dV' \tag{3.49}$$

or

$$q_{lm} = \int Y_{lm}(\theta', \varphi') r'^l \rho(\mathbf{r}') \, dV'$$

are called the *multipole moments* of the charge distribution $\rho(\mathbf{r}')$. In particular, the most important q_{lm} are

$l = 0$	monopole moment
$l = 1$	dipole moment
$l = 2$	quadrupole moment
$l = 3$	octupole moment
$l = 4$	hexadecupole moment

For each l the multipole moments q_{lm} form a tensor of rank l (here in spherical representation with $2l + 1$ components ($m = l, l - 1, \ldots, 0, \ldots, -(l - 1), -l$)). To achieve a unique representation we agree to put the origin of the coordinate system at the center of gravity of the charge distribution. Now, we investigate systematically the first three multipole moments. Due to

$$Y_{lm}^*(\theta', \varphi') = (-1)^m Y_{l-m}^*(\theta', \varphi')$$

we get the following relation:

$$q_{lm}^* = \int \rho(\mathbf{r}') r'^l (-1)^m Y_{l-m}(\theta', \varphi') \, dV' = (-1)^m q_{l-m} \tag{3.50}$$

Hence, for each multipole of order l we have to calculate only $l + 1$ multipole moments q_{lm}.

Monopole moment

Here, the only component is

$$q_{00}^* = \int \rho(\mathbf{r}') r'^0 Y_{00}^*(\theta', \varphi') \, dV' = \frac{1}{\sqrt{4\pi}} \int \rho(\mathbf{r}') \, dV' = \frac{Q}{\sqrt{4\pi}}$$

Observed from a large distance r, any charge distribution acts approximately as if the total charge Q would be concentrated at one point since the dominating term in (3.48) is

$$\phi(\mathbf{r}) = 4\pi q_{00}^* \frac{1}{r} Y_{00} + \cdots = \frac{Q}{r} + \cdots \qquad (3.51)$$

Dipole moment

In Cartesian coordinates the three components of the dipole moment are given by

$$p_x = \int x' \rho(\mathbf{r}') \, dV', \qquad p_y = \int y' \rho(\mathbf{r}') \, dV', \qquad p_z = \int z' \rho(\mathbf{r}') \, dV'$$

The definition $\mathbf{p} = \int \mathbf{r}' \rho(\mathbf{r}') \, dV'$ is identical with the one given previously, after equation (1.80) and in Exercise (1.4). This can be checked easily for the charge distribution consisting of the positive charge q at $\mathbf{a}/2$ and the negative charge $-q$ at $-\mathbf{a}/2$. (See Figure 3.4.) It reads

$$\rho(\mathbf{r}') = q\delta\left(\mathbf{r}' - \frac{\mathbf{a}}{2}\right) - q\delta\left(\mathbf{r}' + \frac{\mathbf{a}}{2}\right)$$

and the dipole moment $\mathbf{p}$ is

$$\mathbf{p} = \int \rho(\mathbf{r}')\mathbf{r}' \, dV' = \int \left(q\delta\left(\mathbf{r}' - \frac{\mathbf{a}}{2}\right) - q\delta\left(\mathbf{r}' + \frac{\mathbf{a}}{2}\right)\right)\mathbf{r}' \, dV'$$

$$= q\frac{\mathbf{a}}{2} - q\left(-\frac{\mathbf{a}}{2}\right) = q\mathbf{a}$$

In spherical representation one obtains with the help of (3.37)

$$q_{11}^* = \int \rho(\mathbf{r}')r'Y_{11}^*(\theta', \varphi') \, dV'$$

$$= -\sqrt{\frac{3}{8\pi}} \int \rho(\mathbf{r}')r'(\sin\theta' \cdot \cos\varphi' - i\sin\theta' \cdot \sin\varphi') \, dV' \qquad (3.52)$$

We may represent the spherical components by Cartesian ones:

$$q_{11}^* = -\sqrt{\frac{3}{8\pi}} \int \rho(\mathbf{r}')(x' - iy') \, dV' = -\sqrt{\frac{3}{8\pi}} \cdot (p_x - ip_y) \qquad (3.53)$$

Also,

$$q_{10}^* = \int \rho(\mathbf{r}')r'Y_{10}^*(\theta', \varphi') \, dV' = \int \rho(\mathbf{r}')r'\sqrt{\frac{4}{3\pi}}\cos\theta' \, dV' = \sqrt{\frac{4}{3\pi}}p_z$$

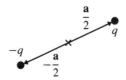

Figure 3.4. Electric dipole.

$$q_{1-1}^* = -q_{11} = \sqrt{\frac{3}{8\pi}}(p_x + ip_y)$$

Hence, the three dipole moments are linear combinations of the Cartesian ones. The two representations (spherical and Cartesian) are equivalent to each other. Which form seems to be more appropriate will depend on the problem.

Quadrupole moment

The Cartesian components of the quadrupole tensor are defined by

$$Q_{ij} = \int \rho(\mathbf{r}')(3x_i'x_j' - r'^2\delta_{ij})\, dV', \qquad i, j = 1, 2, 3 \tag{3.54}$$

In the next section this relation will turn out to be meaningful in the Taylor expansion of the potential. Now, we calculate again the spherical components and express them in terms of the Cartesian ones:

$$
\begin{aligned}
q_{22}^* &= \int \rho(\mathbf{r}')r'^2 Y_{22}^*(\theta', \varphi')\, dV' \\
&= \frac{1}{4}\sqrt{\frac{15}{2\pi}} \int \rho(\mathbf{r}')(r'\sin\theta'(\cos\varphi' - i\sin\varphi'))^2\, dV' \\
&= \frac{1}{4}\sqrt{\frac{15}{2\pi}} \int \rho(\mathbf{r}')(x' - iy')^2\, dV'
\end{aligned} \tag{3.55}
$$

Rewriting $(x' - iy')^2 = \frac{1}{3}[(3x'^2 - r'^2) - 6ix'y' - (3y'^2 - r'^2)]$ we obtain

$$q_{22}^* = \frac{1}{12}\sqrt{\frac{15}{2\pi}}(Q_{11} - 2iQ_{12} - Q_{22}) \tag{3.56}$$

Analogously,

$$
\begin{aligned}
q_{21}^* &= \int \rho(\mathbf{r}')r'^2 Y_{21}^*(\theta', \varphi')\, dV' \\
&= -\sqrt{\frac{15}{8\pi}} \int \rho(\mathbf{r}')z'(x' - iy')\, dV' \\
&= -\frac{1}{6}\sqrt{\frac{15}{2\pi}}(Q_{13} - iQ_{23})
\end{aligned} \tag{3.57}
$$

and

$$q_{20}^* = \int \rho(\mathbf{r}')r'^2 Y_{20}^*(\theta', \varphi')\, dV' = \frac{1}{4}\sqrt{\frac{5}{\pi}}Q_{33}$$

So, with the help of the relation (3.50) we obtain at once the components with negative m:

$$q_{2,-1}^* = \frac{1}{6}\sqrt{\frac{15}{2\pi}}(Q_{13} + iQ_{23})$$

$$q_{2,-2}^* = \frac{1}{12}\sqrt{\frac{15}{2\pi}}(Q_{11} + 2iQ_{12} - Q_{22})$$

(3.58)

Multipole expansions in Cartesian coordinates

The expansion of a function $f(\mathbf{r}')$ in a Taylor series about the point $\mathbf{r}' = 0$ is

$$f(x_1', x_2', x_3') = f(0,0,0) + \sum_{i=1}^{3}\frac{\partial f}{\partial x_i'}(0,0,0)x_i' + \frac{1}{2!}\sum_{i,j=1}^{3}\frac{\partial^2 f}{\partial x_i'\partial x_j'}(0,0,0)x_i'x_j' + \cdots$$

$$= \sum_{n=0}^{\infty}\frac{1}{n!}\left(\sum_{i}x_i'\frac{\partial}{\partial x_i'}\right)^n f(x_1', x_2', x_3')\Bigg|_{x_1'=x_2'=x_3'=0}$$

$$= \sum_{n=0}^{\infty}\frac{(\mathbf{r}'\cdot\nabla')^n}{n!}f(\mathbf{r}')\Bigg|_{\mathbf{r}'=0}$$

(3.59)

Here, the gradient ∇' acts only upon the function $f(\mathbf{r}')$. For the special function

$$f(\mathbf{r}') = \frac{1}{|\mathbf{r} - \mathbf{r}'|} = \frac{1}{\sqrt{(x_1 - x_1')^2 + (x_2 - x_2')^2 + (x_3 - x_3')^2}}$$

$$= \frac{1}{\sqrt{\sum_{i=1}^{3}(x_i - x_i')^2}}$$

(3.60)

one computes

$$f(\mathbf{r}')\big|_{\mathbf{r}'=0} = \frac{1}{\sqrt{x_1^2 + x_2^2 + x_3^2}} = \frac{1}{r}$$

$$\frac{\partial f(\mathbf{r}')}{\partial x_i'}\bigg|_{\mathbf{r}'=0} = \frac{x_i - x_i'}{\sqrt{(x_1 - x_1')^2 + (x_2 - x_2')^2 + (x_3 - x_3')^2}^3}\bigg|_{\mathbf{r}'=0}$$

$$= \frac{x_i}{\sqrt{x_1^2 + x_2^2 + x_3^2}^3} = \frac{x_i}{r^3}$$

$$\frac{\partial^2 f(\mathbf{r}')}{\partial x_j'\partial x_i'}\bigg|_{\mathbf{r}'=0} = \frac{\partial}{\partial x_j'}\frac{x_i - x_i'}{\sqrt{(x_1 - x_1')^2 + (x_2 - x_2')^2 + (x_3 - x_3')^2}^3}\bigg|_{\mathbf{r}'=0}$$

$$= \left\{ \frac{3(x_i - x_i')(x_j - x_j')}{\sqrt{(x_1 - x_1')^2 + (x_2 - x_2')^2 + (x_3 - x_3')^2}^5} \right.$$

$$\left. - \frac{\delta_{ij}}{\sqrt{(x_1 - x_1')^2 + (x_2 - x_2')^2 + (x_3 - x_3')^2}^3} \right\} \Bigg|_{\mathbf{r'}=0}$$

$$= \frac{3x_i x_j - r^2 \delta_{ij}}{r^5} \tag{3.61}$$

Hence, for the potential in the region $r > r'$ we obtain

$$\phi(\mathbf{r}) = \int_{V'} \rho(\mathbf{r}') \frac{1}{|\mathbf{r} - \mathbf{r}'|} dV'$$

$$= \int_{V'} \rho(\mathbf{r}') \left\{ \frac{1}{r} + \sum_{i=1}^{3} \frac{x_i' x_i}{r^3} + \frac{1}{2!} \sum_{i,j=1}^{3} \frac{3x_i x_j - r^2 \delta_{ij}}{r^5} \cdot x_i' x_j' + \cdots \right\} dV'$$

$$= \frac{Q}{r} + \frac{\mathbf{r} \cdot \mathbf{p}}{r^3} + \frac{1}{2!} \sum_{i,j} \frac{3x_i x_j - r^2 \delta_{ij}}{r^5} \int_{V'} \rho(\mathbf{r}') x_i' x_j' dV' + \cdots \tag{3.62}$$

The third term (quadrupole term) can be rewritten further:

$$\frac{1}{1 \cdot 2 \cdot 3} \sum_{i,j} \frac{3x_i x_j - r^2 \delta_{ij}}{r^5} \int_{V'} \rho(\mathbf{r}')(3x_i' x_j' - r'^2 \delta_{ij}) dV'$$

$$+ \frac{1}{1 \cdot 2 \cdot 3} \sum_{i,j} \frac{3x_i x_j - r^2 \delta_{ij}}{r^5} \int_{V'} \rho(\mathbf{r}') r'^2 dV' \delta_{ij}$$

$$= \frac{1}{6} \sum_{i,j} \frac{(3x_i x_j - r^2 \delta_{ij}) Q_{ij}}{r^5} + \frac{1}{6} \sum_{i} \frac{(3x_i^2 - r^2)}{r^5} \int_{V'} \rho(\mathbf{r}') r'^2 dV'$$

$$= \frac{1}{6} \sum_{i,j} \frac{(3x_i x_j - r^2 \delta_{ij}) Q_{ij}}{r^5} \tag{3.63}$$

because

$$\frac{1}{6} \sum_{i} (3x_i^2 - r^2) \frac{\int \rho(\mathbf{r}') r'^2 dV'}{r^5} = \frac{1}{6} (3r^2 - 3r^2) \frac{\int \rho(\mathbf{r}') r'^2 dV'}{r^5} = 0 \tag{3.64}$$

Here, the definition

$$Q_{ij} = \int_{V'} \rho(\mathbf{r}')(3x_i' x_j' - r'^2 \delta_{ij}) dV' \tag{3.65}$$

has been suggested for the Cartesian components of the quadrupole tensor by symmetry reasons. Thus, outside the charge distribution $(r > r')$ the potential is

$$\phi(\mathbf{r}) = \int_{V'} \rho(\mathbf{r}') \frac{1}{|\mathbf{r} - \mathbf{r}'|} dV'$$

$$= \frac{Q}{r} + \frac{\mathbf{r} \cdot \mathbf{p}}{r^3} + \frac{1}{6} \sum_{i,j} \frac{Q_{ij}(3x_i x_j - r^2 \delta_{ij})}{r^5} + \cdots \qquad (3.66)$$

One realizes easily that, due to $\sum_{i=1}^{3} Q_{ii} = 0$ (see (3.67)), the quadrupole term can be written also in the form $\frac{1}{2} \sum_{ij} Q_{ij} x_i x_j / r^5$.

Comparing this series with the corresponding expansion in spherical coordinates, a difference seems to exist between the quadrupole terms. While there are only 5 distinct q_{2m}, 6 independent Q_{ij} seem to exist. That this is not the case one can prove by forming the sum of the $Q_{ii}(i = 1, 2, 3)$ (the trace of the tensor) which vanishes identically:

$$Q_{11} + Q_{22} + Q_{33} = \int \rho(\mathbf{r}') \left[3(x_1'^2 + x_2'^2 + x_3'^2) - 3r'^2 \right] dV' = 0 \qquad (3.67)$$

Consequently, one of the Q_{ii} may be expressed by the other ones.

Interaction of an extended charge distribution with an external field

The multipole expansion of the potential of a charge distribution can also be used to describe the interaction of the charge distribution with an external field. The energy of the charge distribution $\rho(\mathbf{r})$ in an external field $\phi(\mathbf{r})$ (Figure 3.5) is given by

$$W = \int_V \rho(\mathbf{r})\phi(\mathbf{r}) \, dV$$

Compared with the formula for the energy derived previously, the factor $1/2$ is missing here. It was introduced because the interaction energy of two charges appears twice in the integral. Now the double-counting is excluded, since the charge generating the field $\phi(\mathbf{r})$ does not belong to the distribution $\rho(\mathbf{r})$.

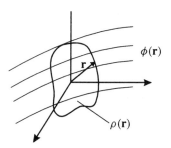

Figure 3.5. Charge distribution $\rho(\mathbf{r})$ in an external field $\phi(\mathbf{r})$.

The external field may be expanded in a Taylor series:

$$\phi(\mathbf{r}) = \phi(0) + \mathbf{r} \cdot \nabla\phi|_0 + \frac{1}{2} \sum_{i,j=1}^{3} x_i x_j \frac{\partial^2}{\partial x_i \partial x_j} \phi \bigg|_0 + \cdots$$

With $\nabla\phi|_0 = -\mathbf{E}(0)$, and therefore also

$$\frac{\partial}{\partial x_i} \frac{\partial \phi}{\partial x_j} \bigg|_0 = -\frac{\partial E_j}{\partial x_i} \bigg|_0$$

the series expansion is

$$\phi(\mathbf{r}) = \phi(0) - \mathbf{r} \cdot \mathbf{E}(0) - \frac{1}{2} \sum_{i,j=1}^{3} x_i x_j \frac{\partial E_j}{\partial x_i} \bigg|_0 + \cdots \tag{3.68}$$

To get the quadrupole moments in the last sum we substract $\frac{1}{6}r^2 \nabla \cdot \mathbf{E}$ from each term,

$$\phi(\mathbf{r}) = \phi(0) - \mathbf{r} \cdot \mathbf{E}(0) - \frac{1}{6} \sum_{i,j=1}^{3} (3x_i x_j - r^2 \delta_{ij}) \frac{\partial E_j}{\partial x_i} \bigg|_0 + \cdots \tag{3.69}$$

This step does not change the potential since $\nabla \cdot \mathbf{E} = 0$ for the external field in the considered region, because the field-producing charges are lying outside of it. Performing the integration, we obtain for the energy

$$W = \int_V \rho(\mathbf{r})\phi(\mathbf{r}) \, dV$$

$$= q\phi(0) - \mathbf{p} \cdot \mathbf{E}(0) - \frac{1}{6} \sum_{ij} Q_{ij} \frac{\partial E_j}{\partial x_i} \bigg|_0 + \cdots \tag{3.70}$$

From this expansion we conclude that different multipoles interact with the outer field in a distinct way: the total charge is connected with the potential, the dipole with the field intensity (i.e., with the gradient of the potential), the quadrupole with the derivative of the field intensity, etc.

One also says: the quadrupole acts upon the gradient of the field intensity, meaning the last term in (3.70). Let us consider subsequently the interaction of two dipoles: the potential of a dipole at the origin, observed at the point $\mathbf{r}$, is

$$\phi(\mathbf{r}) = \frac{\mathbf{p} \cdot \mathbf{r}}{r^3}$$

Hence, for $\mathbf{r} \neq 0$ the field intensity is

$$\mathbf{E}(\mathbf{r}) = -\nabla\phi(\mathbf{r}) = -\frac{\nabla(\mathbf{p} \cdot \mathbf{r})}{r^3} - (\mathbf{p} \cdot \mathbf{r}) \nabla \left(\frac{1}{r^3}\right) = -\frac{\mathbf{p} - 3\mathbf{n}(\mathbf{n} \cdot \mathbf{p})}{r^3} \qquad \text{with } \mathbf{n} = \frac{\mathbf{r}}{r}$$

$$\tag{3.71}$$

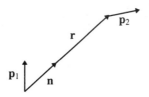

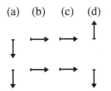

Figure 3.6. Orientation of the dipole moments.

Figure 3.7. Various orientations of the dipoles correspond to different energies.

Then, due to the second term in (3.70) the interaction energy of two dipoles is given by

$$W_{12} = -\mathbf{p}_2 \cdot \mathbf{E}_1 = \frac{\mathbf{p}_1 \cdot \mathbf{p}_2 - 3(\mathbf{n} \cdot \mathbf{p}_1)(\mathbf{n} \cdot \mathbf{p}_2)}{r^3} \tag{3.72}$$

The potential energy, and thus also the interaction force, depend not only on the distance r but also on the orientation of the dipoles with respect to the straight line connecting the dipoles. Such forces exist also between nucleons (e.g., proton–neutron interactions), and they are called *tensor forces*.

The energy reaches its minimum if the dipoles are arranged parallel to each other (along a straight line). If $\mathbf{p}_1 \perp \mathbf{n}$ and $\mathbf{p}_2 \perp \mathbf{n}$, then the antiparallel arrangement is favored energetically. In Figure 3.7, various orientations and the corresponding energies are illustrated:

$$\begin{aligned}
\text{(a)} \quad & W_{12} = \quad -2W \\
\text{(b)} \quad & W_{12} = \quad -W \\
\text{(c)} \quad & W_{12} = \quad +W \\
\text{(d)} \quad & W_{12} = \quad +2W
\end{aligned} \tag{3.73}$$

where $W = |\mathbf{p}_1||\mathbf{p}_2|/r^3$.

The following are examples of higher multipole moments:

Example 3.2: The uniformly charged sphere

Let us consider a sphere of radius R_0 and with constant charge density $\rho(\mathbf{r}) = \rho$. Then, for all values of l and m,

$$q_{lm}^* = \rho \int_{\Omega'} \int_0^{R_0} r'^l Y_{lm}^*(\theta', \varphi') r'^2 \, dr' \, d\Omega' = \rho \frac{R_0^{l+3}}{l+3} \int_{\Omega'} Y_{lm}^*(\theta', \varphi') \, d\Omega'$$

and due to $Y_{00} = 1/\sqrt{4\pi}$ we have from the orthogonality relation

$$q_{lm}^* = \rho \frac{R_0^{l+3}}{l+3} \sqrt{4\pi} \int_{\Omega'} Y_{lm}^* Y_{00} \, d\Omega'$$

$$= \rho \frac{R_0^{l+3}}{l+3} \sqrt{4\pi} \, \delta_{l0} \delta_{m0} \tag{3.74}$$

As to be expected, for a uniformly charged sphere all multipole moments vanish except q_{00}. The higher multipole moments measure the deviation of the charge distribution from the spherical symmetry.

Example 3.3: Deformed sphere with quadrupole moment

Now, we attempt to perform a multipole decomposition of a uniform charge distribution whose surface is a weakly deformed sphere:

$$R = R_0 \left(1 + \sum_{m=-2}^{2} \alpha_{2m}^* Y_{2m}(\theta, \varphi) \right), \qquad |\alpha_{2m}| \ll 1$$

The multipole moments are

$$q_{lm}^* = \int_{\Omega'} \int_0^{R(\theta', \varphi')} \rho r'^l Y_{lm}^* r'^2 \, dr' \, d\Omega'$$

$$= \rho \frac{R_0^{l+3}}{l+3} \int_{\Omega'} Y_{lm}^*(\theta', \varphi') \left(1 + \sum_{\mu=-2}^{2} \alpha_{2\mu}^* Y_{2\mu}^*(\theta', \varphi') \right)^{l+3} d\Omega'$$

$$\cong \rho \frac{R_0^{l+3}}{l+3} \int_{\Omega'} Y_{lm}^*(\theta', \varphi') \left(1 + (l+3) \sum_{\mu=-2}^{2} \alpha_{2\mu}^* Y_{2\mu}^*(\theta', \varphi') \right) d\Omega'$$

Apart from the monopole moment q_{00} from Example 3.1, there are further q_{lm} present:

$$q_{lm}^* = \rho \frac{R_0^{l+3}}{l+3} (l+3) \sum_{\mu=-2}^{2} \alpha_{2\mu}^* \int_{\Omega'} Y_{lm}^* Y_{2\mu} \, d\Omega'$$

$$= \rho R_0^{l+3} \sum_{\mu} \alpha_{2\mu}^* \delta_{m\mu} \delta_{l2}, \qquad l > 0$$

Thus, all multipole moments vanish except for

$$q_{2m}^* = \rho R_0^5 \alpha_{2m}^*$$

Corresponding considerations can be performed also for the higher multipole moments in which we are led to the shapes shown in Figure 3.8.

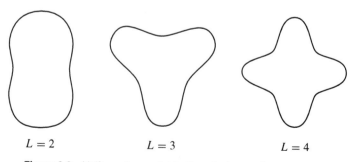

$$L = 2 \qquad\qquad L = 3 \qquad\qquad L = 4$$

Figure 3.8. Uniform charge distributions (cut) with 2^l-pole-moment.

Application in nuclear physics

Many atomic nuclei are deformed; that is, they have large quadrupole moments. The nuclear surface may be represented by $R = R_0(1 + \beta_{20}Y_{20} + \beta_{40}Y_{40})$. The charge distribution inside the surface is by far uniform. β_{20} is called the *quadrupole deformation*, and β_{40} is called the *hexadecupole deformation* of a nucleus. Typical deformations are $\beta_{20} \approx 0.3$ and $\beta_{40} \approx 0.05$. If $\beta_{20} > 0$, the nucleus is deformed cigar-like (*prolate* shape); if $\beta_{20} < 0$, the nucleus looks like a discus (*oblate* shape).

Octupole moments in an exactly static sense do not appear (due to the conservation of parity or the symmetry under time inversion, as we will see in quantum mechanics); but, there are many nuclei containing octupole moments in a symmetrized form. The nucleus exists, so to speak, in both octupole shapes with equal probability (see Figure 3.9b), to

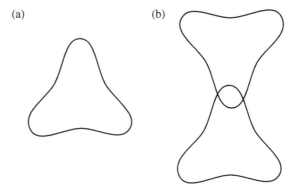

Figure 3.9. (a) Static octupole. (b) Symmetrized octupole.

take a mirror-symmetric configuration (conservation of parity) in this way. In molecular physics, there are similar examples, e.g., the NH_3-molecule (see Figure 3.10).

One can survey the deformed nuclei by drawing up the magic numbers (noble-gas-like, especially compact and tightly bound) as in Figure 3.11.

In nature, nuclei occur only along the *stable valley*. *Magic* and *doubly magic* nuclei are distinguished by closed shells (fully occupied by nucleons).

Due to their high binding energies, they are of highest symmetry; that is, they are

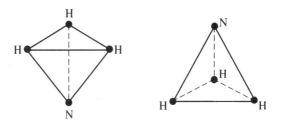

Figure 3.10. Symmetrized form of ammonia.

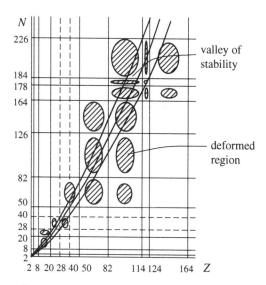

Figure 3.11. On the zoology of atomic nuclei. Magic proton numbers are characterized by vertical lines; the horizontal lines correspond to magic neutron numbers.

spherical. Far away from magic configurations (shadowed region in the figure), the nucleons (protons, neutrons) are less tightly bound: therefore, nuclei in these regions can be deformed easily because of the repulsion between the protons. Typical regions of deformation are the rare earths (Sm to Os) and the actinides (Th, U, transuranic elements).

Beyond uranium, nuclei become unstable: they undergo fission spontaneously due to the strong Coulomb repulsion of the large number of protons in the nucleus. According to Mosel and Greiner (*Zeitschrift für Physik* 217(1968) 256; 222(1969) 261) and S.G. Nielson et al. (*Nucl. Phys.* A131 (1969) 1), near the proton number $Z = 114$ and the neutron number $N = 184$ an island of *superheavy elements* should exist, being relatively stable again. These elements are intended to be produced in the *cold fusion* of two heavy nuclei by means of heavy-ion accelerators. Based on the potential energy surfaces calculated in the framework of the two-center shell model, Sandulescu and Greiner (see W. Greiner, *Int. Journ. of Modern Physics E*, Vol. 5, No. 1 (1995) 1 and references therein) proposed the reactions for the production of these elements. These reactions have been investigated with the heavy-ion mass filter SHIP of GSI (constructed by H. Ewald and G. Münzenberg, Giessen). G. Münzenberg, S. Hofmann, P. Armbruster, and their coworkers succeeded in detecting the elements $Z = 106, 107, 108, 110, 111,$ and 112. They found that the relative stability against nuclear fission increases again. So, one is on the way to the superheavy elements the most stable of which should be the element $Z = 111$ (but with the neutron number $N = 184$, which is much larger than the mentioned, currently synthesized elements, according to the calculations of the Frankfurt school, see, e.g., J. Grumann, U. Mosel, B. Fink, and W. Greiner, *Zeitschrift für Physik* 228 (1969) 371, or J.M. Eisenberg and W. Greiner, *Nuclear Theory 3* (1976) North-Holland, chapter 10).

A model of the uranium nucleus possessing a large quadrupole moment as well as a hexadecupole moment is given in Figure 3.12.

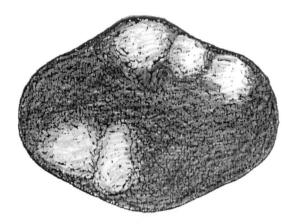

Figure 3.12. Model of the uranium nucleus ^{234}U on the scale $1 : 10^{-11}$. The quadrupole moment is $q_{20} = 2.68$ eb and the hexadecupole moment is $q_{40} = 0.808$ eb^2 (e = electronic charge, b = barn $= 10^{-24}$cm^2).

Example 3.4: Multipoles constructed of point charges

Multipoles can be formed in a simple manner out of point charges by joining multipoles of lower order.

Figure 3.13 shows two possible ways to join two dipoles to form a quadrupole. As for the dipole, we can construct an ideal quadrupple for which all other moments are omitted by a limiting procedure $a^2e \rightarrow 0$ ($a =$ distance, $e =$ charge).

For two antiparallel dipoles on a straight line, the charge distribution is given by

$$\rho(\mathbf{r}) = e\delta(x) \cdot \delta(y) \cdot [\delta(z - a) - 2\delta(z) + \delta(z + a)]$$

The nonvanishing moments are

$$Q_{11} = Q_{22} = -2a^2e, \qquad Q_{33} = 4a^2e$$

In spherical coordinates only $q_{20} = \frac{1}{4}\sqrt{5/\pi} \, Q_{33}$ is different from zero.

Two antiparallel dipoles lying in a plane, as shown in Figure 3.13, are described by

$$\rho(\mathbf{r}) = e\delta(z)\left[\delta(x) \cdot (\delta(y - a) + \delta(y + a)) - \delta(y)(\delta(x - a) + \delta(x + a))\right]$$

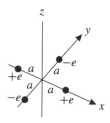

Figure 3.13. Four charges forming a pure quadropole.

The moments are

$$Q_{11} = 6a^2e \qquad \text{and} \qquad Q_{22} = -6a^2e$$

all other ones vanish. The spherical quadrupole moments are $q_{22} = q_{2-2} = \sqrt{15/2\pi}\, a^2e$.

Finally, we give some pure multipoles that follow successively from each other (Figure 3.14): a pure dipole is built up out of two pure monopoles $+q$, $-q$ at the distance d; a pure quadrupole out of two pure opposite dipoles at the distance d, etc. Pure multipoles of the lth order means here that all multipoles of lower order ($l' < l$) vanish and, hence, the first nonvanishing multipole is of the order l. Of course, this is valid exactly only in the limit $a \to 0$!

Finally, we think about how all nonvanishing spherical multipole components of the quadrupole-like charge configuration shown in the figure above should look. Contrary to the previous considerations, now we avoid considering the detour via the Cartesian multipole moments, and write down the charge distribution directly in spherical δ-functions:

$$\rho(\mathbf{r}) = e\frac{\delta(r-a)}{r^2}\frac{\delta(\theta - \pi/2)}{\sin\theta}\left[\delta(\varphi) - \delta\left(\varphi - \frac{\pi}{2}\right) + \delta(\varphi - \pi) - \delta\left(\varphi - \frac{3\pi}{2}\right)\right] \tag{3.75}$$

Now,

$$q_{lm} = \int Y_{lm}(\theta', \varphi')r''\rho(\mathbf{r}')\,d^3r'$$

$$\Leftrightarrow q_{lm} = e\int Y_{lm}(\theta', \varphi')r''\frac{\delta(r'-a)}{r'^2}\frac{\delta(\theta' - \pi/2)}{\sin\theta'}$$

$$\cdot\left[\delta(\varphi') - \delta\left(\varphi' - \frac{\pi}{2}\right) + \delta(\varphi' - \pi) - \delta\left(\varphi' - \frac{3\pi}{2}\right)\right]$$

$$\cdot r'^2\sin\theta'\,d\theta'\,d\varphi'$$

$$\Leftrightarrow q_{lm} = ea^l\sum_{n=0}^{3}(-1)^n Y_{lm}\left(\frac{\pi}{2}, \frac{n\pi}{2}\right) \tag{3.76}$$

We use the definition of the spherical harmonics in terms of associated Legendre polynomials:

$$Y_{lm}\left(\frac{\pi}{2}, \frac{n\pi}{2}\right) = \sqrt{\frac{(2l+1)(l-m)!}{4\pi(l+m)!}}\,P_l^m(0)\,e^{imn\pi/2} \tag{3.77}$$

$$\Rightarrow q_{lm} = ea^l\sqrt{\frac{(2l+1)(l-m)!}{4\pi(l+m)!}}\,P_l^m(0)\sum_{n=0}^{3}(-1)^n e^{imn\pi/2} \tag{3.78}$$

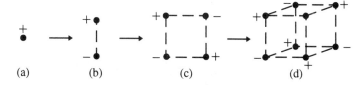

Figure 3.14. (a) Monopole. (b) Dipole. (c) Quadrupole. (d) Octupole.

From equation (3.78), we have immediately $q_{00} = 0$ as well as $q_{l0} = 0$, for all l since for $m = 0$ the sum

$$e^{imn\pi/2} = \left(e^{in\pi/2}\right)^m = i^{mn} \tag{3.79}$$

vanishes. Here, we use the Euler relation $e^{i\varphi} = \cos\varphi + i\sin\varphi$. From (3.79)

$$\sum_{n=0}^{3} (-1)^n e^{imn\pi/2} = \sum_{n=0}^{3} i^{n(2+m)} = 4\delta_{-2m} \tag{3.80}$$

Thus, we see that only the $q_{l-2} \neq 0$. Hence, we obtain from (3.78)

$$q_{l-2} = ea^l 4\sqrt{\frac{(2l+1)(l+2)!}{4\pi(l-2)!}} P_l^{-2}(0), \qquad l = 2, 4, \ldots \tag{3.81}$$

This is the general expression for the spherical multipole moments of our problem. The lowest nonvanishing 2^l-pole reads

$$P_{2-2} = ea^2 4\sqrt{\frac{5 \cdot 4 \cdot 3 \cdot 2 \cdot 1}{4\pi}} P_2^{-2}(0) \tag{3.82}$$

Now, we utilize the relation (see, e.g., Abramowitz and Stegun, (8.6.1)):

$$P_l^m(0) = \frac{2^m}{\sqrt{\pi}} \cos\left(\frac{\pi}{2}(l+m)\right) \frac{\Gamma\left(\frac{1}{2}l + \frac{1}{2}m + \frac{1}{2}\right)}{\Gamma\left(\frac{1}{2}l - \frac{1}{2}m + \frac{1}{2}\right)} \tag{3.83}$$

as well as $\Gamma(n+1) = n!, (n \in \mathbf{N})$

$$\Gamma\left(n + \frac{1}{2}\right) = \sqrt{\pi}\left(\frac{1 \cdot 3 \cdot 5 \cdot 7 \cdots (2n-1)}{2^n}\right)$$

$$\Rightarrow P_2^{-2}(0) = \frac{1}{4\sqrt{\pi}} \cos(0) \frac{\Gamma\left(\frac{1}{2}\right)}{\Gamma(3)} = \frac{1}{8} \tag{3.84}$$

$$\Rightarrow q_{2-2} = ea^2 4\sqrt{\frac{5 \cdot 4 \cdot 3 \cdot 2 \cdot 1}{4\pi}} \frac{1}{8} = \sqrt{\frac{15}{2\pi}} ea^2 \tag{3.85}$$

In particular, $q_{l2}^* = q_{l-2}$, hence,

$$q_{22} = q_{2-2} = \sqrt{\frac{15}{2\pi}} ea^2 \tag{3.86}$$

Equation (3.86) implies that all odd l-terms must vanish. All higher multipole moments can be calculated naturally from equation (3.83), $q_{2-2}, q_{4-2}, q_{6-2}, \ldots$. The next nonvanishing multipole moment would be a hexadecupole. If a is small, then the quadrupole dominates since for $l = 4$, $a^4 \ll a^2$.

Biographical notes

Jean Baptiste Joseph Fourier, b. March 21, 1768, Auxerre–d. May 16, 1830, Paris. The son of a tailor, Fourier attended the local Ecole Militaire. Due to his backgound, he was denied an officer's career. So he decided to become a priest, but the Revolution of 1789 broke out before he took a vow. He became a teacher instead, meddled with politics, and was sent to prison several times. In 1795 he was sent to Paris to study at the Ecole Normale. He soon became a member of the teaching staff at the new Ecole Polytechnique, and in 1798 was appointed head of the Institute d'Egypte in Cairo. In 1801 he returned to Paris and was appointed prefect of Departement Isère by Napoleon. During his period of office from 1802 to 1815 he arranged for the swamps of Bourgoin, which were contaminated with malaria, to be drained. After Napoleon was overthrown, the Bourbons removed Fourier from all of his offices. In 1817, however, the King had to agree to Fourier being elected a member of the Academy of sciences. Five years later, Fourier became the Academy's Standing Secretary. Fourier's most important mathematical achievement is his dealing with the notion of *function*. In 1755 D. Bernoulli had solved the problem of the vibrating string (on which d'Alembert, Euler, and Lagrange had already worked) by a trigonometric series. The ensuing question whether "any" function could be represented by such a series, was answered in the positive by Fourier in 1807–1812. The problem of the conditions for this representation could only be answered by his friend Dirichlet. Fourier became known by his "Theorie analytique de la chaleur" (1822) which mainly discusses the equation of propagation of heat with the means of Fourier series. The work marks the starting point of handling partial differential equations with boundary conditions by trigonometric series. Fourier also made important contributions to the theory of solving equations, and to probability calculus.

Erhard Schmidt, b. Jan. 13, 1876, Dorpat–d. Dec. 6, 1959, Berlin. German mathematician and professor in Zürich, Erlangen, Breslau, and Berlin. Schmidt's fundamental works were on the theory of integral equations. Schmidt was the cofounder of modern functional analysis.

Adrien Marie Legendre, b. Sept. 18, 1752–d. Jan. 10, 1833, Paris. Legendre contributed a lot to the theory of numbers and geodesy. He also made important discoveries in the fields of elliptic integrals, bases and methods of Euclidian geometry, variational calculus, and theoretical astronomy. He was the first to use the method of smallest squares, and calculated extensive tables. Legendre dealt with a lot of problems Gauß was interested in, as well, but always failed to achieve the perfection of the latter. From 1875 Legendre worked as a professor at several Parisian universities, and published excellent textbooks that had a long-lasting impact.

4

Elementary Considerations on Function Theory

MATHEMATICAL SUPPLEMENT

In potential theory as well as later on in the computation of complicated integrals, we need methods coming from function theory, i.e., from the theory of functions with complex arguments. For these reasons, we will arm ourselves with the most important tools in this branch of mathematics.

Complex numbers

Complex numbers can be written in the form

$$z = x + iy =: \operatorname{Re} z + i \cdot \operatorname{Im} z, \qquad i := \sqrt{-1} \tag{4.1}$$

where x and y are real numbers.[1]

The set of all z denoted by $\mathbf{C}$ can be represented by a plane: every point of the plane corresponds to a complex number due to the isomorphism $\mathbf{C} \leftrightarrow \mathbf{R}^2$. Besides (4.1) there is an alternative representation

$$z = |z|e^{i\varphi} \tag{4.2}$$

---

[1] The sign =: means *is defined as*. Depending on whether the colon is on the right or the left side of the equal sign, the quantity to be defined is on the right or the left, respectively. So the first part of equation (4.1) defines the real and the imaginary parts of z after the sign =:, while in the second part of equation (4.1), the sign := defines the quantity i.

with

$$|z| = \sqrt{x^2 + y^2}, \qquad \tan \varphi = \frac{y}{x}$$

It corresponds to the introduction of polar coordinates in the complex plane (see Figure 4.1). Obviously, a comparison of (4.1) and (4.2) yields

$$x = |z| \cos \varphi, \qquad y = |z| \sin \varphi$$

In the arithmetical operations $+$, $-$, $\cdot$, and $/$, i is treated like an algebraic quantity, e.g.,

$$z_1 + z_2 = (x_1 + iy_1) + (x_2 + iy_2) = (x_1 + x_2) + i(y_1 + y_2) = z_3 = x_3 + iy_3$$

Hence, $x_3 = x_1 + x_2$, $y_3 = y_1 + y_2$. Two complex numbers are equal to each other if and only if their real parts, as well as imaginary parts, are equal to each other.

Since i is defined by the (many-valued) root function the exponential law cannot be transferred; that is, the root cannot be solved, as the following example demonstrates:

$$\frac{1}{i} = \frac{1}{\sqrt{-1}} = \frac{\sqrt{1}}{\sqrt{-1}} = \sqrt{\frac{1}{-1}} = \sqrt{-1} = i$$

One is only allowed to rewrite i by taking its powers, e.g., $1/i = i/i^2 = -i$.

In addition to the known basic arithmetic operations, an important operation is the *complex conjugation*, indicated by an asterisk:

$$z^* = (x + iy)^* := x - iy \tag{4.3}$$

One may be convinced easily that

$$\operatorname{Re} z = x = \frac{1}{2}(z + z^*)$$

$$\operatorname{Im} z = y = \frac{1}{2i}(z - z^*) \tag{4.4}$$

$$|z|^2 = x^2 + y^2 = z \cdot z^*$$

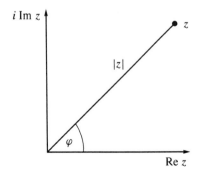

Figure 4.1. A complex number in the complex plane.

Example 4.1:   Computing with complex numbers

We compute $(1 + 4i)/(2 - 3i)$:

$$\frac{1 + 4i}{2 - 3i} = \frac{1 + 4i}{2 - 3i} \cdot \frac{2 + 3i}{2 + 3i} = \frac{2 - 12 + 3i + 8i}{4 + 9} = -\frac{10}{13} + \frac{11}{13}i$$

Cauchy-Riemann differential equations

The significant statements of function theory refer to functions $f(z)$ defined on an open and connected subset (*domain*) G of $\mathbb{C}$:

$$f(z) : G \to \mathbb{C}$$

The notions *continuity* and *differentiability* of complex functions are introduced analogously to the real case, but with a fundamental additional requirement:

Continuity $f(z)$ is called *continuous* in $z_0 \in G$ if for any $\epsilon > 0$ there is a $\delta(\epsilon) > 0$ such that for $|z - z_0| < \delta(\epsilon)$, $|f(z) - f(z_0)| < \epsilon$.

Differentiability f is called *complex differentiable, holomorphic, regular,* or *analytic* in z_0 if the limit

$$\left.\frac{df(z)}{dz}\right| = \lim_{z \to z_0} \frac{f(z) - f(z_0)}{z - z_0} \tag{4.5}$$

exists and is unique.

Here, to be unique means that the limiting value is independent of the path along which one approaches the point z_0. At first, there is an infinite number of possibilities, but they may be reduced to two independent ones: the two limits in the direction of the coordinate axes. With $z = z_0 + \Delta z$, one can rewrite equation (4.5):

$$\frac{df}{dz} = \lim_{\Delta z \to 0} \frac{f(z_0 + \Delta z) - f(z_0)}{\Delta z} \tag{4.6}$$

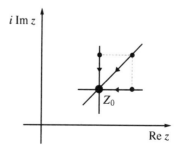

Figure 4.2. On the differentiability of a complex function.

Now, we set (i) $\Delta z = \Delta x$ and (ii) $\Delta z = \Delta y$ and obtain

(i) $\quad \dfrac{df}{dz} = \lim\limits_{\Delta x \to 0} \dfrac{f(x_0 + iy_0 + \Delta x) - f(x_0 + iy_0)}{\Delta x} = \dfrac{\partial f}{\partial x}$

(ii) $\quad \dfrac{df}{dz} = \lim\limits_{\Delta y \to 0} \dfrac{f(x_0 + iy_0 + i\Delta y) - f(x_0 + iy_0)}{i \Delta y} = \dfrac{1}{i} \dfrac{\partial f}{\partial y}$

The requirement that the limiting value has to be independent of the direction leads to

$$\frac{\partial f}{\partial x} = \frac{1}{i} \frac{\partial f}{\partial y} \tag{4.7}$$

Since $f(z) \in \mathbf{C}$ the decomposition

$$f(z) = u(x, y) + iv(x, y), \qquad \text{with} \quad u(x, y), v(x, y) \in R \tag{4.8}$$

is always possible. Substituting equation (4.8) into (4.7), a comparison of the real and the imaginary part leads to the *Cauchy-Riemann differential equations*:

$$\frac{\partial u(x, y)}{\partial x} = \frac{\partial v(x, y)}{\partial y}$$

$$\frac{\partial u(x, y)}{\partial y} = -\frac{\partial v(x, y)}{\partial x} \tag{4.9}$$

Hence, a function is differentiable in the complex domain if it is differentiable in the real domain and satisfies the equations (4.9). The additional requirement (4.9) is the reason for many curious properties of holomorphic functions, as will be shown now.

By the way, the Cauchy-Riemann differential equations may be summarized also in the short form

$$\left(\frac{\partial}{\partial x} + i \frac{\partial}{\partial y} \right) (u + iv) = 0 \tag{4.10}$$

Example 4.2:   Holomorphic functions

Are the functions $f_1(z) = z$, $f_2(z) = \text{Re}\, z$ holomorphic?

$$f_1(z) = z \Rightarrow u = x, \qquad v = y \Rightarrow \frac{\partial u}{\partial x} = 1, \qquad \frac{\partial v}{\partial y} = 1, \qquad \frac{\partial u}{\partial y} = 0, \qquad \frac{\partial v}{\partial x} = 0$$

$\Rightarrow f_1$ is holomorphic (4.9).

$$f_2(z) = \text{Re}\, z \Rightarrow u = x, \qquad v = 0 \Rightarrow \frac{\partial u}{\partial x} = 1 \neq \frac{\partial v}{\partial y} = 0$$

$\Rightarrow f_2$ is not holomorphic because the Cauchy-Riemann differential equations (4.9) are not fulfilled anywhere.

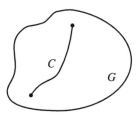

Figure 4.3. Path C in the complex domain G.

Line-integrals

Let $f(z) : G \to \mathbf{C}$ be a function continuous along the path C in G (see Figure 4.3).

To calculate the integral $\int f(z)dz$, defined as the limiting value of a sum, like in the real domain, one has to parametrize the path C; that is, one determines

$$z : [a, b] \to C$$

such that to any $t \in (a, b)$ there corresponds uniquely a $z(t) \in C$. If $z(a) = z(b)$, then the path is closed. Then, the line integral is given by

$$\int_C f(z)\, dz = \int_a^b f(z(t))z'(t)\, dt \tag{4.11}$$

with

$$z'(t) = \frac{dz}{dt}$$

The following statements are valid:

$$\int_{-C} f(z)\, dz = - \int_C f(z)\, dz \tag{4.12}$$

$$\left| \int_C f(z)\, dz \right| \le Ml \tag{4.13}$$

where $-C$ is the path traversed in the opposite sense, M is the maximum of $|f(z)|$ along C, and l is the length of C. The proofs of (4.12) and (4.13) proceed analogously to the real domain. They are so obvious that we omit them here.

Example 4.3: Integration in the complex plane

Let $f(z) = 1/z$ and C be the unit circle about the origin (Figure 4.4). Let the parametrization of C be given by

$$z(t) = \cos t + i \sin t, \qquad t \in [-\pi, \pi]$$

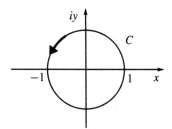

Figure 4.4. On integrating complex functions.

According to (4.11),

$$\int_C f(z)\, dz = \int_{-\pi}^{\pi} \frac{1}{\cos t + i \sin t} (-\sin t + i \cos t)\, dt$$

$$= i \int_{-\pi}^{\pi} (\cos^2 t + \sin^2 t)\, dt = 2\pi i$$

that is,

$$\int_C \frac{1}{z}\, dz = 2\pi i$$

Cauchy's theorem

Let f be a function holomorphic in the complex domain G, $z_1, z_2 \in G$. Then, the line integral

$$\int_C f(z)\, dz$$

has the same value for all paths from z_1 to z_2 in G (see Figure 4.5). This statement, called *Cauchy's integral theorem*, can be formulated equivalently: for any closed path in G (Figure 4.6)

$$\oint f(z)\, dz = 0 \tag{4.14}$$

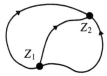

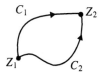

Figure 4.5. Different integration paths between the points Z_1 and Z_2.

Figure 4.6. Illustration of equation (4.15).

if the domain enclosed by C belongs entirely to the region of regularity. Since according to the first statement

$$\int_{C_1} f(z)\,dz = \int_{C_2} f(z)\,dz \tag{4.15}$$

from equation (4.12) follows (4.14). Example 4.3 demonstrates the necessity that any point has to belong to the domain of regularity.

We prove Cauchy's integral theorem in the form of equation (4.14). One has to show that the closed line integral

$$\oint_C f(z)\,dz = 0$$

vanishes if it is extended over an arbitrary, closed path enclosing a domain in which $f(z)$ is regular. To examine this, equation (4.14) is written in the following way:

$$\oint_C f(z)\,dz = \oint_C (u + iv)(dx + i\,dy) = \oint_C (u\,dx - v\,dy) + i \oint_C (u\,dy + v\,dx)$$

The real part, as well as the imaginary part, of this integral can be regarded as a scalar product of the two-dimensional vectors $\mathbf{R} = \{R_x, R_y, R_z\} = \{u, -v, 0\}$ and $\mathbf{I} = \{I_x, I_y, I_z\} = \{v, u, 0\}$ in the $x - y$−plane, with $d\mathbf{r} = \{dx, dy, dz\}$. Hence, we can write

$$\oint_C f(z)\,dz = \oint_C \mathbf{R} \cdot d\mathbf{r} + i \oint_C \mathbf{I} \cdot d\mathbf{r}$$

According to Stoke's integral theorem we can transform the closed line integrals into surface integrals:

$$\oint_C \mathbf{R} \cdot d\mathbf{r} = \int_F \mathrm{curl}\,\mathbf{R} \cdot d\mathbf{F} = \int_F \left\{0, 0, \frac{\partial R_y}{\partial x} - \frac{\partial R_x}{\partial y}\right\} \cdot \{0, 0, dx\,dy\}$$

$$= \int_F \left(\frac{\partial R_y}{\partial x} - \frac{\partial R_x}{\partial y}\right) dx\,dy = \int_F \left(-\frac{\partial v}{\partial x} - \frac{\partial u}{\partial y}\right) dx\,dy$$

$$= 0$$

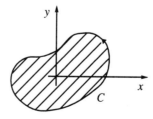

Figure 4.7. In the interior and at the boundary of the curve, $df(z)/dz$ has to exist, that is, the Cauchy-Riemann differential equations (4.9) have to be valid.

and

$$\oint_C \mathbf{I} \cdot d\mathbf{r} = \int_F \mathrm{curl}\,\mathbf{I} \cdot d\mathbf{F} = \int_F \left\{0, 0, \frac{\partial I_y}{\partial x} - \frac{\partial I_x}{\partial y}\right\} \cdot \{0, 0, dx\,dy\}$$

$$= \int_F \left(\frac{\partial I_y}{\partial x} - \frac{\partial I_x}{\partial y}\right) dx\,dy = \int_F \left(\frac{\partial u}{\partial x} - \frac{\partial v}{\partial y}\right) dx\,dy$$

$$= 0$$

Here in the last, we have used the second and the first, respectively, of equations (4.9), and, hence, the regularity of the function $f(z)$.

Cauchy's integral formula

Now, we will derive Cauchy's integral formula as the most important consequence of Cauchy's theorem: if f is regular in G and C is a closed path in G whose interior region G_C lies entirely in G, then for all $z \in G_C$:

$$f(z) = \frac{1}{2\pi i} \oint_C \frac{f(\zeta)}{\zeta - z} d\zeta \tag{4.16}$$

For the proof, we write

$$\frac{1}{2\pi i} \oint_C \frac{f(\zeta)}{\zeta - z} d\zeta = \frac{1}{2\pi i} \oint_C \frac{f(z)}{\zeta - z} d\zeta + \frac{1}{2\pi i} \oint_C \frac{f(\zeta) - f(z)}{\zeta - z} d\zeta$$

The first integral on the right-hand side yields

$$\frac{1}{2\pi i} \oint_C \frac{f(z)}{\zeta - z} d\zeta = \frac{f(z)}{2\pi i} \oint_C \frac{d\zeta}{\zeta - z} = \frac{f(z)}{2\pi i} \oint_{C'} \frac{d\zeta}{\zeta - z} = f(z) \tag{4.17}$$

where C' is a concentric circle about z, and the corresponding integral is solved as in Example 4.3. The transition from C to C' is allowed according to equation (4.15). The second integral vanishes due to the continuity of f: one chooses the radius r of C' so small that for any $\epsilon > 0$ always $|f(\zeta) - f(z)| < \epsilon$. Then according to (4.13)

$$\left| \frac{1}{2\pi i} \oint_C \frac{f(\zeta) - f(z)}{\zeta - z} d\zeta \right| \le \frac{1}{2\pi} \frac{\epsilon}{r} 2\pi r = \epsilon$$

Cauchy's integral formula (4.16) may be understood also in the following way. Let us write equation (4.16) in the form

$$f(z_0) = \frac{1}{2\pi i} \oint_C \frac{f(z)}{z - z_0} dz \tag{4.18}$$

Again we have assumed the differentiability (regularity) of $f(z)$ in the domain with the boundary C. We expand $f(z)$ in a power series about z_0

$$f(z) = f(z_0) + (z - z_0)f'(z_0) + \frac{(z - z_0)^2}{2}f''(z_0) + \cdots$$

and realize that in

$$\frac{1}{2\pi i}\oint_C \frac{f(z)}{z - z_0}dz = \frac{1}{2\pi i}\oint_C \frac{f(z_0)}{z - z_0}dz + \frac{1}{2\pi i}\oint_C \frac{z - z_0}{z - z_0}f'(z_0)\,dz + \cdots \tag{4.19}$$

all higher terms vanish due to (4.14). Only the first term contains an integrand that is not differentiable at $z = z_0$. We consider this in more detail:

$$\frac{1}{2\pi i}\oint_C \frac{f(z)}{z - z_0}dz = \frac{1}{2\pi i}\oint_C \frac{f(z_0)}{z - z_0}dz = \frac{f(z_0)}{2\pi i}\oint_C \frac{dz}{z - z_0} \tag{4.20}$$

The integral over C can be extended along any path C about z_0 since the difference of the integrations over two distinct integration paths C and C' is zero (compare this to Figure 4.8):

$$\oint_C \frac{dz}{z - z_0} - \oint_{C'} \frac{dz}{z - z_0} = \oint_C \frac{dz}{z - z_0} = 0, \qquad C - C' + \uparrow + \downarrow$$

because the function $1/(z - z_0)$ is regular in the shadowed interior region enclosed by the paths $C - C' + \uparrow + \downarrow$. $\uparrow + \downarrow$ means the integration path indicated in the figure forward and backward along a connecting line betweeen C and C', which of course, yields no contribution. Now we can calculate (4.20) easily by choosing C to be simply a circle about z_0, as in Figure 4.9:

$$z - z_0 = re^{i\varphi}, \qquad dz = re^{i\varphi}i\,d\varphi$$

We obtain

$$\frac{f(z_0)}{2\pi i}\oint_C \frac{dz}{z - z_0} = \frac{f(z_0)}{2\pi i}\oint_{circle} i\frac{re^{i\varphi}d\varphi}{re^{i\varphi}} = \frac{f(z_0)}{2\pi i}2\pi i = f(z_0) \tag{4.21}$$

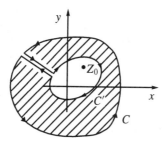

Figure 4.8. On the independence of Cauchy's integral formula of the path.

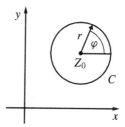

Figure 4.9. Circular curve C about the point z_0.

Substituting (4.21) into (4.20), we can immediately obtain Cauchy's integral formula (4.18). Just as simply we can prove

$$f^{(n)}(z_0) = \frac{n!}{2\pi i} \oint_C \frac{f(z)}{(z - z_0)^{n+1}} dz \tag{4.22}$$

Power series

The surprising statemant of the integral formula (4.16), namely, that it is sufficient to know a function along a closed path to determine any function value in the interior, is still surpassed by the *principle of analytic continuation*. At first, we introduce a *power series* by

$$P(z) = \sum_{n=0}^{\infty} a_n (z - z_0)^n \tag{4.23}$$

whose *radius of convergence* is given by the *Cauchy-Hadamard formula*

$$r = \frac{1}{\overline{\lim} \sqrt[n]{|a_n|}} \tag{4.24}$$

Here $\overline{\lim}$ means the *limes superior* of the sequence. The validity of equation (4.24) is seen immediately starting from the known root test for the convergence of a series: a series $\sum_{n=i}^{\infty} a_n$ is convergent if $\overline{\lim} \sqrt[n]{|a_n|} < 1$. For the power series (4.23) this means

$$\overline{\lim_{n \to \infty}} \sqrt[n]{|a_n z^n|} < 1$$

leading directly to the *radius of convergence* (4.24). See Figure 4.10.

The *identity theorem for power series* states: if two power series

$$P_1(z) = \sum_n a_n (z - z_0)^n, \qquad P_2(z) = \sum_n b_n (z - z_0)^n$$

coincide in z_0 for a sequence $\{z_l\}$ with the limiting point $z_l = z_0$; that is, if $P_1(z_l) = P_2(z_l)$ for all l, then P_1 and P_2 are identical. Really, this means nothing more than a power series which becomes zero with a limiting point at infinitely many distinct points has to vanish identically. This can be seen clearly in the following way: let f be holomorphic in G, $z_0 \in G$. Then, in the neighborhood of z_0 there is a unique representation of $f(z)$ by a power series

$$f(z) = \sum_{n=0}^{\infty} a_n (z - z_0)^n \tag{4.25}$$

and

$$a_n = \frac{1}{n!} f^{(n)}(z_0) \tag{4.26}$$

The power series is convergent within the largest circle (radius r) about z_0 containing only points of G. From the identity theorem for power series and *Taylor's theorem*, the identity

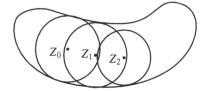

Figure 4.10. On the definition of a radius of convergence of a power series about z_0.

Figure 4.11. Illustration of the chain-of-circles procedure of the analytic continuation.

theorem for holomorphic functions follows: let f_1 and f_2 be two holomorphic functions in G and $z_0 \in G$. Furthermore, let $f_1(z) = f_2(z)$ for an arbitrary neighborhood of z_0, or for a sequence accumulating in z_0, or for an arbitrary path segment emerging from z_0. Then, $f_1 = f_2$ in the entire domain G.

The proof can be given with the aid of the so-called *chain-of-circle procedure*: At first, one expands f_1 and f_2 about z_0. According to *Taylor's theorem* the power series expansion of f_1 and f_2 is unique and convergent in the largest circle about z_0 inside G. Then, from the identity theorem for power series we have $f_1 = f_2$ for all z inside the circle of convergence. Now one chooses a new point z_1 possibly close to the periphery of the first circle and expands again. By the same argument, the identity of f_1 and f_2 is valid in the new circle of convergence. In this manner the entire domain G may be covered by circles within which $f_1 = f_2$.

A direct conclusion of the identity theorem of holomorphic functions is the principle of analytic continuation: if $f(z)$ is holomorphic in G_1 and $G_1 \subset G_2$, then there exists only a single holomorphic continuation of $f(z)$ in the entire domain G_2 (See Figure 4.11). The *chain-of-circles procedure* described above shows in principle, the way in which the analytic continuation of the function may be archieved in terms of a successive power series expansion.

Laurent series

Now we investigate in which way nonholomorphic functions may be represented by series (See Figure 4.12). Let us consider a function $f(z)$ about whose regularity in the shadowed region (in Figure 4.12) nothing can be stated. Furthermore, it is assumed that the two concentric circles K_1 and K_2 (radii r_1, r_2) about z_0 and the region of the circles between them belong to the holomorphic domain of $f(z)$. Introducing an auxiliary path S and using with the integral formula (4.18), one can show at once that for any point z within a circular ring, that is, for any point z with $r_1 < |z - z_0| = r < r_2$:

$$f(z) = \frac{1}{2\pi i} \int_{K_2} \frac{f(\zeta)}{\zeta - z} d\zeta - \frac{1}{2\pi i} \int_{K_1} \frac{f(\zeta)}{\zeta - z} d\zeta \tag{4.27}$$

The cut S is traversed twice in different directions, so that these contributions cancel each other. The inner ring is traversed in a mathematically negative sense. Hence, the second

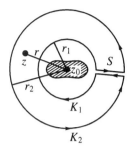

Figure 4.12. On the series expansion of a nonholomorphic function.

term on the right-hand side of equation (4.27) has a negative sign. If ζ lies on the outer circle one can write

$$\frac{1}{\zeta - z} = \frac{1}{\zeta - z_0} \cdot \frac{1}{1 - \dfrac{z - z_0}{\zeta - z_0}} = \sum_{n=0}^{\infty} \frac{(z - z_0)^n}{(\zeta - z_0)^{n+1}}$$

because

$$q := \left| \frac{z - z_0}{\zeta - z_0} \right| = \frac{r}{r_2} < 1$$

and

$$1 + q + \cdots + q^n + \cdots = \frac{1}{1 - q} \qquad \text{for} \quad q < 1$$

If ζ lies on K_1, then

$$\frac{1}{\zeta - z} = -\frac{1}{z - z_0} \cdot \frac{1}{1 - \dfrac{\zeta - z_0}{z - z_0}} = -\sum_{n=0}^{\infty} \frac{(\zeta - z_0)^n}{(z - z_0)^{n+1}}$$

since now

$$\left| \frac{\zeta - z_0}{z - z_0} \right| = \frac{r_1}{r} < 1$$

Substituting both expressions into the corresponding integrals, one obtains:

$$f(z) = \sum_{n=-\infty}^{\infty} a_n (z - z_0)^n$$

$$= -\cdots \frac{a_{-m}}{(z - z_0)^m} + \cdots \frac{a_{-1}}{z - z_0} + a_0 + a_1(z - z_0) + a_2(z - z_0)^2 + \cdots \qquad \textbf{(4.28)}$$

with the definition

$$a_n = \frac{1}{2\pi i} \int_C \frac{f(z)}{(\zeta - z_0)^{n+1}} d\zeta \qquad \textbf{(4.29)}$$

where C may be equal to K_1 or K_2, or any other curve enclosing the shadowed region and lying inside the circular ring. Equations (4.28) and (4.29) are called *Laurent's expansion* of f for the region of the circular ring.

If z_0 is a point for which $f(z)$ is not holomorphic but for which a holomorphic neighborhood exists, one talks about an *isolated singularity*. If z_0 is the only singularity of f inside C, then

$$a_{-1} = \frac{1}{2\pi i} \int_C f(\zeta)\, d\zeta \tag{4.30}$$

is called the *residue* of f at z_0. If the series (4.28) terminates at finite negative n, that is, if

$$a_{n+m} = 0 \tag{4.31}$$

for all $m < 0$ one talks about a *pole of the order n* at z_0; z_0 is called an *essential singularity*. Otherwise, z_0 is called a *nonessential singularity*.

Residue theorem

Let $f(z)$ be a function regular in G, C a closed path within G, in the interior region of which there is a finite number of isolated singular points of $f(z)$. Then

$$\int_C f(z)\, dz = 2\pi i \sum \text{Residues in } G_C \tag{4.32}$$

The proof of the residue theorem (4.32) follows from the definition of the residues (4.30) and the already frequently used decomposition

$$\int_C f(z)\, dz = \int_{C_1} f(z)\, dz + \cdots + \int_{C_N} f(z)\, dz$$

To answer the question of how to calculate residues, we consider the function $f(z)$ with a pole of the order n at z_0. Then, the Laurent's series has the form

$$f(z) = \frac{a_{-n}}{(z-z_0)^n} + \cdots + \frac{a_{-1}}{z-z_0} + a_0 + a_1(z-z_0) + \cdots \tag{4.33}$$

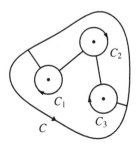

Figure 4.13. On the residue theorem.

The coefficient a_{-1}, the residue, is obtained by

$$a_{-1} = \frac{1}{(n-1)!} \left[\frac{d^{n-1}}{dz^{n-1}} ((z-z_0)^n \cdot f(z)) \right]_{z=z_0} \tag{4.34}$$

Example 4.4: Calculation of a residue

Let $f(z) = (3z^2 + 1)/(z-2)^3$.

At $z_0 = z$, f has a pole of the order 3, hence, according to (4.34),

$$a_{-1} = \frac{1}{2!} \left[\frac{d^2}{dz^2} \left((z-2)^3 \cdot \frac{3z^2+1}{(z-2)^3} \right) \right]_{z=2} = \frac{1}{2!} \cdot 6 = 3$$

The residue theorem is of great importance in the solution of real definite integrals, as demonstrated in the next example.

Example 4.5: Computation of the integral $\int_{-\infty}^{\infty} 1/(1+x^2)\,dx$

We have to compute $I = \int_{-\infty}^{\infty} 1/(1+x^2)\,dx$. At first, the integral is transferred into the complex integral

$$I' = \int_C \frac{dz}{1+z^2}$$

where C is the path indicated in Figure 4.14. Then,

$$\lim_{R \to \infty} I' = I + \lim_{R \to \infty} I_{sc}$$

where I_{sc} is the integral along the semicircle. The transition from the integral I to the integral I' is meaningful only if $I = I'$, that is, if the integral I_{sc} over the infinitely distant semicircle vanishes. As can be seen from

$$\frac{1}{1+z^2} = \frac{1}{2i} \left(\frac{1}{z-i} - \frac{1}{z+i} \right) \tag{4.35}$$

f has only one pole of the first order, with the residue $1/2i$, in the interior region of C. Hence,

$$\lim_{R \to \infty} I' = 2\pi i \cdot \frac{1}{2i} = \pi \tag{4.36}$$

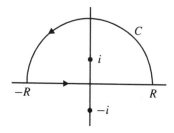

Figure 4.14. Integration path within the complex domain.

The integral over the infinitely distant semicircle I_{sc} may be estimated using equation (4.13):

$$|I_{sc}| = \left| \int_{sc} \frac{dz}{1+z^2} \right| \leq \max_{\phi} \left[\frac{1}{|1+R^2 e^{2i\phi}|} \right] \cdot \pi R \tag{4.37}$$

so that $\lim_{R \to \infty} I_{sc} = 0$. In summary, one obtains

$$I = \int_{-\infty}^{\infty} \frac{dx}{1+x^2} = \pi \tag{4.38}$$

Analytic functions as solution of Laplace's equation

Any analytic function $f(z) = u(x, y) + iv(x, y)$ obeys Laplace's equation in two dimensions, because due to the Cauchy-Riemann differential equations (4.9),

$$\frac{\partial^2 u}{\partial x^2} = \frac{\partial}{\partial x} \frac{\partial v}{\partial y} = \frac{\partial}{\partial y} \frac{\partial v}{\partial x} = -\frac{\partial^2 u}{\partial y^2} \tag{4.39}$$

and

$$\frac{\partial^2 v}{\partial x^2} = \frac{\partial}{\partial x} \left(-\frac{\partial u}{\partial y} \right) = -\frac{\partial}{\partial y} \frac{\partial u}{\partial x} = -\frac{\partial^2 v}{\partial y^2} \tag{4.40}$$

also,

$$\left(\frac{\partial^2}{\partial x^2} + \frac{\partial^2}{\partial y^2} \right) u(x, y) = 0 \qquad \text{and} \qquad \left(\frac{\partial^2}{\partial x^2} + \frac{\partial^2}{\partial y^2} \right) v(x, y) = 0 \tag{4.41}$$

Thus, one has also $\Delta f(z) = (\partial^2/\partial x^2 + \partial^2/\partial y^2) f(x, y) = 0$. If $f(z)$ is a complex solution of the Laplace equation, then also $f^*(z)$, the complex conjugate of $f(z)$, is such a solution, and therefore, so is the sum of both. Hence, both of

$$u(x, y) = \frac{1}{2} (f(x + iy) + f(x - iy))$$

$$v(x, y) = \frac{1}{2i} (f(x + iy) - f(x - iy)) \tag{4.42}$$

are solutions of Laplace's equation, as we know already from (4.41).

Applications in electrostatics

Any analytic function, due to the knowledge just deduced in terms of equations (4.41) and (4.42), corresponds to a solution of a two dimensional boundary problem of the Dirichlet type. To understand this even better, we consider two curves of constant potentials $u(x, y)$ in the (x, y)-plane,

$$u(x, y) = C_1 \qquad \text{and} \qquad u(x, y) = C_2 \tag{4.43}$$

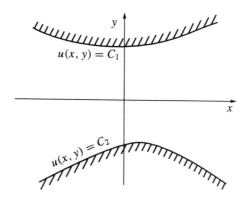

Figure 4.15. On the solution of the capacitor problem by an analytic function.

as illustrated in Figure 4.15. Now the curves characterized by constant $u(x, y)$ are always orthogonal to the family of curves given by constant $v(x, y)$, since

$$\nabla u(x, y) \cdot \nabla v(x, y) = \frac{\partial u}{\partial x}\frac{\partial v}{\partial x} + \frac{\partial u}{\partial y}\frac{\partial v}{\partial y} = -\frac{\partial u}{\partial x}\frac{\partial u}{\partial y} + \frac{\partial u}{\partial y}\frac{\partial u}{\partial x} = 0 \qquad (4.44)$$

The orthogonality follows immediately from (4.44) and from the fact that grad u and grad v are perpendicular to the curves of constant u and v, respectively. This we know already from our general knowledge on gradients but we may also repeat it rapidly: considering two neighboring points $\mathbf{x}$ and $\mathbf{x} + d\mathbf{x}$ in the plane,

$$u(\mathbf{x} + d\mathbf{x}) - u(\mathbf{x}) = \frac{\partial u}{\partial x}dx + \frac{\partial u}{\partial y}dy = \nabla u \cdot d\mathbf{x} \qquad (4.45)$$

If $d\mathbf{x}$ lies in the tangent to the curve, then $u(\mathbf{x})$ is not allowed to change because the curve is determined by $u(\mathbf{x}) = \text{const}$. Hence, along the curve we have $u(\mathbf{x} + d\mathbf{x}) = u(\mathbf{x})$ and

$$\text{grad } u(\mathbf{x}) \cdot d\mathbf{x} = 0 \qquad (4.46)$$

so that everything is clear.

Now we assume that $u(x, y)$ is the electrostatic potential belonging to a plane potential problem. Then, the electric field intensity can be obtained from the regular function $f(z)$ whose real part is $u(x, y)$. Namely,

$$\frac{df}{dz} = \frac{\partial u}{\partial x} + i\frac{\partial v}{\partial x} = \frac{\partial u}{\partial x} - i\frac{\partial u}{\partial y} = -E_x + iE_y \qquad (4.47)$$

Hence, the components of the field intensity E_x and E_y are identical with the real and imaginary part of $-(df(z)/dz)^*$. One calls $f(z) = u(x, y) + iv(x, y)$ the complex potential, $u(x, y)$ the potential, and $v(x, y)$ the stream function.

The electric field $\mathbf{E} = -\text{grad } u(x, y)$ points everywhere in the direction of $v(x, y) = \text{const}$. Obviously, these are the field lines of the problem. In the context discussed here (electrostatics) the physical meaning of the stream function is to represent a measure of the charge density on the boundary surfaces if the boundary surface is a conductor. In this case,

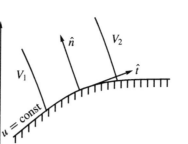

Figure 4.16. On the physical meaning of the stream function $v(x, y)$.

the potential of the boundary surface is $u(x, y) = $ const and the lines $v(x, y) = $ const are perpendicular to the boundary surface. Choosing the normal vector to the boundary surface $\mathbf{n} = \{n_x, n_y\}$ to be directed outward and $\mathbf{t} = \{t_x, t_y\} = \{n_y, -n_x\}$ to lie tangentially in the boundary surface (compare Figure 4.16), then the surface charge is

$$\sigma = \frac{1}{4\pi} \mathbf{E} \cdot \mathbf{n} = -\frac{1}{4\pi} \left[\frac{\partial u}{\partial x} n_x + \frac{\partial u}{\partial y} n_y \right]$$

$$= -\frac{1}{4\pi} \left[\frac{\partial v}{\partial y} n_x - \frac{\partial v}{\partial x} n_y \right] = \frac{1}{4\pi} \left[\frac{\partial v}{\partial x} t_x + \frac{\partial v}{\partial y} t_y \right] = \frac{1}{4\pi} \operatorname{grad} v \cdot \mathbf{t}$$

(4.48)

The charge per unit of length dQ/dl (l means the cylindric coordinate perpendicular to the (x, y)-plane) between v_1 and v_2 on the boundary curve (the image of the boundary surface in the (x, y)-plane) is

$$\frac{dQ}{dl} = \frac{1}{4\pi} \int_1^2 \frac{\partial v}{\partial t} dt = \frac{1}{4\pi} (v_2 - v_1)$$

(4.49)

Example 4.6:   Potential of a charged wire

We consider the complex function

$$f(z) = -\frac{2q}{l} \ln z$$

(4.50)

and introduce cylindric polar coordinates by setting $z = re^{i\theta}$. Now we must note that the function

$$\ln z = \ln r + i\theta + i2\pi n, \qquad n \in \mathbf{Z}$$

(4.51)

is many-valued because all infinitely many values in (4.51) yield the same $z = re^{i(\theta+2\pi n)}$. In order to clarify this, we differentiate (4.51) for fixed n and realize that the derivative $d \ln z/dz$ does not depend on n. It exists everywhere except at the origin ($r = 0$) and at infinity ($r = \infty$). In many

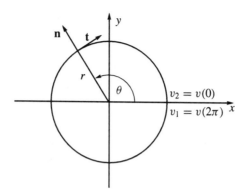

Figure 4.17. Potential and stream function for a line charge (charged wire).

similar cases it is the best to avoid the many-valuedness from the beginning. So, we may define the logarithmic function in a unique manner by prescribing

$$0 \le \text{Im}(\ln z) \le 2\pi \tag{4.52}$$

The derivative of the $\ln z$-function defined uniquely in this way does exist everywhere except at $r = 0, r = \infty$ and along the cuts $\theta = 0$ and $\theta = 2\pi$. With this description, function (4.50) reads

$$f(z) = -\frac{2q}{l}(\ln r + i\theta) \tag{4.53}$$

so that

$$u = -2\frac{q}{l}\ln r \tag{4.54}$$

is the electric potential and

$$v = -2\frac{q}{l}\theta \tag{4.55}$$

is the stream function. Because θ is usually defined in the counterclockwise sense (Figure 4.17), that is, opposite the tangent vector **t** in the figure, the total charge per unit of length is

$$\frac{dQ}{dt} = \frac{1}{4\pi}(v(0) - v(2\pi)) = \frac{q}{l} \tag{4.56}$$

according to equation (4.45). Thus, the potential (4.54) refers to a wire with the line-charge density q/l; **r** is the distance from the wire.

Example 4.7: The potential at the edge of a parallel plate capacitor

Let us consider the function $f(z) = u(z) + iv(z)$ defined implicitly by the following equations:

$$iz = if(z) + e^{if(z)} \qquad ix - y = iu(z) - v(z) + e^{iu-v} \tag{4.57}$$

Hence,

$$x = u + e^{-v}\sin u \qquad y = v - e^{-v}\cos u \tag{4.58}$$

We consider a region in space bounded by the equipotential surfaces $u = +\pi$ and $u = -\pi$. According to equation (4.58), for $u = \pm\pi$ we obtain

$$x = \pm\pi \qquad y = v + e^{-v} \tag{4.59}$$

The function $y(v)$ has an extremum for

$$\frac{dy}{dv} = 1 - e^{-v} = 0 \tag{4.60}$$

giving $v = 0$. Now it is clear that $y = v + e^{-v}$ is always positive and has a minimum with $v = 0$ and $y = 1$. Therefore the boundary curves (boundary surfaces) $u = \pm\pi$ determine the plates of a capacitor as represented in Figure 4.18. For $v \to \infty$ (and therefore $y \to \infty$), one obtains from (4.58)

$$x = u \qquad y = v \tag{4.61}$$

implying that the equipotential curves (equipotential surfaces) are vertical straight lines. Their distance from the x-direction is proportional to u. This corresponds perfectly to the situation in the interior of a capacitor, far from the edge. If $v \to -\infty$ one obtains from (4.58)

$$x = e^{-v} \sin u \qquad y = -e^{-v} \cos u \tag{4.62}$$

if neither $\sin u$ nor $\cos u$ vanish. Then,

$$\frac{x}{y} = -\tan u \tag{4.63}$$

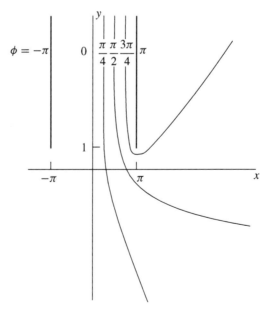

Figure 4.18. Edge effects of a semi-infinite parallel-plate capacitor.

and the equipotential curves (equipotential surfaces) become straight lines again. From (4.58) one can see that $u = 0$ represents the entire y-axis. For $u = \pi/2$ one has

$$x = \frac{\pi}{2} + e^{-v} \qquad y = v \tag{4.64}$$

so that the equation for the equipotential curve reads

$$x = \frac{\pi}{2} + e^{-y} \tag{4.65}$$

This curve never goes into a straight line as in the case of the other equipotential curves on both sides of that belonging to $\pi/2$.

Example 4.8: Potential of a charged wire in front of a conducting surface

As the last example of the solution of the potential problem by means of function theory we discuss the potential of a line charge (charged wire) in the vincinity of a surface as represented in Figure 4.19.

The line charge density (charge per unit of length) is denoted by q/l. Let the wire cross the (x, y)-plane at $x = y = 2c$ (compare Figure 4.19), and let the equation for the hyperbolic surface be $xy = c^2$. We assume the surface to be grounded. In this problem it is suitable to introduce the variables

$$z' = z^2 = x^2 - y^2 + i2xy$$
$$x' = x^2 - y^2$$
$$y' = 2xy \tag{4.66}$$

Then the curve $xy = c^2$ and the point $x = y = 2c$ are represented by $x' = 0$ and $y' = 8c^2$, respectively. So, in the (x', y')-plane the problem is reduced to the problem of a point charge q/l at $z' = i8c^2$ in front of the plane $y' = 2c^2$ which will be solved with the image charge $-q/l$ at $z' = -i4c^2$ (see Example 2.6). Hence, the complex potential (see Example 4.6) is

$$f(z) = -2\frac{q}{l}\ln(z' - i8c^2) + 2\frac{q}{l}\ln(z' + i4c^2) = -2\frac{q}{l}\ln\frac{z' - i8c^2}{z' + i4c^2} \tag{4.67}$$

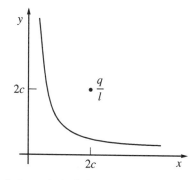

Figure 4.19. A charged wire in front of a hyperbolic grounded surface.

and consequently

$$f(z) = -2\frac{q}{l} \ln \frac{z^2 - i8c^2}{z^2 + i4c^2} \tag{4.68}$$

One should note that $f(z)$ is analytic everywhere except at the points

$$z = \pm 2(1 + i)c$$
$$z = \pm\sqrt{2}(1 - i)c \tag{4.69}$$

Only the point $z = +2(1 + i)c$ lies inside the physical region. Now the complex potential can be written in the following way:

$$f(z) = -2\frac{q}{l} \ln [z + 2(1 + i)c] - 2\frac{q}{l} \ln [z - 2(1 + i)c]$$

$$+ 2\frac{q}{l} \ln \left[z + \sqrt{2}(1 - i)c \right] + 2\frac{q}{l} \ln \left[z - \sqrt{2}(1 - i)c \right] \tag{4.70}$$

Hence, the singularities of $f(z)$ can be interpreted as originating from the original charge in the physical region and the 3 image charges in the unphysical region. The potential itself reads

$$u(x, y) = \frac{1}{2}(f(z) + f^*(z))$$

$$= -\frac{q}{l} \ln \left[\frac{x^2 - y^2 + i2(xy - 4c^2)}{x^2 - y^2 + i2(xy + 2c^2)} \right] - \frac{q}{l} \ln \left[\frac{x^2 - y^2 - i2(xy - 4c^2)}{x^2 - y^2 - i2(xy + 2c^2)} \right]$$

$$= -\frac{q}{l} \ln \left[\frac{(x^2 - y^2)^2 + 4(xy - 4c^2)^2}{(x^2 - y^2)^2 + 4(xy + 2c^2)^2} \right] \tag{4.71}$$

There is no doubt that this is the solution of the problem; in the vicinity of the physical charge it tends to the potential of a line charge (Example 4.6). Obviously, the potential (4.71) is constant at the curve (surface) $xy = c^2$, and it satisfies the Laplace equation everywhere within the physical region except at the position of the charge (of the wire).

Biographical notes

Augustin Louis Cauchy, b. Aug. 21, 1789, Paris–d. May 23, 1857, Scaux. Cauchy was a student of the Ecole Polytechnique and got a chair at the same institution just a few years later. Being a devoted royalist, Cauchy had to leave France after the revolution of July 1830. From 1838 he had lived in Turin and Prague. After his return to Paris, Cauchy taught at a Jesuit school. Cauchy wrote pathbreaking essays on the foundation of analysis, on the theory of differential equations, and on the theory of functions, as well as on physical and astronomical problems.

Pierre Laurent, b. 1813–d. 1854, Paris. Laurent was a student of the Ecole Polytechnique, then an officer in the french corps de génie, and eventually battalion commander in Paris. His most important achievement is the development (given in 1843 and later named for him) of a holomorphic function unique in a circular ring.

PART II

MACROSCOPIC ELECTROSTATICS

5 The Field Equations for Space Filled with Matter

For physical problems which can be described in terms of point charges and simple charge distributions, the two equations of electrostatics

$$\operatorname{div}\mathbf{E} = 4\pi\rho$$
$$\operatorname{curl}\mathbf{E} = 0 \tag{5.1}$$

have to be applied directly. This holds, e.g., on the atomic level for single ions which can be viewed as point charges. These regions are of the dimension 10^{-24} cm^3 (or 10^{-8} cm in linear dimensions).

In macroscopic dimensions (approximately beyond 10^{-2} cm), e.g., in crystal lattices, we are dealing with about 10^{18} charged particles (ions). They cause an electric field in the crystal which would be described by the equation

$$\mathbf{E}(\mathbf{r}) = \int_{\text{crystal}} \rho(\mathbf{r}') \frac{(\mathbf{r} - \mathbf{r}')}{|\mathbf{r} - \mathbf{r}'|^3} \, dV' \tag{5.2}$$

Here, two difficulties arise, namely,

(a) $\rho(\mathbf{r}')$ would have to be determined for about 10^{18} particles.

(b) Since, for example, in a crystal the ions vibrate about their lattice sites, the charge distributions is subject to fluctuations in time.

Thus, we cannot apply the equation (5.2) because it is true only for **E**-fields which are constant in time.

Furthermore, for macroscopic problems this equation yields more information than is of physical interest. Taking into account the fact that in macroscopic measurements the fluctuations of the E-field (caused by the vibrations of the charge carriers) just average to zero, it turns out to be meaningful to form the average values of $\mathbf{E}$ and ρ throughout the volume V':

$$\langle \mathbf{E}(\mathbf{r}) \rangle = \frac{1}{V'} \int V' \mathbf{E}(\mathbf{r} + \mathbf{r}') \, dV'$$

$$\langle \rho(\mathbf{r}) \rangle = \frac{1}{V'} \int V' \rho(\mathbf{r} + \mathbf{r}') \, dV' \qquad\qquad (5.3)$$

where V' means always a volume that contains a large number of particles but that is small compared to the total volume of the material considered. (See Figure 5.1.) The vector $\mathbf{r}'$ covers the entire volume.

These (spatial) average values (5.3) remain constant in time since the motion of the charge carriers is disordered and negligible compared to the magnitude of V'. Therefore, the field and charge distributions given by equation (5.3) can be regarded to be static; this implies that we are allowed to apply the static equations to the average values $\langle \mathbf{E}(\mathbf{r}) \rangle$ and $\langle \rho(\mathbf{r}) \rangle$.

The average charge density $\langle \rho(\mathbf{r}) \rangle$ is composed of two parts: first, we have the uncompensated charges of ions and electrons. Furthermore, there are induced charges (so-called *polarization charges*). These originate from the induced dipole moments since the associated electric dipoles are aligned in an external electric field. To obtain the macroscopic field in a crystal lattice, we start from an individual ion: considering a certain molecule j with the total charge q_j and the charge distribution $\rho_j(\mathbf{r}')$, then at a point P lying outside the molecule it generates the field

$$\mathbf{E}_j(\mathbf{r}) = \int_{\text{Mol}} \rho_j(\mathbf{r}') \frac{(\mathbf{r} - \mathbf{r}_j - \mathbf{r}')}{|\mathbf{r} - \mathbf{r}_j - \mathbf{r}'|^3} \, dV' \qquad (j = 1, \ldots, n)$$

$$= -\nabla \int_{\text{Mol}} \rho_j(\mathbf{r}') \frac{1}{|\mathbf{r} - \mathbf{r}_j - \mathbf{r}'|} \, dV'$$

The coordinate $\mathbf{r}_j$ gives the center of gravity (of the charge) of the molecule j. The index Mol at the integral sign indicates that the integral is extended over the entire molecule.

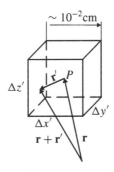

Figure 5.1. The volume $\triangle V'$ (about 10^{-6} cm^3), throughout which the physical quantities are averaged.

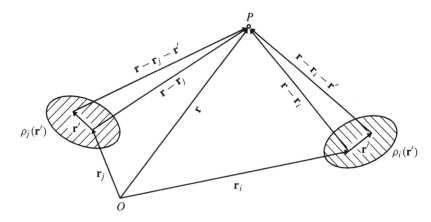

Figure 5.2. On the calculation of the electric field of the molecules i and j at the point P.

Now, we perform a multipole expansion about the center of gravity of the molecule. If ∇_j denotes differentiation with respect to the coordinates of the center of gravity, then

$$\frac{1}{|\mathbf{r} - \mathbf{r}_j - \mathbf{r}'|} = \frac{1}{|\mathbf{r} - \mathbf{r}_j|} + (\mathbf{r}' \cdot \nabla_j) \left(\frac{1}{|\mathbf{r} - \mathbf{r}_j|} \right) + \frac{1}{2}(\mathbf{r}' \cdot \nabla_j)^2 \left(\frac{1}{|\mathbf{r} - \mathbf{r}_j|} \right) + \cdots$$

Since we are interested in the field far outside the molecule, $|\mathbf{r}'| \ll |\mathbf{r} - \mathbf{r}_j|$, the higher terms of the expansion may be neglected. Thus, we obtain

$$\mathbf{E}_j(\mathbf{r}) = -\nabla \left[\int_{\text{Mol}} \rho_j(\mathbf{r}') \left\{ \frac{1}{|\mathbf{r} - \mathbf{r}_j|} + (\mathbf{r}' \cdot \nabla_j) \left(\frac{1}{|\mathbf{r} - \mathbf{r}_j|} \right) \right. \right.$$
$$\left. \left. + \frac{1}{2}(\mathbf{r}' \cdot \nabla_j)^2 \left(\frac{1}{|\mathbf{r} - \mathbf{r}_j|} \right) \right\} dV' + \cdots \right]$$
$$\approx -\nabla \left[\frac{q_j}{|\mathbf{r} - \mathbf{r}_j|} + \nabla_j \left(\frac{1}{|\mathbf{r} - \mathbf{r}_j|} \right) \cdot \mathbf{p}_j + \cdots \right]$$

where $q_j = \int_{\text{Mol}} \rho_j(\mathbf{r}') \, dV'$ denotes the charge of the molecule and

$$\mathbf{p}_j = \int_{\text{Mol}} \rho_j(\mathbf{r}') \mathbf{r}' \, dV'$$

is the dipole moment of the molecule. In general, there are no static dipole moments of a molecule. (The conservation of parity requires $\uparrow\downarrow$-symmetry—compare the discussion of odd multipoles.)

But dipoles may be induced by external electric fields. This important case will be our concern in the following. Usually, the quadrupole moments of the molecules are small;

furthermore, at large distances their field intensities fall off more rapidly than those of the dipoles. Considering the (microscopic) field produced by n molecules we obtain

$$\mathbf{E}(\mathbf{r}) = \sum_{j=1}^{n} \mathbf{E}_j(\mathbf{r}) = -\nabla \sum_{j=1}^{n} \left[\frac{q_j}{|\mathbf{r} - \mathbf{r}_j|} + \mathbf{p}_j \cdot \nabla_j \left(\frac{1}{|\mathbf{r} - \mathbf{r}_j|} \right) \right]$$

This equation is valid only at a fixed time t_1, for which the molecules are at the position $\mathbf{r}_j(t_1)$. The arrangement of the molecules at a later instant of time t_2 does not deviate from the initial arrangement due to the statistical molecular vibrations; the spatial average over all molecules will yield the same result in any case. Hence, the average values over the charge densities are constant in time.

Now we want to perform the averaging according to equation (2.2) to obtain the macroscopic field. It turns out to be suitable to transform from a discrete sum over distinct molecules to an integral over a charge density caused by a very large number of neighboring and thus practically indistinguishable charge carriers (a continuous charge distribution). In the transition from the sum to the integral, the discrete quantity $\mathbf{r}_j$ becomes the continuous integration variable $\mathbf{r}''$ covering the entire charge distribution.

For this purpose, we introduce a *charge density* $\rho_{\text{Mol}}(\mathbf{r})$ and a *polarization density* $\boldsymbol{\pi}_{\text{Mol}}(\mathbf{r})$

$$\rho_{\text{Mol}}(\mathbf{r}) = \sum_j q_j \delta(\mathbf{r} - \mathbf{r}_j)$$

$$\boldsymbol{\pi}_{\text{Mol}}(\mathbf{r}) = \sum_j \mathbf{p}_j \delta(\mathbf{r} - \mathbf{r}_j) \tag{5.4}$$

The index Mol indicates that the charge distribution is produced by the collection of molecules above. With these notations we can perform a transition from the discrete sum to the integral in the formula for the field intensity $\mathbf{E}(\mathbf{r})$:

$$\mathbf{E}(\mathbf{r}) = -\nabla \int_V dV'' \left[\frac{\rho_{\text{Mol}}(\mathbf{r}'')}{|\mathbf{r} - \mathbf{r}''|} + \boldsymbol{\pi}_{\text{Mol}}(\mathbf{r}'') \cdot \nabla'' \left(\frac{1}{|\mathbf{r} - \mathbf{r}''|} \right) \right] \tag{5.5}$$

(∇'' acts on $\mathbf{r}''$ only; $\mathbf{r}''$ has been introduced as the integration variable.) We may demonstrate that equation (5.5) is really equivalent to the original sum, by substituting relation (5.4) for $\rho_{\text{Mol}}(\mathbf{r})$ and $\boldsymbol{\pi}_{\text{Mol}}(\mathbf{r})$ into equation (5.5) and computing the integral where the properties of the δ-function have to be taken into account.

Now, we perform the average value $\langle \mathbf{E}(\mathbf{r}) \rangle$ corresponding to equation (5.3)(in equation (5.5) $\mathbf{r}$ is replaced by $\mathbf{r} + \mathbf{r}'$):

$$\langle \mathbf{E}(\mathbf{r}) \rangle = \frac{1}{V} \int_{V'} \mathbf{E}(\mathbf{r} + \mathbf{r}') \, dV'$$

$$= -\frac{1}{V'} \int_{V'} dV' \, \nabla \int_V dV'' \left[\frac{\rho_{\text{Mol}}(\mathbf{r}'')}{|\mathbf{r} + \mathbf{r}' - \mathbf{r}''|} + \boldsymbol{\pi}_{\text{Mol}}(\mathbf{r}'') \nabla'' \left(\frac{1}{|\mathbf{r} + \mathbf{r}' - \mathbf{r}''|} \right) \right]$$

The volume over which one averages in (5.3) should be the volume V that contains the molecules in (5.5):

$$V' = V$$

To simplify the integrand we substitute

$$\mathbf{z} = \mathbf{r}'' - \mathbf{r}' \quad \text{or} \quad \mathbf{r}'' = \mathbf{z} + \mathbf{r}'$$

and obtain, with $dV'' = d^3z$:

$$\langle \mathbf{E}(\mathbf{r}) \rangle = -\frac{1}{V} \int_V dV' \, \nabla \int_V d^3z \left[\frac{\rho_{\text{Mol}}(\mathbf{z} + \mathbf{r}')}{|\mathbf{r} - \mathbf{z}|} + \pi_{\text{Mol}}(\mathbf{z} + \mathbf{r}') \cdot \nabla_z \left(\frac{1}{|\mathbf{r} - \mathbf{z}|} \right) \right] \quad (5.6)$$

In equation (5.6), we can interchange the averaging procedure and the differentiation and factor out the term $1/|\mathbf{r} - \mathbf{z}|$, which is independent of $\mathbf{r}'$, from under the integral sign in the first term of (5.4):

$$\mathbf{E}^*(\mathbf{r}) := \langle \mathbf{E}(\mathbf{r}) \rangle = \qquad\qquad\qquad\qquad\qquad\qquad\qquad\qquad (5.7)$$

$$-\nabla \int_V d^3z \left[\frac{1}{|\mathbf{r} - \mathbf{z}|} \frac{1}{V} \int_V \rho_{\text{Mol}}(\mathbf{z} + \mathbf{r}') \, dV' + \nabla_z \left(\frac{1}{|\mathbf{r} - \mathbf{z}|} \right) \frac{1}{V} \int_V \pi_{\text{Mol}}(\mathbf{z} + \mathbf{r}') \, dV' \right]$$

According to the definition of ρ_{Mol} in equation (5.4), we have

$$\frac{1}{V} \int_V \rho_{\text{Mol}}(\mathbf{z} + \mathbf{r}') \, dV' = \frac{1}{V} \sum_j q_j$$

Denoting the number of molecules per unit of volume by $N(\mathbf{z})$ and the average charge per molecule at the point $\mathbf{z}$ in V by $\langle q_{\text{Mol}}(\mathbf{z}) \rangle$, we may also write

$$\rho(\mathbf{z}) = \frac{1}{V} \int_V \rho_{\text{Mol}}(\mathbf{z} + \mathbf{r}') \, dV' = N(\mathbf{z}) \langle q_{\text{Mol}}(\mathbf{z}) \rangle = \frac{1}{V} \sum_j q_j \quad (5.8)$$

In the same manner, the second term in equation (5.7) can be transformed:

$$\frac{1}{V} \int_V \pi_{\text{Mol}}(\mathbf{z} + \mathbf{r}') \, dV' = N(\mathbf{z}) \langle \mathbf{p}_{\text{Mol}}(\mathbf{z}) \rangle \equiv \mathbf{P}(\mathbf{z}) \quad (5.9)$$

where $\langle \mathbf{p}_{\text{Mol}}(\mathbf{z}) \rangle$ is the average polarization per molecule in V at the point $\mathbf{z}$. Using the relations (5.8) and (5.9) we can rewrite equation (5.7) in the following way:

$$\mathbf{E}^*(\mathbf{r}) \equiv \langle \mathbf{E}(\mathbf{r}) \rangle = -\nabla \int_V N(\mathbf{z}) \left\{ \frac{\langle q_{\text{Mol}}(\mathbf{z}) \rangle}{|\mathbf{r} - \mathbf{z}|} + \langle \mathbf{p}_{\text{Mol}}(\mathbf{z}) \rangle \cdot \nabla_z \left(\frac{1}{|\mathbf{r} - \mathbf{z}|} \right) \right\} d^3z$$

Introducing the macroscopic charge density

$$\rho(\mathbf{z}) = N(\mathbf{z}) \langle q_{\text{Mol}}(\mathbf{z}) \rangle \quad (5.10)$$

as well as the *polarization vector* (more precisely, the *vector of the polarization density*)

$$\mathbf{P}(\mathbf{z}) = N(\mathbf{z}) \langle \mathbf{p}_{\text{Mol}}(\mathbf{z}) \rangle \quad (5.11)$$

and, to unify the notation, writing $\mathbf{r}'$ everywhere for the integration variable $\mathbf{z}$, in analogy to equation (5.2), we obtain

$$\mathbf{E}^*(\mathbf{r}) = -\nabla \int dV' \left[\frac{\rho(\mathbf{r}')}{|\mathbf{r} - \mathbf{r}'|} + \mathbf{P}(\mathbf{r}') \cdot \nabla' \left(\frac{1}{|\mathbf{r} - \mathbf{r}'|} \right) \right] \quad (5.12)$$

If the macroscopic field $\mathbf{E}^*(\mathbf{r})$ is produced by various species of atoms and molecules, then equation (5.12) can be generalized easily by setting

$$\mathbf{P}(\mathbf{r}) = \sum_i N_i(\mathbf{r})\langle \mathbf{p}_i(\mathbf{r})\rangle \tag{5.13}$$

and

$$\rho(\mathbf{r}) = \sum_i N_k(\mathbf{r})\langle q_i(\mathbf{r})\rangle + \rho_a$$

where ρ_a means a further possible external charge distribution, and the index i describes the species. Equations (5.11) and (5.13) make it clear that the vector of the polarization density $\mathbf{P}(\mathbf{r})$ is equal to the dipole density; that is, the dipole moment/unit volume $\mathbf{P}(\mathbf{r}) =$ *dipole moment/volume*.

The dielectric displacement D

To obtain the macroscopic field equations for material media, equivalent to (5.1) we take the divergence of the field $\mathbf{E}^*$:

$$\nabla \cdot \mathbf{E}^*(\mathbf{r}) = -\int_V dV' \left[\rho(\mathbf{r}')\nabla^2\left(\frac{1}{|\mathbf{r}-\mathbf{r}'|}\right) + \mathbf{P}(\mathbf{r}') \cdot \nabla'\left\{\nabla^2\left(\frac{1}{|\mathbf{r}-\mathbf{r}'|}\right)\right\}\right] \tag{5.14}$$

Since the integrand is continuous for $\mathbf{r} \neq \mathbf{r}'$, differentiation and integration may be interchanged. Using the relation

$$\nabla^2\left(\frac{1}{|\mathbf{r}-\mathbf{r}'|}\right) = -4\pi\delta(\mathbf{r}-\mathbf{r}')$$

we obtain

$$\nabla \cdot \mathbf{E}^*(\mathbf{r}) = 4\pi\int_V dV' \left[\rho(\mathbf{r}')\delta(\mathbf{r}-\mathbf{r}') + \mathbf{P}(\mathbf{r}') \cdot \nabla'(\delta(\mathbf{r}-\mathbf{r}'))\right]$$

Furthermore, since

$$\nabla'\delta(\mathbf{r}-\mathbf{r}') = -\nabla\delta(\mathbf{r}-\mathbf{r}')$$

and we are integrating over $\mathbf{r}'$, we can factor out the nabla operator from under the integral sign:

$$\nabla \cdot \mathbf{E}^*(\mathbf{r}) = 4\pi\int dV'\rho(\mathbf{r}')\delta(\mathbf{r}-\mathbf{r}') - 4\pi\nabla \cdot \int \mathbf{P}(\mathbf{r}')\delta(\mathbf{r}-\mathbf{r}')dV'$$

Now we perform the integration and obtain

$$\nabla \cdot \mathbf{E}^*(\mathbf{r}) = 4\pi\rho(\mathbf{r}) - 4\pi\nabla \cdot \mathbf{P}(\mathbf{r}) \tag{5.15a}$$

or omitting equal arguments

$$\nabla \cdot (\mathbf{E}^* + 4\pi\mathbf{P}) = 4\pi\rho \tag{5.15b}$$

In comparison with equation (5.1) we recognize from equation (5.15a) that, on the macroscopic level, we have to subtract a correction term, resulting from the induced charge, from the electric field.

To obtain an analogy with equation (5.1), we define the *dielectric displacement* **D**

$$\mathbf{D} = \mathbf{E}^* + 4\pi \mathbf{P} \tag{5.15c}$$

So, equation (5.15b) can be rewritten

$$\nabla \cdot \mathbf{D} = \operatorname{div} \mathbf{D} = 4\pi \rho \tag{5.16}$$

Calculating the curl of equation (5.12), we find

$$\operatorname{curl} \mathbf{E}^* = 0 \tag{5.17}$$

since always curl grad $\phi = 0$.

On the macroscopic level, equations (5.16) and (5.17) are the relations analogous to (5.1).

The polarization P

The polarization vector $\mathbf{P}(\mathbf{x})$ describes the displacements of charges in an insulator that results from applying an (external) field. (Positive charges are shifted in the direction of the field; negative charges are shifted in the opposite direction.) Any atom or molecule becomes an electric dipole. Hence, the polarization appears as the electric dipole moment per unit volume, that is, as the dipole density (see equation (5.9)). As an example, we now consider an insulator of cubic-like shape (see Figure 5.3). Due to an applied field, the charges $\pm Q$ are induced on the endface areas.

If V and l are the volume and the edge-length, respectively, then the total dipole moment may be expressed as

$$|\mathbf{P}| V = Q \cdot l = \sigma A \cdot l = \sigma \cdot V$$

where Q denotes the charge on the surface A of the polarized insulator, and $\sigma = Q/A$ is the surface-charge density. Hence,

$$|\mathbf{P}| = \sigma$$

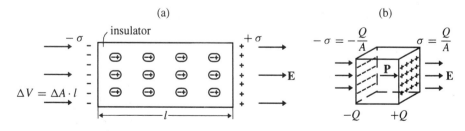

Figure 5.3. On the polarization within a dielectric. (a) Top view. (b) Perspective.

Now, we want to know the general interrelation between σ and $\mathbf{P}$, or between ρ and $\mathbf{P}$. For this purpose, we start from equation (5.15c), which can be regarded to be the gradient of two parts of the potential, ϕ_ρ and ϕ_σ, those parts of the potential (ϕ_p) depending on the polarization $\mathbf{P}$

$$\phi_p(\mathbf{r}) = \int_V \mathbf{P}(\mathbf{r}') \cdot \nabla' \frac{1}{|\mathbf{r} - \mathbf{r}'|} dV'$$

This is the potential ϕ_p generated by the polarized body. Because

$$\nabla' \cdot \left(\mathbf{P}(\mathbf{r}') \frac{1}{|\mathbf{r} - \mathbf{r}'|} \right) = \frac{1}{|\mathbf{r} - \mathbf{r}'|} \nabla' \cdot \mathbf{P}(\mathbf{r}') + \mathbf{P}(\mathbf{r}') \cdot \nabla' \left(\frac{1}{|\mathbf{r} - \mathbf{r}'|} \right)$$

we have for $\phi_p(\mathbf{r})$

$$\phi_p(\mathbf{r}) = \int_V \nabla' \cdot \left(\frac{\mathbf{P}(\mathbf{r}')}{|\mathbf{r} - \mathbf{r}'|} \right) dV' - \int_V \frac{1}{|\mathbf{r} - \mathbf{r}'|} \nabla' \cdot \mathbf{P}(\mathbf{r}') dV'$$

Rewriting the first term according to Gauss' theorem, we obtain

$$\phi_p(\mathbf{r}) = \int_S \frac{\mathbf{P}(\mathbf{r}') \cdot \mathbf{n}}{|\mathbf{r} - \mathbf{r}'|} dA' + \int_V \frac{-\nabla' \cdot \mathbf{P}(\mathbf{r}')}{|\mathbf{r} - \mathbf{r}'|} dV' \equiv \phi_\sigma + \phi_\rho$$

where $\mathbf{n}$ is the unit vector normal to the surface S of the volume V. Due to the relation for the potential of surface and space charges already derived

$$\phi_\sigma = \int_S \frac{\sigma(\mathbf{r}') dA'}{|\mathbf{r} - \mathbf{r}'|}, \qquad \phi_\rho = \int_V \frac{\rho(\mathbf{r}') dV'}{|\mathbf{r} - \mathbf{r}'|}$$

by a comparison of coefficients, we obtain at once

$$\mathbf{P}(\mathbf{r}) \cdot \mathbf{n} = |\mathbf{P}|_n = P_n = \sigma_p \tag{5.18a}$$

The normal component of the polarization vector corresponds to the surface-charge density induced on the surface.

Furthermore,

$$\nabla \cdot \mathbf{P} = \operatorname{div} \mathbf{P} = -\rho_p \tag{5.18b}$$

Hence, the sources of the polarization are the induced space charge densities. In general, the polarization $\mathbf{P}$ is the response of the system to the field $\mathbf{E}$. For some materials there are natural, aligned dipole moments (for example, inside the so-called Weiss domains of ferroelectrics and electrets). But, in general, the dipole moments will be induced. Without an external field, $\mathbf{P}$ would be zero because the dipoles cancel each other. For simplicity, we assume here that the interaction between $\mathbf{P}$ and $\mathbf{E}$ is linear

$$\mathbf{P} = \chi_e \mathbf{E} \tag{5.19}$$

The constant of proportionality χ_e is the *electric susceptibility*. The linear relation (5.19) is valid for fields $\mathbf{E}$ that are not too strong, and it is not valid for ferroelectrics. From (5.15b) and (5.19) we get

$$\mathbf{D} = (1 + 4\pi \chi_e)\mathbf{E} = \epsilon \mathbf{E}$$

$$\epsilon = 1 + 4\pi \chi_e \tag{5.20}$$

This is the relation between the dielectric displacement **D** and the applied field intensity **E** (given from outside). From equation (5.16), for a constant permittivity ϵ we obtain

$$\nabla \cdot \mathbf{E} = 4\pi \frac{\rho}{\epsilon} \tag{5.21}$$

Thus, for media of this kind all problems can be reduced to those of the preceding chapters with the fields of charges are reduced by a factor $1/\epsilon$. In Chapter 6, we will come back to this in more detail.

Example 5.1: Parallel-plate capacitor filled with a dielectric

We calculate the capacitance of a parallel-plate capacitor filled with a dielectric (see Figure 5.4). We have (compare (6.6) and (1.66))

$$D_{1n} - D_{2n} = 4\pi \sigma \tag{5.22}$$

Neglecting the stray fields outside the capacitor, $D_{2n} = 0$, we obtain

$$D = D_{1n} = 4\pi \sigma \tag{5.23}$$

Furthermore, $D = \epsilon E$, and then $E = 4\pi \sigma / \epsilon$.

Then the capacitance is

$$C = \frac{Q}{U} = \frac{\sigma A}{U} = \frac{\sigma A}{dE} = \frac{\epsilon}{4\pi} \cdot \frac{A}{d} \tag{5.24}$$

Here, A is the area of one plate of the capacitor, and d is the distance of the two plates. For some materials the permittivity has the following values:

$$\text{vacuum} \quad \epsilon = 1$$

air $\epsilon = 1.0005$ alcohol $\epsilon = 26$

glass $\epsilon = 5 - 8$ water $\epsilon = 81$

Thus, the capacitance of a capacitor can be increased many times if a material of high permittivity is placed in between the plates.

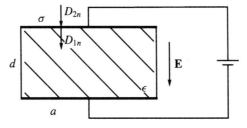

Figure 5.4. Parallel-plate capacitor with a dielectric.

6 Simple Dielectrics and the Susceptibility

In general, without an external electric field there is also no polarization. Concerning the origin of the polarization, we have to distinguish two kinds of polarization. On the one hand, there is the deformation polarization. In this case, in the volume of matter (dielectric) considered there are originally no electric dipole moments; these are generated only by the deformation of the atoms (or molecules) by an applied electric field. An example is shown in Figure 6.1. A neutral atom consists of a negative electronic shell and a positively charged nucleus. Due to the applied external field, the negative charges are displaced against the positive nucleus so that a dipole moment is induced.

In orientation polarization, molecule dipole moments are already present, but they are completely disordered due to the thermal motion, so that no polarization occurs.

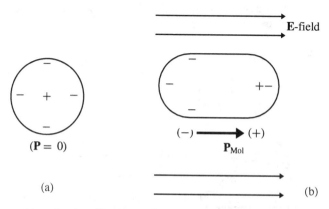

Figure 6.1. Symmetric molecules without inner dipole moments. (a) No external field. (b) In an external field.

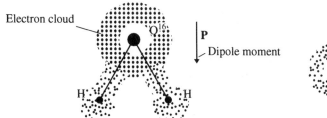

Figure 6.2. The water molecule.

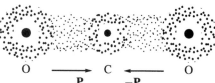

Figure 6.3. The CO_2 molecule.

In the discussion of the multipole moments, we noted that in nulei, atoms, and molecules there are no static dipole moments (odd multipole moments) in the laboratory system. The reason for that is the parity invariance of the electromagnetic interaction; that is, any charge distribution has to be point-symmetric, $(\mathbf{r} \to -\mathbf{r})$. But there are also molecules, e.g., ammonia (NH_3) or water (H_2O) with possessing internal dipole moments. This means that in a (rotating) coordinate system fixed to the molecule a dipole may be present. But this is not the case in the laboratory system, because in this system the various dipole positions are averaged out in space and time due to quantum mechanical rotation.

In the water molecule (Figure 6.2), the higher charge of the oxygen nucleus attracts the common electrons and at the same time repels the two hydrogen nuclei (protons). Hence, the oxygen nucleus gets an excess of negative charge. A dipole moment arises, that is, a *polar molecule*. If only two identical atoms form a molecule (e.g., the hydrogen molecule H_2), the common electrons stay mostly in the middle between the nuclei.

Molecules containing polar bounds are in most cases polar themselves (possessing a dipole moment). If the polarities (dipole moments) of the individual bonds in the molecule cancel each other, such a molecule becomes homopolar (possessing no dipole moments). Thus, the linear molecule CO_2 has no dipole moment, although the C-O bond has a strong dipole moment (Figure 6.3). But molecules have additional degrees of freedom in rotation and vibration. Because of the rotation, the dipole moment is averaged out in the laboratory system (Figure 6.4). The vibration may act similarly, for example, the vibration of the nitrogen (N) nucleus across the H_3-plane in the ammonia molecule (Figure 6.5).

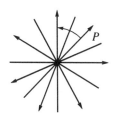

Figure 6.4. The rotation of a molecule leads to a cancellation of the dipole moment.

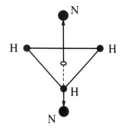

Figure 6.5. Vibration of the N-nucleus across the H_3-plane in the ammonia molecule.

In this manner, a parity conserving pattern of the charge distribution appears in the laboratory system. Frequently, the time for a rotation τ_{curl} is large compared to the interaction time τ_{WW}. Hence, on the scale of the interaction time the polar molecule appears as a quasistatic dipole. This entitles us to the (often not quite correct) statement that molecules have static dipole moments. Using an external field, one can align the molecular dipole moments (see Figure 6.6). If this alignment is performed for liquid dielectrics (liquid wax or resin), then these materials can maintain the polarization after solidification (electret). *The electret is the electrostatic analog to the permanent magnet.* But, since the neighborhood to some extent is always conducting, an electret loses its outward effect after a few hours or days.

But there are also *ferroelectric* crystals that exhibit a finite polarization without an external field below a certain temperature (the Curie point, similar to ferromagnets in the magnetic case). This situation arises because the dipoles of nearby molecules attempt to orient themselves parallel to each other due to the dipole-dipole interaction. A parallel orientation is most favorable for two dipoles (lowest energy), as we have seen in Chapter 3. Within an ensemble of many rotating and vibrating dipoles at a sufficiently low temperature, the free rotations and vibrations of the molecules are changed collectively so that all of them are aligned. One talks about a *phase transition*. The new, oriented phase is simply more favorable energetically (lower energy) than the free rotations and vibrations of the molecules. Although the basic interaction (in our case the electric interaction) exhibits the symmetry of parity and rotational invariance, the many-body system can prefer an energetically more favorable state in which this symmetries are no longer present (i.e., are spontaneously broken). See Figure 6.7.

In general, the permanent polarization of a ferroelectric crystal is compensated at once. The free surface charges produced by the polarization attract compensating charges. Therefore, ferroelectricity can be observed only by a directed compression (*piezoelectric effect*), in a variation of the temperature (*pyroelectric effect*), or by means of a rapidly changing external field. All ferroelectrics are piezoelectric and pyroelectric, but the converse need not be true (e.g., the piezoelectric quartz crystals and the pyroelectric tourmaline crystals are not ferroelectric). The condition for the appearance of piezoelectricity or pyroelectricity is the absence of a symmetry center in the crystal: the directed compression or the heating have to cause a directed displacement of electrons or ions to lead to a finite polarization. But one talks about ferroelectricity only if this polarization is present in the normal state

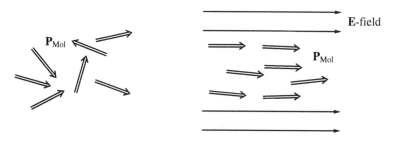

Figure 6.6. Polar molecules are aligned in an external field.

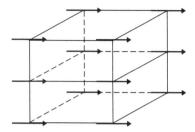

Figure 6.7. Spontaneously broken parity and rotational symmetries in a ferroelectric.

and can be rotated by an external electric field. Typical ferroelectrics are, e.g., the Rochelle salt ($NaKC_4OH_4O_4 + 4H_2O$), the KH_2PO_4-salt, and the ion-crystal $BaTiO_3$.

Hence, the polarization will be a function of the field intensity, $\mathbf{P} = \mathbf{P}(\mathbf{E})$, where $\mathbf{P}(0) = 0$. Because the applied external fields are in general small compared to the fields acting inside the crystal between the individual ions, we may expect that in a series expansion of the polarization in powers of the applied field only the first two terms are of importance. Therefore, we attempt the following ansatz for the components of the polarization:

$$P_i = \sum_{j=1}^{3} a_{ij} E_j + \sum_{j,k=1}^{3} b_{ijk} E_j E_k + \cdots$$

The indices represent the three Cartesian components. The ansatz implies that, in general, components of the polarization in y- and z-directions also occur when a field is applied in x-direction.

The coefficients a_{ij} and b_{ijk} are tensors of the rank two and three, respectively (with 9 and 27 elements, respectively), and are determined by the specific properties of the corresponding dielectric. Their magnitude can be obtained from experiment. It turns out that at normal (not too low) temperatures and the usual field intensities the linear approximation is sufficient; that is, we may set the b_{ijk} and all higher coefficients equal to zero. There are materials (e.g., $BaTiO_3$ and Rochelle salt) for which $\mathbf{P}(\mathbf{E})$ is a complicated function depending on the prehistory of the material. In Figure 6.8, the components of the polarization are drawn as a function of the applied field. Then, the coefficients a_{ij} are the slopes of the straight lines. For $i = j$ the coefficients are greater than for $i \neq j$; that is, the diagonal elements of the tensor are greater than the nondiagonal elements. Also, the tensor is symmetric, $a_{ij} = a_{ji}$; that is, six independent coefficients remain for *anisotropic dielectrics* (e.g., quartz).

In particularly simple cases, for *isotropic dielectrics*,

$$a_{ij} = a\delta_{ij}$$

that is

$$P_i = aE_i$$

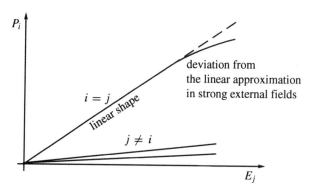

Figure 6.8. Applying an electric field in the j-direction (E_j) creates polarization in the i-direction (P_i).

Then the polarization is parallel to the external field. In this case the *dielectric suscepti-bility* is defined by $\chi_e = a$. Then

$$\mathbf{P} = \chi_e \mathbf{E} \tag{6.1}$$

For such dielectrics the *dielectric susceptibility* is a pure number. In this case the dielectric displacement $\mathbf{D}$ is

$$\mathbf{D} = \mathbf{E} + 4\pi \mathbf{P} = (1 + 4\pi \chi_e)\mathbf{E} = \epsilon \mathbf{E}$$

Here, the *permittivity* is defined by

$$\epsilon = 1 + 4\pi \chi_e \tag{6.2}$$

For a fixed (constant) permittivity we have

$$\operatorname{div} \mathbf{E} = \operatorname{div} \frac{\mathbf{D}}{\epsilon} = \frac{1}{\epsilon} \operatorname{div} \mathbf{D} = \frac{4\pi\rho}{\epsilon} \tag{6.3}$$

Using Gauss' theorem, we want to transform the differential formulation for the field intensity and the dielectric displacement into an integral form. We have

$$\oint_S \mathbf{D} \cdot \mathbf{n}\, dA = \int_V \operatorname{div} \mathbf{D}\, dV = 4\pi \int_V \rho\, dV = 4\pi Q \tag{6.4}$$

Analogously, the equations (5.15b) and (5.17) yield

$$\oint_S \mathbf{E} \cdot \mathbf{n}\, dA = 4\pi(Q + Q_p) \tag{6.5}$$

where the Q_p are the induced polarization charges. With these two relations (6.4) and (6.5), we can now investigate the behavior of the field lines crossing an interface which carries a surface charge σ.

We again single out the disk-shaped element of volume containing the interface. See Figure 6.9. The front areas are taken to be so small that they can be neglected. Then

$$\oint_S \mathbf{D} \cdot \mathbf{n}\, dA = (D_{1n} - D_{2n})\Delta A = 4\pi \Delta Q \quad \text{or} \quad D_{1n} - D_{2n} = 4\pi\sigma \tag{6.6}$$

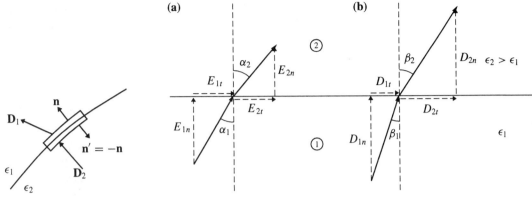

Figure 6.9. On the behavior of the dielectric displacement D at an interface.

Figure 6.10. Refraction of the field lines: (a) The electric field. (b) The dielectric displacement at an interface of two dielectrics without a surface charge.

The normal component of the dielectric displacement jumps by an amount equal to the surface charge density. Corresponding to equation (5.21), surface charges still occur in the electric field intensity:

$$E_{1n} - E_{2n} = 4\pi(\sigma + \sigma_p) \tag{6.7}$$

With the Stokes theorem the continuity of the tangential components follows, as in vacuum, because of $\nabla \times \mathbf{E} = 0$,

$$\oint_{C_A} \mathbf{E} \cdot d\mathbf{s} = \int_A \nabla \times \mathbf{E} \cdot d\mathbf{A} = 0 \quad \Rightarrow \quad E_{1t} = E_{2t} \tag{6.8}$$

With these relations, we can consider the refraction of the electric field lines in a transition between two dielectrics: If the surface charges are absent, then

$$E_{1t} = E_{2t} \quad \text{and} \quad D_{1n} = D_{2n}$$

Introducing the angles corresponding to Figure 6.10, we obtain

$$E_1 \sin\alpha_1 = E_2 \sin\alpha_2 \quad \text{and} \quad \epsilon_1 E_1 \cos\alpha_1 = \epsilon_2 E_2 \cos\alpha_2$$

The division of both equations yields for the **E**-field

$$\frac{\tan\alpha_1}{\epsilon_1} = \frac{\tan\alpha_2}{\epsilon_2} \tag{6.9}$$

and also for the **D**-field

$$\frac{\tan\beta_1}{\epsilon_1} = \frac{\tan\beta_2}{\epsilon_2}$$

In linear dielectrics, $\alpha_1 = \beta_1$ and $\alpha_2 = \beta_2$.

Denoting the medium with the higher permittivity as the electrically denser medium, this refraction law implies that in the transition into the electrically denser medium the electrical field lines are refracted away from the axis of incidence (the opposite of what happens in optics).

Example 6.1:   Potential distribution of a point charge in front of a dielectric half-plane with constant permittivity

Two dielectrics share a planar interface as in Figure 6.11. Introducing cylindrical coordinates (ρ, φ, z), the interface should be described by $z = 0$. In dielectric 1, there is a point charge at the distance $z = d$. The potential distribution $\phi(\rho, \varphi, z)$ in both half-spaces is required. This is a potential problem for the whole space with Dirichlet boundary conditions at infinity ($\phi(r \to \infty) = 0$). The potential ϕ is composed of partial solutions for the half-spaces $z < 0$ and $z > 0$ matched by continuity conditions at $z = 0$. The conditional equations are

$$\epsilon_1 \nabla \cdot \mathbf{E} = 4\pi\rho, \qquad z > 0$$
$$\epsilon_2 \nabla \cdot \mathbf{E} = 0, \qquad z < 0$$
$$\nabla \times \mathbf{E} = 0, \qquad \text{everywhere}$$
$$\rho = q\delta(z - d)\delta(x)\delta(y)$$

Correspondingly, for the potential we have

$$\Delta\phi(\rho, z) = -\frac{4\pi\rho}{\epsilon_1}, \qquad z > 0$$
$$\Delta\phi(\rho, z) = 0, \qquad z < 0$$

If we find a solution in some way, then it is the only one, according to the uniqueness theorem (Chapter 2). The charge lies at $(0, 0, d)$, and the problem is entirely rotationally symmetric about the z-axis; therefore, the angle φ is not needed for the solution:

$$\phi(\rho, \varphi, z) = \phi(\rho, z)$$

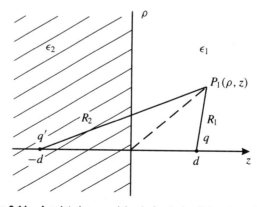

Figure 6.11. A point charge q lying in front of a dielectric half-plane.

Let us take a point $P_1(\rho, z)$ in the right-hand half-space and calculate the potential for it. The effect of dielectric 2 is simulated by the image charge q' at $(0, 0, -d)$. Since in a dielectric with constant ϵ the relation $\nabla \cdot \mathbf{D} = 4\pi\rho = \epsilon \nabla \cdot \mathbf{E}$ is valid, for the potential of the two point charges in the half-space 1 we obtain

$$\phi_1 = \frac{1}{\epsilon_1} \left(\frac{q}{R_1} + \frac{q'}{R_2} \right) \tag{6.10}$$

if $R_1 = \sqrt{\rho^3 + (d - z)^2}$ and $R_2 = \sqrt{\rho^3 + (d + z)^2}$ are the distances of the point P_1 from q and q', respectively. Considering the charge-free half-space 2, then the potential may be calculated there under the assumption that the effect of the dielectrics can be taken into account by placing a point charge q'' at $(0, 0, d)$. Then, the potential is

$$\phi_2 = \frac{1}{\epsilon_2} \frac{q''}{R_1} \tag{6.11}$$

Using the boundary conditions that result for the potential and the field intensity on the interface, the magnitude of the assumed image charges is calculated. The tangential component of the electric field intensity and the normal component of the dielectric displacement are continuous at the interface:

$$E_{1\rho}(z = 0) = E_{2\rho}(z = 0) \tag{6.12}$$

$$D_{1z}(z = 0) = D_{2z}(z = 0) \tag{6.13}$$

Expressing the field intensity by the potential, $E = -\nabla\phi$, then for equation (6.13) we obtain

$$\left. \frac{\partial\phi_1}{\partial\rho} \right|_{z=0} = \left. \frac{\partial\phi_2}{\partial\rho} \right|_{z=0}$$

$$\frac{1}{\epsilon_1} \frac{\partial}{\partial\rho} \left(\frac{q}{\sqrt{\rho^2 + (d - z)^2}} + \frac{q'}{\sqrt{\rho^2 + (d + z)^2}} \right)\bigg|_{z=0} = \frac{1}{\epsilon_2} \frac{\partial}{\partial\rho} \frac{q''}{\sqrt{\rho^2 + (d - z)^2}}\bigg|_{z=0}$$

and performing the differentiation:

$$q + q' = \frac{\epsilon_1}{\epsilon_2} q'' \tag{6.14}$$

A second condition is obtained from equation (6.13):

$$\epsilon_1 \left. \frac{\partial\phi_1}{\partial z} \right|_{z=0} = \epsilon_2 \left. \frac{\partial\phi_2}{\partial z} \right|_{z=0}$$

Substituting ϕ_1 and ϕ_2, then

$$q - q' = q'' \tag{6.15}$$

The image charges can be determined by the equations (6.14) and (6.15)

$$q' = \frac{\epsilon_1 - \epsilon_2}{\epsilon_1 + \epsilon_2} q \qquad \text{and} \qquad q'' = \frac{2\epsilon_2}{\epsilon_1 + \epsilon_2} q$$

So, we obtain the solution for the potential in the half-space 1 $(z > 0)$

$$\phi_1(\rho, z) = \frac{q}{\epsilon_1} \left(\frac{1}{\sqrt{\rho^2 + (d - z)^2}} + \frac{\epsilon_1 - \epsilon_2}{\epsilon_1 + \epsilon_2} \cdot \frac{1}{\sqrt{\rho^2 + (d + z)^2}} \right) \tag{6.16}$$

and in the half-space 2 ($z < 0$):

$$\phi_2(\rho, z) = \frac{2q}{\epsilon_1 + \epsilon_2} \cdot \frac{1}{\sqrt{\rho_2 + (d + z)^2}} \tag{6.17}$$

Discussion of the results

Figure 6.12 shows the field-line diagrams for various ratios ϵ_1/ϵ_2 and a positive charge q. The image charge q'' always has the same sign as q. In the half-space 2, we thus always obtain the potential of a more or less strong (positive) point charge q''. The image charge q' changes its sign if $\epsilon_1 > \epsilon_2$. In the half-space 1 the field lines are repelled or attracted (by the image charge), depending on the ratio of the permittivities.

For $\epsilon_1 = \epsilon_2$, $q' = 0$, and $q'' = q$, we obtain the field of a point charge in the whole space.

For $\epsilon_2 < \epsilon_1$, $q' > 0$, and $0 < q'' < q$, the field lines in the half-space 1 behave as if they are repelled by the image charge q'.

For $\epsilon_2 > \epsilon_1$ and $q' < 0$, the field lines are attracted by the image charge q'.

For $\epsilon_1 \ll \epsilon_2$ and $\epsilon_2 \to \infty$, $q' \to -q$. See Figure 6.13. The dielectric behaves increasingly like a conductor; the field lines are perpendicular in the interface and the potential vanishes in the second half-space (compare equation (6.17)). We will discuss briefly the density of the polarization charge in this problem. As we know, it is given by $\rho_p = -\operatorname{div} \mathbf{P}$; compare equation (5.18b).

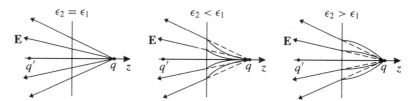

Figure 6.12. Field lines for a point charge in front of a dielectric wall.

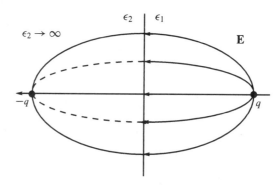

Figure 6.13. Field lines for a point charge in front of a conducting wall ($\epsilon_2 \to \infty$).

Since in the interior of both dielectric media $\mathbf{P} = \chi_e \mathbf{E}$, we have

$$\rho_p = -\operatorname{div} \mathbf{P} = -\chi_e \operatorname{div} \mathbf{E} = \begin{cases} -\chi_e \dfrac{q}{\epsilon_1} 4\pi \delta(x)\delta(y)\delta(z-d) & \text{for } z > 0 \\ 0 & \text{for } z < 0 \end{cases} \tag{6.18}$$

At the position of the charge q, there is a density of the polarization charge with the value

$$\int \rho_p \, dV = \frac{-4\pi q}{\epsilon_1} \chi_e = -\frac{\epsilon_1 - 1}{\epsilon_1} q \tag{6.19}$$

But we have to be careful, because at the interface χ_e changes by

$$\Delta \chi_e = \frac{\epsilon_1 - \epsilon_2}{4\pi} \tag{6.20}$$

Therefore, at the interface there is still a polarization surface-charge density of magnitude

$$\sigma_p = -(\mathbf{P}_1 - \mathbf{P}_2) \cdot \mathbf{n} \tag{6.21}$$

where $\mathbf{n} = \mathbf{e}_z$. Because

$$\mathbf{P}_i \bigg|_{z=0} = \frac{(\epsilon_i - 1)}{4\pi} \mathbf{E}_i \bigg|_{z=0} = -\frac{(\epsilon_i - 1)}{4\pi} \nabla \phi_i \bigg|_{z=0} \tag{6.22}$$

σ_p can be determined easily according to (6.21)

$$\sigma_p = -\frac{q}{2\pi} \frac{(\epsilon_2 - \epsilon_1)}{\epsilon_1(\epsilon_2 + \epsilon_1)} \frac{d}{(\rho^2 + d^2)^{3/2}} \tag{6.23}$$

As expected, the polarization surface-charge density takes the largest value if $\rho = 0$, and it vanishes if $\epsilon_1 = \epsilon_2$, that is, if there is no interface at all. In the case $\epsilon_2 \gg \epsilon_1$ the left-hand side is a conductor, and (6.23) becomes

$$\lim_{\epsilon_2 \to \infty} \sigma_p = \frac{-q}{2\pi} \frac{d}{(\rho^2 + d^2)^{3/2}} \tag{6.24}$$

This is the result for the induced charge that we obtained in Exercise 2.6.

Example 6.2:   Polarization of a sphere in a uniform field E_0

We investigate the potential of an electrically neutral sphere, consisting of a dielectric medium with permittivity ϵ, imbedded in a uniform electric field with the field intensity $\mathbf{E}_0$. Let the permittivity outside the sphere be equal to 1; that is, the sphere is in vacuum (see Figure 6.14).

On the spherical surface, polarization charges are formed. On the other hand, the interior as well as the exterior region of the sphere remain charge-free. Then, the field $\mathbf{E}_i$ in the interior of the sphere and $\mathbf{E}_a = \mathbf{E}_0$, the field in the exterior region satisfy the equations

$$\nabla \times \mathbf{E} = 0, \qquad \mathbf{E} = -\nabla \varphi$$
$$\nabla \cdot \mathbf{D} = 0, \qquad \Delta \varphi = 0 \tag{6.25}$$

because $\mathbf{D} = \epsilon \mathbf{E}$, with $\epsilon = \text{const.}$

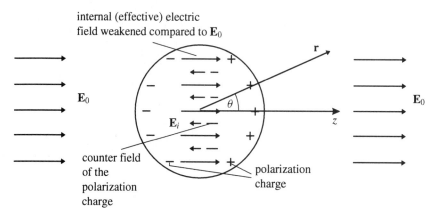

internal (effective) electric
field weakened compared to $\mathbf{E}_0$

$\mathbf{E}_0$ $\mathbf{E}_0$

counter field
of the
polarization
charge

polarization
charge

Figure 6.14. A dielectric sphere in a uniform electric field.

Because we have chosen the center of the sphere to be the origin of the coordinate system and the direction of $\mathbf{E}_0$ to be the z-axis, and since the field at large distance from the sphere is uniform, we obtain as $r \to \infty$:

$$\mathbf{E} = E_0\mathbf{e}_z, \qquad \varphi = -E_0 z, \qquad E_z = -\frac{\partial}{\partial z}\varphi = E_0 \tag{6.26}$$

Further boundary conditions are obtained for the spherical surface:

$$D_r^i\Big|_{r=a} = D_r^a\Big|_{r=a} \tag{6.27}$$

that is, the normal component of $\mathbf{D}$ is continuous across the surface, because due to $\operatorname{div}\mathbf{D} = 4\pi\rho$, we can state that the *sources of* $\mathbf{D}$ *are the true charges only.* Thus, the polarization charges σ on the surface are not sources of $\mathbf{D}$, so that the normal component does not make a jump. Furthermore, for the field intensity we have

$$E_\vartheta^i\Big|_{r=a} = E_\vartheta^a\Big|_{r=a} \tag{6.28}$$

that is, the tangential component remains continuous. Assuming that the polarization is uniform, and parallel and proportional to the external field throughout the sphere, we have

$$\mathbf{D} = \epsilon\mathbf{E} = -\epsilon\operatorname{grad}\varphi$$

Then for the normal component of $\mathbf{D}$ we obtain

$$\mathbf{D}_n = \epsilon\mathbf{E}_n = -\epsilon\frac{\partial\varphi}{\partial r}\mathbf{e}_r$$

Because of (6.27),

$$-\epsilon\frac{\partial\varphi_i}{\partial r}\Big|_{r=a} = -\frac{\partial\varphi_a}{\partial r}\Big|_{r=a} \tag{6.29}$$

Now, we want to investigate the alteration of the field $\mathbf{E}_0$ arising due to the polarization of the sphere. In the interior of the sphere the potential is

$$\varphi_i = -E_i \cdot z \tag{6.30}$$

Outside, the sphere acts like an electric dipole in z-direction whose moment is $\mathbf{p}_z$. The potential originating from this dipole is given by

$$\varphi = \frac{\mathbf{p}_z \cdot \mathbf{r}}{r^3}$$

Thus, in the exterior of the sphere the potential is

$$\varphi_a = -E_0 z + \frac{\mathbf{p}_z \cdot \mathbf{r}}{r^3} \tag{6.31}$$

Since the part of the potential generated by the dipole moment is proportional to $1/r^2$, it vanishes for large r and only the homogeneous part is left according to (6.26). We substitute the potentials (6.30) and (6.31) into the boundary conditions (6.27) and (6.28). Eliminating the dipole moment, we obtain a relation for the electric field in the interior. From (6.29) we get, with $z = r \cos \vartheta$:

$$\epsilon E_i = E_0 + \frac{2 p_z}{a^3} \tag{6.32}$$

On the other hand, from (6.28):

$$\left.\frac{\partial \varphi_i}{\partial \vartheta}\right|_{r=a} = \left.\frac{\partial \varphi_a}{\partial \vartheta}\right|_{r=a}, \qquad E_i = E_0 - \frac{p_z}{a^3} \tag{6.33}$$

Thus, from (6.32) and (6.33) the dielectric field in the interior of the sphere can be calculated:

$$E_i = \frac{3}{\epsilon + 2} E_0 \tag{6.34}$$

The field in the interior of the sphere is weakened by the factor $3/(\epsilon + 2) \leq 1$. Now, we determine p_z, the magnitude of the dipole moment. From (6.33)

$$E_0 = E_i - \frac{p_z}{a^3}$$

Substituting this into (6.34), we can solve for p_z and obtain

$$p_z = \frac{\epsilon - 1}{\epsilon + 2} a^3 E_0 \tag{6.35}$$

Equation (6.35) may also be obtained starting from the polarization (dipole density) $\mathbf{P}$:

$$\mathbf{P} = \chi \mathbf{E}_i = \chi E_i \mathbf{e}_z = \frac{\epsilon - 1}{4\pi} \frac{3 E_0}{\epsilon + 2} \mathbf{e}_z \equiv P \mathbf{e}_z$$

One should note that the polarization $\mathbf{P}$, together with the susceptibility χ_e, are always defined with respect to the macroscopic field in the dielectric. In our case, the field in the dielectric is E_i. Then, we obtain the dipole moment $\mathbf{p}$ by integrating over the volume V:

$$\mathbf{p} = \int_V \mathbf{P} \, dV = \frac{\epsilon - 1}{\epsilon + 2} \frac{3}{4\pi} E_0 \int_V dV \cdot \mathbf{e}_z$$

With $V = \frac{4}{3} \pi a^3$,

$$\mathbf{p} = \frac{\epsilon - 1}{\epsilon + 2} E_0 a^3 \mathbf{e}_z$$

The difference between the internal and external fields can be represented by the polarization

$$\mathbf{E}_i - \mathbf{E}_0 = \left(\frac{3}{\epsilon + 2} - 1\right) \mathbf{E}_0 = -\frac{\epsilon - 1}{\epsilon + 2} \mathbf{E}_0 = \frac{\epsilon - 1}{\epsilon + 2} \frac{\epsilon + 2}{3} \mathbf{E}_i = -\frac{\epsilon - 1}{3} \frac{\mathbf{P}}{\chi}$$

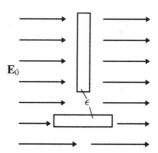

Figure 6.15. Slab and rod in an electric field.

and since $\chi = (\epsilon - 1)/4\pi$,

$$\mathbf{E}_0 - \mathbf{E}_i = +\frac{4\pi}{3}\mathbf{P} \tag{6.36}$$

The weakening of the electric field in the interior of the sphere due to the polarization is called the *deelectrification*. The magnitude of the deelectrification is given by (6.36), the prefactor of the polarization $4\pi/3$ is called the *deelectrification factor*. Apart from the case of the sphere, the deelectrification can be calculated for a shallow slab and a long thin rod. See Figure 6.15.

For a slab of permittivity ϵ positioned perpendicular to the field $\mathbf{E}_0$, the continuity of the normal component of $\mathbf{D}$ yields

$$D_i = D_a \qquad \text{and hence} \qquad \epsilon E_i = E_a = E_0$$

For the difference, we obtain

$$E_0 - E_i = (\epsilon - 1)E_i = 4\pi \chi E_i = +4\pi P$$

The deelectrification factor is equal to 4π.

For the thin rod positioned parallel to the field lines, the continuity of the tangential component of $\mathbf{E}$ yields $E_i = E_0$; therefore, the deelectrification factor is equal to zero.

One could wonder that in the case of the slab we considered the normal component of the dielectric displacement, and in the case of the thin rod we took the tangential component of the field intensity. Of course, the reason for this lies in the geometry. Due to the small thickness of the slab the tangential component of the field intensity is unimportant, apart from the strong fields at the edges of the slab. The same holds for the normal component of the **D**-field in the case of the thin rod.

In general, we can state that in a uniform field an ellipsoid will be polarized uniformly. For a uniform polarization, one defines

$$E_0 - E_i = NP$$

where N is the deelectrification factor. With $\chi E_i = P$,

$$P = \frac{1}{\dfrac{1}{\chi} + N} E_0$$

If the susceptibility is large, the polarization is determined essentially by the geometry-dependent deelectrification factor. This explains the difficulty in measuring the susceptibility for small samples.

Example 6.3: Spherical cavity in a dielectric

A homogeneous sphere of radius a and with permittivity ϵ_2 is imbedded in a region with permittivity ϵ_1. In absence of the sphere, a uniform field $\mathbf{E} = E_0 \mathbf{e}_z$ is in this region. The potential and the shape of the field lines in presence of the field are required. For a spherical cavity in the dielectric ($\epsilon_1 = \epsilon$, $\epsilon_2 = 1$), the density of the polarization charge on the spherical surface has to be determined as well as the electric field produced by the surface charges. See Figure 6.16.

This task can be treated analogously to Exercise 6.2. There, the ansatz for the potential was

$$\phi_1 = -E_0 z + \frac{P_z z}{r^3}, \qquad r > a$$

$$\phi_2 = -E_i \cdot z, \qquad r < a \tag{6.37}$$

From the condition that the normal components of the dielectric displacement are continuous, we obtain

$$\epsilon_2 \frac{\partial \phi_2}{\partial r}\bigg|_{r=a} = \epsilon_1 \frac{\partial \phi_1}{\partial r}\bigg|_{r=a} \tag{6.38}$$

Therefore, we have to replace only $\epsilon \rightarrow \epsilon_2/\epsilon_1$ in equation (6.29). This replacement yields the constants occurring in the ansatz for the potential

$$E_i = \frac{3\epsilon_1}{2\epsilon_1 + \epsilon_2} E_0, \qquad \rightarrow \qquad \phi_2 = -\frac{3\epsilon_1}{2\epsilon_1 + \epsilon_2} E_0 r \cos\vartheta$$

$$p_z = \frac{\epsilon_2 - \epsilon_1}{2\epsilon_1 + \epsilon_2} a^3 E_0, \qquad \rightarrow \qquad \phi_1 = -E_0 r \cos\vartheta + \frac{\epsilon_2 - \epsilon_1}{2\epsilon_1 + \epsilon_2} a^3 E_0 \frac{\cos\vartheta}{r^2} \tag{6.39}$$

The field inside the sphere is always uniform, as per ansatz. The field lines are refracted at the interface according to the relation

$$\textbf{E-field:} \quad \frac{\tan\alpha_1}{\tan\alpha_2} = \frac{\epsilon_1}{\epsilon_2}; \qquad \textbf{D-field:} \quad \frac{\tan\beta_1}{\tan\beta_2} = \frac{\epsilon_1}{\epsilon_2}$$

Thus, we obtain the shape of the field lines shown in Figure 6.17.

The **D**-lines are refracted at the surface (just like the **E**-lines). Their number cannot increase because $\operatorname{div}\mathbf{D} = 4\pi\rho = 0$ since there are no external charges. In contrast, $\operatorname{div}\mathbf{E} = 4\pi\rho - 4\pi\operatorname{div}\mathbf{P} = 4\pi\rho_p$; that is, the number of the **E**-lines may increase because their sources are not only the true charges (ρ) but also the polarization charges $\rho_p = -\operatorname{div}\mathbf{P}$. But a polarization occurs in our example. The polarization density is therefore

$$\sigma_{\text{pol}} = (\mathbf{P}_1 - \mathbf{P}_2) \cdot \mathbf{n} = \mathbf{P}_1 \cdot \mathbf{n} - \mathbf{P}_2 \cdot \mathbf{n}$$

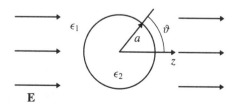

Figure 6.16. Spherical cavity in a dielectric.

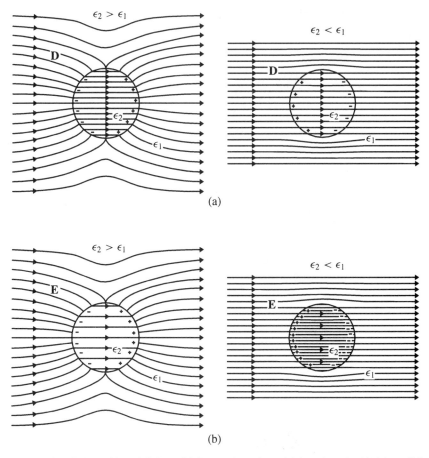

Figure 6.17. The **D**-lines (a) and **E**-lines (b) for a sphere (permittivity ϵ_2) embedded in a dielectric (permittivity ϵ_1).

The interior region of the sphere is the exterior region of the dielectric 1. So, the normal vector **n** has to point outside the sphere, that is, $\mathbf{n} = -\mathbf{r}/r$. For $\mathbf{P}_2$ it is opposite; therefore, the relative minus sign appears. The polarization charges of the both parts superpose on the surface of the sphere. With the relations

$$\mathbf{P}_1 = \chi_1 \mathbf{E}_1 = -\chi_1 \operatorname{grad} \phi_1, \qquad \mathbf{P}_2 = \chi_2 \mathbf{E}_2 = -\chi_2 \operatorname{grad} \phi_2$$

we obtain

$$\sigma_{\text{pol}} = \mathbf{P}_1 \cdot \mathbf{n} - \mathbf{P}_2 \cdot \mathbf{n} = \chi_1 \left. \frac{\partial \phi_1}{\partial r} \right|_{r=a} - \chi_2 \left. \frac{\partial \phi_2}{\partial r} \right|_{r=a}$$

$$= \chi_1 \frac{\partial}{\partial r} \left(-E_0 r \cos \vartheta + \frac{\epsilon_2 - \epsilon_1}{2\epsilon_1 + \epsilon_2} a^3 E_0 \frac{\cos \vartheta}{r^2} \right) \Bigg|_{r=a}$$

$$- \chi_2 \frac{\partial}{\partial r} \left(- \frac{3\epsilon_1}{2\epsilon_1 + \epsilon_2} E_0 r \cos \vartheta \right) \Bigg|_{r=a}$$

$$= \chi_1 \left(-E_0 \cos \vartheta - 2 \frac{\epsilon_2 - \epsilon_1}{2\epsilon_1 + \epsilon_2} E_0 \cos \vartheta \right) + \chi_2 \frac{3\epsilon_1}{2\epsilon_1 + \epsilon_2} E_0 \cos \vartheta$$

$$= E_0 \cos \vartheta \left\{ \chi_1 \left(\frac{-2\epsilon_1 - \epsilon_2 - 2\epsilon_2 + 2\epsilon_1}{2\epsilon_1 + \epsilon_2} \right) + \chi_2 \frac{3\epsilon_1}{2\epsilon_1 + \epsilon_2} \right\}$$

$$= E_0 \cos \vartheta \left\{ \left(\frac{\epsilon_1 - 1}{4\pi} \right) \left(\frac{-3\epsilon_2}{2\epsilon_1 + \epsilon_2} \right) + \left(\frac{\epsilon_2 - 1}{4\pi} \right) \left(\frac{3\epsilon_1}{2\epsilon_1 + \epsilon_2} \right) \right\}$$

$$= \frac{E_0 \cos \vartheta}{4\pi} \frac{3(\epsilon_2 - \epsilon_1)}{2\epsilon_1 + \epsilon_2}$$

If there is a cavity inside the sphere, then $\epsilon_2 = 1$ (vacuum). Hence,

$$\sigma_{\text{pol}} = \frac{1 - \epsilon_1}{1 + 2\epsilon_1} \frac{3}{4\pi} E_0 \cos \vartheta$$

These polarization charges are indicated in Figure 6.17. Their sign changes, depending on the ratio ϵ_2 / ϵ_1. The polarization charge σ_{pol} generates the field in the interior of the sphere. The potential produced by these surface charges is the dipole part of ϕ_1:

$$\phi_D = \frac{1 - \epsilon_1}{2\epsilon_1 + 1} a^3 E_0 \frac{\cos \vartheta}{r^2}$$

This potential corresponds to the field

$$\mathbf{E}_D = - \text{grad} \, \phi_D = - \frac{1 - \epsilon_1}{2\epsilon_1 + 1} a^3 E_0 \left(\mathbf{e}_r \frac{\partial}{\partial r} + \mathbf{e}_\vartheta \frac{1}{r} \frac{\partial}{\partial \vartheta} \right) \frac{\cos \vartheta}{r^2}$$

$$= \frac{1 - \epsilon_1}{2\epsilon_1 + 1} \frac{a^3}{r^3} E_0 (2 \cos \vartheta \mathbf{e}_r + \sin \vartheta \mathbf{e}_\vartheta)$$

Namely, the dipole moment of the polarization charge σ_{pol} is

$$P_z = \int_{\text{sphere}} \sigma_{\text{pol}} z \, dS = \int \sigma_{\text{pol}} a \cos \vartheta a^2 \, d\Omega = \frac{1 - \epsilon_1}{1 + 2\epsilon_1} a^3 E_0$$

just as it should be according to equation (6.39).

A molecular model of the polarizability

In this and in the next sections, we consider the interrelation between the molecular properties (e.g., the polarizability of a molecule) and the macroscopically defined parameter the electric susceptibility χ_e. We discuss this interrelation on a classical basis because thus our understanding is enlarged, although a better, perfect discussion demands quantum mechanics.

At first, we want to calculate the local field E_{loc} in the interior of the dielectric. This field, certainly distinct from the macroscopic average field $\mathbf{E}$, gives rise to the polarization. In fluids and solids the local field is decisively determined by the influence of neighboring molecules. In gases the distance between the molecules is large so that the local field is mainly determined by the applied external field.

The local field at the position of an arbitrarily selected molecule is split into three parts

$$\mathbf{E}_{\text{loc}} = \mathbf{E} + \mathbf{E}_p + \mathbf{E}_i$$

The two fields occurring additionally take into account the effect of the neighboring molecules. The direct electrostatic interaction between the neighboring molecules is given by $\mathbf{E}_i$; $\mathbf{E}_p$ results from the polarization of the molecules of the neighborhood. Both parts will be calculated in the following.

To calculate the polarization field $\mathbf{E}_p$ (Lorentz field), we consider a sphere centered on the selected molecule. The radius of the sphere should be large compared to the molecular distances. In the interior of the sphere we remove all other molecules; the remaining molecules may then be treated as a continuum. See Figure 6.18. The polarization charge on the surface of the hollow sphere creates the field $\mathbf{E}_p$ at the position of the molecule.

Since in the exterior region the field $\mathbf{E}$ acts in the direction of the z-axis, it creates a polarization oriented parallel to it, according to the equation

$$\sigma = P_n$$

a surface-charge density on the spherical surface

$$\sigma = P_n = \mathbf{P} \cdot \mathbf{n} = -P \cos \vartheta$$

One should note that $\mathbf{n}$ points radially inward because the interior of the sphere is the exterior region of the dielectric in the more distant neighborhood of the molecule.

According to Coulomb's law, at the position of the molecule the surface charge $\sigma\, da$ adds a field with the z-component $(\sigma\, da/R^2) \cos \vartheta$; the other components of this additional field vanish due to symmetry reasons. Then, we obtain the Lorentz field by integrating over the entire spherical surface:

$$E_p \mathbf{e}_z = -\oint_s \frac{\sigma\, da}{R^2} \cos \vartheta\, \mathbf{e}_z = +\oint_s \frac{da\, P}{R^2} \cos^2 \vartheta\, \mathbf{e}_z$$

$$= \int_0^{2\pi} \int_0^{\pi} \frac{P}{R^2} \cos^2 \vartheta \sin \vartheta\, R^2\, d\vartheta\, d\varphi\, \mathbf{e}_z$$

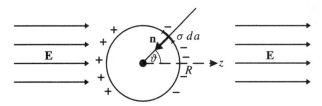

Figure 6.18. On the calculation of the polarization-field intensity at the position of the molecule.

$$= \int_0^{2\pi} \int_0^{\pi} P \cos^2 \vartheta \sin \vartheta \, d\vartheta \, d\varphi \, \mathbf{e}_z$$

$$= P2\pi \left[-\frac{\cos^3 \vartheta}{3} \right]_0^{\pi} \mathbf{e}_z = \frac{4\pi}{3} P \mathbf{e}_z$$

In the first step, the negative sign arises because a positive charge $\sigma \, da$ produces a field component in negative z-direction $(-\mathbf{e}_z)$ at the position of the molecule.

The field $\mathbf{E}_i$ resulting from the molecules in the immediate neighborhood depends on the structure of the crystal. For the sake of simplicity, let us consider a crystal lattice with the lattice parameter a. See Figure 6.19. Let all molecules or atoms have a dipole moment aligned along the z-direction. Now we can show that for such a cubic crystal lattice the contributions of all molecules cancel each other, so that the field $\mathbf{E}_i$ vanishes. The field of a dipole at a distance $\mathbf{r}$ from the origin (see equation equation (1.86)) is

$$\mathbf{E} = \frac{3(\mathbf{p} \cdot \mathbf{r})\mathbf{r} - \mathbf{p}r^2}{r^5}$$

Because $\mathbf{p} = p_z \mathbf{e}_z$ and $\mathbf{r} = la\mathbf{e}_x + ma\mathbf{e}_y + na\mathbf{e}_z$, where l, m, n are integers and $l \neq m \neq n \neq 0$, the total field is obtained by summing over the individual dipoles:

$$\mathbf{E}_i = \sum_{l,m,n} \frac{(3p_z na)(la \, \mathbf{e}_x + ma \, \mathbf{e}_y + na \, \mathbf{e}_z) - p_z \mathbf{e}_z (a^2(l^2 + m^2 + n^2))}{a^5(l^2 + m^2 + n^2)^{5/2}}$$

$$= \sum_{l,m,n} \left(\frac{3p_z nla^2}{a^5(l^2 + m^2 + n^2)^{5/2}} \mathbf{e}_x + \frac{3p_z nma^2}{a^5(l^2 + m^2 + n^2)^{5/2}} \mathbf{e}_y \right.$$

$$\left. + \frac{p_z a^2(3n^2 - l^2 - m^2 - n^2)}{a^5(l^2 + m^2 + n^2)^{5/2}} \mathbf{e}_z \right)$$

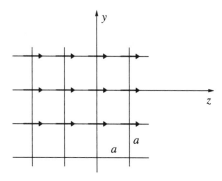

Figure 6.19. Aligned electric dipoles in a cubic lattice.

Because l and n assume negative values as well as positive values, the x- and y-components have to vanish. Also, due to symmetry reasons

$$\sum_{l,m,n} \frac{l^2}{(l^2 + m^2 + n^2)^{5/2}} = \sum_{l,m,n} \frac{m^2}{(l^2 + m^2 + n^2)^{5/2}} = \sum_{l,m,n} \frac{n^2}{(l^2 + m^2 + n^2)^{5/2}}$$

But then also $D_z = 0$. Hence, E_i vanishes in a cubic crystal lattice. In addition to crystals and in media with a random arrangement of molecules, E_i can be set equal to zero on average. So, e.g., in amorphous materials like glass there is no contribution to the field from the nearest molecules.

To obtain the relation between molecular polarizability and the dielectric susceptibility, we start from the definition of the latter. The susceptibility χ_e is defined by

$$\mathbf{P} = \chi_e \mathbf{E}$$

where $\mathbf{E}$ is the macroscopic field in the dielectric. Furthermore, according to (5.15a)

$$\mathbf{P} = N \langle \mathbf{p}_{\text{Mol}} \rangle$$

where $\langle \mathbf{p}_{\text{Mol}} \rangle$ is the average dipole moment of a molecule, and N is the number of molecules per unit of volume. This dipole moment is approximately proportional to the field acting on the molecule:

$$\langle \mathbf{p}_{\text{Mol}} \rangle = \gamma_{\text{Mol}} \cdot \mathbf{E}_{\text{loc}}$$

where γ_{Mol} is the *molecular polarizability*. Due to

$$\mathbf{E}_{\text{loc}} = \mathbf{E} + \frac{4\pi}{3} \mathbf{P}$$

we obtain

$$\langle \mathbf{p}_{\text{Mol}} \rangle = \gamma_{\text{Mol}} \cdot \left(\mathbf{E} + \frac{4\pi}{3} \mathbf{P} \right)$$

$$\mathbf{P} = N \gamma_{\text{Mol}} \left(\mathbf{E} + \frac{4\pi}{3} \mathbf{P} \right) = N \gamma_{\text{Mol}} \left(\frac{1}{\chi_e} + \frac{4\pi}{3} \right) \mathbf{P}$$

Since the vectors have to be identical on both sides of the equation, a comparison of coefficients yields

$$1 = N \gamma_{\text{Mol}} \left(\frac{1}{\chi_e} + \frac{4\pi}{3} \right)$$

This equation is solved for χ_e,

$$\chi_e = \frac{N \gamma_{\text{Mol}}}{1 - \frac{4\pi}{3} N \gamma_{\text{Mol}}}$$

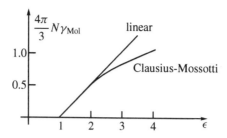

Figure 6.20. The relation between $N\gamma_{\mathrm{Mol}}$ and permittivity according to Clausius-Mossotti.

Because $\chi_e = (\epsilon - 1)/4\pi$, we obtain the *Clausius Mossotti formula* for the molecular polarizability:[1]

$$\gamma_{\mathrm{Mol}} = \frac{3}{4\pi N}\frac{\epsilon - 1}{\epsilon + 2}$$

See Figure 6.20. In the vacuum ($\epsilon = 1$), there are no molecules, and therefore, also $\gamma_{\mathrm{Mol}} = 0$. The measurement of the permittivity ϵ of a medium allows for the determination of the molecular polarizability if the density (particle number N per cm^3) is known.

Because γ_{Mol} is independent of the particle number N by definition, $(\epsilon - 1)/(\epsilon + 2)$ should be proportional to the particle density of the material. This relation is best for gases; it is least fulfilled for fluids and solids, in particular for materials having a high permittivity. This results from the fact that for high ϵ the Clausius-Mossotti formula gives a maximal polarizability of matter:

$$N\gamma_{\mathrm{Mol}} = \frac{3}{4\pi}\frac{\epsilon - 1}{\epsilon + 2} = \frac{3}{4\pi}\frac{1 - \dfrac{1}{\epsilon}}{1 + \dfrac{2}{\epsilon}} = \frac{3}{4\pi} \qquad \text{as} \qquad \epsilon \to \infty$$

which is physically meaningless, of course. In this case ($\epsilon \to \infty$) nonlinear relations between P and E as well as the tensor character of ϵ play a role.

Models for the molecular polarizability

Since all material bodies consist of positive charge carriers that are hard to move (atomic cores, ions, nuclei) and negative ones that are easy to move (electrons), all bodies respond to a penetrating external field by a displacement of these charge carriers. These displacements, which in matter cause a charge current in direction of the field lines, produce only a *local distortion of the atomic structure*, the *dielectric polarization*, in an insulator. The variety of the atomic structure gives rise, of course, also to a variety of possibilities for such

---

[1] As we will see later, $\epsilon = n^2$ for optical frequencies, where n is the index of refraction. The corresponding modified formula is sometimes called the Lorentz-Lorenz equation (1880).

polarizations. In principle, we may distinguish two kinds of polarizability. On the one hand, there is the displacement of charges in the dielectric. Thus, a deformation of the electron clouds, as well as a displacement of the ions, can occur. The other kind of polarization is the *orientation polarization*, which occurs for polar molecules (e.g., H_2O) possessing a permanent intrinsic dipole moment. We want to examine both kinds of polarization in greater detail.

Displacement polarization

We will consider an isolated atom; hence, we will neglect its interaction with the environment in the molecule or in the crystal lattice. Within the simplest model, let the electron be bound harmonically to the atom (see Figure 6.21). It vibrates about its equilibrium position with the frequency ω_0. Now, applying a constant electric field $\mathbf{E}$ to the atom, the center of oscillation is shifted due to the additional force $e\mathbf{E}$. The differential equation

$$m\ddot{\mathbf{r}} = -m\omega_0^2\mathbf{r} + e\mathbf{E}$$

is valid. The solution of this inhomogeneous differential equation is

$$\mathbf{r} = \mathbf{r}_0 e^{i\omega_0 t} + \frac{e}{m\omega_0^2}\mathbf{E}$$

Thus, for the dipole moment $\mathbf{p}$ we obtain

$$\mathbf{p} = e\mathbf{r} = e\mathbf{r}_0 e^{i\omega_0 t} + \frac{e^2}{m\omega_0^2}\mathbf{E}$$

Averaging over time, we obtain the *average dipole moment* caused by the shift of the center of oscillation of an electron:

$$\langle\mathbf{p}\rangle = \frac{e^2}{m\omega_0^2}\mathbf{E} = \gamma_{\text{el}}\cdot\mathbf{E}$$

Thus, $\gamma_{\text{el}} = e^2/(m\omega_0^2)$, or if there are Z electrons,

$$\gamma_{\text{el}} = \sum_{i=1}^{Z}\frac{e^2}{m\omega_i^2}$$

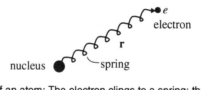

Figure 6.21. Naive model of an atom: The electron clings to a spring; the nucleus fixes the other end of the spring.

The dimension of $\gamma_{el} = p/E$ can be determined easily:

$$[\gamma_{el}] = \frac{[p]}{[E]} = \frac{\text{charge} \cdot \text{length}}{\text{charge} \cdot \text{length}^{-2}} = \text{length}^3$$

We want to obtain an estimate of the susceptibility. The quantity γ_{el} describes an atomic process. The length occurring in the dimensional consideration is set equal to one Ångström, corresponding approximately to the atomic order of magnitude, that is, $\gamma_{el} \approx 1 \text{Å}^3 = 10^{-24} \text{cm}^3$. The order of magnitude of the dielectric susceptibility is obtained by multiplying this quantity by the number of atoms per unit volume. For gases, $N \approx 10^{19} \text{cm}^{-3}$; thus, $\chi \approx 10^{-5}$; while for solids, $N \approx (10^{22} - 10^{23}) \text{cm}^{-3}$; thus, $\chi \approx 1$.

Orientation polarization

If the elementary units of a material have a permanent dipole moment, the dipoles will be oriented randomly in the normal case due to the thermal motion, so that the dipole fields cancel on the average. Applying an external field selects a direction along which the dipoles may align. See Figure 6.22. The parallel orientation of all dipoles in the external field is the energetically most favorable one. But thermal motion acts against this. A thermodynamic *equilibrium of partial order* is reached in which a resultant dipole moment per unit of volume arises in the material; the magnitude of the dipole moment is proportional to the external field. Here, we neglect the interactions between the molecules.

If a unit volume of the material contains N molecules, each having a dipole moment **p** at an angle ϑ to the field **E**, then a polarization (a resultant dipole moment per volume) arises in the direction of the field. The polarization is

$$\mathbf{P} = Np\langle\cos\vartheta\rangle\mathbf{e}_z$$

Now, we can calculate the average resulting angle. To do so, we must determine the ratio of the potential energy in the field to the thermal energy. The potential energy is $V(\vartheta) = -\mathbf{p} \cdot \mathbf{E} = -pE\cos\vartheta$. According to *Boltzmann's distribution law* the probability that an orientation corresponding to the potential energy has been taken, is given by the factor $\exp(-V/kT)$.

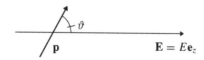

Figure 6.22. Orientation of a dipole with respect to an electric field.

Here, k and T are Boltzmann's constant and the absolute temperature, respectively. The average value for $\cos\vartheta$ is obtained if it is weighted by this probability and integrated over all angles. Furthermore, one has to normalize for the total probability:

$$\langle\cos\vartheta\rangle = \frac{\int e^{-V/kT}\cos\vartheta\,d\Omega}{\int e^{-V/kT}\,d\Omega}$$

For the computation we consider the magnitude of the exponent: we have $V/kT = (-pE/kT)\cos\vartheta \equiv -a\cos\vartheta$. The dipole moment is on the order of $p = e \cdot 10^{-8}$ cm (e is the electron charge, 10^{-8} cm corresponds to the atomic radius). At room temperature, the thermal energy is $kT \approx \frac{1}{40}$ eV, and for the field, we substitute $E = 4000$ V/cm. Hence, the factor is $a \approx 2 \cdot 10^{-3}$. It is small enough to perform a power expansion of the exponential function. We obtain

$$\langle\cos\vartheta\rangle = \frac{\int_0^{2\pi} d\varphi \int_0^{\pi}(1 + a\cos\vartheta)\cos\vartheta\sin\vartheta\,d\vartheta}{\int_0^{2\pi} d\varphi \int_0^{\pi}(1 + a\cos\vartheta)\sin\vartheta\,d\vartheta}$$

and after integration

$$\langle\cos\vartheta\rangle = \frac{a}{3} = \frac{pE}{3kT}$$

This implies that an average molecular dipole moment

$$\langle p\rangle = p\langle\cos\vartheta\rangle = \frac{p^2 E}{3kT}$$

results from the *orientation polarization*. Then, the molecular polarizability is composed of two parts: the displacement polarization γ_{el} and the orientation polarizarion $\gamma_{or} = p^2/3kT$.

The orientation polarization depends linearly on the reciprocal temperature. The displacement polarization is independent of the temperature. Both parts are shown in Figure 6.23.

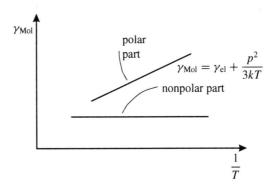

Figure 6.23. The atomic polarizability as a function of the reciprocal temperature.

Since the slope of the straight line γ_{Mol} yields the dipole moment, the molecular dipole moment can be determined from measurements of the polarizability.

Biographical notes

Rudolf Julius Emanuel Clausius, b. Jan. 2, 1822, Köslin (Pomerania)–d. Apr. 24, 1888, Bonn, physicist and professor in Zürich and Würzburg, and from 1869 in Bonn. Clausius was one of the founders of "classical thermodynamics," i.e., both of thermodynamics and of the kinetic theory of gases. In 1850 he proved the compatibility of Carnot's principle with the law of conservation of energy. In 1865 he established the notion of entropy, and for the first time formulated the second law of thermodynamics. It was L. Boltzmann who interpreted this law on the basis of probability calculus. The most noticeable of Clausius' works on electricity is the theory of electrolytic conduction, which can be seen as an early stage of S. Arrhenius' theory of electrolytic dissociation.

7 Electrostatic Energy and Forces in a Dielectric

To calculate the electrostatic energy of a charge distribution in vacuum, we have brought the charges q_i to the positions $\mathbf{r}_i$ in a gedanken experiment, and the necessary work has been calculated. At once, we see that in a dielectric work has to be done not only against the potential but also to build up the polarization of the dielectric.

Let a charge distribution $\rho(\mathbf{r})$ be present in an infinitely extended dielectric. The charge distribution produces a displacement field in the dielectric. We have

$$\operatorname{div} \mathbf{D}(\mathbf{r}) = 4\pi\rho(\mathbf{r})$$

At the point $\mathbf{r}$, the charge distribution $\rho(\mathbf{r})$ is to be altered by introducing a small finite charge density $\delta\rho(\mathbf{r})$ from infinity. The dielectric properties should not be altered. Then, the energy

$$\delta W = \int \delta\rho\,\phi\,dV \tag{7.1}$$

has to be spent against the potential $\phi(\mathbf{r})$.

The general expression for the energy is $W = \frac{1}{2}\int_V \rho\phi\,dV$ (compare equation (1.106)). Therefore, $\delta W = \frac{1}{2}\int(\delta\rho\,\phi + \rho\,\delta\phi)\,dV$. In equation (7.25), we will show that the two expressions in the integrand are equal to each other if the dielectric (i.e., $\epsilon(\mathbf{x})$) is not changed. So, equation (7.1) is derived.

The change in the charge density is connected with a change in the displacement density

$$\operatorname{div} \delta\mathbf{D} = 4\pi\,\delta\rho$$

So, we obtain for the change in energy (7.1)

$$\delta W = \frac{1}{4\pi}\int_V \phi\operatorname{div}\delta\mathbf{D}\,dV$$

With the relation

$$\phi \operatorname{div} \delta \mathbf{D} = \operatorname{div}(\delta \mathbf{D} \phi) - \operatorname{grad} \phi \cdot \delta \mathbf{D} \tag{7.2}$$

we obtain further

$$\delta W = \frac{1}{4\pi} \int_V \operatorname{div}(\delta \mathbf{D}\, \phi)\, dV - \frac{1}{4\pi} \int_V \operatorname{grad} \phi \cdot \delta \mathbf{D}\, dV$$

Applying Gauss' theorem to the first integral with the bounding surface at infinity yields

$$\int_V \operatorname{div}(\delta \mathbf{D}\, \phi)\, dV = \oint_S \delta \mathbf{D} \cdot \phi\, \mathbf{n}\, da = 0$$

since the variation $\delta \mathbf{D}$ tends to zero as $\mathbf{r} \to \infty$.

Then the energy variation becomes

$$\delta W = \frac{-1}{4\pi} \int_V \operatorname{grad} \phi \cdot \delta \mathbf{D}\, dV$$

and with $\mathbf{E} = -\operatorname{grad} \phi$ it follows that

$$\delta W = \frac{1}{4\pi} \int_V \mathbf{E} \cdot \delta \mathbf{D}\, dV$$

The total energy W is obtained by integrating over the dielectric displacement

$$W = \frac{1}{4\pi} \int_V dV \int_0^D \mathbf{E} \cdot \delta \mathbf{D} \tag{7.3}$$

where the charge density $\delta \rho$ is transported from infinity with $\mathbf{D} = 0$ to $\mathbf{r}$ with $\mathbf{D} \neq 0$.

If a linear relation exists in the dielectric between the displacement density $\mathbf{D}$ and the electric field $\mathbf{E}$:

$$\mathbf{D} = \epsilon \mathbf{E}$$

then

$$\mathbf{E} \cdot \delta \mathbf{D} = \mathbf{E} \epsilon \cdot \delta \mathbf{E} = \frac{1}{2} \epsilon\, \delta \mathbf{E}^2 = \frac{1}{2}\, \delta (\mathbf{D} \cdot \mathbf{E})$$

Hence, the total energy is

$$W = \frac{1}{8\pi} \int_V \mathbf{E} \cdot \mathbf{D}\, dV \tag{7.4a}$$

The field then has the *energy density*

$$w = \frac{W}{V} = \frac{1}{8\pi} \mathbf{E} \cdot \mathbf{D} \tag{7.4b}$$

Now, we will calculate the electrostatic energy in a dielectric which contains n conductors with constant potentials ϕ_v, and, furthermore, contains a charge density ρ (see Figure 7.1). The total energy of the electrostatic field is

$$W = \frac{1}{8\pi} \int \mathbf{E} \cdot \mathbf{D}\, dV \tag{7.5}$$

where V is the volume of the dielectric (without the volume of the conductors). It is important to make this point clear. While in equations (7.1) through (7.4a) the volume V included also the conductors, the conductors are now explicitly excluded. Inside the conductors, the field intensity vanishes ($\mathbf{E} = 0$), and consequently, also the dielectric displacement ($\mathbf{D} = 0$). Therefore, it makes no difference whether the volume integration in the equations (7.2) to (7.4a) includes the conductors. With $\mathbf{E} = -\,\text{grad}\,\phi$ and equation (7.2), we obtain

$$W = \frac{1}{8\pi} \int_V \mathbf{E} \cdot \mathbf{D}\, dV = -\frac{1}{8\pi} \int_V \text{grad}\,\phi \cdot \mathbf{D}\, dV$$

$$= -\frac{1}{8\pi} \int_V (\text{div}(\phi\mathbf{D}) - \phi\,\text{div}\,\mathbf{D})\, dV$$

The displacement density is linked to the charge density because

$$\text{div}\,\mathbf{D} = 4\pi\rho$$

and thus,

$$W = \frac{-1}{8\pi} \int_V \text{div}(\phi\mathbf{D})\, dV + \frac{1}{2} \int_V \phi\rho\, dV$$

Applying Gauss' theorem to the first integral for n conductors, one obtains

$$\int_V \text{div}(\phi\mathbf{D})\, dV = \sum_{\nu=1}^{n} \oint_{S_\nu} \phi_\nu \mathbf{D} \cdot \mathbf{n}_i\, da + \oint_S \phi\mathbf{D} \cdot \mathbf{n}_a\, da$$

The volume integral is taken over the entire dielectric. The first surface integral runs over the surface separating the dielectric from the conductors ($\mathbf{n}_i$ points into the conductor). The second integral runs over the surface enclosing the dielectric at infinity ($\mathbf{n}_a$ points outward).

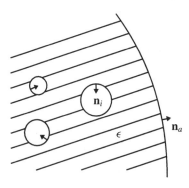

Figure 7.1. Dielectric (hatched) with conductors. The conductors are marked as circles (e.g., cross-sectional areas of wires).

The last integral vanishes because the integration surface lies at infinity, $\phi \sim 1/r$, and $D \sim 1/r^2$; as $r \to \infty$ the integrand tends to zero more rapidly than the area element tends to infinity.

Therefore, the total energy is

$$W = \frac{1}{2} \int_V \phi \rho \, dV - \frac{1}{8\pi} \sum_{\nu=1}^{n} \oint_{S_\nu} \phi_\nu \mathbf{D} \cdot \mathbf{n}_i \, da$$

The potential of the conductors is $\phi_\nu = $ constant because there the charges move until the field intensity vanishes inside the conductors, that is, until $\phi_\nu = $ constant is reached.

The last integral can be written

$$-\oint_{S_\nu} \phi_\nu \mathbf{D} \cdot \mathbf{n}_i \, da = +\phi_\nu \oint_{S_\nu} \mathbf{D} \cdot n_a \, da = 4\pi q_\nu \phi_\nu$$

q_ν is the total charge in the νth conductor:

$$q_\nu = \int_{V_\nu} \rho \, dV = \frac{1}{4\pi} \int_{V_\nu} \operatorname{div} \mathbf{D} \, dV = \frac{1}{4\pi} \int_{S_\nu} \mathbf{D} \cdot d\mathbf{a} = \frac{1}{4\pi} \int_{S_\nu} \mathbf{D}_\nu \cdot \mathbf{n}_a \, da$$

The normal vector to the surface $\mathbf{n}_a = -\mathbf{n}_i$ is directed from the conductor outward, that is, into the dielectric. Then, the total energy is

$$W = \frac{1}{2} \int \rho \phi \, dV + \frac{1}{2} \sum_{\nu=1}^{n} q_\nu \phi_\nu \tag{7.6}$$

Here, the integral yields the electrostatic energy of the charge density ρ, and the sum yields the contribution of the conductors to the energy. The integral is extended over the volume *without* the conductors. If the volume of the conductors is included in the integral, then the sum drops out.

The minimum property of the electrostatic field energy

Let a conductor be embedded in a dielectric. We show that for a constant potential ϕ on the conductor, that is, for a corresponding distribution of the charges on the the surface of the conductor, the field energy

$$W = \frac{1}{8\pi} \int_V \mathbf{E} \cdot \mathbf{D} \, dV$$

is minimal. The field intensity $\mathbf{E}$ satisfies curl $\mathbf{E} = 0$ and the dielectric displacement satisfies Poisson's equation div $\mathbf{D} = 4\pi\rho$.

For the energy W' corresponding to the fields $\mathbf{E}'$ and $\mathbf{D}'$, with curl $\mathbf{E}' = 0$ and div $\mathbf{D}' = 4\pi\rho$, the following relation must be true:

$$W' = \frac{1}{8\pi} \int_V \mathbf{E}' \cdot \mathbf{D}' \, dV \geq W$$

Therefore, the difference $W' - W$ must be greater than or equal to zero

$$\Delta W = W' - W = \frac{1}{8\pi} \int_V (\mathbf{E}' \cdot \mathbf{D}' - \mathbf{E} \cdot \mathbf{D}) \, dV \geq 0$$

We introduce the difference of the fields:

$$\mathbf{E}'' = \mathbf{E}' - \mathbf{E}, \qquad \mathbf{D}'' = \mathbf{D}' - \mathbf{D}$$

Hence,

$$\Delta W = \frac{1}{8\pi} \int_V ((\mathbf{E} + \mathbf{E}'') \cdot (\mathbf{D} + \mathbf{D}'') - \mathbf{E} \cdot \mathbf{D}) \, dV$$

$$= \frac{1}{8\pi} \int_V \mathbf{E}'' \cdot \mathbf{D}'' \, dV + \frac{1}{8\pi} \int_V (\mathbf{E}'' \cdot \mathbf{D} + \mathbf{E} \cdot \mathbf{D}'') \, dV$$

Furthermore, we substitute $\mathbf{D} = \epsilon \mathbf{E}$ and $\mathbf{D}'' = \epsilon \mathbf{E}''$, so that

$$\Delta W = \frac{1}{8\pi} \int_V \mathbf{E}'' \cdot \mathbf{D}'' \, dV + \frac{1}{4\pi} \int_V \mathbf{E} \cdot \mathbf{D}'' \, dV$$

The second integral vanishes. This can be seen in the following way: with $\mathbf{E} = -\,\mathrm{grad}\,\phi$ the second integral becomes

$$\int_V \mathbf{E} \cdot \mathbf{D}'' \, dV = - \int_V \mathrm{grad}\,\phi \cdot \mathbf{D}'' \, dV$$

Utilizing the relation (7.2) and the fact that the true charges do not change, $\mathrm{div}\,\mathbf{D}'' = \mathrm{div}\,\mathbf{D}' - \mathrm{div}\,\mathbf{D} = 0$, we get

$$\int_V \mathbf{E} \cdot \mathbf{D}'' \, dV = - \int_V \mathrm{div}(\phi \mathbf{D}'') \, dV$$

By Gauss' theorem, the volume integral can be split up into an integral over the outer surface s_∞ and an integral over the conductor surface s_L:

$$\int_V \mathbf{E} \cdot \mathbf{D}'' \, dV = - \int_{s_\infty} \phi \mathbf{D}'' \cdot \mathbf{n}_a \, da - \int_{s_L} \phi_L \mathbf{D}'' \cdot \mathbf{n}_i \, da$$

The first integral over the outer boundary vanishes again. With $\mathbf{D}'' = \mathbf{D}' - \mathbf{D}$,

$$\int_V \mathbf{E} \cdot \mathbf{D}'' \, dV = - \int_{s_L} \phi_L \mathbf{D}''_L \cdot d\mathbf{a} = -\phi_L \int_{s_L} (\mathbf{D}'_L - \mathbf{D}_L) \cdot d\mathbf{a} = 0$$

because for the conductor $\mathbf{D}' = \mathbf{D}$: the tangential components of $\mathbf{D} = \epsilon \mathbf{E}$ vanish on the conductor surface because the charges redistribute themselves to compensate for the field. The normal components $D_n = 4\pi \sigma_n$ are the same because σ_n is the same. The energy difference ΔW is therefore greater than or equal to zero, since

$$\Delta W = \frac{1}{8\pi} \int_V \mathbf{E}'' \cdot \mathbf{D}'' \, dV = \frac{\epsilon}{8\pi} \int_V (\mathbf{E}' - \mathbf{E})^2 \, dV \geq 0 \tag{7.7}$$

Therefore, the field energy W is minimal for fields with constant potentials on the conductor surfaces. Now, we may repeat this derivation step by step by interchanging $\mathbf{E}'$ with $\mathbf{E}$. Then we get $\Delta W' = -\Delta W = (e/8\pi) \int_V (\mathbf{E} - \mathbf{E}')^2 dV \geq 0$. A contradiction can be avoided only by taking the equality sign. Thus, $\mathbf{E}' = \mathbf{E}$, $\mathbf{D}' = \mathbf{D}$. This is another proof of the uniqueness of the solution of the field equation with the boundary conditions of constant potentials ϕ_v on conductors having a charge q_v.

Change of energy by a dielectric object

Let the sources of an electric field be fixed in space. Let the electric field $\mathbf{E}_0$ thus be determined by $\rho_0(\mathbf{r})$. Consider a material having the permittivity $\epsilon_0(\mathbf{r})$. Then, the electrostatic energy is

$$W_0 = \frac{1}{8\pi} \int_V \mathbf{E}_0 \cdot \mathbf{D}_0 \, dV \qquad \text{with} \quad \mathbf{D}_0 = \epsilon_0 \mathbf{E}_0$$

Now, let an object of volume V, having a different permittivity, be placed in the field. See Figure 7.2. The field and the permittivity are now $\mathbf{E}(\mathbf{r})$ and $\epsilon(\mathbf{r})$, respectively. For ϵ we have

$$\epsilon = \begin{cases} \epsilon_1(\mathbf{r}) & \text{inside } V_1 \\ \epsilon_0(\mathbf{r}) & \text{outside} \end{cases}$$

Now, the energy is

$$W_1 = \frac{1}{8\pi} \int_V \mathbf{E} \cdot \mathbf{D} \, dV$$

Hence, the energy of the field has changed by

$$\Delta W = W_1 - W_0 = \frac{1}{8\pi} \int_V (\mathbf{E} \cdot \mathbf{D} - \mathbf{E}_0 \cdot \mathbf{D}_0) \, dV$$

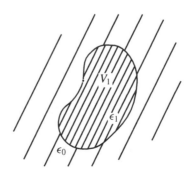

Figure 7.2. A dielectric object (ϵ_1, V_1) is imbedded in a dielectric (ϵ_0).

which we can rewrite as

$$\Delta W = \frac{1}{8\pi} \int_V (\mathbf{E} \cdot \mathbf{D}_0 - \mathbf{D} \cdot \mathbf{E}_0) \, dV + \frac{1}{8\pi} \int_V (\mathbf{E} + \mathbf{E}_0) \cdot (\mathbf{D} - \mathbf{D}_0) \, dV$$

The electric fields can now be represented by the gradient of a potential; we write $\mathbf{E} + \mathbf{E}_0 = -\operatorname{grad} \phi$. So, for the second integral we obtain

$$I = \frac{1}{8\pi} \int_V -\operatorname{grad} \phi \cdot (\mathbf{D} - \mathbf{D}_0) \, dV$$

We use the relation $(\operatorname{grad} \phi) \cdot (\mathbf{D} - \mathbf{D}_0) = \operatorname{div}(\phi(\mathbf{D} - \mathbf{D}_0)) - \phi \operatorname{div}(\mathbf{D} - \mathbf{D}_0)$, and since $\operatorname{div} \mathbf{D} = 4\pi\rho_0 = \operatorname{div} \mathbf{D}_0$ (no new charges have been brought into the field), we have $\phi \operatorname{div}(\mathbf{D} - \mathbf{D}_0) = 0$. Therefore, $(\operatorname{grad} \phi) \cdot (\mathbf{D} - \mathbf{D}_0) = \operatorname{div}(\phi(\mathbf{D} - \mathbf{D}_0))$, and thus,

$$I = -\frac{1}{8\pi} \int_V \operatorname{div}(\phi(\mathbf{D} - \mathbf{D}_0)) \, dV$$

With the help of Gauss' theorem, we may convert this integral into a surface integral,

$$I = -\frac{1}{8\pi} \int_s \phi(\mathbf{D} - \mathbf{D}_0) \cdot \mathbf{n} \, da$$

As has been frequently shown, this integral tends to zero as $r \to \infty$. Hence,

$$\Delta W = \frac{1}{8\pi} \int_V (\mathbf{E} \cdot \mathbf{D}_0 - \mathbf{D} \cdot \mathbf{E}_0) \, dV \qquad \text{(7.8a)}$$

So, for isotropic media, (6.1):

$$\Delta W = -\frac{1}{8\pi} \int_V (\epsilon_1 - \epsilon_0) \, \mathbf{E} \cdot \mathbf{E}_0 \, dV \qquad \text{(7.8b)}$$

If $\epsilon_0 = 1$ and $\epsilon_1 = \epsilon$, then

$$\Delta W = -\frac{1}{2} \int_{V_1} \mathbf{P} \cdot \mathbf{E}_0 \, dV \qquad \text{(7.9)}$$

where the polarization $\mathbf{P}$ is given by $\mathbf{P} = [(\epsilon - 1)/4\pi] \, \mathbf{E}$, (6.1). One should note that $d\mathbf{p} = \mathbf{P} \, dV$ represents the dipole moment contained in the volume element, and according to equation (3.70), $-d\mathbf{p} \cdot \mathbf{E}_0$ is the interaction energy of the dipole with the field $\mathbf{E}_0$.

The integration can be restricted to V_1 because the sample is now in vacuum, so that the integrand vanishes outside V_1. The relation (7.8a) makes it clear that the change of energy due to the introduction of a dielectric comes from the polarization of the dielectric.

General remarks on ponderomotive forces

Since the charges in dielectrics are tied to carriers, the latter are accelerated by forces acting only upon charges. These ponderomotive forces have to be determined. For sake of consistency, we will consider the bulk distribution. Later we will take into consideration also the surface distribution, by taking a simple limit. The ponderomotive forces are determined in the following way: we assume that charged bodies and dielectrics make small changes in their positions and are subject, eventually, to deformations. An element of volume dV will be displaced by the shift $\delta\mathbf{r}$. But, for sake of simplicity, we will assume that there are no compressions or expansions of the volume element dV during the displacement. That is, let dilatation be excluded so that *electrostriction is not taken into account*, and the representation becomes simpler. Electrostriction leads to a condition for $\delta\mathbf{r}$. The variation of the volume V bounded by the surface a can be written as an integral in the following way (see Figure 7.3):

$$\delta V = \int \delta\mathbf{r} \cdot \mathbf{n}\, da = \int_V \operatorname{div} \delta\mathbf{r}\, dV = 0$$

Since this has to be true for any volume,

$$\operatorname{div} \delta\mathbf{r} = 0 \tag{7.10}$$

During its displacement, a volume element should continue to carry its initial charge $\rho\, dV$. Due to the displacement, ρ as well as ϵ (due to the positional change in the dielectric) will change at any spatial position. Namely, we admit inhomogeneous dielectrics; therefore, ϵ is also a function of the position $\mathbf{r}$, $\epsilon = \epsilon(\mathbf{r})$.

The permittivities depend also on the density of the dielectric. But, because we have excluded dilatation and thus density variations in our derivations, these changes of ϵ do not occur.

Denoting the *force density*, that is, the force acting upon the unit volume, by $\mathbf{k}$, the force acting on the volume element is $\mathbf{k}\, dV$. The work done in the displacement is

$$\delta A = \int_V \mathbf{k} \cdot \delta\mathbf{r}\, dV \tag{7.11a}$$

From the physical point of view, the displacements take place because the field energy has not yet reached its static minimal value. Let the work be done entirely at the expense of the energy U of the electric field in space:

$$\delta A = -\delta U \tag{7.11b}$$

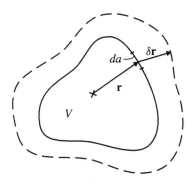

Figure 7.3. Variation of the volume δV by the displacement $\delta\mathbf{r}$ of the surface.

In the following, we will attempt to bring the energy variation δU to the form (7.11a) to identify the expressions replacing **k** with this **k**.

A change in U may occur for two reasons: at the position **r**, there is a change of ϵ, first, and of ρ, second. Due to the displacement, the position **r** is taken by that volume element that had the position $\mathbf{r} - \delta\mathbf{r}$ before, involving the quantities $\epsilon(\mathbf{r} - \delta\mathbf{r})$ and $\rho(\mathbf{r} - \delta\mathbf{r})$. By a series expansion the changes taking place at the position **r** are obtained in the form

$$\delta\epsilon(\mathbf{r}) = \epsilon(\text{new}) - \epsilon(\text{old}) = \epsilon(\mathbf{r} - \delta\mathbf{r}) - \epsilon(\mathbf{r}) = -\operatorname{grad}\epsilon \cdot \delta\mathbf{r} \tag{7.12}$$

$$\delta\rho(\mathbf{r}) = \rho(\text{new}) - \rho(\text{old}) = \rho(\mathbf{r} - \delta\mathbf{r}) - \rho(\mathbf{r}) = -\operatorname{grad}\rho \cdot \delta\mathbf{r}$$
$$= -\operatorname{div}(\rho\delta\mathbf{r}) + \rho\operatorname{div}\delta\mathbf{r} = -\operatorname{div}(\rho\,;\delta\mathbf{r})$$

The quantities $\delta\epsilon$ and $\delta\rho$ correspond to the variations of ϵ and ρ, respectively, that result from the displacement. The related variations of U are denoted by $\delta_\epsilon U$ and $\delta_\rho U$, respectively:

$$\delta U = \delta_\epsilon U + \delta_\rho U$$

where

$$U = \frac{1}{8\pi}\int \frac{\mathbf{D}^2}{\epsilon}\,dV$$

In the first case, the density ρ remains unchanged at any point.

From $\operatorname{div}\mathbf{D} = 4\pi\rho$, we obtain

$$\operatorname{div}\delta_\epsilon\mathbf{D} = 0 \tag{7.13}$$

This is obvious, since the displacement fields $\mathbf{D}_\epsilon$ and $\mathbf{D}_{\epsilon+\delta\epsilon}$ have the same sources ρ everywhere. Hence, their difference $\delta_\epsilon\mathbf{D}$ represents a source-free field. By differentiation one obtains

$$\delta_\epsilon U = -\frac{1}{8\pi}\int \frac{\mathbf{D}^2}{\epsilon^2}\delta\epsilon\,dV + \frac{1}{4\pi}\int \frac{\mathbf{D}}{\epsilon}\cdot\delta_\epsilon\mathbf{D}\,dV$$

It is easy to show that the second integral vanishes. One writes $\mathbf{D}/\epsilon = \mathbf{E} = -\operatorname{grad}\Phi$ and obtains according to the frequently used conversions

$$-\operatorname{grad}\Phi \cdot \delta_\epsilon\mathbf{D} = -\operatorname{div}(\Phi\delta_\epsilon\mathbf{D}) + \Phi\operatorname{div}\delta_\epsilon\mathbf{D}$$

The second term is zero due to (7.13). By applying Gauss' theorem, the space integral of the first term becomes the surface integral $-\int \Phi\delta_\epsilon\mathbf{D} \cdot d\mathbf{a}$. Extending this surface integral over the infinite spherical surface, it becomes zero. Thus, one has

$$\delta_\epsilon U = -\frac{1}{8\pi}\int \mathbf{E}^2\delta\epsilon\,dV \tag{7.14a}$$

Substituting in expression (7.12) for $\delta\epsilon$ yields

$$\delta_\epsilon U = \frac{1}{8\pi}\int \mathbf{E}^2(\operatorname{grad}\epsilon \cdot \delta\mathbf{r})\,dV \tag{7.14b}$$

In the second case, the variation of the equation div $\mathbf{D} = 4\pi\rho$ yields

$$\text{div }\delta_\rho \mathbf{D} = 4\pi \,\delta\rho \qquad (7.15)$$

The corresponding energy variation becomes

$$\delta_\rho U = \frac{1}{4\pi} \int \frac{\mathbf{D}}{\epsilon} \cdot \delta_\rho \mathbf{D}\, dV = -\frac{1}{4\pi} \int \text{grad }\Phi \cdot \delta_\rho \mathbf{D}\, dV \qquad (7.16)$$

As before,

$$- \text{grad }\Phi \cdot \delta_\rho \mathbf{D} = -\text{div}(\Phi \delta_\rho \mathbf{D}) + \Phi \,\text{div}(\delta_\rho \mathbf{D})$$

The space integral of the first term on the right-hand side vanishes again. One has

$$\delta_\rho U = \int \Phi \delta\rho \, dV$$

Using equation (7.12), the integrand becomes

$$\Phi\,\delta\rho = -\Phi\,\text{div}(\rho\delta\mathbf{r}) = -\text{div}(\Phi\rho\delta\mathbf{r}) + \rho\,\text{grad }\Phi \cdot \delta\mathbf{r}$$

Because the space integral over the divergence vanishes again, one obtains

$$\delta_\rho U = -\int (\rho\mathbf{E} \cdot \delta\mathbf{r})\, dV$$

Now, we may write down equation (7.11a) in the required form:

$$\int \{\mathbf{k} \cdot \delta\mathbf{r}\}\, dV = -\delta_\epsilon U - \delta_\rho U = \int \left\{ \left[\rho\mathbf{E} - \frac{1}{8\pi}\mathbf{E}^2(\text{grad }\epsilon) \right] \cdot \delta\mathbf{r} \right\}\, dV$$

A comparison of the left-hand and the right-hand sides demonstrates that one has to set

$$\mathbf{k} = \rho\mathbf{E} - \frac{1}{8\pi}\mathbf{E}^2\,\text{grad }\epsilon \qquad (7.17)$$

The occurrence of the first term on the right-hand side is obvious and follows from the definition of the field intensity. The second term vanishes in the interior of uniform dielectrics. However, in regions of nonuniformity and discontinuity it has to be taken into account, and it then answers a question of practical importance about the pressure at the interface of two dielectrics.

In practice, the electric charges appear nearly exclusively as surface-like distributions. But then the spatial force density $\mathbf{k}$ loses its meaning, and we have to consider those tensions and pressures which are exerted on a surface element by a field. It is clear that the term of the force density $\rho\mathbf{E}$ has to be replaced by $\sigma\mathbf{E}$. The conversion of the term $-(1/8\pi)\mathbf{E}^2 \cdot \text{grad }\epsilon$ is somewhat more complicated in the case when ϵ has a discontinuity at the interface. We proceed in such a way that at the interface we assume a thin boundary layer within which ϵ varies continuously from ϵ_1 to ϵ_2. Finally, we let the thickness of the layer converge to zero. In Figure 7.4, dielectric 2 is below the boundary layer, and the dielectric 1 is above

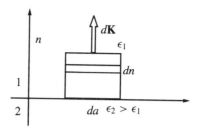

Figure 7.4. Detailed model of a boundary layer. The base of a small cylinder through the boundary layer is *da*, and its height *n* corresponds to the thickness of the boundary layer.

it. The normal vector **n** points into the first insulator. In the boundary layer, we enclose an elementary cylinder of base da and height dn. Its volume is $dV = da\,dn$.

$$d\mathbf{k} = -\frac{1}{8\pi}\mathbf{E}^2\frac{\partial\epsilon}{\partial n}\,dn\,da\,\mathbf{n}$$

According to equation (7.17), the direction of the force is given by $-\,\mathrm{grad}\,\epsilon$, in our case by $-\partial\epsilon/\partial n$, so that the force points in direction of **n** if $\epsilon_2 > \epsilon_1$. In any case, it is perpendicular to the tangential plane. Then, one talks of pressure rather than tension. The pressure acting on the interface is denoted by $p = dk/da$, and we have

$$p = -\frac{1}{8\pi}\int_2^1\mathbf{E}^2\frac{\partial\epsilon}{\partial n}\,dn \tag{7.18}$$

Of course, the integration can be performed only if $\mathbf{E}^2$ is known as a function of n. But, there is the equation

$$\mathbf{E}^2 = \mathbf{E}_t^2 + \mathbf{E}_n^2 = \mathbf{E}_t^2 + \frac{\mathbf{D}_n^2}{\epsilon^2} \tag{7.19}$$

$\mathbf{E}_t$ as well as $\mathbf{D}_n$ are independent of n if the boundary layer is infinitely thin. Substituting (7.19) in (7.18), one obtains

$$p = -\frac{1}{8\pi}\mathbf{E}_t^2\int_2^1\frac{\partial\epsilon}{\partial n}\,dn - \frac{1}{8\pi}\mathbf{D}_n^2\int_2^1\frac{1}{\epsilon^2}\frac{\partial\epsilon}{\partial n}\,dn$$

$$= \frac{1}{8\pi}\mathbf{E}_t^2(\epsilon_2 - \epsilon_1) + \frac{1}{8\pi}\mathbf{D}_n^2\left(\frac{1}{\epsilon_1} - \frac{1}{\epsilon_2}\right) \tag{7.20}$$

Obviously, the limit has been taken already, due to keeping $\mathbf{E}_t$ and $\mathbf{D}_n$ constant. In final form the equation reads

$$p = \frac{1}{8\pi}\mathbf{E}_t^2(\epsilon_2 - \epsilon_1) + \frac{1}{8\pi}\mathbf{D}_n^2\frac{\epsilon_2 - \epsilon_1}{\epsilon_2 \cdot \epsilon_1} \tag{7.21}$$

Each term characterizes a pressure according to which the insulator with the higher permittivity tries to push the other one out of the field.

Example 7.1: Kelvin's absolute electrometer

We wish to measure the force between the charged plates of a capacitor to determine their difference of voltage. See Figure 7.5. We have

$$K_z = -\frac{\partial}{\partial z}\left(\frac{1}{8\pi} \cdot E^2 az\right) = -\frac{E^2 a}{8\pi} \tag{7.22}$$

With $Q = \sigma a = Ea/4\pi$,

$$K_z = -\frac{EQ}{2} \tag{7.23}$$

Therefore, the voltage V also has been determined, because in this case $V = Ez$. This is a particular example of a ponderomotive force.

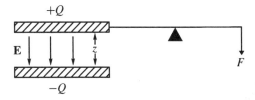

Figure 7.5. Kelvin's absolute electrometer.

Example 7.2: The force on a dielectric in a parallel-plate capacitor

Now we will apply our result of Example 7.1 to the following case: at the edge of a rectangular vacuum capacitor, a dielectric plate is added into the gap. See Figure 7.6. The electrostatic field pulls the plate into the gap, performing work at the same time. We want to calculate this work. Let the distance of the plate be d, and let their areas—taken separately—be equal to a. We consider a transistent state in which the insulator plate covers the area a' of the capacitor. The area still free is $a - a'$. Let the electric charge density be σ; the charge density on the covered one is σ'. It follows that in the still empty gap $|\mathbf{E}| = 4\pi\sigma$, but in the filled space, $|\mathbf{E}'| = 4\pi\sigma'/\epsilon$. Since $\mathbf{E}$ and $\mathbf{E}'$ are parallel to the front area of the insulator plate, $\mathbf{E} = \mathbf{E}'$, and furthermore, $\sigma' = \epsilon\sigma$. The charge Q is no

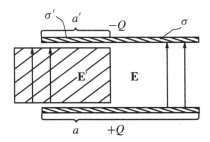

Figure 7.6. A plate is pulled into the capacitor. The charge Q on the capacitor plates remains constant.

longer distributed uniformly over the plates of capacitor. Its density is ϵ-times larger on the covered part than on the uncovered part. The charge Q is now the sum of the two partial charges:

$$Q = \sigma' a' + \sigma(a - a') = \sigma\{(\epsilon - 1)a' + a\}$$

and therefore

$$\sigma = \frac{Q}{(\epsilon - 1)a' + a} \quad |\mathbf{E}| = E_t = 4\pi Q \frac{1}{(\epsilon - 1)a' + a}$$

The pressure pushing the front areas into the gap is ($\mathbf{D}_n = 0$ in this case)

$$p = \frac{1}{8\pi}\mathbf{E}^2(\epsilon - 1) = 2\pi Q^2 \frac{\epsilon - 1}{\{(\epsilon - 1)a' + a\}^2}$$

Now, the element of work done by the field intensity pulling the plate into the capacitor by the space element dV can be expressed as

$$dA = p\, dV = p\, da'd$$

The entire work is obtained by integration

$$A = 2\pi Q^2(\epsilon - 1)d \int_0^a \frac{da'}{\{(\epsilon - 1)a' + a\}^2} = 2\pi Q^2 d\left(\frac{1}{a} - \frac{1}{\epsilon a}\right)$$

Taking into account the capacitances of the empty and the dielectrically filled capacitor, respectively,

$$C = \frac{a}{4\pi d}, \quad C' = \frac{\epsilon a}{4\pi d}$$

then

$$A = \frac{Q^2}{2C} - \frac{Q^2}{2C'} = (U - U')\frac{Q}{2}$$

The energy of the dielectric-filled capacitor is obviously smaller than that of the empty one. The dielectric attempts to creep into the state of minimal energy. During the displacement, the work A is done. A convenient measurement process to determine the dielectric constant of liquids is the *capillary rise method*, which is based on the action of ponderomotive forces. See Figure 7.7. A U-shaped glass tube is filled with the liquid in question, and one of the legs is positioned between the plates of a vertical capacitor. The field intensity $\mathbf{E}$ is directed horizontally, parallel to the surface of the liquid. Due to the pressure

$$p = \frac{1}{8\pi}\mathbf{E}^2(\epsilon - 1)$$

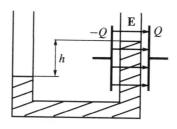

Figure 7.7. On the capillary rise method: A dielectric liquid in an electric field.

the surface rises in this leg. If ρ is the density of the liquid, and h is the level distance, then the hydrostatic pressure is $\rho g h$. The equilibrium condition is

$$\frac{1}{8\pi} \mathbf{E}^2(\epsilon - 1) = \rho g h \tag{7.24}$$

The field intensity E is obtained from the potential difference of the plates in the form $(U_1 - U_2)/d$. So, ϵ may be calculated easily from the measured height h.

Example 7.3: Action of a force for potentials kept constant

In practice, electrodes are frequently kept at a constant potential by using batteries. Shifting the electrodes between the dielectrics, currents are flowing such that the potentials are conserved. Therefore, energy is supplied to or taken away from the field. For a linear dielectric the *energy of a charge density in free space* is

$$W = \frac{1}{2} \int \rho \phi \, dV \rightarrow \delta W = \frac{1}{2} \int (\phi \, \delta \rho + \rho \, \delta \phi) \, dV$$

The two contributions to δW are equal to each other if the dielectrics are not disturbed, that is, if $\epsilon(\mathbf{r})$ is fixed in variations of ρ and ϕ.

This can be proved in the following way:

$$\int_V \delta \rho \, \phi \, dV = \int_V \frac{\operatorname{div} \delta \mathbf{D}}{4\pi} \phi \, dV$$

$$= \frac{1}{4\pi} \int_V \operatorname{div}(\phi \delta \mathbf{D}) \, dV - \frac{1}{4\pi} \int_V \delta \mathbf{D} \cdot \operatorname{grad} \phi \, dV$$

$$= \frac{1}{4\pi} \int_{S_\infty} \phi \delta \mathbf{D} \cdot \mathbf{n} \, da - \frac{1}{4\pi} \int_V \epsilon(\mathbf{r}) \delta \mathbf{E} \cdot \operatorname{grad} \phi \, dV$$

$$= \frac{1}{4\pi} \int \epsilon(\mathbf{r}) \delta \mathbf{E} \cdot \mathbf{E} \, dV = -\frac{1}{4\pi} \int_V \operatorname{grad} \delta \phi \cdot \mathbf{D} \, dV$$

$$= -\frac{1}{4\pi} \int_V \operatorname{div}(\mathbf{D} \delta \phi) \, dV + \frac{1}{4\pi} \int \delta \phi \operatorname{div} \mathbf{D} \, dV$$

$$= -\frac{1}{4\pi} \int_{S_\infty} \mathbf{D} \delta \phi \cdot \mathbf{n} \, da + \frac{1}{4\pi} \int \delta \phi \operatorname{div} \mathbf{D} \, dV = \int \delta \phi \rho \, dV \tag{7.25}$$

As usual, we have neglected the contributions of the surface integrals. The identity of the two contributions $\int \delta \rho \phi \, dV = \int \rho \delta \phi \, dV$ is lost when the dielectric properties are changing $\epsilon(\mathbf{r}) \rightarrow \epsilon(\mathbf{r}) + \delta \epsilon(\mathbf{r})$.

If the dielectric property of an arrangement is changed for fixed potentials, this can proceed in two steps:

(1) Closing the connection between the battery and the electrode ($\delta \rho_1 = 0$), displacing of the dielectric. In this way, the external charge density (e.g., on the plates of the capacitor) is most changed, but the potential is also altered: $\phi \rightarrow \phi + \delta \phi$ because the dielectric has been shifted. Then

$$dW_1 = \frac{1}{2} \int \rho \delta \phi_1 \, dV$$

is the work due to the change in the potential.

(2) The electrodes are again connected with the battery: $\delta\rho_2$ flows from the battery to them and causes a change in the potential $\delta\phi_2 = -\delta\phi_1$ (the old potential ϕ is restored), thus,

$$dW_2 = \frac{1}{2}\int (\delta\rho_2\phi + \delta\phi_2\rho)\,dV = -\int \rho\delta\phi_1\,dV = -2\,dW_1$$

since both contributions are of the same magnitude according to (7.25).

Hence, the total change for constant potentials is

$$dW_\phi = dW_1 + dW_2 = -dW_1$$

For the charge kept constant, thus without an external energy supply, the energy conservation law has to be valid:

$$K\,ds + dW_Q = 0$$

where dW_Q is our dW_1; that is, $dW_Q = dW_1$. The work $K\,ds$ has to be done against a force (e.g., a spring; compare Example 7.4). Then

$$K = -\left(\frac{dW}{ds}\right)_{Q=\text{const}} = \left(\frac{dW}{ds}\right)_{\phi=\text{const}} \tag{7.26}$$

The last step is clear due to the result obtained above.

If a body moves, because of $\epsilon > 1$, into a region of higher field intensity then its energy increases.

Example 7.4: Force on a dielectric in a capacitor at constant potential

For the arrangement shown in Figure 7.8, where $F = al$ is the total area of the plates, $f = ax$ is the area of the plates covered by the dielectric, and d is the separation of the plates, we obtain from the equation for the field energy the relation $W = (1/8\pi)\int \mathbf{E}\cdot\mathbf{D}\,dV$

$$W = \frac{1}{8\pi}E^2\left((F-f) + \epsilon f\right)d = \frac{1}{8\pi}E^2 d\,(F + (\epsilon - 1)f)$$

Then, if the voltage V remains fixed, and f is increased by δf,

$$\delta W = \frac{\epsilon - 1}{8\pi}E^2 d\cdot\delta f$$

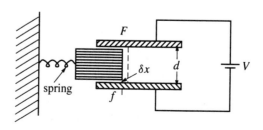

Figure 7.8. Dielectric plate in a parallel-plate capacitor at constant voltage.

For the charge Q, we have

$$Q = \frac{1}{4\pi} E(F - f + \epsilon f)$$

Then the battery does the work (taking into account $V = Ed$)

$$\delta A = V \delta Q = \frac{\epsilon - 1}{4\pi} E^2 d \cdot \delta f$$

Obviously, only half of this energy is needed to increase the field energy. The other half is the work the field has to do against the external force, e.g., a spring, keeping the insulator at equilibrium if it is pulled into the capacitor by δx. Therefore,

$$K = \frac{\epsilon - 1}{8\pi} E^2 d \frac{\partial f}{\partial x}$$

With $f = ax$

$$K = \frac{\epsilon - 1}{8\pi} E^2 da$$

The area $A = d \cdot a$ is the cross sectional area of the insulator (or of the capacitor), and the force is proportional to this area:

$$K = \frac{\epsilon - 1}{8\pi} E^2 A$$

Example 7.5:   Exercise: Liquid in a cylindrical capacitor

Two long, coaxial cylindrical conductors of the radii a and b are submerged perpendicular to a dielectric liquid, as in Figure 7.9. The voltage V is maintained between the two conductors. Show that the liquid rises into the gap by the height

$$h = \frac{2V^2 \chi_e}{\rho g (b^2 - a^2) \ln \dfrac{b}{a}}$$

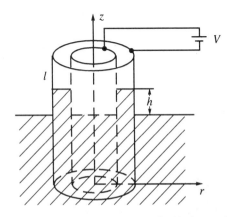

Figure 7.9. Dielectric liquid in a cylindrical capacitor.

The mass density of the liquid is ρ, g is the acceleration of gravity, and χ is the dielectric susceptibility. Let the radius of the inner cylinder be a, the radius of the outer cylinder be b, and the length of the cylinder be l.

Solution   We set up the energy balance, $dW_B = dW_E + dW_M$, with dW_B the change of the work of the battery, dW_E the change of the field energy, and dW_M the change of the mechanical work, if the liquid is lifted by dz. We have determined the electric field in the interior of a coaxial cylindrical capacitor to be $E = 2Q/rl$. Since the potential of the cylindrical capacitor is

$$V = \frac{2Q}{l} \ln \frac{b}{a}$$

we obtain

$$E = \frac{V}{r} \left(\ln \frac{b}{a} \right)^{-1}$$

The electrostatic energy is given by ($d\tau$ = volume element)

$$W_E = \frac{1}{8\pi} \int_0^l \mathbf{E}\,\mathbf{D}\,d\tau \qquad \text{with} \quad \mathbf{D} = \epsilon \mathbf{E}$$

Because the cylinders are partially submerged in the liquid,

$$W_E = \frac{\epsilon}{8\pi} \int_0^{2\pi} d\varphi \int_0^z dz \int_a^b E^2 r\,dr + \frac{1}{8\pi} \int_0^{2\pi} d\varphi \int_z^l dz \int_a^b E^2 r\,dr$$

Substituting for E^2 and integrating, we obtain

$$W_E = \frac{\epsilon V^2}{4 \ln \dfrac{b}{a}} z + \frac{V^2}{4 \ln \dfrac{b}{a}} (l - z)$$

So, the energy change as a function of the height of lift z is

$$dW_E = \frac{\epsilon - 1}{4} \frac{V^2}{\ln \dfrac{b}{a}}\, dz$$

If the liquid is wetting the conductor, the change of the charge density on the inner conductor ($r = a$) is given by

$$\sigma = \frac{D_\epsilon - D_0}{4\pi}$$

with $D_\epsilon = \epsilon E$ and $D_0 = E$. Substituting $E = V/[r \ln(b/a)]$ for $r = a$ yields

$$\sigma = \frac{\epsilon - 1}{4\pi} \frac{V}{a \ln \dfrac{b}{a}}$$

The work of the battery is $dW_B = V\,dQ = V\sigma 2\pi a \cdot dz$, since $2\pi a \cdot dz$ is the area of the inner cylinder. Substituting σ,

$$dW_B = \frac{\epsilon - 1}{2 \ln \dfrac{b}{a}} V^2\, dz$$

The mechanical work is given by $dW_M = gz \cdot dm$ with $dm = \rho(b^2 - a^2)\pi \cdot dz$. Since according to the energy balance $dW_M = dW_B - dW_E$, we obtain

$$dW_M = \frac{\epsilon - 1}{4 \ln \dfrac{b}{a}} V^2 dz$$

Equating the two yields

$$\int_0^h \frac{\epsilon - 1}{4 \ln \dfrac{b}{a}} V^2 dz = \int_0^h zg\rho(b^2 - a^2)\pi \, dz$$

We solve for h:

$$h = \frac{\epsilon - 1}{4\pi} \frac{2V^2}{\ln \dfrac{b}{a}(b^2 - a^2)g\rho}$$

since $\chi = (\epsilon - 1)/4\pi$.

Example 7.6: Exercise: Point charge in a uniform but anisotropic medium

Calculate the potential ϕ of a point charge q in a uniform anisotropic medium with the symmetric dielectric tensor ϵ_{ik} ($\mathbf{D} = \hat{\epsilon}\mathbf{E}$). Determine the modified point charge $q' = q/\sqrt{\det \hat{\epsilon}}$ for the dielectric tensor

$$(\epsilon_{ik}) = \frac{1}{4} \begin{pmatrix} 29 & 3 & -3\sqrt{2} \\ 3 & 29 & 3\sqrt{2} \\ -3\sqrt{2} & 3\sqrt{2} & 26 \end{pmatrix}$$

the transformation to principal axes as well as the excentricity of the equipotential surfaces. Think about materials that may be described by dielectric tensors.

Solution We must find a solution of the system of equations

$$\mathbf{D}(\mathbf{x}) = \hat{\epsilon}\mathbf{E}(\mathbf{x})$$
$$\nabla \cdot \mathbf{D}(\mathbf{x}) = 4\pi q\delta(\mathbf{x})$$
$$\nabla \times \mathbf{E}(\mathbf{x}) = \mathbf{0}$$

Furthermore, $\mathbf{D}$ and $\mathbf{E}$ must vanish for large distances. Introducing the potential according to

$$\mathbf{E}(\mathbf{x}) = -\nabla\phi(\mathbf{x})$$

the last equation is fulfilled automatically, and the first two equations may be combined as

$$\sum_{ij} \epsilon_{ij} \frac{\partial}{\partial x_i} \frac{\partial}{\partial x_j} \phi(\mathbf{x}) = -4\pi q\delta(\mathbf{x}) \tag{7.27}$$

One can assume that this equation takes a simpler form if we pass from the coordinate system $(\mathbf{x})$ to the principal coordinate system $(\mathbf{x}')$ of the matrix $\hat{\epsilon}$. Let $\hat{S} = (S_{ij})$ be an orthogonal matrix which diagonalizes $\hat{\epsilon}$

$$\hat{\epsilon}' = \hat{S}\hat{\epsilon}\hat{S}^{-1} \quad \text{or} \quad \epsilon'_{kl} = \sum_{ij} S_{ki}\epsilon_{ij}S_{lj} =: \epsilon_k \delta_{kl} = \epsilon_k \tag{7.28}$$

Here, we have used the facts that for orthogonal matrices $\hat{S}^{-1} = \hat{S}^T$ and that $\hat{\epsilon}$ is symmetric. Thus, the ϵ_k are the eigenvalues of the matrix $\hat{\epsilon}$. We want to assure that all of them are positive. Substituting,

$$\mathbf{x}' = \hat{S}\mathbf{x}, \qquad x'_i = \sum_j S_{ij}x_j \tag{7.29}$$

and

$$\phi(\mathbf{x}) = \phi'(\hat{S}\mathbf{x}) = \phi'(\mathbf{x}'), \qquad \text{since } \phi(\mathbf{x}) \text{ is a scalar.} \tag{7.30}$$

According to the chain rule

$$\frac{\partial}{\partial x_i} = \sum_k \frac{\partial x'_k}{\partial x_i} \frac{\partial}{\partial x'_k} = \sum_k S_{ki} \frac{\partial}{\partial x'_k} \tag{7.31}$$

because

$$x'_k = \sum_j S_{kj}x_j, \qquad \frac{\partial x'_k}{\partial x_i} = \sum_j S_{kj}\frac{\partial x_j}{\partial x_i} = \sum_j S_{kj}\delta_{ij} = S_{ki}$$

Substituting this into (7.27), we get $\left(\partial/\partial x_j = \sum_l S_{lj}\partial/\partial x'_l\right)$

$$\sum_{ij}\sum_{kl} \epsilon_{ij}S_{ki}S_{lj}\frac{\partial}{\partial x'_k}\frac{\partial}{\partial x'_l}\phi'(\mathbf{x}') = \sum_{kl} \epsilon'_{kl}\frac{\partial}{\partial x'_k}\frac{\partial}{\partial x'_l}\phi'(\mathbf{x}') = \sum_k \epsilon_k \frac{\partial^2}{\partial x'^2_k}\phi'(\mathbf{x}')$$

$$= -4\pi q\delta(\mathbf{x}') \tag{7.32}$$

since the relation $\delta(\mathbf{x}) = \delta(\hat{S}^{-1}\mathbf{x}') = \delta(\mathbf{x}')$ is valid due to $|\det \hat{S}| = 1$.

By the substitution

$$x'_k = \sqrt{\epsilon_k}\, y_k$$

$$\frac{\partial}{\partial x'_k} = \frac{dy_k}{dx'_k}\frac{\partial}{\partial y_k} = \frac{1}{\sqrt{\epsilon}}\frac{\partial}{\partial y_k} \tag{7.33}$$

$$\phi'(\mathbf{x}') = \phi'(x'_k) = \psi\left(\frac{x'_k}{\sqrt{\epsilon_k}}\right) = \psi(y_k)$$

this equation may now be simplified further

$$\sum_k \epsilon_k \frac{1}{\epsilon_k}\frac{\partial^2}{\partial y^2_k}\psi(\mathbf{y}) = -4\pi q\delta(\mathbf{x}') = -4\pi \delta(\sqrt{\epsilon_1}y_1)\delta(\sqrt{\epsilon_2}y_2)\delta(\sqrt{\epsilon_3}y_3) \tag{7.34}$$

Because

$$\delta(\sqrt{\epsilon_1}y_1)\delta(\sqrt{\epsilon_2}y_2)\delta(\sqrt{\epsilon_3}y_3) = \frac{1}{\sqrt{\epsilon_1\epsilon_2\epsilon_3}}\delta(y_1)\delta(y_2)\delta(y_3) = \frac{1}{\sqrt{\det \hat{\epsilon}}}\delta(\mathbf{y}) \tag{7.35}$$

we obtain

$$\nabla^2\psi(\mathbf{y}) = -4\pi \frac{q}{\sqrt{\det \hat{\epsilon}}}\delta(\mathbf{y}) \tag{7.36}$$

This is the ordinary Poisson equation with the charge $q' = q/\sqrt{\det \hat{\epsilon}}$. Therefore, the solution is

$$\psi(\mathbf{y}) = \frac{q}{\sqrt{\det \hat{\epsilon}}} \frac{1}{|\mathbf{y}|} \tag{7.37}$$

and the solution in the principal coordinate system is

$$\phi'(\mathbf{x}') = \frac{q}{\sqrt{\det \hat{\epsilon}}} \frac{1}{\sqrt{x_1'^2/\epsilon_1 + x_2'^2/\epsilon_2 + x_3'^2/\epsilon_3}} \tag{7.38}$$

The potential is constant, if

$$R^2 = \frac{x_1'^2}{\epsilon_1} + \frac{x_2'^2}{\epsilon_2} + \frac{x_3'^2}{\epsilon_3} \tag{7.39}$$

is constant. Hence, the equipotential surfaces are ellipsoids. This expression can be written as

$$R^2 = \mathbf{x}'\hat{\epsilon}'^{-1}\mathbf{x}' = \mathbf{x}\hat{S}^T\hat{S}\hat{\epsilon}^{-1}\hat{S}^{-1}\hat{S}\mathbf{x} = \mathbf{x}\hat{\epsilon}^{-1}\mathbf{x} = \sum_{ij} x_i (\epsilon^{-1})_{ij} x_j \tag{7.40}$$

In the original coordinate system we have

$$\phi(\mathbf{x}) = \frac{q}{\sqrt{\det \hat{\epsilon}}} \frac{1}{\sqrt{\mathbf{x}\hat{\epsilon}^{-1}\mathbf{x}}} = \frac{q}{\sqrt{\det \hat{\epsilon}}} \frac{1}{\sqrt{\sum_{ij} x_{ij} (\epsilon^{-1})_{ij} x_j}} \tag{7.41}$$

Therefore, the equipotential surfaces are ellipsoids. Their principal axes point in the direction of the eigenvectors of $\hat{\epsilon}$. To determine the transformation matrix $\hat{S}$ such that $\hat{\epsilon}' = \hat{S}\hat{\epsilon}\hat{S}^{-1}$ lies in the principal coordinate system, that is, $\hat{\epsilon}'$ is diagonal, we calculate the eigenvalues of the dielectric tensors $\hat{\epsilon}$:

$$(\epsilon_{ik}) = \frac{1}{4} \begin{pmatrix} 29 & 3 & -3\sqrt{2} \\ 3 & 29 & 3\sqrt{2} \\ -3\sqrt{2} & 3\sqrt{2} & 26 \end{pmatrix} \tag{7.42}$$

The eigenvalues are computed from

$$\det(\hat{\epsilon} - \lambda\hat{1}) = \|\hat{\epsilon} - \lambda\hat{1}\| = \begin{Vmatrix} \dfrac{29}{4} - \lambda & \dfrac{3}{4} & -\dfrac{3\sqrt{2}}{4} \\ \dfrac{3}{4} & \dfrac{29}{4} - \lambda & \dfrac{3\sqrt{2}}{4} \\ -\dfrac{3\sqrt{2}}{4} & \dfrac{3\sqrt{2}}{4} & \dfrac{26}{4} - \lambda \end{Vmatrix} = 0 \tag{7.43}$$

$$\Longleftrightarrow \quad -4\lambda^3 + 84\lambda^2 - 576\lambda + 1280 = 0 \tag{7.44}$$

Using the division by polynomials shows that $(\lambda - 5)$ works

$$(-4\lambda^3 + 84\lambda^2 - 576\lambda + 1280) : (\lambda - 5) = -4\lambda^2 + 64\lambda - 256$$

$$\underline{-4\lambda^3 + 20\lambda^2}$$

$$64\lambda^2 - 576\lambda$$

$$\underline{64\lambda^2 - 320\lambda}$$

$$-256\lambda + 1280$$

$$-256\lambda + 1280$$

So, the equation (7.44) becomes equivalent to

$$(\lambda - 5)(-4\lambda^2 + 64\lambda - 256) = 0 \tag{7.45}$$

$\lambda_1 = 5$ is an eigenvalue, or if the second bracket vanishes

$$\lambda^2 - 16\lambda + 64 = 0 \qquad \Longleftrightarrow \qquad \lambda_{2,3} = 8 \pm \sqrt{8^2 - 64}$$

and the remaining eigenvalues are:

$$\lambda_1 = 5, \qquad \lambda_2 = 8, \qquad \lambda_3 = 8 \tag{7.46}$$

Now, for the eigenvalues we calculate the eigenvectors $\mathbf{u}_1$, $\mathbf{u}_2$, $\mathbf{u}_3$, from which we may build up the transformation matrix $\hat{S}$

$$\lambda_1 = 5, \qquad \mathbf{u}_1 = \begin{pmatrix} x_1 \\ x_2 \\ x_3 \end{pmatrix}, \qquad \text{without loss of generality} \quad x_1 = 1$$

$$\hat{\epsilon}\mathbf{u}_1 = 5\mathbf{u}_1 \tag{7.47}$$

$$\Longleftrightarrow \quad \begin{aligned} \frac{29}{4} + \frac{3}{4}x_2 - \frac{3}{4}\sqrt{2}x_3 &= 5 \\ \frac{3}{4} + \frac{29}{4}x_2 + \frac{3}{4}\sqrt{2}x_3 &= 5x_2 \\ -\frac{3\sqrt{2}}{4} + \frac{3\sqrt{2}}{4}x_2 + \frac{26}{4}x_3 &= 5x_3 \end{aligned} \tag{7.48}$$

$$\Longleftrightarrow \quad \begin{aligned} 9 + 3x_2 - 3\sqrt{2}x_3 &= 0 \\ 3 + 9x_2 + 3\sqrt{2}x_3 &= 0 \\ -3\sqrt{2} + 3\sqrt{2}x_2 + 6x_3 &= 0 \end{aligned} \tag{7.49}$$

$\Rightarrow \quad x_2 = -1, x_3 = \sqrt{2}$ and since the eigenvalues are determined only up to a constant:

$$\mathbf{u}_1 = c_1 \begin{pmatrix} 1 \\ -1 \\ \sqrt{2} \end{pmatrix} \tag{7.50}$$

We proceed analogously for the eigenvalue $\lambda = 8$:

$$\begin{aligned} & -3x_1 + & 3x_2 - 3\sqrt{2}x_3 = 0 \\ \Rightarrow \quad & 3x_1 - & 3x_2 + 3\sqrt{2}x_3 = 0 \\ & -3\sqrt{2}x_1 - 3\sqrt{2}x_2 - & 6x_3 = 0 \end{aligned} \tag{7.51}$$

Since the first two equations are linearly dependent it is not sufficient to give $x_1 = 1$. We set $x_3 = 0 \Rightarrow x_2 = 1$ so that

$$\mathbf{u}_2 = c_2 \begin{pmatrix} 1 \\ 1 \\ 0 \end{pmatrix} \tag{7.52}$$

Because $\lambda = 8$ is a zero of multiplicity two of the polynomial (7.44), we use the orthogonality relation between the eigenvectors to determine $\mathbf{u}_3$:

$$\mathbf{u}_1 \cdot \mathbf{u}_3 = 0, \qquad \mathbf{u}_2 \cdot \mathbf{u}_3 = 0$$

$$\Rightarrow \quad \begin{aligned} x_1 - x_2 + \sqrt{2}x_3 = 0 \\ x_1 + x_2 \qquad\quad = 0 \end{aligned}$$

$$\Rightarrow \quad x_1 = -x_2, \qquad x_3 = -\sqrt{2}$$

$$\Rightarrow \quad \mathbf{u}_3 = c_3 \begin{pmatrix} 1 \\ -1 \\ -\sqrt{2} \end{pmatrix} \tag{7.53}$$

Besides the normalization factors we obtain for the transformation matrix $\hat{S}$:

$$\hat{S} = \begin{pmatrix} c_1 & c_2 & c_3 \\ -c_1 & c_2 & -c_3 \\ \sqrt{2}c_1 & 0 & -\sqrt{2}c_3 \end{pmatrix} \tag{7.54}$$

The c_i are determined from $\|\hat{S}\| = 1$ and the fact that the $\mathbf{u}_i$ are normalized to unity:

$$\|\hat{S}\| = c_1 \cdot c_2 \cdot c_3 [-\sqrt{2} - \sqrt{2} - (\sqrt{2} + \sqrt{2})] = 1 \quad \Rightarrow \quad c_1 \cdot c_2 \cdot c_3 = -\frac{1}{4\sqrt{2}} \tag{7.55}$$

$$4c_1^2 = 1 \quad \Rightarrow \quad c_1 = \frac{1}{2}$$

$$2c_2^2 = 1 \quad \Rightarrow \quad c_2 = \frac{1}{\sqrt{2}} \tag{7.56}$$

$$4c_3^2 = 1 \quad \Rightarrow \quad c_3 = -\frac{1}{2}$$

Hence, we obtain for the transformation matrix

$$
\hat{S} = \begin{pmatrix} \dfrac{1}{2} & \dfrac{1}{\sqrt{2}} & -\dfrac{1}{2} \\[2ex] -\dfrac{1}{2} & \dfrac{1}{\sqrt{2}} & \dfrac{1}{2} \\[2ex] \dfrac{1}{\sqrt{2}} & 0 & \dfrac{1}{\sqrt{2}} \end{pmatrix}
\tag{7.57}
$$

One may check easily that

$$
\hat{S}\hat{\epsilon}\hat{S}^{-1} = \begin{pmatrix} 5 & 0 & 0 \\ 0 & 8 & 0 \\ 0 & 0 & 8 \end{pmatrix} \qquad \text{and therefore} \quad \|\hat{\epsilon}\| = 320
$$

The modified point charge is

$$
q' = \frac{q}{8\sqrt{5}}
\tag{7.58}
$$

The equipotential surfaces are described by the ellipsoid $R^2 = \text{constant} = x_1'^2/5 + x_2'^2/8 + x_3'^2/8$ rotationally symmetric about the x_1'-axis. The excentricity in the (x_1', x_2')-plane is calculated according to

$$
c = \sqrt{a^2 - b^2}, \qquad \epsilon = \frac{c}{a} \quad \Rightarrow \quad \epsilon \simeq .78
\tag{7.59}
$$

Introducing plane polar coordinates in the (x_1', x_2')-axis, the equation of the ellipse is

$$
r = \frac{a(1 - \epsilon^2)}{1 + \epsilon \cos\theta} = \frac{3.125}{1 + .78 \cdot \cos\theta}
\tag{7.60}
$$

A rotation about the x_1'-axis yields the equipotential surface. The dielectric tensor

$$
\hat{\epsilon} = \begin{pmatrix} 5 & 0 & 0 \\ 0 & 8 & 0 \\ 0 & 0 & 8 \end{pmatrix}
$$

describes a medium which has a permittivity in the $\mathbf{u}_1$-direction that is different from the permittivity perpendicular to this direction (optically preferred orientation for the propagation of light). Such a medium would be, e.g., quartz glass.

PART III

MAGNETOSTATICS

8 Foundations of Magnetostatics

In this section on magnetostatics, we will treat the phenomena caused by steadily flowing charges. The restriction on *stationary* processes implies that the *partial derivatives* with respect to time vanish in magnetostatics, hence, entirely $\partial/\partial t = 0$. That statement means that there are no time variations; thus, e.g., $\partial\rho/\partial t = 0$ or $\partial\mathbf{j}/\partial t = 0$, etc. Here, ρ and $\mathbf{j}$ are the charge density and the current density, respectively.

On the magnetic field intensity and the magnetic moment

The starting point of electrostatics is the discovery that there are electric charges exerting forces on each other according to Coulomb's law. In magnetostatics there is a similar situation. The striking difference is the absence of a magnetic charge; that is, the magnetic

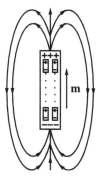

Figure 8.1. Bar magnet and its magnetic dipole field. The assembly out of elementary magnets is shown schematically.

monopole (analogous to the electric charge) does not exist.[1] Magnetic charges occur always only in combination with the opposite magnetic charges, that is, in the form of a *magnetic dipole*. A magnetic rod (Figure 8.1) has a magnetic dipole moment **m** and generates a magnetic field **B**. For the latter, we assume that for a magnetic dipole **m** it will be completely analogous to the electric dipole field (see also Figure 8.2 and exercise 1.4, in particular equation (1.86)), that is,

$$\mathbf{B(r)} = \frac{3(\mathbf{m} \cdot \mathbf{r})\mathbf{r} - r^2 \mathbf{m}}{r^5} \tag{8.1}$$

In a magnetic field **B**, a torque **N** is exerted on a magnetic dipole **m**:

$$\mathbf{N} = \mathbf{m} \times \mathbf{B} \tag{8.2}$$

analogous to the torque of an electric dipole

$$\mathbf{N} = \mathbf{d}_+ \times e\mathbf{E} + \mathbf{d}_- \times (-e\mathbf{E})$$
$$= e(\mathbf{d}_+ - \mathbf{d}_-) \times \mathbf{E} = \mathbf{p} \times \mathbf{E}$$

See Figures 8.3 and 8.4. Based on the postulated relations (8.1) and (8.2), we will show now how to measure magnetic fields, their dimensions in cgs-units, and the dimensions of the magnetic dipole moment **m**. Then, with the help of these methods of measurement, we will empirically and precisely formulate the Biot-Savart law and the force law for current-carrying coils. After these theorems are known, we will derive (8.1) for the magnetic dipole field and (8.2) for the torque of a magnetic dipole in a magnetic field. So, our approach becomes self-consistent, corresponding to practical research. Now we turn to the measurement of $|\mathbf{m}|$ and $|\mathbf{B}_0|$, which becomes possible by determining the products $|\mathbf{m}||\mathbf{B}_0|$ and $|\mathbf{m}||\mathbf{B}_0|^{-1}$ or $|\mathbf{B}_0||\mathbf{m}|^{-1}$:

(a) Reduction of $|\mathbf{m}||\mathbf{B}_0|$ to mechanical units by studying the *oscillations of a magnetic needle* (with magnetic moment **m**) in the field $\mathbf{B}_0$. We have $|\mathbf{N}| = N = mB_0 \sin\varphi$; $\partial \mathbf{L}/\partial t = \mathbf{N}$, where the angular momentum $\mathbf{L} = \theta\dot{\varphi}\mathbf{e}_z$ points out of the paper's plane, while **N** points into it. θ is the moment of inertia. Hence,

$$\frac{d}{dt}\theta\dot{\varphi} = -mB_0 \sin\varphi$$

and for small angles

$$\ddot{\varphi} + \omega^2\varphi = 0, \qquad \omega^2 = \frac{mB_0}{\theta}$$

---

[1]The search for single magnetic charges, i.e., magnetic monopoles, continues to this day. Up to now, however, it has been unsuccessful. We will discuss in detail in Chapter 13 the changes in the theory brought about by the assumption of the existence of magnetic monopoles.

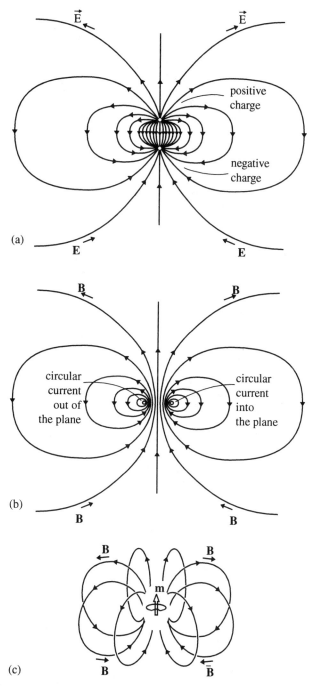

Figure 8.2. A magnetic dipole may be realized also by a current-carrying conductor loop. (Compare Example 9.1.) (a) Electric field lines of a positive and a negative charge. At large distances from the double-charge, the field becomes purely dipole-like. (b) Magnetic field lines of a circular current. At large distances from the circle, this field also becomes purely dipole-like. (c) The same as (b), seen in perspective.

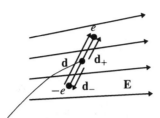

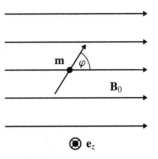

Center of rotation (center of gravity)   ⊛ e_z

Figure 8.3. Electric dipole in an elec-   **Figure 8.4.** Magnetic dipole in
tric field **E**. Here **d** = **d**$_+$ − **d**$_−$ is the   a magnetic induction field **B**$_0$.
vector from the negative charge to the
positive one.

This is the equation for the oscillation of the magnetic needle. If the moment of inertia
θ is known, then together with ω, one knows also the product $m B_0$, whose dimension
is obviously

$$[m B_0] = [\omega^2 \theta] = \frac{g \cdot cm^2}{sec^2} \tag{8.3}$$

(b) The reduction of m/B to mechanical units by the investigation of the relative orientation
of two magnetic needles (Figure 8.5). We consider one magnetic needle **m** itself to be
the origin of the magnetic field **B** corresponding to equation (8.1). At the position $\mathbf{r}_0$ a
magnetic field **B** arises (in the direction of **m**),

$$\mathbf{B} = \frac{3mr_0^2 - mr_0^2}{r_0^5} \frac{\mathbf{r}_0}{r_0} = \frac{2m}{r_0^3} \frac{\mathbf{r}_0}{r_0}$$

Everything is arranged such that **B** is perpendicular to the external field **B**$_0$. This is
possible since the direction of **B**$_0$ alone can be easily determined by means of a magnetic

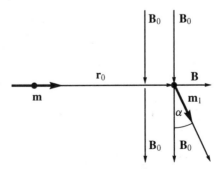

Figure 8.5. On the measurement of the ratio m/B_0. The special geometry **m** $\perp$ **B**$_0$ corresponds
to one of the so-called *Gaussian positions*.

test needle m_1. This test needle orients itself in the combined $(B + B_0)$-field along the resulting field intensity. Then,

$$\tan \alpha = \frac{B}{B_0} = \frac{2m}{r_0^3 B_0} \quad \text{or} \quad \frac{m}{B_0} = \frac{r_0^3 \tan \alpha}{2}$$

and therefore $[m/B] = \text{cm}^3$. Together with (8.3) this leads to

$$[B^2] = \frac{\text{g} \cdot \text{cm}^2}{\text{sec}^2 \text{cm}^3} = \frac{\text{g}}{\text{sec}^2 \text{cm}}$$

$$[m^2] = \frac{\text{g} \cdot \text{cm}^5}{\text{sec}^2}$$

By comparison with our result of Chapter 1, we recognize

$$[B] = [E]$$

The static unit of the magnetic field intensity is called the *Gauss*. (1 Gauss $= 1\sqrt{\text{g}/\text{sec}^2\text{cm}}$). We note that the combined method of measurement of the magnetic fields and the dipole moment is used, e.g., for the measurement of Earth's field B_0 (*Gaussian method*).

Current density and continuity equation

The flow of a quantity of charge dq through the area da in time dt is denoted the *current density*:

$$j = |\mathbf{j}| = \frac{dq}{dt\, da} \tag{8.4a}$$

This time derivative should be the only one which does not vanish, so that a steady current may exist. The current density is a vector whose direction characterizes the direction of motion of the (positive) charge carriers. The integral over the area ($\mathbf{n}$ is the surface normal)

$$I = \int_a \mathbf{j} \cdot \mathbf{n}\, da \tag{8.4b}$$

yields the electric current.

Let us consider a volume V with the charge density ρ. Because we assume *charge conservation* to hold, a variation of the charge density in time is caused by a current flow through the surface of the volume:

$$-\frac{d}{dt} \int_V \rho\, dV = \int_s \mathbf{j} \cdot \mathbf{n}\, da = \int_V \text{div}\, \mathbf{j}\, dV \quad \text{or} \quad \int_V \left(\frac{\partial \rho}{\partial t} + \text{div}\, \mathbf{j} \right) dV = 0$$

where the current $I = a|\mathbf{j}| = a\sigma E$ (more precisely, $I = \int_a |\mathbf{j}|\, da = |\mathbf{j}|a$, if $\mathbf{j}$ is constant over a); the voltage $V = V_1 - V_2 = EL$; the resistance $V/I = EL/(a\sigma E) = L/(\sigma a) =$

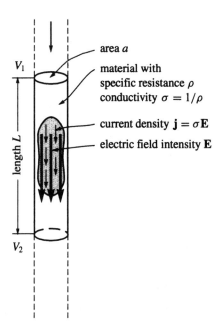

Figure 8.6. Current flow in a conduction wire. The relation between the current density **j** and the electric field intensity **E** ($j = \sigma E$) is implied by Ohm's law $V = IR$.

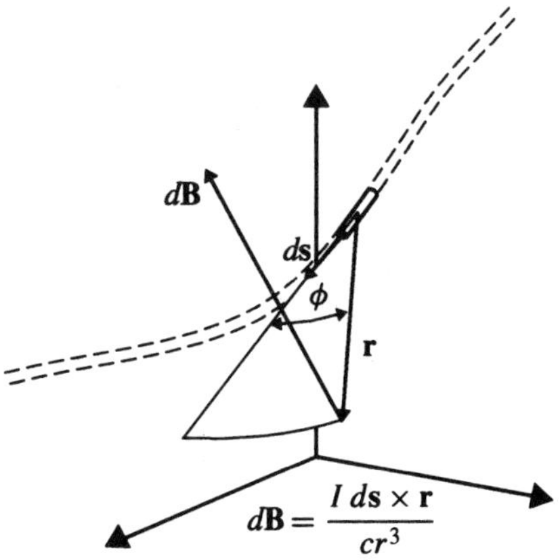

Figure 8.7. The magnetic field at the position **r** produced by a conductor element $d\mathbf{s}$ carrying the current I.

$\rho L/a = R$, with cgs-units esu/(sec · cm^3) = (1/sec)(statvolt/cm); and the current density $\mathbf{j} = \sigma\mathbf{E}$, with rationalized units amp/cm^2 = [1/(ohm · cm)](volt/cm). See Figure 8.6.

Because this is valid for any volume V, we obtain the *continuity equation*

$$\frac{\partial\rho}{\partial t} + \text{div}\,\mathbf{j} = 0 \tag{8.5}$$

Obviously, the continuity equation is equivalent to charge conservation. Since in magnetostatics all partial derivatives with respect to time are set equal to zero,

$$\text{div}\,\mathbf{j} = 0 \tag{8.6}$$

While equation (8.5) is valid in general, (8.6) is true only in magnetostatics. The current flow in a conducting wire, and the definitions and notions occurring in this context are explained and illustrated in Figure 8.7.

The Biot-Savart law

Biot and Savart (1820), Oerstedt (1819), and Ampère (1820–1825) investigated the magnetic fields induced by currents and the correspondingly exerted forces. In general, such measurements are rather complicated. But, after long efforts the observations were summarized in two laws. The first of them reads: A current element $I d\mathbf{s}$ generates a magnetic field at the point P. For this magnetic field the relation

$$d\mathbf{B}(\mathbf{r}) = k \frac{I \cdot d\mathbf{s} \times \mathbf{r}}{r^3} \tag{8.7}$$

has been found *empirically*; that is, the magnetic field produced by an element of a conductor is proportional to the current and (for $d\mathbf{s} \perp \mathbf{r}$) is inversely proportional to the square of the distance. This relation is called the Biot-Savart law. $\mathbf{B}$ is the so-called *magnetic induction* or induction flux density. As we will see later, in vacuum it is identical to the magnetic field intensity (which we will call $\mathbf{H}$ later on).

An external $\mathbf{B}$-field exerts a force on a current-carrying conductor. This force is proportional to the product of the current element and the field intensity (*Ampère's first law*[2]).

$$d\mathbf{F}(\mathbf{r}) = kI \, d\mathbf{s}(\mathbf{r}) \times \mathbf{B}(\mathbf{r}) \tag{8.8}$$

This law also has been found empirically. Both laws (8.7) and (8.8) are, so to speak, the mathematically precise summary of many difficult, detailed experiments. In the Gaussian system of units which we are using and which has been the foundation of our measurement of the magnetic induction, the two constants of proportionality are equal to each other. In this system of units, k is found experimentally to be equal to the reciprocal of the speed of light: $k = 1/c$. The first experiment to determine the constant of proportionality k if q, F, and B are measured in cgs-units was performed by Weber and Kohlrausch in 1856. It gave the remarkable value $k = 1/c$, $c = 3.1 \cdot 10^{10}$ cm/sec, in fact, the velocity of light. Then, equations (8.7) and (8.8) have the form

$$d\mathbf{B}(\mathbf{r}) = \frac{1}{c} \frac{I \, d\mathbf{s} \times \mathbf{r}}{r^3}$$

---

[2]Biot and Savart (1820), and later Ampère (1825), determined the forces between current-carrying coils (cf. the following pages). The laws (8.7) and (8.8) were deduced from this. Apparently the force between conductors is proportional to k^2. In (8.7) and (8.8), k is split up proportionally. The basis for the calculation of magnetic fields was (and still is) the undistorted beat of the individual magnetic fields which are generated by different current-carrying conductors (principle of superposition). According to this, the vector of magnetic induction $\mathbf{B}$ (drainage density) is obtained by representing the total field as a beat of those individual fields that belong to short parts of a conductor thought of as a straight line (see equation (8.7)). In most of such calculations, the mathematical effort required is considerable. This method goes back to the three french physicists J.B. Biot (1774–1862), his assistant F. Savart (1791–1841), and P. de Laplace (1749–1827). Biot and Savart scrutinized the magnetic field of a long, straight wire with the help of a rotating (oscillating) magnetic needle. For magnetic fields with horizontal lines of force, the amount of magnetic induction $\mathbf{B}$ is inversely proportional to the second power of the period of oscillation T; i.e., $B \propto 1/T^2$. In 1821 Biot and Savart found out that the magnetic induction field of a straight wire is inversely proportional to the distance between point under consideration and the wire. They showed the dependency of magnetic induction from the current on a qualitative level only, however, for in 1821 the means for measuring current were not yet in existence. It was about a decade afterwards that useful measuring procedures became known. It was Laplace who calculated the fields of current-carrying wires and formulated (8.7) in a mathematically precise manner. Thus, he could deduce the findings of Biot and Savart. So the authorship for the law which is named after Biot and Savart belongs to Laplace, really.

$$dF(r) = \frac{I}{c}ds(r) \times B(r) \tag{8.9}$$

with $c = 2.998 \cdot 10^{10}$cm/sec. The fact that the value $k = 1/c$ is obtained seems to be rather arbitrary. Later on, in the relativistic formulation of Maxwell's equations this, and therefore, our system of units (the Gaussian system of units) will turn out to be convenient.[3] If, instead of the current, we take a charge q moving at the position $\mathbf{r}' = \mathbf{r}_0$ at the velocity $\mathbf{v}$, then with

$$dB = \frac{I\,ds(r')}{c} \times \frac{(r - r')}{|r - r'|^3}$$

$$= \frac{j(r')\,da\,ds}{c} \times \frac{(r - r')}{|r - r'|^3}$$

$$= \frac{j(r')}{c} \times \frac{(r - r')}{|r - r'|^3}dV'$$

and

$$j(r') = qv(r')\,\delta(r - r_0) \tag{8.10}$$

we obtain

$$B(r) = \frac{q}{c}\int v(r') \times \frac{(r - r')}{|r - r'|^3}\delta(r' - r_0)\,dV'$$

$$= \frac{q}{c}v(r_0) \times \frac{(r - r_0)}{|r - r_0|^3}$$

$$= \frac{1}{c}v(r_0) \times E(r)$$

because the electric field of a point charge q at the position $\mathbf{r}_0$ is given by $E = q(\mathbf{r}-\mathbf{r}_0)/|\mathbf{r}-\mathbf{r}_0|^3$. Note in the last form that the velocity vector $\mathbf{v}$ lies at the position $\mathbf{r}_0$ of the charge, while the vector of the electric field intensity $\mathbf{E}$ is at the position $\mathbf{r}$. The field lines of the B-field are concentric around the conductor. Looking in the direction of the current, they proceed in a clockwise sense (right-hand rule). For that we refer to the following example.

At the position $\mathbf{r}$, a current I is flowing in the length element $\mathbf{s}$, generating a magnetic induction field $d\mathbf{B}$ at the position $\mathbf{r}$.

---

[3]In equations (8.7) and (8.8), $k = \mu_0/4\pi$ or $k = 1$ in the rationlized unit system. Here μ_0 is the magnetic permeability of the vacuum with the value

$$\mu_0 = 4\pi \cdot 10^{-7}\frac{Vs}{Am} = 1.26 \cdot 10^{-6}\frac{Vs}{Am}$$

Example 8.1: Magnetic field of a straight conductor

We will calculate the field of the infinitely long, straight conductor shown in Figure 8.8. From equation (8.9), we have

$$\mathbf{B} = \frac{I}{c} \int_{-\infty}^{+\infty} \frac{d\mathbf{s} \times \mathbf{r}}{r^3}$$

or in scalar form

$$B = \frac{I}{c} \int_{-\infty}^{+\infty} \frac{ds \cdot r \sin\theta}{r^3} = \frac{IR}{c} \int_{-\infty}^{+\infty} \frac{ds}{(s^2 + R^2)^{3/2}}$$

With the substitution $s = R \cdot \sinh u$, with $\cosh^2 u - \sinh^2 u = 1$, the value $2/R^2$ follows for the integral.

Thus, the field of a straight charge-carrying conductor is

$$B = \frac{2I}{cR}$$

Because of the cross product, the field lines point always in the φ-direction (cylindrical coordinates). See Figure 8.9. This can be summarized in vector form in the following way:

$$\mathbf{B} = \frac{2I}{cR} \mathbf{e}_\varphi$$

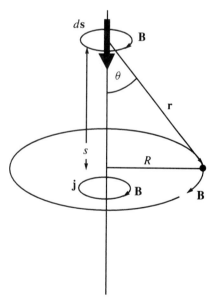

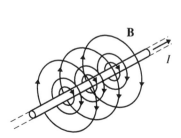

Figure 8.8. Field lines of a linear wire with constant current I.

Figure 8.9. Straight current-carrying conductor and its magnetic induction.

Force between two closed currents

Consider the arrangement shown in Figure 8.10.

We want to calculate the force between the two current-carrying conductors. At the position $\mathbf{r}_1$, the current element $I_2\, d\mathbf{s}_2$ generates a magnetic field, which according to the Biot-Savart law is given by

$$d\mathbf{B}_2 = \frac{I_2\, d\mathbf{s}_2 \times (\mathbf{r}_1 - \mathbf{r}_2)}{c \quad |\mathbf{r}_1 - \mathbf{r}_2|^3}$$

According to Ampère's force law, $\mathbf{B}_2$ exerts a force $d\mathbf{F}_{12}$ on the conductor element $d\mathbf{s}_1$:

$$d\mathbf{F}_{12} = \frac{I_1}{c}(d\mathbf{s}_1 \times d\mathbf{B}_2) = \frac{I_1 I_2}{c^2} \frac{d\mathbf{s}_1 \times (d\mathbf{s}_2 \times (\mathbf{r}_1 - \mathbf{r}_2))}{|\mathbf{r}_1 - \mathbf{r}_2|^3}$$

The total force is obtained by integrating over both conductor loops, that is,

$$\mathbf{F}_{12} = \frac{I_1 I_2}{c^2} \oint_1 \oint_2 \frac{d\mathbf{s}_1 \times (d\mathbf{s}_2 \times (\mathbf{r}_1 - \mathbf{r}_2))}{|\mathbf{r}_1 - \mathbf{r}_2|^3}$$

From the vector formula $\mathbf{a} \times (\mathbf{b} \times \mathbf{c}) = (\mathbf{a} \cdot \mathbf{c})\mathbf{b} - (\mathbf{a} \cdot \mathbf{b})\mathbf{c}$, and with $\mathbf{r}_{12} = \mathbf{r}_1 - \mathbf{r}_2$, we have $d\mathbf{s}_1 \times (d\mathbf{s}_2 \times \mathbf{r}_{12}) = (d\mathbf{s}_1 \cdot \mathbf{r}_{12})d\mathbf{s}_2 - (d\mathbf{s}_1 \cdot d\mathbf{s}_2)\mathbf{r}_{12}$. The first term then can be transformed to

$$d\mathbf{s}_2 \frac{d\mathbf{s}_1 \cdot \mathbf{r}_{12}}{r_{12}^3} = -d\mathbf{s}_2 \left(d\mathbf{s}_1 \cdot \nabla_{12} \frac{1}{r_{12}} \right)$$

where ∇_{12} is the gradient with respect to the difference vector,

$$\nabla_{12} \equiv \left[\frac{\partial}{\partial(x_1 - x_2)}, \frac{\partial}{\partial(y_1 - y_2)}, \frac{\partial}{\partial(z_1 - z_2)} \right]$$

Hence, we obtain for the force

$$\mathbf{F}_{12} = \frac{I_1 I_2}{c^2} \oint_2 d\mathbf{s}_2 \oint_1 d\mathbf{s}_1 \cdot \nabla_{12} \frac{1}{r_{12}} - \frac{I_1 I_2}{c^2} \oint_2 \oint_1 \frac{\mathbf{r}_{12}}{r_{12}^3}(d\mathbf{s}_1 \cdot d\mathbf{s}_2)$$

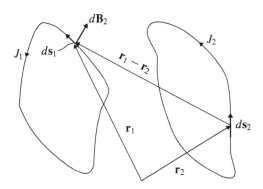

Figure 8.10. Two current-carrying conductor loops.

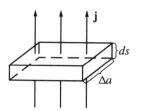

Figure 8.11. $j\Delta V = j\Delta a\Delta s = I\Delta\mathbf{s}$.

The integral over a gradient is path-independent; because the integrals are performed over the closed conductor loop the first line integral vanishes, and the force of the conductor loop 2 on the conductor loop 1 remains:

$$\mathbf{F}_{12} = -\frac{I_1 I_2}{c^2} \oint_2 \oint_1 \frac{\mathbf{r}_1 - \mathbf{r}_2}{|\mathbf{r}_1 - \mathbf{r}_2|^3}(d\mathbf{s}_1 \cdot d\mathbf{s}_2) \tag{8.11}$$

According to (8.8), in an external magnetic field $\mathbf{B}$ an arbitrary conductor carrying the current-density distribution $\mathbf{j}(\mathbf{r})$ experiences the force (see Figure 8.11)

$$\mathbf{F} = \frac{1}{c} \int_{\text{volume}} \mathbf{j}(\mathbf{r}') \times \mathbf{B}(\mathbf{r}')\,dV' \tag{8.12}$$

For a point charge q moving at the velocity $\mathbf{v}(\mathbf{r})$, $\mathbf{j}(\mathbf{r}) = q\mathbf{v}(\mathbf{r}')\delta(\mathbf{r} - \mathbf{r}')$. With equation (8.12), we obtain the *Lorentz force*

$$\mathbf{F} = \frac{q}{c}\mathbf{v} \times \mathbf{B} \tag{8.13}$$

experienced by the charge during the motion in the $\mathbf{B}$-field. The force (8.13) exerts the torque $\mathbf{N}$ on the body:

$$\mathbf{N} = \int \mathbf{r} \times d\mathbf{F} = \frac{1}{c} \int_{\text{volume}} \mathbf{r} \times (\mathbf{j} \times \mathbf{B})\,dV \tag{8.14}$$

Example 8.2: The force between two parallel conductors

We shall calculate the force per unit length exerted by two infinitely long, straight conductors on each other. The conductors are arranged parallel to each other at the distance d. The field $\mathbf{B}_2$ generated by conductor 2 has been calculated in the Example 8.1;

$$B_2 = \frac{2I_2}{cd}$$

From equation (8.6), we obtain for the force

$$d\mathbf{F} = \frac{I_1}{c}d\mathbf{s}_1 \times \mathbf{B}_2$$

Since the two conductors are parallel to each other, the line element $d\mathbf{s}_1$ and the field $\mathbf{B}_2$ are perpendicular to each other. Then, the magnitude of the force per unit length is

$$\frac{dF}{ds_1} = \frac{I_1}{c}\frac{|d\mathbf{s}_1 \times \mathbf{B}_2|}{ds_1} = \frac{I_1}{c}\frac{ds_1|\mathbf{B}_2|}{ds_1} = \frac{2I_1I_2}{c^2d}$$

This force is attractive (repulsive) if the currents are flowing in the same (opposite) direction. This follows immediately from the cross product in Ampère's force law or from equation (8.11). As mentioned in the introduction, an experimental arrangement of this kind was used to deduce the Biot-Savart law and the first Ampère's law via a measurement of the force.

Ampère's second law

Let the current density in a conductor be given by $\mathbf{j}(\mathbf{r}')$. Then, the generated magnetic induction $\mathbf{B}(\mathbf{r})$ is (see Figure 8.12)

$$\mathbf{B}(\mathbf{r}) = \int d\mathbf{B} = \frac{1}{c}\int \frac{\mathbf{j}(\mathbf{r}') \times (\mathbf{r} - \mathbf{r}')}{|\mathbf{r} - \mathbf{r}'|^3}\,dV'$$

Here, (8.7) or (8.9) together with $I\,d\mathbf{s} = \mathbf{j}\,dV$ (Figure 8.11) has been used. This expression is analogous to the equation for the electric field terms of the charge density

$$\mathbf{E}(\mathbf{r}) = \int \rho(\mathbf{r}')\frac{\mathbf{r} - \mathbf{r}'}{|\mathbf{r} - \mathbf{r}'|^3}\,dV'$$

Now, we substitute as follows: $(\mathbf{r} - \mathbf{r}')/|\mathbf{r} - \mathbf{r}'|^3 = -\nabla(1/|\mathbf{r} - \mathbf{r}'|) = \nabla'(1/|\mathbf{r} - \mathbf{r}'|)$. Inserting the gradient yields

$$\mathbf{B}(\mathbf{r}) = \nabla \times \frac{1}{c}\int \frac{\mathbf{j}(\mathbf{r}')}{|\mathbf{r} - \mathbf{r}'|}\,dV'$$

$$\mathbf{B}(\mathbf{r}) = \operatorname{curl}\left(\frac{1}{c}\int \frac{\mathbf{j}(\mathbf{r}')}{|\mathbf{r} - \mathbf{r}'|}\,dV'\right) \tag{8.15}$$

Since the divergence of a rotation always vanishes, we can also write

$$\operatorname{div}\mathbf{B} = 0 \tag{8.16}$$

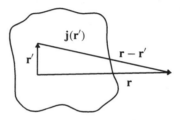

Figure 8.12. For the understanding of Ampère's second law.

analogous to the relation curl $\mathbf{E} = 0$ found previously. This relation implies that there are no isolated sources of the magnetic field, or in other words: *there are no magnetic monopoles.* For the electric field, the situation is different, since div $\mathbf{E} = 4\pi\rho$, where ρ is the electric charge density. The charges ρ represent the electric monopoles. According to the laws of vector calculus, we have for any vector $\mathbf{A}$: $\nabla \times (\nabla \times \mathbf{A}) = \nabla(\nabla \cdot \mathbf{A}) - \Delta\mathbf{A}$. With the help of this relation curl $\mathbf{B}$ can be rewritten as follows:

$$\operatorname{curl} \mathbf{B}(\mathbf{r}) = \nabla \times \left(\nabla \times \frac{1}{c} \int \frac{\mathbf{j}(\mathbf{r}')}{|\mathbf{r} - \mathbf{r}'|} \, dV' \right)$$

$$= \frac{1}{c} \int \nabla \left(\nabla \cdot \frac{\mathbf{j}(\mathbf{r}')}{|\mathbf{r} - \mathbf{r}'|} \right) dV' - \frac{1}{c} \int \Delta \frac{\mathbf{j}(\mathbf{r}')}{|\mathbf{r} - \mathbf{r}'|} \, dV'$$

For the operator Δ, we have $\Delta(1/|\mathbf{r} - \mathbf{r}'|) = -4\pi\delta(\mathbf{r} - \mathbf{r}')$, so that

$$\operatorname{curl} \mathbf{B}(\mathbf{r}) = \frac{1}{c} \int \nabla \left(\nabla \cdot \frac{\mathbf{j}(\mathbf{r}')}{|\mathbf{r} - \mathbf{r}'|} \right) dV' + \frac{4\pi}{c} \int \delta(\mathbf{r} - \mathbf{r}')\mathbf{j}(\mathbf{r}') \, dV'$$

Now, we consider both summands separately. The second one can be integrated at once:

$$\frac{4\pi}{c} \int \delta(\mathbf{r} - \mathbf{r}') \, \mathbf{j}(\mathbf{r}') \, dV' = \frac{4\pi}{c} \mathbf{j}(\mathbf{r})$$

Because ∇ operates on the $\mathbf{r}$-coordinate only, the first term can be rewritten

$$\frac{1}{c} \int \nabla \left(\nabla \cdot \frac{\mathbf{j}(\mathbf{r}')}{|\mathbf{r} - \mathbf{r}'|} \right) dV' = -\frac{\nabla}{c} \int \mathbf{j}(\mathbf{r}') \cdot \nabla' \frac{1}{|\mathbf{r} - \mathbf{r}'|} \, dV'$$

After integration by parts

$$\mathbf{B}(\mathbf{r}) = \frac{\nabla}{c} \int \frac{\nabla' \cdot \mathbf{j}(\mathbf{r}') \, dV'}{|\mathbf{r} - \mathbf{r}'|} - \frac{\nabla}{c} \int \nabla' \cdot \left(\frac{\mathbf{j}(\mathbf{r}')}{|\mathbf{r} - \mathbf{r}'|} \right) dV'$$

$$= \frac{\nabla}{c} \int \frac{\nabla' \cdot \mathbf{j}(\mathbf{r}') \, dV'}{|\mathbf{r} - \mathbf{r}'|} - \frac{\nabla}{c} \int_a \frac{\mathbf{j}(\mathbf{r}') \cdot \mathbf{n}}{|\mathbf{r} - \mathbf{r}'|} \, da' \qquad (8.17)$$

Both terms vanish; the second term because the current distribution is bounded; the first integral because in magnetostatic problems $\nabla \cdot \mathbf{j} = 0$ (equation (8.6)). Thus, we have

$$\operatorname{curl} \mathbf{B}(\mathbf{r}) = \frac{4\pi}{c} \mathbf{j}(\mathbf{r}) \qquad (8.18)$$

This relation is called *Ampère's second law* or *Oerstedt's law*. It is the analog to Gauss' law in electrostatics: div $\mathbf{E} = 4\pi\rho$. In Chapter 13 we will see that equation (8.18) nearly represents one of the Maxwell equations. On the right-hand side of (8.18), only the displacement current vanishing in magnetostatics is still missing. Applying Stokes' theorem and denoting by $\mathbf{n}$ the normal vector to the current-carrying surface, we arrive at another form of Ampère's law:

$$\oint_C \mathbf{B} \cdot d\mathbf{s} = \int_a \operatorname{curl} \mathbf{B} \cdot \mathbf{n} \, da = \frac{4\pi}{c} \int_a \mathbf{j} \cdot \mathbf{n} \, da = \frac{4\pi}{c} I$$

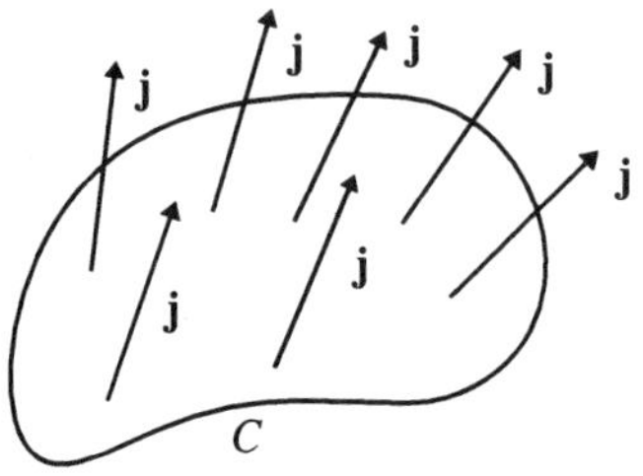

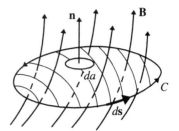

Figure 8.13. **j** is the current density. The surface integral $\int \mathbf{j} \cdot d\mathbf{a} = I$ yields the current flowing through the surface bounded by the curve C.

Figure 8.14. Illustration of the curl $\oint_C \mathbf{B} \cdot d\mathbf{s}$ of the magnetic field.

where I is the total current flowing through the surface. (See Figures 8.13 and 8.14.) Hence,

$$\oint_C \mathbf{B} \cdot d\mathbf{s} = \frac{4\pi}{c} I \tag{8.19}$$

This is the *integral form of Ampère's second law*. Compare with exercise 8.4.

Motion of charged particles in a magnetic field

If a particle of electric charge q is moving at the velocity $\mathbf{v}$ in a uniform magnetic field of intensity $\mathbf{B}$ (Figure 8.15), then according to equation (8.13), the force acting on the particle is

$$\mathbf{F} = \frac{q}{c} \mathbf{v} \times \mathbf{B}$$

It is sufficient to take into account only the velocity component perpendicular to the magnetic field, since the component parallel to $\mathbf{B}$ yields no contribution in the cross product:

$$F = \frac{q}{c} v_s B$$

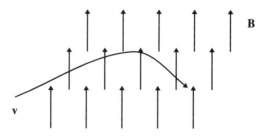

Figure 8.15. A particle in a magnetic field.

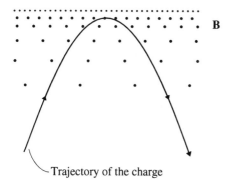

Figure 8.16. Deflection of a charged particle from a magnetic wall.

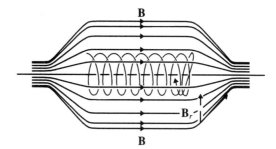

Figure 8.17. Schematic drawing of a magnetic bottle.

At any point of the trajectory the particle experiences the centrifugal force

$$F = \frac{m v_s^2}{a}$$

where a is the radius of curvature of the trajectory. For the forces to balance, we have

$$\frac{m v_s^2}{a} = \frac{q}{c} v_s B$$

From this equation we may calculate the radius of the trajectory as a function of particle momentum, charge, and magnetic field intensity:

$$a = \frac{m v_s c}{q B}$$

Because the force $\mathbf{F} = (q/c)\mathbf{v}_s \times \mathbf{B}$ is perpendicular to the velocity $\mathbf{v}_s$ only the direction of the velocity changes, and not its magnitude. Hence, in a uniform magnetic field the radius of curvature of the trajectory remains constant: $a = \mathrm{const}$.

Thus, in a uniform magnetic field, charged particles move along circular trajectories, and in spirals if the velocity still has a component parallel to $\mathbf{B}$. If the magnetic field is not uniform, the curvature of the trajectory changes from point to point. When a charged particle moves into a magnetic field of increasing intensity, as shown in Figure 8.16, it is reflected. Therefore, a magnetic field exhibiting this shape is called a magnetic mirror. Such mirrors are important for the suspension of plasmas (highly ionized gases at high temperature), because in this manner contact between the plasma particles and the solid outer walls can be avoided. Hence, frequently coils are used to generate the magnetic field, and are twisted in the form of a bent eight, due to stability reasons. No breakthrough has yet been achieved in plasma physics. The longest suspension times are 10^{-13} sec. This principle is utilized also in so-called magnetic bottles (Figure 8.17), e.g., for the confinement of charged particles.

Example 8.3:　Exercise: Helmholtz coils

Two coaxial, parallel, circular conductors of radius R are carrying the current I. See Figure 8.18. At what distance a from each other must the conductors be placed so that in series connection the magnetic field in between them becomes as uniform as possible?

Solution　　According to the Biot-Savart law,

$$\mathbf{B}(\mathbf{r}) = \frac{I}{c} \oint d\mathbf{s}_1 \times \frac{\mathbf{r} - \mathbf{r}_1}{|\mathbf{r} - \mathbf{r}_1|^3} + \frac{I}{c} \oint d\mathbf{s}_2 \times \frac{\mathbf{r} - \mathbf{r}_2}{|\mathbf{r} - \mathbf{r}_2|^3}$$

The origin of the coordinate system is placed at the center of the two coils. The Helmholtz coil is a special current circuit for which the dipole approximation is not valid.

We have (compare Figure 8.18)

$$\mathbf{r}_1 = \left(R \cdot \cos\varphi,\ R \cdot \sin\varphi,\ -\frac{a}{2} \right)$$

$$\mathbf{r}_2 = \left(R \cdot \cos\varphi,\ R \cdot \sin\varphi,\ +\frac{a}{2} \right)$$

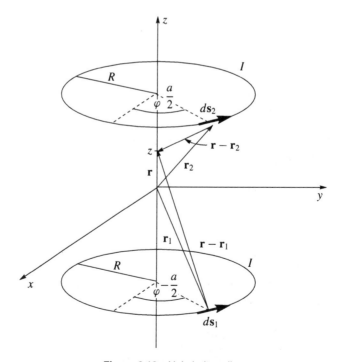

Figure 8.18.　Helmholtz coils.

where $\mathbf{r} = (0, 0, z)$, $\mathbf{j}\,dV' = I\,d\mathbf{s}$, $d\mathbf{s} = ds(-\sin\varphi, \cos\varphi, 0)$, $ds = R\,d\varphi$, and

$$\mathbf{B} = \mathbf{B}_1 + \mathbf{B}_2 = \frac{1}{c}\int\frac{\mathbf{j}(\mathbf{r}_1)\times(\mathbf{r}-\mathbf{r}_1)\,dV_1'}{|\mathbf{r}-\mathbf{r}_1|^3} + \frac{1}{c}\int\frac{\mathbf{j}(\mathbf{r}_2)\times(\mathbf{r}-\mathbf{r}_2)\,dV_2'}{|\mathbf{r}-\mathbf{r}_2|^3}$$

We consider the component along the z-axis. From the definition of the cross product it is clear that for $\mathbf{r} = z\mathbf{e}_z$ the z-components of the cross products $(d\mathbf{s}_1 \times \mathbf{r})_z$ and $(d\mathbf{s}_2 \times \mathbf{r})_z$ vanish. Introducing cylindrical coordinates,

$$\mathbf{r}_{1/2} = \left(R\cos\varphi, R\sin\varphi, \mp\frac{a}{2}\right), \qquad |\mathbf{r}-\mathbf{r}_1| = \sqrt{R^2 + \left(z - \frac{a}{2}\right)^2}$$

$$d\mathbf{s}_1 = d\mathbf{s}_2 = R(-\sin\varphi, \cos\varphi, 0)\,d\varphi, \qquad |\mathbf{r}-\mathbf{r}_2| = \sqrt{R^2 + \left(z + \frac{a}{2}\right)^2}$$

After performing the cross products $d\mathbf{s}_1 \times (\mathbf{r} - \mathbf{r}_1) = -d\mathbf{s}_1 \times \mathbf{r}_1$ and $d\mathbf{s}_2 \times (\mathbf{r} - \mathbf{r}_2) = -d\mathbf{s}_2 \times \mathbf{r}_2$, we have

$$B_z = \frac{I}{c}R^2\int_0^{2\pi}\frac{d\varphi}{\left(R^2 + (z - a/2)^2\right)^{3/2}} + \frac{I}{c}R^2\int_0^{2\pi}\frac{d\varphi}{\left(R^2 + (z + a/2)^2\right)^{3/2}}$$

$$= \frac{I}{c}R^2 2\pi\left[\left(R^2 + \left(z - \frac{a}{2}\right)^2\right)^{-3/2} + \left(R^2 + \left(z + \frac{a}{2}\right)^2\right)^{-3/2}\right]$$

To get a uniform magnetic field we require $B_z(z) \approx B_z(0)$. An expansion of B_z about $z = 0$ yields

$$B_z(z) \approx B_z(0) - \frac{2\pi I R^2}{c}\frac{6(R^2 - a^2)}{(R^2 + (a/2)^2)^{7/2}}\frac{z^2}{2!}$$

So, $R = a$. The pair of coils yields a uniform field along the axis if it is positioned at a distance equal to the radius of the coils. Coils of this kind are called Helmholtz coils.

Example 8.4: Exercise: Magnetic field of a long coil

Calculate the field of a long coil (solenoid, Figure 8.19).

Solution   For a very long coil the field exists practically only in the interior. There it is uniform and parallel to the axis of the coil.

We have

$$\oint \mathbf{B}\cdot d\mathbf{s} = \frac{4\pi}{c}NI \tag{8.20}$$

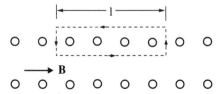

Figure 8.19. Cross section through a coil with a magnetic field.

We use the path of integration shown in the Figure. It comprises N turns of the conductor on a length l; that is, the current flows N times through the domain of integration. Then, for equation (8.20) we obtain

$$\oint \mathbf{B} \cdot d\mathbf{s} = \frac{4\pi}{c} NI$$

Neglecting the magnetic field in the exterior, the integral yields

$$\oint \mathbf{B} \cdot d\mathbf{s} = lB$$

or, in summary,

$$B = \frac{4\pi}{c} \frac{NI}{l} = \frac{4\pi}{c} In$$

$n = N/l$ is the density of turns (the number of turns per cm).

Example 8.5:   Exercise: Magnetic field in a semicircular loop

A long wire is bent in a hairpin-like shape as in Figure 8.20. Determine the magnetic field (exactly) at the point P lying at the center of curvature at the semicircle.

Solution   We calculate the magnetic field with the help of the Biot-Savart law:

$$\mathbf{B}(\mathbf{r}) = \frac{I}{c} \int_{\text{conductor}} \frac{d\mathbf{s}' \times (\mathbf{r} - \mathbf{r}')}{|\mathbf{r} - \mathbf{r}'|^3}$$

Let the coordinate system be chosen so that P is its origin. Then,

$$\mathbf{B}(P) = -\frac{I}{c} \left(\int_{-\infty}^{0} \frac{dx[\mathbf{e}_x \times (x, 0, -R)]}{\sqrt{x^2 + R^2}^3} + \int_{-\pi/2}^{+\pi/2} \frac{d\varphi[R\mathbf{e}_\varphi \times R\mathbf{e}_r]}{R^3} \right)$$

$$-\frac{I}{c} \left(\int_{0}^{-\infty} \frac{dx[\mathbf{e}_x \times (x, 0, +R)]}{\sqrt{x^2 + R^2}^3} \right) = \mathbf{B}_{\rightarrow} + \mathbf{B}_{\uparrow} + \mathbf{B}_{\leftarrow}$$

$$\mathbf{B}_{\rightarrow} = -\frac{I}{c} \int_{-\infty}^{0} \frac{\mathbf{e}_y R}{\sqrt{x^2 + R^2}^3} dx = -\frac{IR}{c} \mathbf{e}_y \left[\frac{x}{R^2 \sqrt{x^2 + R^2}} \right]_{-\infty}^{0} = -\frac{I}{cR} \mathbf{e}_y$$

$$\mathbf{B}_{\leftarrow} = +\frac{IR}{c} \mathbf{e}_y \int_{0}^{-\infty} \frac{dx}{\sqrt{x^2 + R^2}^3} dx = -\frac{I}{cR} \mathbf{e}_y$$

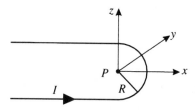

Figure 8.20. Geometry of the current-carrying wire.

$$\mathbf{B}_\uparrow = -\frac{I}{cR}\mathbf{e}_y \int_{-\pi/2}^{+\pi/2} d\varphi = -I/cR\pi\mathbf{e}_y$$

Thus,

$$\mathbf{B}(P) = -\frac{I}{cR}(2+\pi)\mathbf{e}_y$$

Example 8.6: Exercise: Magnetic field in a wire having a hollow cylindrical hole

A long, straight, cylindrical conductor of radius a carries a current I. The conductor has a cylindrical hole (radius R) whose axis is shifted by the distance d from the axis of the conductor. See Figure 8.21. Calculate the magnetic field intensity in the hole with the aid of the principle of superposition.

Solution The cylindrical cavity in the conductor can be described by replacing it by an oppositely directed current of equal density, so that the currents cancel each other at the position of the cavity. Because of the symmetry of the problem we use cylindrical coordinates. Let the position vectors from the center of the conductor (cavity) be $\boldsymbol{\rho}(\boldsymbol{\rho}')$, as in Figure 8.22. The distance between the centers is

$$\boldsymbol{\rho} - \boldsymbol{\rho}' = \mathbf{d} = d\mathbf{e}_x$$

The current density in the conductor is

$$\mathbf{j} = \frac{I}{\pi(a^2 - b^2)}\mathbf{e}_z$$

At first, we calculate the magnetic field generated by this current density in a cylinder of radius a. The field is written in terms of cylindrical coordinates

$$\mathbf{B}(\mathbf{r}) = B_\rho \mathbf{e}_\rho + B_\varphi \mathbf{e}_\varphi + B_z \mathbf{e}_z$$

From the integral form of the Biot-Savart law we infer at once that $B_z = 0$. Also, $B_\rho = B_\rho(\rho)$, $B_\varphi = B_\varphi(\rho)$ for reasons of symmetry. Furthermore, the **B**-field has to vanish along the z-axis: the

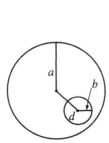

Figure 8.21. Cross section through a wire with a hollow cylindrical hole.

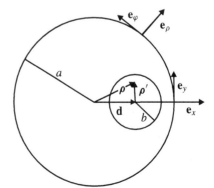

Figure 8.22. On the definition of the coordinates.

current density may be viewed to be built up out of stream filaments, but for any stream filament there is another one symmetric to the z–axis canceling the field there. Then the Biot-Savart law gives

$$\nabla \cdot \mathbf{B} = \frac{1}{\rho}\frac{\partial}{\partial \rho}(\rho B_\rho) = 0 \quad \Rightarrow \quad \rho B_\rho = c_1 \quad \Rightarrow \quad B_\rho = \frac{c_1}{\rho}$$

and from $B_\rho(\rho = 0) = 0$, at once $B_\rho = 0$. Thus, we have $\mathbf{B}(\mathbf{r}) = B_\varphi(\rho)\mathbf{e}_\varphi$. From Ampère's law we obtain

$$\nabla \times \mathbf{B} = \frac{4\pi}{c}\mathbf{j} = \frac{1}{\rho}\frac{\partial}{\partial \rho}(\rho B_\varphi)\mathbf{e}_z \quad \Rightarrow \quad B_\varphi = \frac{2\pi}{c}j\rho$$

So, the field in the interior would be

$$\mathbf{B} = \frac{2\pi j}{c}\rho\mathbf{e}_\varphi = \frac{2\pi j}{c}\rho(\mathbf{e}_z \times \mathbf{e}_\rho) = \frac{2\pi}{c}j(\mathbf{e}_z \times \boldsymbol{\rho}) = \frac{2\pi}{c}\mathbf{j} \times \boldsymbol{\rho}, \qquad (\rho < a)$$

In the hole the field of the fictitious countercurrent is

$$\mathbf{B}' = \frac{2\pi}{c}(-\mathbf{j}) \times \boldsymbol{\rho}', \qquad (\rho' < b)$$

Then, according to the principle of superposition the total field in the hole is

$$\mathbf{B}_L = \mathbf{B} + \mathbf{B}' = \frac{2\pi}{c}\mathbf{j} \times (\boldsymbol{\rho} - \boldsymbol{\rho}') = \frac{2\pi}{c}\mathbf{j} \times \mathbf{d} = \frac{2I \cdot d}{c(a^2 - b^2)}\mathbf{e}_z \times \mathbf{e}_x$$

that is,

$$\mathbf{B}_L = \frac{2I \cdot d}{c(a^2 - b^2)}\mathbf{e}_y$$

The field inside the hole is constant and uniform, and it vanishes if the cylinders are concentric $(d = 0)$.

Example 8.7: Exercise: Particles in a magnetic field—The mass spectrometer

Particles of mass M are singly ionized in an ion source Q and accelerated by the voltage U. They are entering the magnetic field B through a slit S perpendicular to the plane of paper, as in Figure 8.23.

(a) Where do they hit the photoplate?

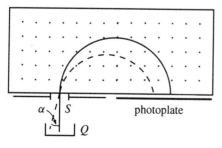

Figure 8.23. Particles from the ion source enter a magnetic field.

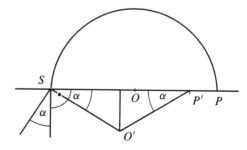

Figure 8.24. Definition of the coordinates.

(b) Where do the particles entering the magnetic field at an angle $\alpha \ll 1$ with respect to the axis hit the photoplate?

(c) How can the mass of the particles be determined by this arrangement?

Solution   Since all particles have the same charge e and have been accelerated by the same voltage U, they also all have the *same energy* $E = eU$. Therefore, their velocity is $v = \sqrt{2E/M}$, and depends on their masses only. The magnetic field exerts a force $\mathbf{F} = ev/c \times \mathbf{B}$ on the particles directed perpendicular to the velocity. Because $\mathbf{v}$ and $\mathbf{B}$ are perpendicular to each other the force has the magnitude $F = (eB/c)v$. Due to the geometry of the arrangement the particles are guided along a circular trajectory. The radius R follows from the balance condition with the centrifugal force $F_z = M\omega^2 R = Mv^2/R$, $R = [c/(eB)]Mv$. So, particles with *equal momentum* travel along circles of the same radii.

(a) The particles entering the magnetic field perpendicularly hit the phototplate in the distance $Y = 2R = [2c/(eB)]Mv = [\sqrt{8}c/(eB)]\sqrt{E}\sqrt{M}$ from the slit S.

(b) The particles entering the magnetic field at an angle α to the perpendicular travel along a circle of radius R whose center O' is shifted from O, the center of the circle for perpendicularly entering particles. See Figure 8.24. As can be inferred from the construction, these particles hit the photoplate in the distance $\Delta' = 2\cos\alpha R$. Thus, the point P' lies closer to the slit than P. For small α, $\Delta' = \Delta - \alpha^2 R + O(\alpha^4)$; that is, an inclination becomes apparent in quadratic deviations only. So, the arrangement acts as focussing.

(c) As we stated in (a) for constant values of U and B the position of the impact point depends on the mass of the particle only. So M may be determined from the impact point. Such an arrangement is called a mass spectrometer.

Biographical notes

Jean Baptiste Biot, b. Apr. 21, 1774, Paris–d. Feb. 3, 1862, Paris. Physicist and astronomer, professor for mathematics in Beauvais, professor for physics at the College de France. Together with F.D. Arago, Biot extended French gauging to the Balearic Islands (1807–1808) and the Shetlands (1817–1818). Later, Biot became professor for astronomy at the Paris Faculté des Sciences and member of all three Parisian academies. He made an important contribution to the findings of rotary polarization (discovered by Arago) by introducing the notion of the specific rotation of the polarization plane. Together with S. Savart, he developed in 1820 Biot-Savart's law for the effect of an electric current on a magnetic pole. Throughout his whole life, he was a supporter and defender of the emission theory of light. (see also p. 516)

Félix Savart, b. June 30, 1791, Mézières–d. March 16, 1841, Paris. French physician and professor of physics at the Collège de France. Savart did a lot of experiments in the field of acoustics. In 1820 he classified the oscillation frequency of tones and (together with Biot) deduced from experiments the basic law of electromagnetism, which is named for them. (see also p. 517) [BR]

Hans Christian Oerstedt, b. Aug. 14, 1777, Rudkobing–d. March 9, 1851. Danish chemist and physicist. Having studied pharmacy at first, Oerstedt received a doctorate in 1799 for his "Dissertatio de forma metaphysica elementaris naturae exterioris," and traveled through Germany and France from 1801 to 1803. In 1806 he became professor for physics, and in 1829 head of the Polytechnical Institute in Copenhagen. In Jena he was won over to Schelling's philosophy of nature by J.W. Ritter. In its spirit he published "Ansicht der chemischen Naturgesetze, durch die neueren Entdeckungen gewonnen" during a second journey (Berlin, 1812, French edition published in Paris together with M. De Serres, titled "Recherches sur l'identité des forces Électriques et chimiques"). In 1819–1820, he discovered the alcaloid piperin in pepper, and the deviation of a magnetic needle by an electric current. Thus, he became a founder of the science of electromagnetism, which was broadened the same year by A.M. Ampère. In 1822 Oerstedt was the first to prove beyond doubt the compressibility of water. Francis Bacon had already tried to find this proof. During a third stay in Paris, Oerstedt constructed the first thermopile together with J. Fourier (1822–1823). His works on philosophy of nature (German: Gesammelte Schriften, 6 volumes, 1850–1853) were much read in the mid-19th century. The unit of measurement of magnetic field strength was named for Oerstedt, both in the electromagnetic, Gaussian system of measures and in the new international system of units; 1 Oe $= (1/4\pi)10^3$ A/m. (see also p. 514) [HE]

Georg Simon Ohm, b. March 16, 1789, Erlangen–d. July 6, 1854, Munich. Physicist. Ohm was a mathematics teacher in Bamberg, Cologne, and at the Berlin war school, and head of the polytechnical school in Nürnberg. In 1849 he became a professor in Munich. Through a series of experiments, in 1826 he found the basic law of electric conduction. The treatise "Die galvanische Kette, mathematisch bearbeitet" (1827, repr. 1965) gives a theory of electric conduction analogous to Fourier's theory of heat conduction. This theory includes the laws of branching points which very often are named after R. Kirchhoff. In 1843 Ohm defined the single tone as a purely sinusoidal oscillation. In 1852–1853, he made important contributions to the problem of interference of light which is polarized in straight lines when going through uniaxial crystals. (see also p. 517) [HE]

Wilhelm Eduard Weber, b. Oct. 24, 1804, Wittenberg–d. June 23, 1891, Göttingen. Together with his brother, Weber published the fundamental "Wellenlehre auf Experimente gegründet" in 1825. In 1831 he was appointed to a chair in Göttingen. Together with C. F. Gauß, he developed both the first electromagnetic telegraph and a number of measuring instruments, e.g., inductorium (1837), electrodynamometer (1840 and 1846), and reflecting galvanometer (1846 and 1852). In 1835 Weber discovered the elastic after-effect, and one year later he and his younger brother, the anatomist Eduard Friedrich (b. March 10, 1806, Wittenberg–d. May 10, 1871, Leipzig), published a work titled "Mechanik der menschlichen Gehwerkzeuge." Being one of the "Göttinger Sieben," he was dismissed in 1837. In 1843 he went to Leipzig as a professor and six years later he was again called to Göttingen. Since 1846 Weber published electrodynamical mensurations which became the basis for international units of measurement. (see also p. 521) [HE]

Friedrich Wilhelm Georg Kohlrausch, b. Oct. 14, 1840, Rinteln–d. Jan. 17 1910, Marburg. This physicist was the son of Rudolf Kohlrausch (b. 1809–d. 1858), who had carried out the "Zurückführung der Stromintensitätsmessungen auf mechanische Maße" (measuring the intensity of current via mechanical measurements) together with Wilhelm Weber in 1856, and so had done basic work for the development of the science of electricity. Kohlrausch had chairs in Göttingen (1866), Zürich (1870), and Strassburg (1888). After Helmholtz' death in 1895, Kohlrausch became principal of the "Physikalisch-Technische Reichsanstalt" in Berlin-Charlottenburg. He did important experiments on the conductivity of electrolytes, found the law of the independent migration of ions (which is named for him), and precisely determined basic electromagnetic quantities. In 1870 Kohlrausch wrote "Leitfaden der praktischen Physik" (Introduction to practical physics). Entitled "Lehrbuch der praktischen Physik" (3 volumes, 1968), this work served as a model for all similar textbooks.

André Marie Ampère, b. Jan. 20, 1775, Lyon–d. June 10, 1836, Marseilles. Mathematician and physicist. At first, Ampère dealt with mathematics. In 1802 he published "Considérations sur la théorie mathématique du jeu." He was a physics teacher in Bourg and Lyon. He then had a chair at the Ecole polytechnique and at the College du France. Later on, he became inspector general of the universities. Starting from crystallography, he reflected on the structure of molecules and arrived at assumptions similar to those Avogadro had uttered three years earlier. After Oersted's discovery of electromagnetism had become known, Ampère, in the autumn of 1820, devoted himself to this new field. He discovered the interaction between current-carrying conductors and named it *electrodynamic*, developed the float rule for the deflection of a magnetic needle by this current, and reduced the phenomena of magnetism to the summation effect of hypothetic "molecular currents." In the treatise "Exposé des nouvelles découvertes sur les magnétisme et l'électricité," published together with J. Babinet in 1822, Ampère proposed the use of an electromagnetic telegraph instead of the electrochemical one suggested by Sömmering. Ampère's fundamental treatise "Sur la théorie mathématique des phènoménes électrodynamiques" (1827) described electrodynamic phenomena as actions at a distance, and served as a model to all similar works. After that, he devoted himself to philosophical topics and published "Essai sur la philosophie des sciences ou Exposition analytique d'une classification naturelle de toutes les connaissances humaines" (2 volumes, 1834). (see also p. 516) [HE]

Hermann Ludwig Ferdinand von Helmholtz, b. Aug. 31, 1821, Potsdam–d. Sept. 8, 1894, Berlin-Charlottenburg. This physicist and physiologist was a medical officer in Potsdam, then an anatomy teacher at Berlin art college, and professor for physiology and anatomy in Königsberg. From 1870 he was professor for physics in Berlin, where he became the first president of the new "Physikalisch-Technische Reichsanstalt" in 1888. Helmholtz was one of the founders of antivitalistic physiology. In 1842 he discovered the nerve fibres to originate in gangliocytes, and in 1850 he measured the rate of speed of nerve stimulation. In the field of music, Helmholtz proved the importance of overtones for the color of a tone and discovered the summation tones (1856). His work "Die Lehre von den Tonempfindungen als physiologische Grundlage für die Theorie der Musik" (1863, 1913) led to his reputation as a founder of modern musical-acoustic research. Without any knowledge of Ch. Babbage's achievements, Helmholtz invented the ophthalmoscope in 1850 (the respective treatise being published in 1851). The same year saw the development of the ophthalmometer, and the year 1857 that of the telestereoscope. Moreover, Helmholtz improved Young's theory of trichromatic vision. In 1873, simultaneously to E. Abbe, Helmholtz determined the theoretical limit of the capacity of microscopes. In the field of physics, Helmholtz explained the impact of the energy principle, and dealt with hydrodynamics of rotational flows (1858). With his experiments in electrodynamics since 1870 he became a pioneer of maxwellian theory, and clearly demonstrated the importance of the principle of least action (1884 to 1894). In 1881 he introduced both the term *free energy* and the notion of elementary charge, and in 1882–1883 he published the important treatise "Zur Thermodynamik chemischer Vorgänge." Helmholtz also dealt with meteorological phenomena. He developed ideas which epistemologically took up Kant's views in several treatises, including "Über die Tatsachen, die der Geometrie zugrunde liegen" (1868), "Das Denken in der Medizin" (1877), "Über die Tatsachen in der Wahrnehmung" (1878), and "Zählen und Messen erkenntnistheoretisch betrachtet" (1887). (see also p. 527)

9 The Vector Potential

Up to now, we have met the description of the magnetic field by two differential equations. These two equations are

$$\text{div } \mathbf{B} = 0 \tag{9.1}$$

$$\text{curl } \mathbf{B} = \frac{4\pi}{c}\mathbf{j} \tag{9.2}$$

If the current density $\mathbf{j}$ vanishes, equation (9.2) demonstrates that the magnetic field can be represented as the gradient of a scalar potential:

$$\mathbf{B} = \text{grad } \phi_M \tag{9.3}$$

since curl grad ($\nabla \times \nabla$) is always equal to zero. From equation (9.1) we conclude that the magnetic field $\mathbf{B}$ may be represented as well in the form ($\nabla \cdot \nabla \times = 0$)

$$\mathbf{B} = \text{curl } \mathbf{A} \tag{9.4}$$

The quantity $\mathbf{A}$ is called a *vector potential*. A scalar potential is determined always only up to an additive constant. The vector potential is determined only up to the gradient of a scalar function, since we can always perform the gauge transformation:

$$\mathbf{A} \rightarrow \mathbf{A} + \text{grad } \varphi$$

without changing the magnetic field. This is similar to the case of the electric potential ϕ, which is also determined only up to a constant. But the electric field intensity is always the same whether the potential is ϕ or $\phi + C$. To achieve uniqueness of the vector potential, we need an additional constraint. We choose div $\mathbf{A} = 0$. This is called the *Coulomb gauge*. The general problems connected with the gauge will be discussed in more detail in another context (Chapter 20). With the vector equation

$$\nabla \times \mathbf{B} = \nabla \times \nabla \times \mathbf{A} = \nabla(\nabla \cdot \mathbf{A}) - \Delta\mathbf{A}$$

a further relationship between the magnetic field and the vector potential can be derived. Due to the Coulomb gauge the first term on the right-hand side vanishes, and we obtain

$$\text{curl } \mathbf{B} = -\Delta\mathbf{A} \tag{9.5}$$

With equation (9.2) we have for the vector potential

$$\Delta \mathbf{A} = -\frac{4\pi}{c}\mathbf{j} \tag{9.6}$$

In equation (8.15) the magnetic field was represented in terms of the curl of a vector field. Hence,

$$\mathbf{A}(\mathbf{r}) = \frac{1}{c}\int \frac{\mathbf{j}(\mathbf{r}')}{|\mathbf{r} - \mathbf{r}'|}\,dV' \tag{9.7}$$

Also, div $\mathbf{A}(\mathbf{r}) = 0$. This follows directly from equation (9.7) by applying Gauss' theorem. This expression for the vector potential corresponds also to a solution of Poisson's equation (9.6) encountered analogously for the scalar potential of electrostatics (see equations (1.31), (1.60), (1.61)). Therefore, the uniqueness theorems discussed in Chapter 2 are also valid.

Example 9.1:   The field of a current-carrying circular conductor

We place a conducting loop in the (x, y)-plane as in Figure 9.1. Let the center coincide with the origin. For the calculation we will use spherical coordinates (r, θ, φ). Then, the current has a component in φ-direction only, as in Figure 9.2.

For the current density, we write

$$\mathbf{j}(\mathbf{r}) = (j_\varphi)\mathbf{e}_\varphi = (I\delta(\cos\theta)\frac{1}{a}\delta(r - a))\mathbf{e}_\varphi \tag{9.8}$$

where a is the radius of the circular conductor. The factor $1/a$ comes from the volume element in spherical coordinates. Note that according to our considerations on the properties of the δ-function in Chapter 1, $\delta(\cos\theta) = (1/|-\sin\theta|)\delta(\theta - \pi/2) = (1/\sin\theta)\delta(\theta - \pi/2)$. So, we can see that $\int j \cdot dV = 2\pi aI$. Due to equation (9.7), we can write for the vector potential

$$\mathbf{A}(\mathbf{r}) = \frac{1}{c}\int \frac{j_\varphi(\mathbf{r}')\mathbf{e}_\varphi}{|\mathbf{r} - \mathbf{r}'|}\,dV' \tag{9.9}$$

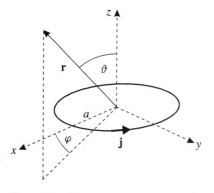

Figure 9.1. The geometry of the problem.

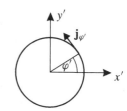

Figure 9.2. Top view of the circular conductor.

Since the unit vectors in spherical coordinates are not constant, for the integration it is suitable to go to Cartesian coordinates. We have placed the conductor in the (x, y)-plane so that the current density has only (x, y)-components. Because of equation (9.8), the same holds also for the vector potential,

$$A_x(\mathbf{r}) = \frac{1}{c} \int \frac{j_x(\mathbf{r}')}{|\mathbf{r} - \mathbf{r}'|} dV'$$

$$A_y(\mathbf{r}) = \frac{1}{c} \int \frac{j_y(\mathbf{r}')}{|\mathbf{r} - \mathbf{r}'|} dV' \tag{9.10}$$

$$A_z(\mathbf{r}) = 0$$

Cartesian coordinates and spherical coordinates are related according to

$$A_x = -A_\varphi \sin\varphi, \qquad j_x = -j_\varphi(\mathbf{r}') \sin\varphi'$$

$$A_y = A_\varphi \cos\varphi, \qquad j_y = j_\varphi(\mathbf{r}') \cos\varphi'$$

We are free to choose the origin of the angle φ: setting $\varphi = 0$, then $A_x = 0$, and we need to consider only the y–component. Since the problem is cylindrically symmetric, for the observation point we can set always $\varphi = 0$. Thus,

$$A_y = A_\varphi = \frac{1}{c} \int \frac{j_\varphi \cos\varphi'}{|\mathbf{r} - \mathbf{r}'|} dV' \tag{9.11}$$

The unit vectors $\mathbf{r}/r$ and $\mathbf{r}'/r'$ can be expressed by the angles of the polar coordinates, $\mathbf{r}/r = \{\sin\theta\cos\varphi, \sin\theta\sin\varphi, \cos\theta\}$ and correspondingly, $\mathbf{r}'/r' = \{\sin\theta'\cos\varphi', \sin\theta'\sin\varphi', \cos\theta'\}$. Substituting equation (9.8) into (9.11), then (with $\varphi = 0$)

$$A_\varphi = \frac{I}{ac} \int \frac{\delta(\cos\theta')\delta(r' - a)\cos\varphi' \cdot r'^2 \sin\theta' dr' d\theta' d\varphi'}{\sqrt{r^2 + r'^2 - 2rr'(\sin\theta\sin\theta'\cos\varphi' + \cos\theta\cos\theta')}}$$

and after the integration over the two δ-functions

$$A_\varphi = \frac{Ia}{c} \int \frac{\cos\varphi' d\varphi'}{\sqrt{r^2 + a^2 - 2ar\sin\theta\cos\varphi'}}$$

Restricting our discussion to large distances from the conducting loop, $r \gg a$, we may expand the square root:

$$A_\varphi \approx \frac{Ia}{cr} \int_0^{2\pi} \cos\varphi' \left(1 - \frac{1}{2}\left(\frac{a^2}{r^2} - \frac{2a}{r}\sin\theta\cos\varphi'\right)\right) d\varphi'$$

$$A_\varphi \approx \frac{Ia^2\pi}{c} \frac{\sin\theta}{r^2} \tag{9.12}$$

The two other components A_θ and A_r vanish (because of the special choice of $\varphi = 0$ and equations (9.10)). Now, from $\mathbf{B} = \text{curl}\,\mathbf{A}$ we calculate the components of the $\mathbf{B}$-field. With $A_\varphi = A_r = 0$ we obtain for the components of the curl in spherical coordinates, according to the formula known from the lectures on mechanics (Chapter 11),

$$\nabla \times \mathbf{A} = \frac{1}{r^2 \sin\theta} \begin{vmatrix} \mathbf{e}_r & r\mathbf{e}_\varphi & r\sin\theta\mathbf{e}_\varphi \\ \partial/\partial r & \partial/\partial\theta & \partial/\partial\varphi \\ A_r & rA_\theta & r\sin\theta A_\varphi \end{vmatrix}$$

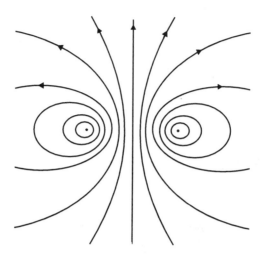

Figure 9.3. B-field of a circular conducting loop.

Hence,

$$(\text{curl }\mathbf{A})_r = \frac{1}{r\sin\theta}\frac{\partial}{\partial\theta}(\sin\theta\, A_\varphi), \qquad (\text{curl }\mathbf{A})_\theta = -\frac{1}{r}\frac{\partial}{\partial r}(r A_\varphi), \qquad (\text{curl }\mathbf{A})_\varphi = 0$$

Substituting equation (9.12), then

$$B_r = 2\frac{I\pi a^2}{c}\frac{\cos\theta}{r^3}, \qquad B_\theta = \frac{I\pi a^2}{c}\frac{\sin\theta}{r^3}, \qquad B_\varphi = 0 \tag{9.13}$$

This field is analogous to the field of the electrostatic dipole (compare this equation to Exercise 1.4). See Figure 9.3. The circular conductor corresponds to a (magnetic) dipole with dipole moment

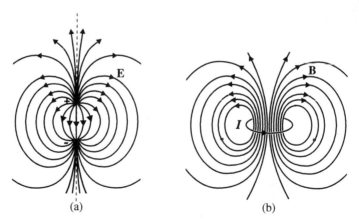

Figure 9.4. Comparison of the electric (a) and magnetic (b) dipole fields: Far from the dipoles both fields are identical.

$m = (I/c)\pi a^2$, where πa^2 is the area enclosed by the conductor. This fact is of general importance, as we will see in Chapter 10.

From this exercise we learn that a conducting loop represents a magnetic dipole, at least at large distance from the loop. See Figure 9.4. There, the field of the loop is identical with the field of a bar magnet having the moment $m = (I/c)\pi a^2$.

Example 9.2: Exercise: Magnetic field of a charged rotating sphere

A sphere of radius R with a charge q uniformly distributed on its surface is rotating about a diameter at a constant angular velocity $\boldsymbol{\omega}$ as in Figure 9.5. Calculate the vector potential $\mathbf{A}$ and the magnetic induction inside and outside the sphere.

Solution The vector potential is given by $\mathbf{A}(\mathbf{r}) = (1/c) \int \mathbf{j}(\mathbf{r}')/|\mathbf{r} - \mathbf{r}'| \, dV'$. The current density and the charge density are given by $\mathbf{j} = \rho\mathbf{v}$, $\mathbf{v} = \boldsymbol{\omega} \times \mathbf{r}$, and $\rho = \sigma\delta(r - R)$, respectively, with $\sigma = q/(4\pi R^2)$.

Thus,

$$\mathbf{j}(\mathbf{r}') = \frac{q}{4\pi R^2}\delta(r' - R)(\boldsymbol{\omega} \times \mathbf{r}')$$

and the vector potential is

$$\mathbf{A}(\mathbf{r}) = \frac{q}{c4\pi R^2}\boldsymbol{\omega} \times \mathbf{F}(\mathbf{r})$$

with the abbreviation

$$\mathbf{F}(\mathbf{r}) = \int \frac{\delta(r' - R)}{|\mathbf{r} - \mathbf{r}'|}\mathbf{r}' dV'$$

Since the integral over the sphere is rotationally symmetric about the direction $\mathbf{r}$, $\mathbf{F} = f \cdot \mathbf{r}/r$. We take the $\mathbf{r}$-direction as the z'-axis; then θ' is the angle between $\mathbf{r}$ and $\mathbf{r}'$. (See Figure 9.6.) With dV' in spherical coordinates and integration over φ',

$$F = \int_0^\pi d\theta' \int_0^\infty dr' \frac{\delta(r' - R)r' \cos\theta' 2\pi r'^2 \sin\theta'}{(r^2 + r'^2 - 2rr'\cos\theta')^{1/2}}$$

$$= \int_0^\pi d\theta' \frac{2\pi R^3 \cos\theta' \sin\theta'}{(r^2 + R^2 - 2rR\cos\theta')^{1/2}}$$

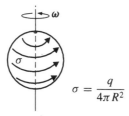

$$\sigma = \frac{q}{4\pi R^2}$$

Figure 9.5. Rotating charged sphere.

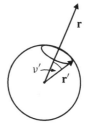

Figure 9.6. On the definition of the coordinates.

We substitute $\cos\theta' = \xi$, leading to

$$F = -2\pi R^3 \int_{+1}^{-1} \frac{\xi \, d\xi}{(r^2 + R^2 - 2rR\xi)^{1/2}}$$

This is an integral of the type

$$I = \int \frac{\xi \, d\xi}{(A - B\xi)^{1/2}}$$

where $A = r^2 + R^2$; $B = 2r \cdot R$, with the solution

$$I = -\frac{2}{3B^2}(2A + B\xi)(A - B\xi)^{1/2} + c$$

We obtain

$$F = \frac{2\pi R}{3r^2}((r^2 + R^2 - rR)(r + R) - (r^2 + R^2 + rR)|r - R|)$$

Inside the sphere, $r < R$, and hence, $|r - R| = R - r$ and

$$F = \frac{4\pi}{3}Rr$$

Outside the sphere,

$$F = \frac{4\pi}{3}\frac{R^4}{r^2}$$

Thus, the vector potential is

$$\mathbf{A} = \begin{cases} \dfrac{q}{3cR}\boldsymbol{\omega} \times \mathbf{r}, & r < R \\[2mm] \dfrac{qR^2}{3c}\boldsymbol{\omega} \times \dfrac{\mathbf{r}}{r^3}, & r > R \end{cases}$$

We have $\mathbf{B} = \operatorname{curl} \mathbf{A}$, and for a constant angular velocity $\boldsymbol{\omega}$

$$\mathbf{B} = \frac{q}{3cR}(\boldsymbol{\omega}(\nabla \cdot \mathbf{r}) - \boldsymbol{\omega} \cdot (\nabla \mathbf{r})), \qquad r < R$$

Let us consider the expression $\boldsymbol{\omega} \cdot (\nabla \mathbf{r})$ in more detail. Written in components, the kth component of this term reads $\sum_i \omega_i (\partial/\partial x_i)x_k$. The 3×3 matrix $A_{ik} = (\partial/\partial x_i)x_k$ appearing in this expression is written frequently in dyadic form (without dot and without cross!), $A_{ik} = (\nabla \mathbf{r})_{ik} = (\partial/\partial x_i)x_k = \delta_{ik}$.

Since the dyad (a tensor of the rank 2) $\nabla \mathbf{r} = \mathbf{1}$, or more explicitly,

$$\begin{pmatrix} \dfrac{\partial x}{\partial x} & \dfrac{\partial y}{\partial x} & \dfrac{\partial z}{\partial x} \\[2mm] \dfrac{\partial x}{\partial y} & \dfrac{\partial y}{\partial y} & \dfrac{\partial z}{\partial y} \\[2mm] \dfrac{\partial x}{\partial z} & \dfrac{\partial y}{\partial z} & \dfrac{\partial z}{\partial z} \end{pmatrix} = \begin{pmatrix} 1 & 0 & 0 \\ 0 & 1 & 0 \\ 0 & 0 & 1 \end{pmatrix}$$

and $\nabla \cdot \mathbf{r} = 3$, for $r < R$ we have

$$\mathbf{B} = \frac{2q}{3cR}\boldsymbol{\omega}$$

For $r > R$ we obtain

$$\mathbf{B} = \frac{qR^2}{3c}\left(\boldsymbol{\omega}\left(\nabla \cdot \frac{\mathbf{r}}{r^3}\right) - (\boldsymbol{\omega} \cdot \nabla)\frac{\mathbf{r}}{r^3}\right)$$

Since $\nabla(\mathbf{r}/r^3) = 1/r^3 - 3\mathbf{r}/r^5$ and $\nabla \cdot \mathbf{r}/r^3 = (3-3)/r^3 = 0$ for $r > R$:

$$\mathbf{B} = \frac{qR^2}{3c} \cdot \frac{3\left(\boldsymbol{\omega} \cdot \dfrac{\mathbf{r}}{r}\right)\dfrac{\mathbf{r}}{r} - \boldsymbol{\omega}}{r^3} \equiv \frac{3(\mathbf{m} \cdot \mathbf{r})\mathbf{r} - r^2\mathbf{m}}{r^5}$$

So, inside the sphere the **B**-field is uniform and proportional to ω; this is plausible. In the exterior the sphere behaves like a dipole with the moment

$$\mathbf{m} = \frac{qR^2\boldsymbol{\omega}}{3c}$$

This serves as a simple model for the magnetic field of the Earth.

Example 9.3: Exercise: The vector potential of parallel conductors

Calculate the vector potential of two straight, infinitely long conductors carrying steady antiparallel currents I. Consider also the case of parallel currents. How can a well-defined convergent expression be achieved?

Solution   The vector potential is

$$\mathbf{A}(\mathbf{r}) = \frac{1}{c}\int_{\text{conductor}} \frac{\mathbf{j}(\mathbf{r}')}{|\mathbf{r} - \mathbf{r}'|}dV' \tag{9.14}$$

We utilize the symmetries of the arrangement and choose $\mathbf{r} = (x, y, z = 0)$. With $\mathbf{j}\,dV' \rightarrow I\,dz'\mathbf{e}_z$, equation (9.14) becomes

$$\mathbf{A}(\mathbf{r}) = \frac{I}{c}\left[\int_{-L}^{L} \frac{dz'}{\sqrt{(x+d)^2 + y^2 + z'^2}} - \int_{-L}^{L} \frac{dz'}{\sqrt{(x-d)^2 + y^2 + z'^2}}\right]\mathbf{e}_z \tag{9.15}$$

Taking into account that the integrand is an even function of z', and introducing the abbreviations

$$\begin{aligned}
\varrho_1^2 &= (x+d)^2 + y^2 \\
\varrho_2^2 &= (x-d)^2 + y^2
\end{aligned} \tag{9.16}$$

we can write

$$\mathbf{A}(\mathbf{r}) = \frac{2I}{c}\left[\int_{0}^{L} \frac{dz'}{\sqrt{\varrho_1^2 + z'^2}} - \int_{0}^{L} \frac{dz'}{\sqrt{\varrho_2^2 + z'^2}}\right]\mathbf{e}_z \tag{9.17}$$

By a clever substitution the integral can be calculated simply. Setting $z' = \varrho_1 \sinh u$, together with $\sqrt{\varrho_2^2 + z'^2} = \varrho_1 \cosh u$, the unpleasant root expression vanishes, and we obtain for the integral

$$\int_{0}^{L} \frac{dz'}{\sqrt{\varrho_i^2 + z'^2}} = \operatorname{arcsinh}\left(\frac{z}{\varrho_i}\right)\Bigg|_{0}^{L}$$

$$= \ln\left(\frac{1}{\varrho_i}\left(L + \sqrt{L^2 + \varrho_i^2}\right)\right) \tag{9.18}$$

Substituting this into (9.17), we obtain

$$\mathbf{A}(\mathbf{r}) = \frac{2I}{c} \left[\ln \left(\frac{\varrho_2}{\varrho_1} \left(\frac{L + \sqrt{L^2 + \varrho_1^2}}{L + \sqrt{L^2 + \varrho_2^2}} \right) \right) \right] \mathbf{e}_z \tag{9.19}$$

Taking the limit $L \to \infty$ we obtain the following result for the vector potential of infinitely long conductors:

$$\lim_{L \to \infty} \mathbf{A}(\mathbf{r}) = \frac{2I}{c} \ln \left(\frac{\varrho_2}{\varrho_1} \right) \mathbf{e}_z \tag{9.20}$$

In the (x, y)-plane the equipotential lines are defined by $\varrho_2/\varrho_1 = \text{const}$. The vector potential vanishes on the y-axis. In the case of parallel currents the "minus" in (9.15) becomes a "plus." With this replacement, from (9.18) and (9.19) we obtain for the vector potential of two parallel, current-carrying conductors

$$\mathbf{A}_{\|}(\mathbf{r}) = \frac{2I}{c} \left[\ln \left(\frac{1}{\varrho_2 \varrho_1} \right) + \ln \left[\left(L + \sqrt{L^2 + \varrho_1^2} \right) \left(L + \sqrt{L^2 + \varrho_2^2} \right) \right] \right] \mathbf{e}_z \tag{9.21}$$

Obviously, this expression is divergent for infinitely long conductors ($L \to \infty$). We rewrite this expression so that the divergent part is split off.

$$\mathbf{A}_{\|}(\mathbf{r}) = \left[\frac{2I}{c} \left[\ln \left(\frac{1}{\varrho_2 \varrho_1} \right) + \ln \left[\frac{\left(L + \sqrt{L^2 + \varrho_1^2} \right)}{2L} \frac{\left(L + \sqrt{L^2 + \varrho_2^2} \right)}{2L} \right] \right] \right.$$

$$\left. + \frac{2I}{c} \ln(4L^2) \right] \mathbf{e}_z \tag{9.22}$$

The last, divergent term has no position dependence. Taking the curl of $\mathbf{A}_{\|}(\mathbf{r})$, this term plays no role. Hence, we may subtract the divergent term without changing something in the physically relevant field intensity. In physics, with these or similar methods, one can frequently derive physically relevant results from formally divergent expressions. Taking the limit $L \to \infty$ in the corrected expression, the vector potential of parallel currents is

$$\lim_{L \to \infty} \mathbf{A}_{\|}(\mathbf{r}) = -\frac{2I}{c} \ln (\varrho_1 \varrho_2) \, \mathbf{e}_z \tag{9.23}$$

Now, the equipotential lines in the (x, y)-plane are determined by $\varrho_1 \varrho_2 = \text{const}$.

10 Magnetic Moment

We wish to consider a narrowly bounded region in which the steady current distribution $\mathbf{j}(\mathbf{r}')$ is given (Figure 10.1).

For the vector potential generated by this current distribution, we obtain

$$\mathbf{A}(\mathbf{r}) = \frac{1}{c} \int \frac{\mathbf{j}(\mathbf{r}')}{|\mathbf{r} - \mathbf{r}'|} \, dV' \tag{10.1}$$

Let the point of observation $\mathbf{r}$ lie far outside the current distribution, so that $|\mathbf{r}| \gg |\mathbf{r}'|$. Then, the denominator $1/|\mathbf{r} - \mathbf{r}'|$ can be expanded in a Taylor series retaining only the first two terms:

$$\frac{1}{|\mathbf{r} - \mathbf{r}'|} = \frac{1}{r} + \frac{\mathbf{r} \cdot \mathbf{r}'}{r^3} + \cdots$$

We pick out one of the (Cartesian) components of the current and the potential, e.g., the ith one:

$$A_i = \frac{1}{c} \int j_i(\mathbf{r}') \left(\frac{1}{r} + \frac{\mathbf{r} \cdot \mathbf{r}'}{r^3} \right) dV'$$

$$= \frac{1}{cr} \int j_i(\mathbf{r}') \, dV' + \frac{1}{c} \int j_i(\mathbf{r}') \frac{\mathbf{r} \cdot \mathbf{r}'}{r^3} \, dV'$$

Since the volume integrals cover the entire current distribution, the first integral vanishes because $\operatorname{div} \mathbf{j} = 0$. Although this is obvious, one can see it by the following trick:

$$0 = \int_V (x \operatorname{div} \mathbf{j}) \, dV = \int \operatorname{div}(x\mathbf{j}) \, dV - \int j_x \, dV = \int_a (x\mathbf{j}) \cdot \mathbf{n} \, da - \int j_x \, dV \tag{10.2a}$$

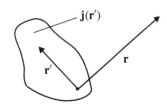

Figure 10.1. Current distribution at $\mathbf{r}'$ and observation point $\mathbf{r}$.

The surface integral vanishes since a is far outside the region of the current. Hence, $\int j_x\,dV = 0$, and analogously for the other components. So, the closed current region has no monopole. Therefore,

$$A_i = \frac{1}{cr^3}\int j_i(\mathbf{r}')\mathbf{r}\cdot\mathbf{r}'dV'$$

or written in vector form,

$$\mathbf{A} = \frac{1}{cr^3}\int \mathbf{j}(\mathbf{r}')[\mathbf{r}\cdot\mathbf{r}']dV'$$

The integrand is rewritten using the relation

$$(\mathbf{r}\cdot\mathbf{r}')\mathbf{j} = (\mathbf{r}\cdot\mathbf{j})\mathbf{r}' - \mathbf{r}\times(\mathbf{r}'\times\mathbf{j})$$

Now, we will demonstrate, that apart from a sign, the integrals over the left-hand side and the first term on the right-hand side are equal to each other, so that

$$\int (\mathbf{r}\cdot\mathbf{r}')\mathbf{j}(\mathbf{r}')\,dV' = -\int [\mathbf{r}\cdot\mathbf{j}(\mathbf{r}')]\mathbf{r}'dV' \tag{10.2b}$$

For this purpose, we treat the x-component first. Starting from one of the terms, e.g., $\int yy'j_x'\,dV'$, we factor out the primed component from the integral sign and consider further ($\mathbf{j}(\mathbf{r}') \equiv \mathbf{j}'$):

$$\int y'j_x'dV' = \int \left[\nabla'\cdot(x'\mathbf{j}')\right]y'dV$$

because

$$\nabla'\cdot(x'\mathbf{j}') = (\nabla'x')\cdot\mathbf{j}' + x'\nabla'\cdot\mathbf{j}' = j_x' + x'\nabla'\cdot\mathbf{j}' = j_x'$$

and the divergence $\nabla'\cdot\mathbf{j}(\mathbf{r}') = 0$. An integration by parts yields

$$\int \left[\nabla'\cdot(x'\mathbf{j}')\right]y'dV' = -\int (x'\mathbf{j}')\cdot\nabla'y'dV' = -\int x'j_y'\,dV'$$

thus,

$$\int y'j_x'\,dV' = -\int x'j_y'\,dV'$$

or in general,

$$\int x_i'j_k'\,dV' = -\int x_k'j_i'\,dV'$$

In the integration by parts, the constant term vanishes (Gauss' theorem); furthermore, $\operatorname{grad} x = \nabla x = \mathbf{e}_x$ has been used. So, we have

$$\int \sum x_i x_i'j_k'\,dV' = -\int \sum x_i j_i'x_k'\,dV' \tag{10.3}$$

This is the kth component of (10.2b). Finally, we have

$$2 \int \mathbf{j}(\mathbf{r}')[\mathbf{r} \cdot \mathbf{r}']dV' = -\mathbf{r} \times \int (\mathbf{r}' \times \mathbf{j}(\mathbf{r}'))dV' \tag{10.4}$$

For the vector potential this implies

$$\mathbf{A} = -\frac{\mathbf{r}}{2cr^3} \times \int (\mathbf{r}' \times \mathbf{j}(\mathbf{r}'))dV'$$

The integral is the magnetic moment $\mathbf{m}$; we define

$$\mathbf{m} = \frac{1}{2c} \int (\mathbf{r}' \times \mathbf{j}(\mathbf{r}'))dV' \tag{10.5}$$

hence,

$$\mathbf{A} = \frac{\mathbf{m} \times \mathbf{r}}{r^3}$$

is the vector potential of a magnetic dipole $\mathbf{m}$. The magnetic field corresponding to this vector potential results from $\mathbf{B} = \operatorname{curl} \mathbf{A}$ outside the current distribution. After a short calculation, one finds (Exercise 9.2):

$$\mathbf{B} = \frac{3\mathbf{n}(\mathbf{n} \cdot \mathbf{m}) - \mathbf{m}}{r^3} \tag{10.6}$$

where $\mathbf{n} = \mathbf{r}/r$ is a unit vector in $\mathbf{r}$-direction. Really, with $\mathbf{r} = \{x_1, x_2, x_3\}$ we calculate directly

$$(\operatorname{curl} \mathbf{A})_j = \left(\nabla \times \mathbf{m} \times \frac{\mathbf{r}}{r^3} \right)_j = \nabla \cdot \frac{\mathbf{r}}{r^3} m_j - \mathbf{m} \cdot \nabla \frac{x_j}{r^3}$$

Now

$$\nabla \cdot \frac{\mathbf{r}}{r^3} = \frac{\nabla \cdot \mathbf{r}}{r^3} + \mathbf{r} \cdot \nabla \frac{1}{r^3} = 0$$

and

$$(\nabla)_i \frac{x_j}{r^3} = \frac{\delta_{ij}}{r^3} - \frac{3x_i x_j}{r^5}$$

Hence,

$$(\operatorname{curl} \mathbf{A})_j = \frac{r^2 m_j - 3x_j(\mathbf{m} \cdot \mathbf{r})}{r^5}$$

so that

$$\operatorname{curl} \mathbf{A} = \frac{3(\mathbf{m} \cdot \mathbf{r})\mathbf{r} - r^2\mathbf{m}}{r^5}$$

This is the desired result. It confirms our assumption for the magnetic dipole field (as being completely analogous to the electric dipole field) and demonstrates the consistency of our considerations. In other words, the assumption in the original measurement of $\mathbf{m}$ and $\mathbf{B}$ that the field of a magnetic dipole can be determined in complete analogy to the field of an

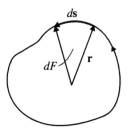

Figure 10.2. Plane conducting loop.

electric dipole follows directly from the Biot-Savart law. On the other hand, the latter has been deduced from experiments utilizing (10.6).

For an arbitrarily shaped conducting loop lying in a plane (Figure 10.2) the magnetic moment is given by

$$\mathbf{m} = \frac{1}{2c} \int (\mathbf{r}' \times \mathbf{j}(\mathbf{r}')) dV' = \frac{1}{2c} \int \mathbf{r}' \times I \, d\mathbf{s}'$$

$$= \frac{I}{2c} \int \mathbf{r}' \times d\mathbf{r}' = \frac{I}{c} a\mathbf{n} \tag{10.7}$$

where $\mathbf{n}$ is the normal vector to the plane. The cross product gives the area element $\frac{1}{2}|\mathbf{r} \times d\mathbf{s}| = da$, and the magnetic moment is $m = (I/c)a$ if a is the area enclosed by the loop as in Exercise 9.1 for the circular conductor.

Force and torque on a magnetic dipole in a magnetic field

Consider an external magnetic field $\mathbf{B}(\mathbf{x})$ (see Figure 10.3). The ith component $B_i(\mathbf{x})$ may be expanded in a Taylor series about the origin:

$$B_i(\mathbf{r}) = B_i(0) + (\mathbf{r} \cdot \text{grad}) B_i(0) + \cdots$$

We assume that the B-field depends only weakly on the position; that is, it is essentially uniform. Then, all terms with higher powers of $\mathbf{r}$ can be neglected. If a body with the current-density distribution $\mathbf{j}(\mathbf{r})$ is placed in the field, then a force and a torque act on it. According to (3.9) the force on a current-density distribution is

$$\mathbf{F} = -\frac{1}{c} \int \mathbf{B}(\mathbf{r}') \times \mathbf{j}(\mathbf{r}') \, dV' \tag{10.8}$$

Substituting in the Taylor expansion given above we get

$$\mathbf{F} = -\frac{1}{c} \mathbf{B}(0) \times \int \mathbf{j}(\mathbf{r}') \, dV' - \frac{1}{c} \int (\mathbf{r}' \cdot \nabla) \mathbf{B}(0) \times \mathbf{j}(\mathbf{r}') \, dV'$$

The volume integral $\int \mathbf{j}(\mathbf{r}') \, dV'$ (that is, the monopole of the current distibution) is equal to zero again; see (10.2a).

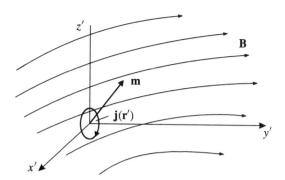

Figure 10.3. Magnetic dipole **m** in a magnetic field.

Thus, the force is

$$\mathbf{F} = \frac{1}{c} \int \mathbf{j}(\mathbf{r}') \times (\mathbf{r}' \cdot \nabla)\mathbf{B}(0) \, dV' \tag{10.9}$$

Note the tensor character of the term $\mathbf{r}' \cdot \nabla\mathbf{B}$ whose, e.g., x-component is

$$x'\frac{\partial B_x}{\partial x} + y'\frac{\partial B_x}{\partial y} + z'\frac{\partial B_x}{\partial z}$$

Since for an external magnetic field $\nabla \times \mathbf{B} = 0$,[1] one obtains (writing **B** for **B**(0)):

$$\mathbf{r}' \times (\nabla \times \mathbf{B}) = \nabla(\mathbf{r}' \cdot \mathbf{B}) - (\mathbf{r}' \cdot \nabla)\mathbf{B} = 0$$

Substituting this into the equation for the force, we obtain

$$\mathbf{F} = \frac{1}{c} \int \mathbf{j}(\mathbf{r}') \times \nabla(\mathbf{r}' \cdot \mathbf{B}) \, dV' = -\frac{1}{c}\nabla \times \int (\mathbf{r}' \cdot \mathbf{B})\mathbf{j}(\mathbf{r}') \, dV'$$

To convince ourselves that this transformation is correct we check, e.g., the x-component. Here, we will refrain from doing so. Comparing the integrand with equation (10.4),

$$\int (\mathbf{r} \cdot \mathbf{r}')\mathbf{j}(\mathbf{r}') \, dV' = -\frac{1}{2}\mathbf{r} \times \int \mathbf{r}' \times \mathbf{j}(\mathbf{r}') \, dV'$$

and identifying **r** with **B(r)** (both are not integrated), then the force can be written in the form

$$\mathbf{F} = \frac{1}{2c}\nabla \times \left(\mathbf{B} \times \int \mathbf{r}' \times \mathbf{j}(\mathbf{r}') \, dV' \right) = \nabla \times \left(\mathbf{B} \times \int \frac{\mathbf{r}' \times \mathbf{j}(\mathbf{r}') \, dV'}{2c} \right)$$

But the integral is just equal to the magnetic dipole moment **m** so that

$$\mathbf{F} = \nabla \times (\mathbf{B} \times \mathbf{m}) = (\mathbf{m} \cdot \nabla)\mathbf{B} - (\nabla \cdot \mathbf{B})\mathbf{m}$$

---

[1] Generally, curl $\mathbf{B} = (4\pi/c)\mathbf{j}$ holds. Outside the sources **j** of the outer field curl $\mathbf{B} = 0$.

and because div $\mathbf{B} = 0$

$$\mathbf{F} = (\mathbf{m} \cdot \nabla)\mathbf{B} = \nabla(\mathbf{m} \cdot \mathbf{B}) \tag{10.10}$$

The last step can be checked by computing, e.g., the x-component:

$$[(\mathbf{m} \cdot \nabla)\mathbf{B}]_x = m_x \frac{\partial B_x}{\partial x} + m_y \frac{\partial B_x}{\partial y} + m_z \frac{\partial B_x}{\partial z}$$

$$= m_x \frac{\partial B_x}{\partial x} + m_y \frac{\partial B_y}{\partial x} + m_z \frac{\partial B_z}{\partial x}$$

$$= [\nabla(\mathbf{m} \cdot \mathbf{B})]_x$$

Here, because curl $\mathbf{B} = 0$, the relations

$$\frac{\partial B_x}{\partial y} = \frac{\partial B_y}{\partial x} \qquad \text{and} \qquad \frac{\partial B_z}{\partial x} = \frac{\partial B_x}{\partial z}$$

have been used.

The energy of the magnetic dipole $\mathbf{m}$ in the external field can be calculated easily from

$$U = -\int \mathbf{F} \cdot d\mathbf{r} = -\int \nabla(\mathbf{m} \cdot \mathbf{B}) \cdot d\mathbf{r} = -\mathbf{m} \cdot \mathbf{B} \tag{10.11}$$

The torque of a current-density distribution in a magnetic induction field $\mathbf{B}$ is given by

$$\mathbf{N} = \int \mathbf{r}' \times d\mathbf{K}' = \int \mathbf{r}' \times \frac{(\mathbf{j}(\mathbf{r}') \times \mathbf{B})}{c} dV'$$

$$= \frac{1}{c} \int \mathbf{r}' \times (\mathbf{j}(\mathbf{r}') \times \mathbf{B}) \, dV'$$

$$= \frac{1}{c} \int [(\mathbf{r}' \cdot \mathbf{B})\mathbf{j}(\mathbf{r}') - (\mathbf{r}' \cdot \mathbf{j}(\mathbf{r}'))\mathbf{B}] \, dV'$$

The identity $\nabla \cdot (r^2 \mathbf{j}) = 2(\mathbf{r} \cdot \mathbf{j}) + r^2 \nabla \cdot \mathbf{j}$ is valid. In magnetostatics, div $\mathbf{j} = 0$. The second part of the integral is ($\mathbf{B} = \mathbf{B}(0)$ can be factored out from under the integral sign):

$$\int \mathbf{r}' \cdot \mathbf{j}(\mathbf{r}') \, dV' = \frac{1}{2} \int \nabla' \cdot (r'^2 \mathbf{j}') \, dV' = \frac{1}{2} \oint_a r'^2 \mathbf{j} \cdot \mathbf{n} \, da = 0$$

because the current distribution is bounded and the surface can be placed arbitrarily far outside the current distribution. Thus, we obtain for the torque

$$\mathbf{N} = \frac{1}{c} \int (\mathbf{r}' \cdot \mathbf{B})\mathbf{j}(\mathbf{r}') \, dV'$$

Taking into account equation (10.4),

$$\mathbf{N} = -\mathbf{B} \times \int \frac{\mathbf{r}' \times \mathbf{j}(\mathbf{r}')}{2c} dV' \tag{10.12}$$

Again, the integral gives the magnetic moment, and finally we can write

$$\mathbf{N} = \mathbf{m} \times \mathbf{B} \qquad\qquad (10.13)$$

Relations (10.10), (10.11), and (10.13) are valid generally in an external field, and they are analogous to the relations between electric dipoles in an electric field. They are needed frequently. Also, this result demonstrates the consistency of our considerations, because (10.13) is identical with (8.2), which is one of the starting points of magnetostatics, in addition to (8.1).

Example 10.1: Illustration of the magnetic moment

We consider a plane, rectangular current loop (Figures 10.4 and 10.5), which is placed in a uniform induction field **B**. The figure displays this situation in perspective as well as in a top view. We can immediately clarify the direction of the forces drawn (Lorentz force!). Obviously,

$$\mathbf{F}_1 = -\mathbf{F}_2 \qquad \text{and} \qquad \mathbf{F}_3 = -\mathbf{F}_4$$

The forces $\mathbf{F}_1$ and $\mathbf{F}_2$ try to squash the loop; on the other hand, the forces $\mathbf{F}_3$ and $\mathbf{F}_4$ exert a torque on the rectangle. We denote the torque by **D**. Its magnitude is (compare the figure)

$$|\mathbf{D}| = |\mathbf{F}_3| d = |\mathbf{F}_3| a \cos \Theta = |\mathbf{F}_3| a \sin \phi$$

Now,

$$|\mathbf{F}_3| = \left| \frac{1}{c} \int \mathbf{j} \times \mathbf{B} \, dV \right| = \left| \frac{I}{c} \int d\mathbf{s} \times \mathbf{B} \right| = \frac{I}{c} B \int_0^b ds = \frac{I}{c} b B$$

and thus

$$|\mathbf{D}| = \frac{I}{c} a b B \sin \phi = |\mathbf{m}| B \sin \phi$$

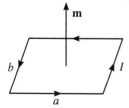

Figure 10.4. The magnetic moment of a rectangular, plane current loop is $|\mathbf{m}| = (I/c)a \cdot b$.

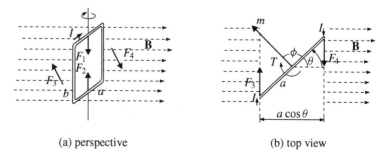

(a) perspective   (b) top view

Figure 10.5. Flux of magnetic induction through a conduction loop.

where $|\mathbf{m}| = (I/c)a \cdot b$ is the magnitude of the magnetic dipole moment. Defining its direction as indicated in the figure, we can write in vector notation

$$\mathbf{D} = \mathbf{m} \times \mathbf{B}$$

for the torque. This relation is the exact analog to the torque of an electric dipole moment in an electric field,

$$\mathbf{D} = \mathbf{p} \times \mathbf{E}$$

Example 10.2:   Exercise: Magnetic moment and angular momentum of a charged particle

Derive the relationship between the magnetic moment $\mathbf{m}$ and the angular momentum of a charged particle. Let the particle move in a central field. Refer to Figure 10.6.

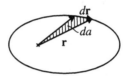

Figure 10.6.   Trajectory of a particle in a central force field.

Solution   The area a of a particle's closed orbit is

$$a = \int da = \int \frac{|\mathbf{L}|}{2m} dt = \frac{|\mathbf{L}|}{2m} \int dt = \frac{|\mathbf{L}|}{2m} T$$

where T is the period. We have used the fact that $\mathbf{L}$ is a constant vector (central force field). With the charge q the current is

$$I = \frac{q}{T}$$

For a positive charge q, $\mathbf{m}$ and $\mathbf{L}$ have the same direction. Therefore,

$$\mathbf{m} = \frac{I}{c} a \frac{\mathbf{L}}{|\mathbf{L}|} = \frac{q}{cT} \frac{|\mathbf{L}|}{2m} T \frac{\mathbf{L}}{|\mathbf{L}|} = \frac{q}{2mc} \mathbf{L}$$

In quantum mechanics, $|\mathbf{L}| = n\hbar$, where $\hbar = h/2\pi = 1.05 \cdot 10^{-34}$ Js, and h is Planck's constant. $n = 0, 1, 2 \ldots$ are the quantum numbers of the angular momentum. Hence, the magnetic moment is a multiple of

$$m_B = \frac{q\hbar}{2mc}$$

This is called the Bohr magneton for the particle of charge q and mass m.

Example 10.3:   Exercise: Force and torque between two circular conductors

Determine the torque and the force between two circular conductors with parallel axes carrying equal and equidirectional currents I if the distance between the centers of the loops is large ($L \gg R$). Express the force and the torque as a function of the angle ϑ between the tie line and the conductor plane. Refer to Figure 10.7.

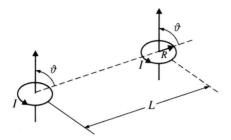

Figure 10.7. Two parallel circular conductors.

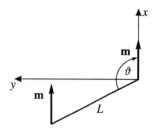

Figure 10.8. Definition of the coordinates in the given problem.

Solution

At large distance $L \gg R$ the field of a small conductor loop of radius R is given by (see equation (10.6))

$$\mathbf{B}(\mathbf{r}) = \frac{3(\mathbf{m_1} \cdot \mathbf{r})\mathbf{r}}{r^5} - \frac{\mathbf{m_1}}{r^3} \qquad (r = L) \tag{10.14}$$

where

$$\mathbf{m_1} = \frac{\pi I R^2}{c}\mathbf{n}, \qquad \mathbf{m_2} = \frac{\pi I R^2}{c}\mathbf{n} \tag{10.15}$$

I is the current; $\mathbf{n}$ is a unit vector. The coordinate system can always be chosen such that the dipoles lie in the (x, y)-plane, and $\mathbf{n} = \mathbf{e}_x$ (Figure 10.8).

Then, corresponding to equation (10.11) the interaction energy of the two magnetic dipoles is given by

$$U = -\mathbf{m_2} \cdot \mathbf{B}(\mathbf{r}) = -\left(\frac{3(\mathbf{m_1} \cdot \mathbf{r})(\mathbf{m_2} \cdot \mathbf{r})}{r^5} - \frac{\mathbf{m_1} \cdot \mathbf{m_2}}{r^2}\right) = -\frac{m^2}{r^3}(3\cos^2\vartheta - 1) \tag{10.16}$$

and

$$\mathbf{F} = -\nabla U = \nabla(\mathbf{m_2} \cdot \mathbf{B}) = \left(\mathbf{e}_r\frac{\partial}{\partial r} + \frac{1}{r}\mathbf{e}_\vartheta\frac{\partial}{\partial \vartheta}\right)(\mathbf{m} \cdot \mathbf{B})$$

$$= -\mathbf{e}_r\frac{3m^2}{r^4}(3\cos^2\vartheta - 1) - 6\frac{m^2}{r^4}\cos\vartheta\sin\vartheta\,\mathbf{e}_\vartheta \tag{10.17}$$

$$= \frac{3m^2}{L^4}\{(1 - 3\cos^2\vartheta)\mathbf{e}_r - \sin 2\vartheta\,\mathbf{e}_\vartheta\}$$

Here, we have set $\mathbf{m_1} = \mathbf{m_2} = \mathbf{m}$, as given, and noted in equation (10.15). We recognize that equation (10.16) is completely analogous to the case of the interacting electric dipoles (see equation (3.72)):

$$\mathbf{D} = \mathbf{m} \times \mathbf{B} = m^2[\mathbf{e}_x \times \mathbf{r}]\frac{3\cos\vartheta}{r^4} = -\frac{3m^2\cos\vartheta}{L^3}\mathbf{e}_z \tag{10.18}$$

11 The Magnetic Field in Matter

Up to now, we have started always from precisely known current density distribution $\mathbf{j}(\mathbf{r})$. When we intend to calculate the magnetic field in space filled with matter we can no longer assume this. There are molecular currents, and magnetic moments of atoms and ions that are not known in detail, and whose average values are of interest only in macroscopic considerations. So, we will proceed as in the treatment of the electrostatic field in matter. The total current density is split into a part arising from the macroscopic charge transport and a part that takes into account the circular currents of electrons in atoms.

The microscopic vector potential depending on all currents and describing exactly also the atomic region is

$$\mathbf{A} = \mathbf{A}_{\text{macr}} + \mathbf{A}_{\text{mol}}$$

or, expressed in terms of the current density distribution,

$$\mathbf{A} = \frac{1}{c} \int \frac{\mathbf{j}(\mathbf{r}')\,dV'}{|\mathbf{r} - \mathbf{r}'|} + \frac{1}{c} \int \frac{\mathbf{j}_{\text{mol}}(\mathbf{r}')\,dV'}{|\mathbf{r} - \mathbf{r}'|} \tag{11.1}$$

where the second integral represents the contribution of the atomic circular currents. Independent of $\mathbf{j}$, the vector potential at the position $\mathbf{r}$, belonging to a molecule at the point $\mathbf{r}_i$, can be approximated by

$$\mathbf{A}_{\text{mol}_i}(\mathbf{r}) = \frac{\mathbf{m}_{\text{mol}} \times (\mathbf{r} - \mathbf{r}_i)}{|\mathbf{r} - \mathbf{r}_i|^3}$$

where $\mathbf{m}_{\text{mol}}$ is the total magnetic moment of the molecule. Now, the total vector potential is

$$\mathbf{A} = \frac{1}{c} \int \frac{\mathbf{j}(\mathbf{r}')\,dV'}{|\mathbf{r} - \mathbf{r}'|} + \sum_i \frac{\mathbf{m}_{\text{mol}} \times (\mathbf{r} - \mathbf{r}_i)}{|\mathbf{r} - \mathbf{r}_i|^3} \tag{11.2}$$

An average value $\langle \mathbf{m} \rangle$ is assumed for $\mathbf{m}_{\text{mol}}$. If N is the number of molecules per volume, then with the magnetic dipole density $\mathbf{M} = \langle \mathbf{m}_{\text{mol}} \rangle$ we can go from the sum to the integral over the volume. The quantity $\mathbf{M}$ is called the *(macroscopic) magnetization*. Corresponding to the polarization in the electrostatic case it is a *density of magnetic dipoles*. Hence, we obtain for the vector potential

$$\mathbf{A}(\mathbf{r}) = \frac{1}{c} \int \frac{\mathbf{j}(\mathbf{r}') \, dV'}{|\mathbf{r} - \mathbf{r}'|} + \int \frac{\mathbf{M}(\mathbf{r}') \times (\mathbf{r} - \mathbf{r}') \, dV'}{|\mathbf{r} - \mathbf{r}'|^3} \tag{11.3}$$

This equation can be rewritten in the following way:

$$\mathbf{A}(\mathbf{r}) = \frac{1}{c} \int \frac{\mathbf{j}(\mathbf{r}') \, dV'}{|\mathbf{r} - \mathbf{r}'|} + \int \mathbf{M}(\mathbf{r}') \times \nabla' \frac{1}{|\mathbf{r} - \mathbf{r}'|} dV' \tag{11.4}$$

Now,

$$\nabla \times (\varphi \mathbf{M}) = (\text{grad } \varphi) \times \mathbf{M} + \varphi \, \text{curl} \, \mathbf{M}$$

Therefore, with $\varphi = 1/|\mathbf{r} - \mathbf{r}'|$, the integral over the magnetization is

$$\int \mathbf{M}(\mathbf{r}') \times \nabla' \frac{1}{|\mathbf{r} - \mathbf{r}'|} dV' = -\int \nabla' \times \frac{\mathbf{M}(\mathbf{r}')}{|\mathbf{r} - \mathbf{r}'|} dV' + \int \frac{\nabla' \times \mathbf{M}(\mathbf{r}')}{|\mathbf{r} - \mathbf{r}'|} dV'$$

The first integral on the right-hand side may be transformed into a surface integral which becomes zero under the assumption that $\mathbf{M}$ is bounded in space:

$$\int \nabla' \times \frac{\mathbf{M}(\mathbf{r}')}{|\mathbf{r} - \mathbf{r}'|} dV' = \int_a \frac{d\mathbf{a}' \times \mathbf{M}}{|\mathbf{r} - \mathbf{r}'|}$$

(see the mathematical consideration at the end of this section). Therefore,

$$\mathbf{A}(\mathbf{r}) = \frac{1}{c} \int \frac{\mathbf{j}(\mathbf{r}') + c \, \text{curl}' \, \mathbf{M}(\mathbf{r}')}{|\mathbf{r} - \mathbf{r}'|} dV' \tag{11.5}$$

The macroscopic magnetization corresponds to a current, the so-called *magnetizing current*

$$\mathbf{j}_M(\mathbf{r}) = c \, \text{curl} \, \mathbf{M}(\mathbf{r})$$

So, the effective current is the sum of the conduction current $\mathbf{j}$ and the magnetizing current $\mathbf{j}_M$. Really, because $\text{curl} \, \mathbf{B} = \text{curl} \, \text{curl} \, \mathbf{A} = \nabla(\nabla \cdot \mathbf{A}) - \Delta \mathbf{A} = -\Delta \mathbf{A} = -i/c \int (\mathbf{j}(\mathbf{r}') + \mathbf{j}_M(\mathbf{r}')) \Delta (1/|\mathbf{r} - \mathbf{r}'|) dV' = 4\pi/c \int (\mathbf{j}(\mathbf{r}') + \mathbf{j}_M(\mathbf{r}')) \delta(\mathbf{r} - \mathbf{r}') \, dV'$, we obtain

$$\text{curl} \, \mathbf{B} = \frac{4\pi}{c} (\mathbf{j} + \mathbf{j}_M) = \frac{4\pi}{c} (\mathbf{j} + c \, \text{curl} \, \mathbf{M}) \tag{11.6}$$

A conversion yields

$$\text{curl}(\mathbf{B} - 4\pi \mathbf{M}) = \frac{4\pi}{c} \mathbf{j} \tag{11.7}$$

The quantity

$$\mathbf{H} = \mathbf{B} - 4\pi \mathbf{M} \tag{11.8}$$

is denoted the *magnetic field intensity*. It has to be regarded in analogy to the electric field intensity, for which in a dielectric $\mathbf{E} = \mathbf{D} - 4\pi \mathbf{P}$.

For the magnetic field intensity, the relation

$$\text{curl } \mathbf{H} = \frac{4\pi}{c}\mathbf{j} \tag{11.9}$$

holds. Hence, it does not depend on the molecular dipoles. In the presence of matter the magnetic field intensity $\mathbf{H}$ replaces the magnetic induction $\mathbf{B}$. In vacuum, these field quantities are equal to each other, $\mathbf{H} = \mathbf{B}$.

Mathematical consideration

For a vector field $\mathbf{B}(\mathbf{r})$, we can prove generally

$$\iiint_V \nabla \times \mathbf{B}\, dV = \iint_a \mathbf{n} \times \mathbf{B}\, da \tag{11.10}$$

by forming the vector field $\mathbf{A} = \mathbf{B} \times \mathbf{C}$ with an arbitrary constant vector $\mathbf{C}$. From Gauss' theorem,

$$\iiint_V \nabla \cdot \mathbf{A}\, dV = \iiint_V \nabla \cdot (\mathbf{B} \times \mathbf{C})\, dV = \iint_a (\mathbf{B} \times \mathbf{C}) \cdot \mathbf{n}\, da$$

Since

$$\nabla \cdot (\mathbf{B} \times \mathbf{C}) = \mathbf{C} \cdot (\nabla \times \mathbf{B})$$

and

$$(\mathbf{B} \times \mathbf{C}) \cdot \mathbf{n} = \mathbf{B} \cdot (\mathbf{C} \times \mathbf{n}) = (\mathbf{C} \times \mathbf{n}) \cdot \mathbf{B} = \mathbf{C} \cdot (\mathbf{n} \times \mathbf{B})$$

we obtain

$$\iiint_V \mathbf{C} \cdot (\nabla \times \mathbf{B})\, dV = \iint_a \mathbf{C} \cdot (\mathbf{n} \times \mathbf{B})\, da$$

or, since $\mathbf{C}$ is constant

$$\mathbf{C} \cdot \iiint_V (\nabla \times \mathbf{B})\, dV = \mathbf{C} \cdot \iint_a (\mathbf{n} \times \mathbf{B})\, da$$

Because $\mathbf{C}$ is an arbitrary constant vector,

$$\iiint_V (\nabla \times \mathbf{B})\, dV = \iint_a (\mathbf{n} \times \mathbf{B})\, da = \iint_a d\mathbf{a} \times \mathbf{B}$$

Susceptibility and permeability

Now, the relationship between $\mathbf{M}$ and $\mathbf{H}$ has to be investigated. Experience shows that in many cases the ansatz

$$\mathbf{M} = \chi_m \mathbf{H}, \qquad \chi_m = \text{const.} \tag{11.11}$$

is true. The factor χ_m is denoted the *magnetic susceptibility*. But, in some cases to be discussed below, χ_m depends on **H**; if **M** and **H** are not parallel to each other, then χ_m is a tensor. The ansatz given above yields for the induction **B**

$$\mathbf{B} = \mathbf{H} + 4\pi\mathbf{M} = (1 + 4\pi\chi_m)\mathbf{H} = \mu\mathbf{H} \tag{11.12}$$

with the *permeability* μ:

$$\mu = (1 + 4\pi\chi_m) \tag{11.13}$$

Depending on the value of χ_m or μ the following materials are distinguished:

Diamagnetics : $\mu < 1$, $\chi_m < 0$

Paramagnetics : $\mu > 1$, $\chi_m > 0$

Ferromagnetics : $\mu \gg 1$, $\mu = \mu(H)$

The values of the susceptibility for some materials can be taken from the following list:

$$
\left.
\begin{aligned}
\text{H}_2\text{:} & \quad \chi_m = -2.3 \cdot 10^{-9} \\
\text{H}_2\text{O:} & \quad \chi_m = -1.2 \cdot 10^{-5} \\
\text{N}_2\text{:} & \quad \chi_m = -0.7 \cdot 10^{-8} \\
\text{Ag:} & \quad \chi_m = -2.5 \cdot 10^{-5}
\end{aligned}
\right\} \quad \text{Diamagnetics } (T = 0°\text{C})
$$

$$
\left.
\begin{aligned}
\text{O}_2\text{:} & \quad \chi_m = 1.8 \cdot 10^{-8} \\
\text{Pt:} & \quad \chi_m = 2.7 \cdot 10^{-5} \\
\text{Al:} & \quad \chi_m = 2.1 \cdot 10^{-5}
\end{aligned}
\right\} \quad \text{Paramagnetics } (T = 20°\text{C})
$$

$$
\left.
\begin{aligned}
\text{Fe:} & \quad \chi_m \approx 10^6 \\
\text{Co:} & \quad \chi_m \approx 10^6 \\
\text{Ni:} & \quad \chi_m \approx 10^6
\end{aligned}
\right\} \quad \text{Ferromagnetics}
$$

Paramagnetism occurs for materials whose atoms or molecules have a magnetic moment. This magnetic moment originates from electrons lying outside closed electron shells. The magnetic moments are completely disordered due to the thermal motion. Applying an external magnetic field, the average angle between the moments and the field direction depends on the ratio of the potential energy of the moments in the field and the thermal energy (analogous to the considerations for dielectrics). Thus, the susceptibility is temperature-dependent, and it is given by *Curie's law*:

$$\chi_m = c\frac{\sigma}{T} \tag{11.14}$$

where σ is the density of the material and T is the absolute temperature. The state in which all moments possess the smallest possible angle with respect to the field direction is called *saturation magnetization*. For paramagnetics the moments are so small that for normal temperatures saturation cannot be reached.

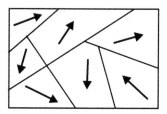

Figure 11.1. Weiss-domains in a nonmagnetized ferromagnetic.

For *ferromagnetics*, the saturation magnetization occurs always inside microscopic domains, called *Weiss domains*. But in the normal case the directions of the magnetization in the Weiss domains are statistically distributed, so that no magnetization appears outside. However, the directions of the Weiss domains will align themselves with respect to an applied field. Figure 11.1 shows the Weiss domains in a nonmagnetized ferromagnet. For ferromagnetics there is no linear relation between **M** and **H**. Some materials even exhibit the phenomenon that **M** is not a unique function of **H**. A change in **M** depends on a change in **H** as well as on the magnetization already present. This effect is called *hysteresis*. If one starts from a high magnetic field **H** and decreases it continuously, then even for **H** = 0 a remanent magnetization is left. A material is called magnetically hard or magnetically soft depending on whether a *remanence* occurs (Figure 11.2).

Decreasing the field intensity **H**, the coercive force, which directed opposite to **M**, is needed to make the magnetization **M** vanish.

As shown in Figure 11.2, for magnetically hard ferromagnets a part of the magnetization is left even if the external force is zero. These materials are suited for permanent magnets.

Above a certain temperature, the so-called *Curie temperature T_c*, ferromagnets lose their ferromagnetic properties and behave like paramagnetics. In this case, their behavior is described also by Curie's law, but the Curie temperature has to be taken into account:

$$\chi_m = \frac{c\sigma}{T - T_c}, \qquad T > T_c \tag{11.15}$$

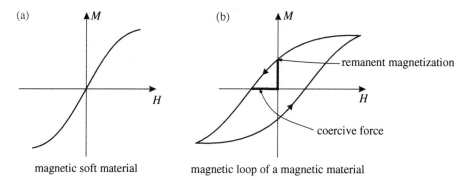

Figure 11.2. Magnetization in (a) a magnetically soft material; and (b) a magnetically hard material.

Example of *Curie temperatures* are the following:

Fe : $T_c = 774°C$

Co : $T_c = 1131°C$

Ni : $T_c = 372°C$

For *diamagnetic materials* the magnetization **M** is directed opposite to the field **H**. This can be explained in the following way: in switching on a magnetic field, circular currents are induced in the individual atoms whose magnetic moments are oriented opposite to the field. Therefore, atoms and molecules without a magnetic moment are diamagnetic. As already stated, the reason for the magnetization is merely the occurrence of circular currents of the atomic electrons. In particular, atoms and ions having closed rare gas-like electron shells, e.g., the ions in salts, have no own moment. Diamagnetism is independent of the temperature, as is understandable from this model.

The behavior of B and H at boundary surfaces

In Figure 11.3, two materials with permeability μ_1 and μ_2, respectively, are displayed, separated by the boundary surface S. In the following we will investigate how the **B**-field changes in the transition from one material to the other. To solve this problem we start from the equation div **B** $= 0$.

This equation implies that the flux into the volume element is equal to the flux out of the volume element. Rewriting

$$\int \text{div} \, \mathbf{B} \, dV = 0$$

with the help of Gauss' theorem as

$$\oint \mathbf{B} \cdot \mathbf{n} \, da = 0$$

we obtain for a flat volume whose side faces may be neglected:

$$((B_1)_n - (B_2)_n)\Delta a = 0$$

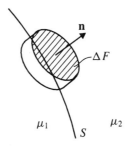

Figure 11.3. Boundary surface between two magnetic materials.

Since this equation is valid for arbitrary Δa, we obtain that the normal components are equal on both sides

$$(B_1)_n = (B_2)_n \tag{11.16}$$

A corresponding condition can be derived for $\mathbf{H}$ from the equation $\text{curl}\,\mathbf{H} = (4\pi/c)\mathbf{j}$.

Choosing the integration path in such a way (Figure 11.4) that the normal vector $\mathbf{n}'$ to Δa is tangential to the boundary surface the equation can be written in integral form

$$\int (\text{curl}\,\mathbf{H}) \cdot \mathbf{n}'\,da = \frac{4\pi}{c} \int_{\Delta a} \mathbf{j} \cdot \mathbf{n}'\,da$$

This equation can be rewritten according to the Stokes theorem

$$\oint \mathbf{H} \cdot d\mathbf{s} = \frac{4\pi}{c} \int_{\Delta a} \mathbf{j} \cdot \mathbf{n}'\,da = \frac{4\pi}{c} K \tag{11.17}$$

The left-hand side of the equation is equal to $[(H_2)_{\text{tan}} - (H_1)_{\text{tan}}]L$. The right-hand side represents a *surface current of magnitude K*. Normally, $K = 0$.

Therefore, the tangential component of $\mathbf{H}$ is conserved in the transition from one medium to another one.

$$(H_1)_{\text{tan}} = (H_2)_{\text{tan}} \tag{11.18}$$

Superconductors, for example, in which no magnetic fields may exist since they are canceled by surface currents, may be viewed as an exceptional case. If there are *surface currents* $\mathbf{K}$ the boundary-layer condition

$$\mathbf{n} \times (\mathbf{H}_2 - \mathbf{H}_1) = \frac{4\pi}{c} \frac{\mathbf{K}}{L} = \frac{4\pi}{c} \mathbf{k} \tag{11.19}$$

is valid instead of equation (11.18). Here, $\mathbf{k} = \mathbf{K}/L$ has the dimension current per unit of length; hence, it represents the density of the *surface current*. L is the length of the considered area element $\Delta a = L \cdot \Delta x$. From these considerations one obtains, altogether, the following conditions:

(a) The normal component of $\mathbf{B}$ is continuous on the boundary surface.

(b) The tangential component of $\mathbf{H}$ is in general continuous on the boundary surface.

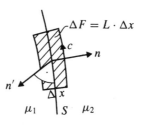

Figure 11.4. The integration path in the boundary surface. L and Δx are the length and width, respectively of the traversed area element Δa.

Because $\mathbf{B} = \mu \mathbf{H}$, we have from $\mathbf{B}_1 \cdot \mathbf{n} = \mathbf{B}_2 \cdot \mathbf{n}$ that $\mu_1(\mathbf{H}_1 \cdot \mathbf{n}) = \mu_2(\mathbf{H}_2 \cdot \mathbf{n})$, or for the normal component

$$(\mathbf{H}_1)_n = \frac{\mu_2}{\mu_1}(\mathbf{H}_2)_n \tag{11.20}$$

Example 11.1: The magnetic field of a uniformly magnetized sphere

A sphere is uniformly magnetized (Figure 11.5) in the direction of the z-axis if $\mathbf{M} = M_0 \mathbf{e}_3$, where M_0 is constant.

For the external field, we have

$$\mathrm{curl}\,\mathbf{B}_a = \mathrm{div}\,\mathbf{B}_a = 0$$

Therefore, the external field may be represented as the gradient of a magnetic field $\mathbf{B}_a = -\,\mathrm{grad}\,\phi_M$, where $\Delta \phi_M = 0$ is valid in the exterior region. The solution of Laplace's equation is known already from electrostatics. Expressed in terms of Legendre polynomials it reads:

$$\phi_M = \sum_{l=0}^{\infty} \alpha_l \frac{P_l(\cos \vartheta)}{r^{l+1}} \tag{11.21}$$

Inside the sphere, $\mathbf{H} = \mathbf{B} - 4\pi \mathbf{M}$ where we have assumed that $\mathbf{B}$ is parallel to $\mathbf{M}$:

$$\mathbf{B}_i = B_0 \cdot \mathbf{e}_3 \tag{11.22}$$
$$\mathbf{H}_i = B_0 \cdot \mathbf{e}_3 - 4\pi M_0 \cdot \mathbf{e}_3 \tag{11.23}$$

According to the conditions derived above, we have at the boundary surface:

$$\mathbf{B}_i \cdot \mathbf{e}_r = \mathbf{B}_a \cdot \mathbf{e}_r$$
$$\mathbf{H}_i \cdot \mathbf{e}_\vartheta = \mathbf{H}_a \cdot \mathbf{e}_\vartheta$$

Substituting for $\mathbf{B}_a = -\,\mathrm{grad}\,\phi_M$ and expressing the first component in spherical coordinates, the following relationship between ϕ_M and B_0 is obtained:

$$B_0 \cos \vartheta = -\frac{\partial}{\partial r}\phi_M \bigg|_{r=a}$$

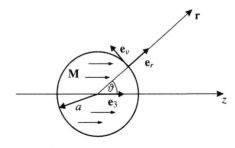

Figure 11.5. A uniformly magnetized sphere.

Taking the expansion (11.21) for ϕ_M, then, with $P_1(\cos \vartheta) = \cos \vartheta$:

$$B_0 P_1(\cos \vartheta) = \sum_{l=0}^{\infty} \alpha_l (l+1) \left. \frac{P_l(\cos \vartheta)}{r^{l+2}} \right|_{r=a}$$

Correspondingly, we have for **H** at the boundary surface:

$$(B_0 - 4\pi M_0) \sin \vartheta = \sum_{l=0}^{\infty} \frac{\alpha_l}{r^{l+2}} \left. \frac{d P_l(\cos \vartheta)}{d\vartheta} \right|_{r=a}$$

With the relation $\sin \vartheta = -(d/d\vartheta) P_1(\cos \vartheta)$ a comparison of coefficients yields the equations

$$B_0 = \frac{\alpha_1 \cdot 2}{a^3} \qquad \text{and} \qquad (B_0 - 4\pi M_0) = -\frac{\alpha_1}{a^3}$$

We obtain

$$B_0 = \frac{8\pi}{3} M_0, \qquad \alpha_1 = \frac{4\pi}{3} M_0 a^3$$

Substituting this result into equations (11.22) and (11.22) for $\mathbf{B}_i$ and $\mathbf{H}_i$, then

$$\mathbf{B}_i = \frac{8\pi}{3} M_0 \mathbf{e}_3 = \frac{8\pi}{3} \mathbf{M}, \qquad \mathbf{H}_i = -\frac{4\pi}{3} M_0 \mathbf{e}_3 = -\frac{4\pi}{3} \mathbf{M} \qquad (11.24)$$

The field in the exterior region becomes

$$\mathbf{B}_a = -\operatorname{grad} \phi_M = -\operatorname{grad} \left(\alpha_1 \frac{\cos \vartheta}{r^2} \right)$$

For the various components, we obtain

$$B_r = \frac{8\pi}{3} M_0 a^3 \frac{\cos \vartheta}{r^3}$$

$$B_\vartheta = \frac{4\pi}{3} M_0 a^3 \frac{\sin \vartheta}{r^3}$$

$$B_\phi = 0$$

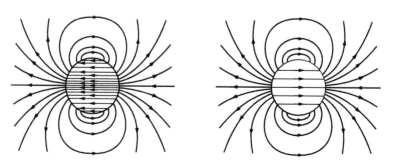

Figure 11.6. B-field and H-field.

This is just the field of a magnetic dipole (compare this result to equation (8.1) and Exercise 1.4), with

$$\mathbf{m} = \frac{4\pi}{3} a^3 \mathbf{M}$$

For **H**

$$\mathbf{H} = \mathbf{B} - 4\pi \mathbf{M}$$

Because div **B** = 0, we obtain div **H** = −4π div **M**.

The quantity div **M** is called the *magnetic charge*. It represents the sources of the magnetic field.

The field of the magnetized sphere is represented in Figure 11.6. For the **H**-field one has to note the following: in the interior there are only half as many field lines as for the **B**-field; moreover, they are oriented in the opposite direction. Because $\mu = 1$, the external field is equal to the **B**-field. This expresses the result (11.24), according to which $|B_i| = 2|H_i|$ and $\mathbf{B}_i = -2\mathbf{H}_i$.

Example 11.2: Example: Magnetizable sphere in an external field

A magnetizable sphere of permeability μ is in an external field $\mathbf{B}_0$. The magnetization of the sphere has to be calculated. Due to the linearity of the field equations, the internal fields are, according to Example 11.1:

$$\mathbf{B}_i = \mathbf{B}_0 + \frac{8\pi}{3} \mathbf{M} \tag{11.25}$$

$$\mathbf{H}_i = \mathbf{B}_0 - \frac{4\pi}{3} \mathbf{M} \tag{11.26}$$

Supposing that the material is not ferromagnetic there is a linear relationship between $\mathbf{B}_i$ and $\mathbf{H}_i$.

$$\mathbf{B}_i = \mu \mathbf{H}_i$$

Thus, from equations (11.25) and (11.26),

$$\mu \left(\mathbf{B}_0 - \frac{4\pi}{3} \mathbf{M} \right) = \mathbf{B}_0 + \frac{8\pi}{3} \mathbf{M}$$

Solving this equation we obtain for the magnetization:

$$\mathbf{M} = \frac{3}{4\pi} \cdot \frac{\mu - 1}{\mu + 2} \mathbf{B}_0$$

The polarization in electrostatics (Exercise 6.2) is its analog:

$$\mathbf{P} = \frac{3}{4\pi} \cdot \frac{\epsilon - 1}{\epsilon + 2} \mathbf{E}_0$$

For ferromagnetics there is no simple proportionality of **B** to **H**. Rather, $\mathbf{B}_i$ is a complicated function of $\mathbf{H}_i$:

$$\mathbf{B}_i = \mathbf{B}_i(\mathbf{H}_i)$$

Now, we may solve the equations

$$\mathbf{B}_i = \mathbf{B}_0 + \frac{8\pi}{3} \mathbf{M}, \qquad \mathbf{H}_i = \mathbf{B}_0 - \frac{4\pi}{3} \mathbf{M}$$

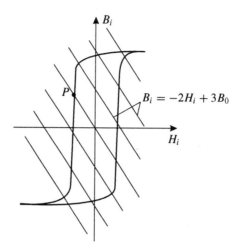

Figure 11.7. Hysteresis loop of a material.

for a relation between $\mathbf{H}_i$ and $\mathbf{B}_i$ by eliminating $\mathbf{M}$. The result reads

$$\mathbf{B}_i + 2\mathbf{H}_i = 3\mathbf{B}_0$$

For various $\mathbf{B}_0$ this equation corresponds to a family of straight lines with the slope -2.

The hysteresis curve of the material yields a further relation between $\mathbf{B}_i$ and $\mathbf{H}_i$. With these two relations special values for any external field may be found.

For example, if we want to calculate the residual magnetization of a ferromagnetic sphere after $\mathbf{B}_0$ has been increased first and then has been taken back to zero the values for $\mathbf{B}_i$ and $\mathbf{H}_i$ result graphically as the intersection point P of the hysteresis curve and the straight line with $\mathbf{B}_0 = 0$ (Figure 11.7)

$$\mathbf{B}_i = \frac{8\pi}{3}\mathbf{M}, \qquad \mathbf{H}_i = -\frac{4\pi}{3}\mathbf{M}$$

From these equations, $\mathbf{M}$ may be determined.

Example 11.3: Exercise: Energy loss and hysteresis

In a ferromagnet the connection of the fields $\mathbf{B}$ and $\mathbf{H}$ depends on the history of the system (hysteresis). Such a body is placed in a uniform magnetic field $-\mathbf{B}_0$ the intensity of which is high enough so that the internal fields are unique. Now, the polarity of the external field is slowly alternated ($-\mathbf{B}_0 \to +\mathbf{B}_0$) and then brought back to the original state ($+\mathbf{B}_0 \to -\mathbf{B}_0$). Prove that the energy loss is given by $1/4\pi \int S(\mathbf{r})d^3r$ where $S(\mathbf{r})$ is the area enclosed by the hysteresis loop (Figure 11.8).

Note: It will be shown later that the change of the energy density $w(\mathbf{r})$ for the transition ($\mathbf{B}_1 \to \mathbf{B}_2$) along the path C is given by

$$w = \frac{1}{4\pi} \int_C \mathbf{H}(\mathbf{B}) \cdot d\mathbf{B}$$

This expression for the magnetic energy density is completely analogous to the expression for the electric energy density.

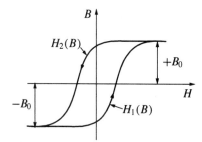

Figure 11.8. Hysteresis loop.

Solution The variation of the energy density is given by

$$w(\mathbf{r}) = \frac{1}{4\pi} \int_{-B_0}^{+B_0} \mathbf{H}_1(\mathbf{B}) \cdot d\mathbf{B} + \int_{+B_0}^{-B_0} \mathbf{H}_2(\mathbf{B}) \cdot d\mathbf{B}$$

$$= -\frac{1}{4\pi} \int_{-B_0}^{+B_0} (\mathbf{H}_2(\mathbf{B}) - \mathbf{H}_1(\mathbf{B})) \cdot d\mathbf{B} = -\frac{1}{4\pi} S(\mathbf{r}) \qquad (11.27)$$

As can be seen, the integrand is just the area enclosed by the hysteresis loop.

Outside the ferromagnetic volume, $H_1 = H_2$ and therefore $S(\mathbf{r}) = 0$. The total energy change is

$$\Delta W = -\frac{1}{4\pi} \int_V S(\mathbf{r}) \, d^3 x$$

Example 11.4: Exercise: Measurement of E, D, B, and H

Point out a possible method for measuring the fields **E** and **D**, and **B** and **H**.

Solution In vacuum, **E** and **B** may be determined by the force acting on a test charge at rest or moving. In space filled with matter, this procedure would run into difficulties; furthermore, also **D** and **H** have to be measured. For sake of simplicity, we restrict ourselves to uncharged insulators, so that no surface charges or surface currents occur. We distinguish two cases:

(a) There is a linear relationship $\mathbf{D} = \epsilon \mathbf{E}$ and $\mathbf{B} = \mu \mathbf{H}$ with known values of ϵ and μ. Then, it is sufficient to determine **E** and **H**; both other quantities can be calculated. According to Exercise 6.3, the relationship between the field $\mathbf{E}_{\text{int}}$ inside a spherical cavity and the asymptotically constant exterior field $\mathbf{E}_0$ is given by $\mathbf{E}_{\text{int}} = 3\epsilon/(2\epsilon + 1)\mathbf{E}_0$.

Taking into account the comparison,

Figure 11.9. Fields $\mathbf{E}_{\text{int}}$ and $\mathbf{H}_{\text{int}}$ inside a spherical cavity.

$$\nabla \times \mathbf{E} = 0, \qquad \nabla \cdot \mathbf{D} = 0, \qquad \nabla \times \mathbf{H} = 0, \qquad \nabla \cdot \mathbf{B} = 0$$

$$\mathbf{D}_{\text{ext}} = \epsilon \mathbf{E}_{\text{ext}}, \qquad \mathbf{D}_{\text{int}} = \mathbf{E}_{\text{int}}, \qquad \mathbf{B}_{\text{ext}} = \mu \mathbf{H}_{\text{ext}}, \qquad \mathbf{B}_{\text{int}} = \mathbf{H}_{\text{int}}$$

$$D_{\text{ext}\perp} = D_{\text{int}\perp}, \qquad E_{\text{ext}\parallel} = E_{\text{int}\parallel}, \qquad B_{\text{ext}\perp} = B_{\text{int}\perp}, \qquad H_{\text{ext}\parallel} = H_{\text{int}\parallel}$$

$$\mathbf{E}_{\text{ext}} \to \mathbf{E}_0 (r \to \infty), \qquad\qquad \mathbf{H}_{\text{ext}} \to \mathbf{H}_0 (r \to \infty)$$

then, for the analogous problem of a spherical cavity in a magnetizable material, we obtain $\mathbf{H}_{\text{int}} = 3\mu/(2\mu + 1)\mathbf{H}_0$. When a spherical cavity is introduced into our sample whose dimensions are small enough to regard the fields to be uniform, then

$$\mathbf{E}_0 = \frac{2\epsilon + 1}{3\epsilon} \mathbf{E}_{\text{int}}, \qquad \mathbf{D}_0 = \frac{2\epsilon + 1}{3} \mathbf{E}_{\text{int}}$$

$$\mathbf{B}_0 = \frac{2\mu + 1}{3} \mathbf{B}_{\text{int}}, \qquad \mathbf{H}_0 = \frac{2\mu + 1}{3\mu} \mathbf{B}_{\text{int}}$$

(b) The assumptions for the method described in (a) do not hold. Then, all four quantities have to be determined separately. We restrict ourselves to the discussion of the measurements of **E** and **D**. The method can be used *mutatis mutandis* to determine **B** and **H**.

We place a flat hollow cylinder (beer-coaster-like) in the dielectric. The symmetry axis points in z-direction. Due to the boundary conditions $D_{\text{int}\perp} = D_{\text{ext}\perp}$, $E_{\text{ext}\parallel} = E_{\text{int}\parallel}$, and $\mathbf{D}_{\text{int}} = \mathbf{E}_{\text{int}}$, we have $D_{0z} = E_{\text{int}z}$ and $E_{0x} = E_{\text{int}x}$, $E_{0y} = E_{\text{int}y}$. The other field components can be determined by a rotation of the hollow cylinder.

PART IV

ELECTRODYNAMICS

12 Faraday's Law of Induction

The first quantitative studies of time-dependent electric and magnetic fields were performed by Faraday in 1831. He discovered that an electric current arises in a closed wire loop when it is moved through a magnetic field (compare Figure 12.1).

Faraday's discovery may be formulated mathematically in the following way: If C is the curve of the current loop, then an arbitrary surface can be placed through C and the *magnetic flux*

$$\phi = \int_{\text{area}} \mathbf{B} \cdot \mathbf{n} \, da \tag{12.1}$$

can be calculated. Because div $\mathbf{B} = 0$, ϕ is independent of the choice of the surface a over C (Figures 12.2 and 12.3).

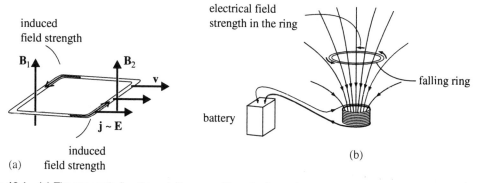

Figure 12.1. (a) The magnetic flux through the current loop is directed upward and decreases in time for the motion indicated. In this case the current flows, due to an induced E-field, as indicated. (b) If the ring is falling, the magnetic flux through it increases. Lenz's law tells us that due to the induced electromotive force a circular current is induced in such a way that the external flux is diminished. The system responds such that the causing changes are hindered: The field intensity induced in the ring induces the current, which creates an E-field opposing the external B-field, thus weakening the external B-field.

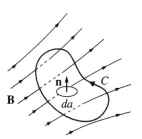

Figure 12.2. On the induction law.

The *electromotive force* (voltage) V acting on the wire loop C is

$$V = \oint_C \mathbf{E} \cdot \mathbf{ds} \tag{12.2}$$

Here, $\mathbf{E}$ is the electric field intensity acting at the position of the conductor element $\mathbf{ds}$. Faraday's discovery can now be formulated in the following way:

$$V = -k\frac{d\phi}{dt} \quad \text{or} \quad \oint_C \mathbf{E} \cdot \mathbf{ds} = -k\frac{d}{dt} \int_{\text{area}} \mathbf{B} \cdot d\mathbf{a} \tag{12.3}$$

The induced voltage is proportional to the rate of change of the magnetic flux. The sign is fixed by the *Lenz law*, implying that the induced currents and the magnetic flux associated with it are directed such that they oppose the change of the external flux. The law is generally valid. The change of flux may come about in different ways, for example, by modification or motion of the current loop in an external magnetic field; by alteration of the magnetic field (e.g., by displacement of the generating magnet); or both. The constant of proportionality k in equation (12.3) can be fixed by considering the following case: At first, we move a metallic wire C_1 carrying no charge at the velocity $\mathbf{v}$ to the new position C_2 (Figure 12.4), then the charges contained in the wire take part in this motion. The Lorentz force acting on these charges is

$$\mathbf{F} = \frac{q}{c}\mathbf{v} \times \mathbf{B} \tag{12.4}$$

If the force $\mathbf{F}$ has a component in direction of the wire, it will set the charge in motion: a current will flow. The field intensity due to the Lorentz force is

$$\mathbf{E}_{\text{ind}} = \frac{\mathbf{F}}{q} = \frac{1}{c}\mathbf{v} \times \mathbf{B} \tag{12.5}$$

The voltage in the wire loop is

$$V = \oint_{C_1} \mathbf{E}_{\text{ind}} \cdot \mathbf{ds} = \frac{1}{c}\oint_{C_1} (\mathbf{v} \times \mathbf{B}) \cdot \mathbf{ds} \tag{12.6}$$

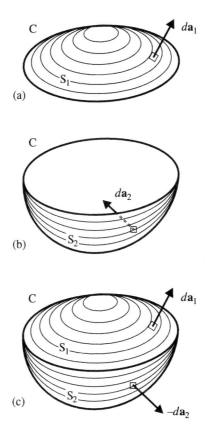

Figure 12.3. (a) The flux through C is $\phi = \int_{\text{area}_1} \mathbf{B} \cdot d\mathbf{a}_1$. (b) a_2 is another surface over C. The flux through a_2 is equal to the flux through a_1. (c) The combination of both areas a_1 and a_2 leads to a closed surface for which $\int_{\text{area}} \mathbf{B} \cdot d\mathbf{a} = \int_{\text{area}_1} \mathbf{B} \cdot d\mathbf{a}_1 - \int_{\text{area}_2} \mathbf{B} \cdot d\mathbf{a}_2 = \int \text{div} \, \mathbf{B} \, dV = 0$.

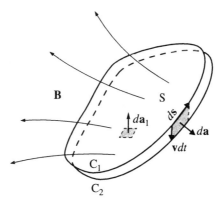

Figure 12.4. The conducting loop moves with the velocity $\mathbf{v}$ from the position C_1 to the position C_2.

Now, we will demonstrate that

$$\oint_{C_1} (\mathbf{v} \times \mathbf{B}) \cdot \mathbf{ds} = -\frac{d}{dt} \int_{\text{area}(C_1)} \mathbf{B} \cdot \mathbf{n}\, da$$

that is, it is equal to the negative time rate of change of the magnetic flux through a surface over C_1: The two positions of the wire loop C_1 (at time t) and C_2 (at time $t + \Delta t$) arise from each other by the displacement vectors $\mathbf{v}\Delta t$. The vector product

$$-\mathbf{ds} \times \mathbf{v}\Delta t = d\mathbf{a}$$

describes the surface element $d\mathbf{a}$ directed outward (Figure 12.4). Therefore,

$$\oint_{C_1} \mathbf{ds} \cdot (\mathbf{v} \times \mathbf{B}) = \frac{1}{\Delta t} \oint_{C_1} (\mathbf{ds} \times \mathbf{v}\Delta t) \cdot \mathbf{B} = -\frac{1}{\Delta t} \int_{\text{area}_M} \mathbf{B} \cdot d\mathbf{a} = -\frac{\Delta \phi_M}{\Delta t} \tag{12.7}$$

The last integral represents obviously the flux $\Delta \phi_M$ crossing the lateral surface Δa_M connecting the curves C_1 and C_2. The flux through the closed surface $a(C_1) + a(C_2) + \Delta a_M$ has to vanish since div $\mathbf{B} = 0$; that is,

$$\phi_1 + \phi_2 + \Delta \phi_M = 0 \tag{12.8}$$

where

$$\phi_1 = \int_{\text{area}_1(C_1)} \mathbf{B} \cdot d\mathbf{a}_1$$

and

$$\phi_2 = \int_{\text{area}_1(C_1)} \mathbf{B} \cdot d\mathbf{a}_2$$

The normal direction is always chosen to point outward (seen from the interior of the cylinder determined by the two curves C_1 and C_2). But, choosing it on the surface, that is, $\mathbf{n}_2 = \mathbf{n}_1$, then $\phi_2' = -\phi_2$ and therefore from (12.8)

$$\phi_1 - \phi_2' + \Delta \phi_M = 0$$

Hence,

$$\Delta \phi_M = (\phi_2' - \phi_1) \equiv +\Delta \phi \tag{12.9}$$

where

$$\Delta \phi = \phi_2' - \phi_1 = \phi(t + \Delta t) - \phi(t)$$

So, $\Delta \phi_M$ is equal to the change of flux through the endface, area_1. Together with (12.7) and (12.6)

$$\oint_{C_1} \mathbf{E}_{\text{ind}} \cdot \mathbf{ds} = -\frac{1}{c}\frac{d\phi}{dt} = -\frac{1}{c}\frac{d}{dt} \int_{\text{area}_1} \mathbf{B} \cdot d\mathbf{a} \tag{12.10}$$

this is exactly Faraday's law. Obviously, the constant of proportionality in (12.3) is given by

$$k = \frac{1}{c}$$

so that generally

$$\oint \mathbf{E} \cdot d\mathbf{s} = -\frac{1}{c} \frac{d}{dt} \int_{\text{area}} \mathbf{B} \cdot d\mathbf{a}. \tag{12.11}$$

The *universality* of the induction law (12.3) has been assumed in this argument.

We note that, starting from equation (12.11), any single step can be inverted, and then we may infer equation (12.6) for a fixed **B**-field, but with moving conductors. Since equation (12.6) has to be valid for arbitrary closed conductors, equation (12.11) and thus the Lorentz force can be derived. It is interesting to realize that the Lorentz force follows from Faraday's induction law.

Using this law, e.g., the intensity of a uniform magnetic field can be determined: According to Ohm's law the induced voltage corresponds to the current $I = V/R$, where R is the resistance. The charge which is moved in a wire loop during the motion of a wire loop through a uniform magnetic field is

$$q = \int_0^T I \, dt = \frac{1}{R} \int_0^T \left(-\frac{1}{c} \frac{d\phi}{dt} \right) dt = -\frac{1}{Rc} \int_0^T d\phi = -\frac{1}{Rc} (\phi(T) - \phi(0))$$

Moving the loop out of the field, then $\phi(T) = 0$. In a uniform field $\phi(0) = B \cdot a$ (a is the area of the loop) so that by substitution one obtains finally

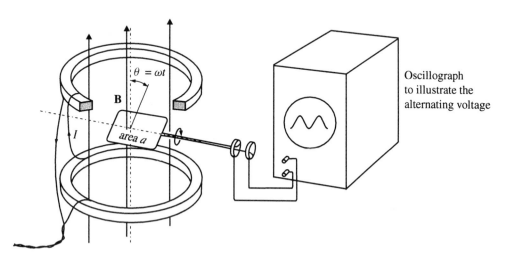

Figure 12.5. The two rings generate a magnetic induction field **B**, which is approximately uniform in the vicinity of the conducting loop (compare Exercise 8.3). If the conducting loop rotates with the angular velocity ω, then a cosine-like voltage $V = -(1/c)d\phi/dt = -(1/c)d/dt\, B \cdot a \sin \omega t = -(Ba\omega/c) \cos \omega t$, and hence, a corresponding alternating current is induced.

$$q = \frac{B \cdot a}{R \cdot c}$$

The charge q may be measured directly by a ballistic galvanometer, so that B can be calculated. It is worth mentioning that the generation of electric power (and the electric motor in its inversion) is based on the law (12.11). This is obvious from Figure 12.5. Compare also Example 12.4.

The betatron

The betatron serves for the generation of high-energetic electron beams. It consists of a highly evacuated cyclic discharge tube arranged symmetrically between the pole pieces of an electromagnet. Electrons in the discharge tube will constantly follow a circular orbit if the acting magnetic field (control field) B_{con} just cancels the centrifugal acceleration:

$$\frac{mv^2}{r} = e\frac{v}{c}B_{con} \qquad \text{or} \qquad mv = e\frac{r}{c}B_{con}$$

This condition between the control field B_{con} and the momentum of the electron has to be met to maintain a stable orbit. In addition to the control field, there is an accelerating field. Changing this field, a circulation voltage is induced in the tube that accelerates the electrons.

For the force (we consider magnitudes only) we have

$$\frac{d}{dt}(mv) = eE = \frac{e}{c}\frac{1}{2\pi r}\frac{d}{dt}\int B\,da$$

since

$$E2\pi r = \frac{1}{c}\frac{d}{dt}\int B\,da$$

By integration we obtain a condition between the momentum of the electron and the accelerating field:

$$mv = \frac{e}{c}\frac{1}{2\pi r}\int B\,da$$

Hence, we have two conditions, one for the control field and one for the accelerating field. Eliminating the momentum yields

$$e\frac{r}{c}B_{con} = \frac{e}{c}\frac{1}{2\pi r}\int B\,da = \frac{e}{c}\frac{1}{2\pi r}B\pi r^2$$

The two fields have to fulfill always the 1:2-ratio if the electron must stay on a stable orbit during the acceleration, $B_{con} = \frac{1}{2}B$. See Figure 12.6.

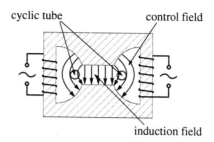

Figure 12.6. A betatron.

Example 12.1: Exercise: Induction of a current in a conducting loop

A rectangular conducting loop is taken out of a magnetic induction field of intensity $\mathbf{B}_0$ (Figure 12.7). In which direction does the current flow? The induction field points into the plane of the page, as indicated in the figure.

Solution   Obviously, the induction flux decreases during this motion. According to Faraday's induction law, for a circulation in the mathematically positive sense the voltage is given by

$$V = \oint \mathbf{E} \cdot \mathbf{ds} = -\frac{1}{c}\frac{d\phi}{dt} = -\frac{1}{c}(-B)\frac{da}{dt}$$

where a is the area crossed by the induction field. Now, da/dt is negative, hence V is negative, so the current flows in a mathematically negative sense, that is, clockwise.

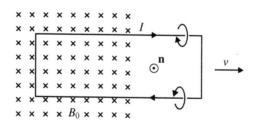

Figure 12.7. A loop is pulled out of a magnetic field.

Example 12.2: Excercise: Voltage in a conducting loop

A conducting loop is connected to a parallel-plate capacitor (Figure 12.8). For the region indicated in the figure the magnetic induction field $\mathbf{B}$ points into the plane of the page. Its magnitude is increasing in time. Which of the two plates of the capacitor is charged positively?

Solution   According to Faraday's induction law the circulation voltage measured in a mathematically positive sense is

$$V = -\frac{1}{c}\frac{d\phi}{dt} = -\frac{1}{c}\frac{d}{dt}\mathbf{B} \cdot \mathbf{n}a = -\frac{a}{c}\left(-\frac{dB}{dt}\right)$$

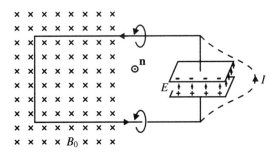

Figure 12.8. Conducting loop with parallel-plate capacitor in a magnetic field.

because $\mathbf{B} \cdot \mathbf{n} = -B$. a is the area threaded by the magnetic induction. Since $dB/dt > 0$, a positive circulation voltage is induced; the current flows in a mathematically positive sense; that is, the current flows away from the upper plate (then, it is charged negatively).

Example 12.3:   Exercise: Induction in a coil by a time-varying magnetic field

A coil with N turns is placed into an electromagnet (Figure 12.9). The magnetic field increases linearly in time, reaching a value of $B_0 = 10^4$ Gauss after T=10 s. Then it remains constant. The coil has a cross sectional area of $a = 100 \text{cm}^2$, and it is oriented perpendicular to the magnetic field.

(a) What voltage is induced in the coil?

(b) Let the coil have a resistance of $R = 0.5\Omega$. What is the magnitude of the current in the coil if the ends of the coil are closed?

(c) How much energy is dissipated during the switching-on of the magnet?

Solution   (a) The induction flux through the individual turns of the coil is

$$\phi(t) = B(t) \cdot a$$

where the magnetic induction is increasing according to

$$B = B_0 \frac{t}{T}$$

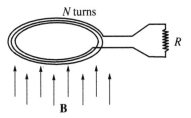

Figure 12.9. A coil with N turns and with the resistance R in a magnetic field **B**.

Hence, the induced voltage is

$$V = -\frac{N}{c}\frac{d\phi}{dt} = -\frac{Na}{c}\frac{dB}{dt} = \frac{-NaB_0}{cT}$$

$$= -10 \cdot 10^{-2}\text{m}^2 \cdot \frac{10^4\text{Gauss}}{10\text{sec}}$$

$$= -10 \cdot 10^{-2} \cdot 10^{-1}\left[\text{m}^2 \cdot \frac{V\text{s}}{\text{m}^2} \cdot \frac{1}{\text{s}}\right] = -0.01 V$$

calculated in the rationalized system of units. In electromagnetic units (Chapter 8) we have

$$V = -10 \cdot 10^{-2}[\text{m}^2] \cdot 10^4[\text{Gauss}]/10[\text{sec}]$$
$$= -10 \cdot 10^{-6}[\text{cm}^2] \cdot 10^4[\text{g}^{1/2}\text{cm}^{-1/2}\text{sec}^{-1}] \cdot 10^{-1}[\text{sec}^{-1}]$$
$$= -0.01[\text{g}^{1/2}\text{cm}^{3/2}\text{sec}^{-2}] = -0.01[V]$$

(b) If the current circuit is closed, the current flowing follows from

$$V = IR \Longrightarrow I = \frac{V}{R} = 0.02A = 20 \text{ mA}$$

(c) The energy dissipated in heat may be calculated in the following way: The current is flowing during the time T. Therefore, the charge flowing through an arbitrary cross sectional area of the wire will be

$$Q = IT = \frac{V}{R}T = 0.2 \text{ Coul}$$

This charge Q moved through the potential difference V and thus gained the energy

$$W = V \cdot Q = 0.01\text{Volt} \cdot 0.2 \text{ Coul} = 0.002 \text{ Joule}$$

This energy is converted to heat. We may think about it also in the following way. The energy released per unit of time at the resistor is

$$P = V \cdot I = RI^2$$

and the energy converted to heat during the time T is

$$W = \int_0^T P(t)\,dt = RI^2T = VIT = VQ$$

as above.

Example 12.4: Electric generators and motors

The principle of an electric generator is represented in Figure 12.10. A coil (N conducting loops) is revolved in an external magnetic field at the frequency ω by a steam turbine or a water turbine. In one loop with the area a, the induction flux

$$\Phi = Ba\sin\varphi = Ba\sin\omega t, \qquad \omega = \frac{2\pi}{T} \tag{12.12}$$

Because the flux is varying in time, in N loops the voltage

$$V = -N\frac{d\Phi}{c\,dt} = -N\frac{d}{c\,dt}(Ba\sin\omega t) = \frac{-NaB\omega}{c}\cos\omega t \tag{12.13}$$

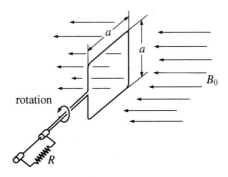

Figure 12.10. Principle of an electric generator. Only one conducting loop is drawn.

is induced (Figure 12.11). An alternating voltage occurs with the maximal values $NaB\omega/c$. While the coil is open, no current flows. If the current circuit is closed, the alternating current (let the resistance be R)

$$I = \frac{V}{R} = \frac{-NaB\omega}{cR} \cos \omega t \tag{12.14}$$

flows. The heat output of the device is

$$P = RI^2 = \frac{N^2 a^2 B^2 \omega^2}{c^2 R} \cos^2 \omega t \tag{12.15}$$

It fluctuates also periodically with the maximal values

$$P_{\text{max}} = \frac{N^2 a^2 B^2 \omega^2}{c^2 R}$$

The flux generated on the average over one period is

$$\langle P \rangle = \frac{1}{T} \int_0^T P(t) \, dt = \frac{N^2 a^2 B^2 \omega^2}{c^2 R T} \int_0^T \cos^2 \omega \, dt$$

$$= \frac{N^2 a^2 B^2 \omega^2}{R T c^2} \frac{1}{2\omega} (\sin \omega t \cos \omega t + \omega t) \Big|_0^T = \frac{P_{\text{max}}}{2} \tag{12.16}$$

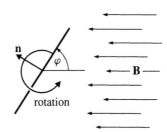

Figure 12.11. On the calculation of the flux in the loop. $\varphi = \omega t$ is the rotation angle of the plane of the loop against the magnetic induction field **B**.

Of course, this heat output must come from the turbine. We want to be convinced about this in more detail and at first state that the coil has a magnetic dipole moment of magnitude

$$|\mathbf{m}| = \frac{NIa}{c} \tag{12.17}$$

So, a torque $\mathbf{N}$ is exerted on the current-carrying coil,

$$\mathbf{N} = \mathbf{m} \times \mathbf{B} \tag{12.18}$$

This torque has to be canceled from outside, by the turbine, to keep the coil running (revolving). The external torque is

$$\mathbf{N}' = -\mathbf{N} = -\mathbf{m} \times \mathbf{B} \tag{12.19}$$

From the drawing we can see that

$$|\mathbf{N}| = |\mathbf{m}|B\cos\varphi = \frac{NIaB}{c}\cos\omega t \tag{12.20}$$

The mechanical work that has to be done in order to revolve the coil by the small angle $d\varphi$ is

$$dW = |\mathbf{N}|d\varphi \tag{12.21}$$

So, the mechanical power is

$$P_{\text{mech}} = \frac{dW}{dt} = |\mathbf{N}|\frac{d\varphi}{dt} = |\mathbf{N}| \cdot \omega = \frac{N^2 a^2 B^2 \omega^2}{Rc^2}\cos^2\omega t \tag{12.22}$$

This is just the electric power (12.15) converted to heat at the resistor. The electric power for the generator is exactly equal to the mechanical power of the turbine driving the generator, as it must be, due to the energy conservation.

Example 12.5: Exercise: Linear motor

A rod of length $L = 10\,\text{cm}$ is lying on two ideally conducting rails (Figure 12.12). The potential difference between the rails is $V_0 = 15$ volts. Let the resistance of the rod be $R = 0.1\ \Omega$ (ohm). The rod is connected to a mass $m = 1.2\,\text{kg}$ by a rope and a pulley.

Calculate the velocity of the rod if a magnetic field B_0 is applied in the direction shown in the figure and is 10^4 Gauss. What fraction of the power supplied by the battery is converted to mechanical power?

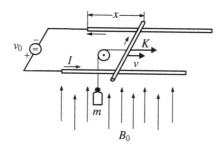

Figure 12.12. A linear motor.

Solution   This example is the linear (one-dimensional) analog of a rotating coil in an external magnetic field. If the current I flows through the conduction, a force $\mathbf{F}$ acts on the rod (see the figure):

$$|\mathbf{F}| = \left| \frac{1}{c} \int \mathbf{j} \times \mathbf{B}\, dV \right| = \left| \frac{I}{c} \int_0^L d\mathbf{r} \times \mathbf{B} \right| = \left| \frac{IB}{c} \int_0^L d\mathbf{r} \right| = \frac{IBL}{c} \qquad (12.23)$$

For a balance,

$$|\mathbf{F}| = mg \qquad (12.24)$$

But if the rod is moving the total area $a = Lx$ enclosed by the current circuit changes; hence, also the induction flux ϕ changes. Then,

$$\frac{d\phi}{dt} = B \frac{d}{dt}(Lx) = BL \frac{dx}{dt} = BLv \qquad (12.25)$$

where v is the velocity of the rod sliding on the rails. So, according to Faraday's induction law a voltage is induced in the circuit. This voltage is

$$V = -\frac{1}{c}\frac{d\phi}{dt} = -\frac{BLv}{c} \qquad (12.26)$$

The total voltage in the current circuit is $V_0 + V$. It determines the current flowing according to

$$RI = V_0 + V = V_0 - \frac{BLv}{c} \qquad (12.27)$$

and the balance condition (12.24) becomes

$$mg = |\mathbf{F}| = \frac{IBL}{c} = \frac{BL}{c}\frac{1}{R}\left(V_0 - \frac{BLv}{c} \right) \qquad (12.28)$$

hence, the constant velocity is

$$v = \frac{c}{LB}\left(V_0 - \frac{Rmgc}{LB} \right) \qquad (12.29)$$

The numerical values of our example yield

$$v = 32.3 \left[\frac{\mathrm{m}}{\sec} \right] \qquad (12.30)$$

This is a high velocity. But, we note that a slight change of the resistance, $R = 0.128\ \Omega$, would have given $v = 0$. We see that the constant velocity sensitively depends on the resistance R . It determines the maximal current and thus also the maximal strengh of the electromagnetic force which sets the rod in motion.

The total power supplied by the battery is

$$P_1 = V_0 I = V_0 \frac{mgc}{LB} \qquad (12.31)$$

where we have substituted the current following from (12.23) and (12.24). This power must equal the sum of the mechanical power $Fv = mgv$ and the thermal power RI^2 at the resistor, thus, with (12.29) and (12.27)

$$P_2 = mgv + RI^2 = \frac{mgc}{LB}\left(V_0 - \frac{Rmgc}{LB} \right) + R\left(\frac{mgc}{LB} \right)^2 \qquad (12.32)$$

In fact, we see, $P_1 = P_2$ .The efficiency of this motor can be calculated as the ratio of the mechanical power and the total power

$$\eta = \frac{mgv}{P_1} = \frac{LB}{V_0 c}v = 1 - \frac{mgRc}{B_0 L V_0}$$

(12.33)

The numerical values yield

$$\eta = 1 - \frac{1.2 \cdot 9.81}{15}\left[\frac{\text{Nm}}{\text{Ws}}\right] = 22\%$$

So, only $\approx 1/5$ of the electrical power is converted to mechanical work; 4/5 of the electric power is converted to heat in the resistor. Obviously this motor is not efficient.

Biographical notes

Michael Faraday, b. Sept. 22, 1791, Newington Butts (n. London)–d. Aug. 25, 1867, Hampton Court-Green (n. London). English physicist and chemist. The son of a blacksmith, he served his apprenticeship in a bookbindery. He educated himself, and became lab assistant to Humphry Davy in 1813. In 1825 he was promoted head of the Royal Institution Laboratories. By 1824 he was elected Fellow of the Royal Society (F.R.S.). During his lifetime he became a member of all the important academies of science. In 1821 Faraday proved the rotation of a current-carrier around a magnetic pole and vice versa. Between 1820 and 1822 he unsuccessfully tried to produce stainless steel. In 1823 he discovered that gaseous chlorine liquefied under pressure. While working on the distillation of the condensate of illuminating gas bottles in 1824, he found the matters now known as benzole and butylene. At the same time, he, J. Herschel, and the optician G. Dolland worked at the production of improved optical glasses. In 1856 he researched properties of colloidal solution of gold. One of Faraday's biggest discoveries, namely, that of induction, was made on August 29, 1831. The report on it was to become the first published in a series called "Experimental Researches in Electricity" which dealt with all fields of electricity known at that time. In 1833–1834 Faraday etablished his basic electrochemical laws, and (advised by W. Whewell) introduced the electrochemical nomenclature still in use today. Faraday studied in detail spark discharge and glow discharge, as well as the action in a dielectric (a term coined by Whewell), and the phenomenon of diamagnetism (another term by Whewell). Faraday described all these actions very vividly, by using the terms "magnetic field of force" and "electric field of force." James Clerk Maxwell united these ideas in the notion of an electromagnetic field, and developed them into an electromagnetic theory of light. This theory was mainly based on the rotation of the polarization plane of light by a magnetic field, as proved by Faraday in 1845. Faraday disapproved both of the theory of action at a distance (supported by French and German physico-mathematicians at that time) and the theory of the atom as revived by J. Dalton. (see also p. 518)

Heinrich Fridrich Emil Lenz, b. Feb. 12, 1804, Dorpat–d. Feb. 10, 1865, Paris. This physicist accompanied the German-Russian lieutenant-commander Otto von Kotzebue on the latter's third voyage around the world (1823–1826). In 1834 Lenz established the rule for the direction of an induced current. During the following years (1835–1838), he established a law for the dependency of an electrical resistance on temperature, and succeeded in freezing water by making use of the Peltier effect.

13 Maxwell's Equations

Maxwell's equations are based on the following empirical facts:

(1) The electric charges are the sources and sinks of the vector field of the dielectric displacement density **D**. Hence, for the flux of the dielectric displacement through a surface enclosing the charge we have

$$\frac{1}{4\pi} \oint_{\text{area}} \mathbf{D} \cdot \mathbf{n}\, da = Q = \int_V \rho\, dV$$

Here, **n** is the outward pointing unit normal vector. This relation can be derived from Coulomb's force law.

(2) Faraday's induction law:

$$V = \oint \mathbf{E} \cdot \mathbf{dr} = -\frac{1}{c}\frac{\partial \phi}{\partial t} \qquad \text{with} \quad \phi = \int_{\text{area}} \mathbf{B} \cdot \mathbf{n}\, da$$

(3) The fact that there are no isolated monopoles implies

$$\oint_{\text{area}} \mathbf{B} \cdot \mathbf{n}\, da = 0$$

that is, the magnetic induction is source-free; its field lines are closed curves.

(4) Oersted's or Ampère's law:

$$\oint \mathbf{H} \cdot \mathbf{dr} = \frac{4\pi}{c} I = \frac{4\pi}{c}\int \mathbf{j} \cdot \mathbf{n}\, da$$

From these observations (1)–(4), Maxwell's equations in integral representation can be inferred directly:

$$\oint_{\text{area}} \mathbf{D} \cdot \mathbf{n}\, da = 4\pi \int_V \rho\, dV \tag{13.1}$$

$$\oint \mathbf{E} \cdot \mathbf{dr} = -\frac{1}{c}\frac{\partial}{\partial t} \int_{\text{area}} \mathbf{B} \cdot \mathbf{n}\, da \tag{13.2}$$

$$\oint_{\text{area}} \mathbf{B} \cdot \mathbf{n}\, da = 0 \tag{13.3}$$

$$\oint \mathbf{H} \cdot \mathbf{dr} = \frac{4\pi}{c}\left(\int_{\text{area}} \mathbf{j} \cdot \mathbf{n}\, da + \frac{1}{4\pi}\frac{d}{dt}\int_{\text{area}} \mathbf{D} \cdot \mathbf{n}\, da \right) \tag{13.4}$$

where the last term in equation (13.4) has the dimension of a current (compare equation (13.1)) introduced by Maxwell as the *displacement current*. This additional term is necessary to fulfill the continuity equation

$$\frac{d}{dt}\int_V \rho\, dV = -\oint_{\text{area}} \mathbf{j} \cdot \mathbf{n}\, da \tag{13.5}$$

This relation implies the conservation of the electric charge: The electric charge in a region of space V can change in time only if a current flows through its surface.

A more detailed justification of the displacement current may be given after transforming the equations (13.1)–(13.5) into the associated differential form, which can be done by means of Gauss' and Stokes' theorems. As an example, we consider equation (13.1):

$$\oint_{\text{area}} \mathbf{D} \cdot \mathbf{n}\, da = \int_V \operatorname{div}\mathbf{D}\, dV = 4\pi \int_V \rho\, dV$$

or

$$\int_V \operatorname{div}\mathbf{D}\, dV - 4\pi \int_V \rho\, dV = \int_V (\operatorname{div}\mathbf{D} - 4\pi\rho)\, dV = 0$$

Immediately, $\operatorname{div}\mathbf{D} = 4\pi\rho$. Analogously, we transform the remaining equations (13.2)–(13.5). Then, we obtain *Maxwell's equations in differential form*:

$$\operatorname{div}\mathbf{D} = 4\pi\rho \tag{13.6}$$

$$\operatorname{curl}\mathbf{E} = -\frac{1}{c}\frac{\partial \mathbf{B}}{\partial t} \tag{13.7}$$

$$\operatorname{div}\mathbf{B} = 0 \tag{13.8}$$

$$\operatorname{curl}\mathbf{H} = \frac{4\pi}{c}\mathbf{j} + \frac{1}{c}\frac{\partial \mathbf{D}}{\partial t} = \frac{4\pi}{c}\left(\mathbf{j} + \frac{1}{4\pi}\frac{\partial \mathbf{D}}{\partial t}\right) \tag{13.9}$$

as well as the *continuity equation*

$$\operatorname{div}\mathbf{j} + \frac{\partial \rho}{\partial t} = 0 \tag{13.10a}$$

In addition, there is the *force law*

$$\mathbf{f} = \rho\mathbf{E} + \frac{1}{c}\mathbf{j} \times \mathbf{B} \tag{13.10b}$$

The force law links electrodynamics and mechanics. At this point, *Ampère's law*, given in (13.9), reads (compare equation (3.10))

$$\operatorname{curl}\mathbf{H} = \frac{4\pi}{c}\mathbf{j} \tag{13.11}$$

In this form, it holds only for *stationary current distributions*. Under the assumption that the same assignment exists also for time-varying currents, equation (13.11) requires a crucial completion in the case that the conduction current is interrupted, e.g., by a capacitor (Figure 13.1): At the plates of the capacitor, the divergence of $\mathbf{j}$ is not equal to zero, while the left-hand side of equation (13.11) is always source-free (div curl $\mathbf{H} = 0$). After introducing the displacement current $(1/4\pi)\partial\mathbf{D}/\partial t$ the vector

$$\mathbf{Z} = 4\pi \left(\mathbf{j} + \frac{1}{4\pi} \frac{\partial\mathbf{D}}{\partial t} \right)$$

is always source-free due to the eqations (13.6) and (13.10a):

$$\operatorname{div}\mathbf{Z} = 4\pi \left(\operatorname{div}\mathbf{j} + \frac{1}{4\pi} \frac{\partial}{\partial t} \operatorname{div}\mathbf{D} \right) = 4\pi \left(\operatorname{div}\mathbf{j} + \frac{\partial\rho}{\partial t} \right) = 0$$

that is, after introducing Maxwell's displacement current there are only closed currents. Without the displacement current $(1/4\pi)\partial\mathbf{D}/\partial t$ in equation (13.9), the system of Maxwell's equations (13.6)–(13.9) would not be consistent if the continuity equation (13.10a) was taken into account. Only the equations completed in this way are consistent.

Maxwell's equations (13.6)–(13.9) are completed by the so-called *connecting equations* relating the vectors $\mathbf{D}$ and $\mathbf{B}$ to the field intensities $\mathbf{E}$ and $\mathbf{H}$:

$$\mathbf{D} = \mathbf{E} + 4\pi\mathbf{P}, \qquad \mathbf{B} = \mathbf{H} + 4\pi\mathbf{M} \tag{13.12}$$

where $\mathbf{P}$ is the vector of the electric polarization, and $\mathbf{M}$ is the vector of the magnetization. The link to mechanics is realized by the force equation (13.10b), which describes the Lorentz force.

In steady fields, for isotropic materials which may be polarized and magnetized in normal manner, there are *constitutive equations* (a special case of (13.12)):

$$\mathbf{D} = \epsilon\mathbf{E} = \mathbf{E} + 4\pi\mathbf{P} = (1 + 4\pi\chi_e)\mathbf{E}$$

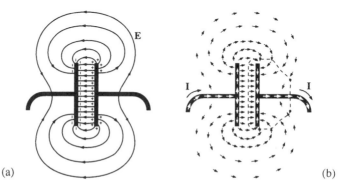

(a)   (b)

Figure 13.1. (a) The electric field of a capacitor connected to a current circuit at a certain instant. The intensity of **E** may change anywhere as a function of time. (b) Illustration of the conducting current (white arrows) and the displacement current (black arrows) at the capacitor. The displacement current guarantees the validity of the continuity equation (13.10a) also at the plates of the capacitor.

$$\mathbf{B} = \mu\mathbf{H} = \mathbf{H} + 4\pi\mathbf{M} = (1 + 4\pi\chi_m)\mathbf{H} \qquad (13.13)$$

The connection of $\mathbf{j}$ and $\mathbf{E}$ yields the *generalized Ohm's law*:

$$\mathbf{j} = \sigma\mathbf{E} \qquad (13.14)$$

where σ is the electric conductivity. In nonlinear media, ϵ, μ, and σ are tensors depending on the field intensities; moreover, in time-varying fields ϵ, μ, and σ are frequency dependent.

To express equation (13.9) in terms of the quantities $\mathbf{E}$ and $\mathbf{B}$, we use the connecting equations (13.12) :

$$\text{curl}\,\mathbf{H} = \frac{4\pi}{c}\mathbf{j} + \frac{1}{c}\frac{\partial\mathbf{D}}{\partial t} \qquad (13.15)$$

$$\text{curl}(\mathbf{B} - 4\pi\mathbf{M}) = \frac{4\pi}{c}\mathbf{j} + \frac{1}{c}\frac{\partial}{\partial t}(\mathbf{E} + 4\pi\mathbf{P}) \qquad (13.16)$$

$$\text{curl}\,\mathbf{B} = \frac{4\pi}{c}(\mathbf{j} + \frac{\partial}{\partial t}\mathbf{P} + c\,\text{curl}\,\mathbf{M}) + \frac{1}{c}\frac{\partial}{\partial t}\mathbf{E} \qquad (13.17)$$

Obviously, the curl of the $\mathbf{B}$-field depends on the densities of the *conducting current* $\mathbf{j}$, the *magnetization current* $\mathbf{j_M} = c\,\text{curl}\,\mathbf{M}$, and the *polarization current* $\mathbf{j_P} = \partial\mathbf{P}/\partial t$. Since in vacuum the current densities $\mathbf{j}$, $\mathbf{j}_M$, and $\mathbf{j}_P$ vanish, only the term $(1/c)\partial\mathbf{E}/\partial t$ in connection with equation (13.7) may explain the propagation of electromagnetic radiation in vacuum. (See Chapter 14.)

Remark Maxwell's equations (13.6)–(13.9) are partial, linear, coupled differential equations of the first order. Due to linearity the *principle of superposition* is valid. Hence, the interference fringes of electromagnetic waves, in particular, and also those of light as special electromagnetic waves are guaranteed.

We have inferred Maxwell's equations (13.6)–(13.10b) including the connecting relations (13.13) and (13.14), step by step from empirical observations: equation (13.6) follows, e.g., from the Coulomb interaction of charges, equation (13.7) from Faraday's induction law, and equation (13.8) from the fact that there are no magnetic monopole charges. Without the displacement current, equation (13.9) represents the second Ampère's law being again equivalent to the Biot-Savart law. The conservation of charge is related to (13.10a) and the first Ampère's law (the Lorentz force) is fixed in (13.10b). So, we have oriented ourselves according to the results of experiments to find, finally, the fundamental equations of electrodynamics, Maxwell's equations. We found also new, unexpected things like the displacement current, which was not contained in the experimental phenomena by which we have been guided.

Now, we will accept Maxwell's equations as fundamental; we will use them as axioms from which to deduce everything that follows. These fundamental equations are further supported by the success of these deductions.

In mechanics everything was much simpler: Newton's force law $\dot{\mathbf{p}} = \mathbf{F}$ could be grasped rapidly; soon we could place it at the top of our theory. The fundamental electromagnetic equations (the Maxwell equations) are much more complicated. Therefore, the way has been much more difficult.

Example 13.1:   Exercise: Ohm's law

We want to demonstrate that from Ohm's law in its differential form

$$\mathbf{j} = \sigma \mathbf{E}$$

under certain assumptions the known integral form

$$\phi_{12} = I R_{12}$$

can be derived. The constant of proportionality σ is the *conductivity* of the material, ϕ_{12} is the difference of the voltage between two points of the conductor, R_{12} is the corresponding resistance, and I is the current.

Solution   We restrict ourselves to fields $\mathbf{E}$ that are constant in time and to uniform charge distributions ρ so that the current distributions are stationary. Now, we consider a cylindrical conductor between two cross-sectional areas a_1 and a_2 having the potentials $\phi(a_1)$ and $\phi(a_2)$, respectively (equipotential surfaces). See Figure 13.2.

In this case, the electric voltage ϕ_{12} between the two surfaces is

$$\phi_{12} = \phi(a_1) - \phi(a_2) = -\int_{\phi_1}^{\phi_2} d\phi = -\int_1^2 (\nabla\phi \cdot \mathbf{ds})$$

$$= \int_1^2 \mathbf{E} \cdot \mathbf{ds} = \int_1^2 \frac{1}{\sigma}(\mathbf{j} \cdot \mathbf{ds}) = I \int_1^2 \frac{ds}{\sigma a}$$

Hence, the integral form of Ohm's law for the voltage is

$$V = \phi_{12} = I R_{12}$$

with $R_{12} = \int_1^2 (1/\sigma a)\, ds$ as the ohmic resistance. Due to the specific resistance $1/\sigma$, the resistance depends on the properties of the material, and due to s and a, it depends on the spatial dimensions of the conductor. For a cylindric conductor composed of a homogeneous material with length l and the constant cross sectional area a, the ohmic resistance is

$$R_{12} = \int_1^2 \frac{ds}{\sigma a} = \frac{l}{\sigma a}$$

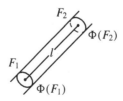

Figure 13.2. Conducting wire with the cross sectional areas a_1 and a_2.

The energy conservation law of electrodynamics: The Poynting vector

The *Poynting theorem* represents the energy conservation law of electrodynamics; it can be derived with the help of Maxwell's equations. For this purpose, we perform the scalar product of equation (13.7) and $-\mathbf{H}$, and of equation (13.9) and $\mathbf{E}$, and then we add both equations. We obtain

$$
\mathbf{E} \cdot \operatorname{curl} \mathbf{H} - \mathbf{H} \cdot \operatorname{curl} \mathbf{E} = \frac{4\pi}{c} \mathbf{j} \cdot \mathbf{E} + \frac{1}{c} \left(\mathbf{E} \cdot \frac{\partial \mathbf{D}}{\partial t} + \mathbf{H} \cdot \frac{\partial B}{\partial t} \right)
\tag{13.18}
$$

With the identity

$$
-\mathbf{E} \cdot \operatorname{curl} \mathbf{H} + \mathbf{H} \cdot \operatorname{curl} \mathbf{E} = \operatorname{div}(\mathbf{E} \times \mathbf{H})
$$

The equation (13.15) can be rewritten

$$
-\operatorname{div}(\mathbf{E} \times \mathbf{H}) = \frac{4\pi}{c} \mathbf{j} \cdot \mathbf{E} + \frac{1}{c} \left(\mathbf{E} \cdot \frac{\partial \mathbf{D}}{\partial t} + \mathbf{H} \cdot \frac{\partial \mathbf{B}}{\partial t} \right)
$$

Multiplying this equation by $-c/4\pi$, then it can be brought to the form

$$
\frac{c}{4\pi} \operatorname{div}(\mathbf{E} \times \mathbf{H}) + \mathbf{j} \cdot \mathbf{E} = -\frac{1}{4\pi} \left(\mathbf{E} \cdot \frac{\partial \mathbf{D}}{\partial t} + \mathbf{H} \cdot \frac{\partial \mathbf{B}}{\partial t} \right)
\tag{13.19}
$$

Now, we must interpret the various terms. At first, based on our experiences, e.g., the electrostatics (see equation (7.3)), we recognize the following: the right-hand side of equation (13.19) is the rate of change of the total energy density of the electromagnetic field:

$$
-\frac{\partial}{\partial t}(w_{\mathrm{el}} + w_{\mathrm{m}}) = -\frac{1}{4\pi} \left(\mathbf{E} \cdot \frac{\partial \mathbf{D}}{\partial t} + \mathbf{H} \cdot \frac{\partial \mathbf{B}}{\partial t} \right)
\tag{13.20a}
$$

This can be seen by introducing the energy density of the electric field,

$$
w_{\mathrm{el}} = \frac{1}{4\pi} \int_0^{\mathbf{D}} \mathbf{E} \cdot d\mathbf{D} = \frac{1}{8\pi} \cdot \frac{\mathbf{D}^2}{\epsilon} = \frac{\epsilon}{8\pi} \cdot \mathbf{E}^2 = \frac{1}{8\pi} \mathbf{E} \cdot \mathbf{D}
\tag{13.20b}
$$

and the energy density of the magnetic field

$$
w_{\mathrm{m}} = \frac{1}{4\pi} \int_0^{\mathbf{B}} \mathbf{H} \cdot d\mathbf{B} = \frac{1}{8\pi} \mathbf{H} \cdot \mathbf{B}
\tag{13.20c}
$$

by substituting these expressions for w_{el} and w_{m} into equation (13.20a) and performing the partial differentiation with respect to the time. Here, we have assumed that $\mathbf{D} = \epsilon \mathbf{E}$ and $\mathbf{B} = \mu \mathbf{H}$, and hence, also that

$$
\frac{\partial \mathbf{E}}{\partial t} \cdot \mathbf{D} = \frac{\partial \mathbf{D}}{\partial t} \cdot \mathbf{E}
$$

$$
\frac{\partial \mathbf{H}}{\partial t} \cdot \mathbf{B} = \frac{\partial \mathbf{B}}{\partial t} \cdot \mathbf{H}
$$

We already have encountered the relations (13.20a)–(13.20c) for the energy density of the electric and the magnetic field in special cases in electrostatics and magnetostatics, respectively. Now, the expressions (13.20a)–(13.20c) follow from the fundamental equations. So, in the first form (13.20a) they are valid generally. Then, we can write the equation (13.19) in the first form

$$\frac{c}{4\pi}\operatorname{div}(\mathbf{E}\times\mathbf{H})+\mathbf{j}\cdot\mathbf{E}=-\frac{\partial}{\partial t}(w_{\mathrm{el}}+w_{\mathrm{m}}) \qquad \textbf{(13.21a)}$$

Equation (13.21a) describes the transfer of energy during a decrease of the total energy density of the electrodynamic field in time. The term $\mathbf{j}\cdot\mathbf{E}$ represents the *work done by the field on the electric current density*. The energy put into the current in this way can be found as the *Joule heat generated per second per unit volume*; it is the thermal power of the field. For example, for the ohmic wire corresponding to Exercise 13.1 we have

$$\int\mathbf{j}\cdot\mathbf{E}dV=\int\sigma E^2\int a\cdot dl=\sigma E^2 a\cdot l=\frac{\sigma V^2 a l}{l^2}=V^2/R=I^2 R$$

This is the *ohmic power of a resistor*. A further reason for the decrease of the energy of the field is the divergence of the vector

$$\mathbf{S}=\frac{c}{4\pi}(\mathbf{E}\times\mathbf{H}) \qquad \textbf{(13.21b)}$$

denoted the *Poynting vector*. This vector describes the intensity of the energy flow per unit time through a unit area perpendicular to the direction of the flow. Strictly, only the surface integral extended over a closed surface

$$\oint_{\mathrm{surface}}\mathbf{S}\cdot\mathbf{n}da$$

has a physical meaning. This can be seen by integrating equation (13.20a) over an arbitrary volume and applying Gauss' theorem

$$-\frac{d}{dt}\int_V (w_{\mathrm{el}}+w_{\mathrm{m}})\,dV=\int_V \mathbf{j}\cdot\mathbf{E}dV+\oint_{\mathrm{surface}}\mathbf{S}\cdot\mathbf{n}da$$

This equation represents the *energy conservation law of electrodynamics* (*Poynting theorem*); the term $\mathbf{S}\cdot\mathbf{n}$ gives the energy flow crossing the area element da per second in direction of the outer normal $\mathbf{n}$. As we will see later on in equation (15.30), for a plane electromagnetic wave the Poynting vector points in direction of the propagation of the energy. The quantity $\mathbf{S}/c^2$ represents the momentum density transferred by a propagating electromagnetic wave that leads to a radiation pressure on the material boundary surfaces. (These results will be derived in the next section.)

The Maxwellian stress tensor: The conservation of the linear momentum in the electromagnetic field

The Poynting theorem in the form

$$-\frac{d}{dt} \int_V (w_{el} + w_m) \, dV = \int_V \mathbf{j} \cdot \mathbf{E} \, dV + \oint_{\text{surface}} \mathbf{S} \cdot \mathbf{n} \, da$$

offers the possibility to derive the momentum of the electromagnetic field. The work done by the electric field $\mathbf{E}$ per unit volume and time, $\mathbf{j} \cdot \mathbf{E}$, implies the conversion of electromagnetic energy into mechanical energy or thermal energy:

$$\int \mathbf{j} \cdot \mathbf{E} \, dV = \frac{d}{dt} E_{\text{mech}}$$

there, E_{mech} is the total mechanical energy of the particles inside the volume V. The energy conservation in a closed system may then be represented as

$$\frac{dE}{dt} = \frac{d}{dt}(E_{\text{mech}} + E_{\text{field}}) = -\oint_{\text{surface}} \mathbf{S} \cdot \mathbf{n} \, da$$

E_{field} is the total field energy per volume, which is given by the equation

$$E_{\text{field}} = \frac{1}{8\pi} \int_V (\mathbf{E} \cdot \mathbf{D} + \mathbf{B} \cdot \mathbf{H}) \, dV$$

The momentum conservation law in electrodynamics

Now that we have found a formulation of the conservation of energy in the electromagnetic field in form of the Poynting theorem we want to set up a corresponding conservation law for the linear momentum. For this purpose, we consider the case that, apart from the electromagnetic field, there are moving charges in the whole space. The total force on these charges follows from the Lorentz-force density

$$\mathbf{f} = \rho \mathbf{E} + \frac{\mathbf{j}}{c} \times \mathbf{B}$$

by volume integration:

$$\mathbf{F} = \frac{d}{dt} \mathbf{P}_{\text{mech}} = \int_V \mathbf{f}(\mathbf{r}) \, dV = \int_V (\rho \mathbf{E} + \frac{\mathbf{j}}{c} \times \mathbf{B}) \, dV \tag{13.22}$$

where $\mathbf{P}_{\text{mech}}$ means the total mechanical (linear) momentum of the particle. The great importance of equation (13.22) lies in the fact that it links *electrodynamics to mechanics* It has to be viewed as the mechanical supplementary equation to Maxwell's equation.

Now, with the help of Maxwell's equations (13.6) and (13.9) we express $\mathbf{j}$ and ρ by the fields. We get

$$\nabla \cdot \mathbf{D} = 4\pi\rho \Rightarrow \rho = \frac{1}{4\pi} \nabla \cdot \mathbf{D}$$

$$\nabla \times \mathbf{H} = \frac{4\pi}{c}\mathbf{j} + \frac{1}{c}\frac{\partial \mathbf{D}}{\partial t} \Rightarrow \mathbf{j} = \frac{c}{4\pi}\left(\nabla \times \mathbf{H} - \frac{1}{c}\frac{\partial \mathbf{D}}{\partial t}\right)$$

And equation (13.19) becomes

$$\frac{d}{dt}\mathbf{P}_{\text{mech}} = \frac{1}{4\pi}\int_V \left(\mathbf{E}(\nabla \cdot \mathbf{D}) + (\nabla \times \mathbf{H}) \times \mathbf{B} - \frac{1}{c}\frac{\partial \mathbf{D}}{\partial t} \times \mathbf{B}\right) dV$$

Adding further the terms $(\nabla \cdot \mathbf{B})\mathbf{H} = 0$ (Maxwell equation (13.8): div $\mathbf{B} = 0$) and utilizing the relation:

$$\frac{\partial \mathbf{D}}{\partial t} \times \mathbf{B} = \frac{\partial}{\partial t}(\mathbf{D} \times \mathbf{B}) - \mathbf{D} \times \frac{\partial \mathbf{B}}{\partial t}$$

our equation becomes

$$\frac{d}{dt}\mathbf{P}_{\text{mech}} = \frac{1}{4\pi}\int_V \Big((\nabla \cdot \mathbf{D})\mathbf{E} + (\nabla \cdot \mathbf{B})\mathbf{H} + (\nabla \times \mathbf{H}) \times \mathbf{B}$$
$$- \frac{1}{c}\frac{\partial}{\partial t}(\mathbf{D} \times \mathbf{B}) + \frac{1}{c}\mathbf{D} \times \frac{\partial \mathbf{B}}{\partial t}\Big)dV \tag{13.23}$$

Substituting the Maxwell equation (13.7), $\nabla \times \mathbf{E} = -(1/c)\partial \mathbf{B}/\partial t$, into equation (13.22), we obtain

$$\frac{d}{dt}\mathbf{P}_{\text{mech}} = \frac{1}{4\pi}\int_V \Big((\nabla \cdot \mathbf{D})\mathbf{E} + (\nabla \cdot \mathbf{B})\mathbf{H} + (\nabla \times \mathbf{H}) \times \mathbf{B}$$
$$- \frac{1}{c}\frac{\partial}{\partial t}(\mathbf{D} \times \mathbf{B}) - \mathbf{D} \times (\nabla \times \mathbf{E})\Big)dV \tag{13.24}$$

or, transferring the term

$$-\frac{1}{c}\frac{\partial}{\partial t}(\mathbf{D} \times \mathbf{B})$$

under the integral sign to the other side

$$\frac{d}{dt}\mathbf{P}_{\text{mech}} + \frac{d}{dt}\left(\frac{1}{4\pi c}\int_V (\mathbf{D} \times \mathbf{B})\, dV\right)$$
$$= \frac{1}{4\pi}\int_V ((\nabla \cdot \mathbf{D})\mathbf{E} + (\nabla \cdot \mathbf{B})\mathbf{H} + (\nabla \times \mathbf{H}) \times \mathbf{B} + (\nabla \times \mathbf{E}) \times \mathbf{D})\, dV \tag{13.25}$$

To simplify the remaining derivation, we consider the electromagnetic field in vacuum, with $\epsilon = \mu = 1$; that is, $\mathbf{E} = \mathbf{D}$ and $\mathbf{B} = \mathbf{H}$. Then, equation (13.25) reads

$$\frac{d}{dt}\left(\mathbf{P}_{\text{mech}} + \frac{1}{4\pi c}\int_V (\mathbf{E} \times \mathbf{B})\, dV\right)$$
$$= \frac{1}{4\pi}\int_V ((\nabla \cdot \mathbf{E})\mathbf{E} + (\nabla \cdot \mathbf{B})\mathbf{B} + (\nabla \times \mathbf{B}) \times \mathbf{B} + (\nabla \times \mathbf{E}) \times \mathbf{E})\, dV \tag{13.26a}$$

The volume integral on the left-hand side of equation (13.26a) may be interpreted as the electromagnetic field momentum in the volume V. The expression

$$\frac{1}{4\pi c}\mathbf{E} \times \mathbf{B} = \frac{\mathbf{S}}{c^2} \qquad (13.26b)$$

has to be regarded as the spatial density of the field momentum $\mathbf{P}_{\text{field}}$. It is proportional to the density of the energy flow, the Poynting vector $\mathbf{S}$, with the factor $1/c^2$

$$\frac{1}{4\pi c}\int_V (\mathbf{E} \times \mathbf{B})\, dV = \mathbf{P}_{\text{field}} = \frac{1}{c^2}\int_V \mathbf{S}\, dV$$

Substituting $\mathbf{P}_{\text{field}}$ into equation (13.26a)

$$\frac{d}{dt}(\mathbf{P}_{\text{mech}} + \mathbf{P}_{\text{field}}) = \frac{1}{4\pi}\int_V [(\nabla \cdot \mathbf{E})\mathbf{E} + (\nabla \cdot \mathbf{B})\mathbf{B} + (\nabla \times \mathbf{E}) \times \mathbf{E} + (\nabla \times \mathbf{B}) \times \mathbf{B}]\, dV$$

$$(13.27)$$

The integrand of equation (13.27) may be expressed as the divergence of a tensor, the *Maxwellian stress tensor*.

To demonstrate this, for the representation of divergence and curl we use Einstein's summation convention, according to which one sums automatically over indices appearing twice. Then, the summation sign is supressed. For example, in Cartesian coordinates:

$$\text{div}\, \mathbf{F} = \mathbf{e}_i \frac{\partial}{\partial x_i} \cdot \mathbf{F} = \mathbf{e}_i \cdot \mathbf{e}_j \frac{\partial F_j}{\partial x_i}$$

$$\text{curl}\, \mathbf{F} = \mathbf{e}_i \frac{\partial}{\partial x_i} \times \mathbf{F} = \mathbf{e}_i \times \mathbf{e}_j \frac{\partial F_j}{\partial x_i}$$

where $\mathbf{F}$ is any differentiable vector field. We now rewrite the part of the integrand of equation (13.27) that depends on the $\mathbf{E}$-field:

$$(\nabla \cdot \mathbf{E})\mathbf{E} + (\nabla \times \mathbf{E}) \times \mathbf{E}$$

$$= \left(\mathbf{e}_i \cdot \mathbf{e}_j \frac{\partial E_j}{\partial x_i}\right) E_k \mathbf{e}_k + \left(\mathbf{e}_i \times \mathbf{e}_j \frac{\partial E_j}{\partial x_i}\right) \times E_k \mathbf{e}_k$$

$$= \left(\delta_{ij} \frac{\partial E_j}{\partial x_i}\right) E_k \mathbf{e}_k + (\mathbf{e}_i \cdot \mathbf{e}_k E_k)\frac{\partial E_j}{\partial x_i}\mathbf{e}_j - \left(\frac{\partial E_j}{\partial x_i}\mathbf{e}_j \cdot \mathbf{e}_k E_k\right)\mathbf{e}_i$$

$(i, j, k = 1, 2, 3)$ and write further

$$(\nabla \cdot \mathbf{E})\mathbf{E} + (\nabla \times \mathbf{E}) \times \mathbf{E} = \left(\frac{\partial E_i}{\partial x_i}\right) E_k \mathbf{e}_k + \delta_{ik} E_k \frac{\partial E_j}{\partial x_i}\mathbf{e}_j - \left(E_k \frac{\partial E_j}{\partial x_i}\delta_{jk}\right)\mathbf{e}_i$$

$$= \left(\frac{\partial E_i}{\partial x_i}\right) E_k \mathbf{e}_k + E_i \left(\frac{\partial E_j}{\partial x_i}\right)\mathbf{e}_j - E_j \left(\frac{\partial E_j}{\partial x_i}\right)\mathbf{e}_i$$

$$= \left(E_i \left(\frac{\partial E_j}{\partial x_j}\right) + E_j \left(\frac{\partial E_i}{\partial x_j}\right) - E_j \left(\frac{\partial E_j}{\partial x_i}\right)\right)\mathbf{e}_i$$

because we can rewrite the indices in the various terms due to the summation over all indices, e.g.,

$$E_j \left(\frac{\partial E_i}{\partial x_i} \right) \mathbf{e}_j = E_i \left(\frac{\partial E_j}{\partial x_j} \right) \mathbf{e}_i$$

In the following, we consider only the ith component of our expression in order to rewrite further

$$[(\nabla \cdot \mathbf{E})\mathbf{E} + (\nabla \times \mathbf{E}) \times \mathbf{E}]_i$$

$$= E_i \left(\frac{\partial E_j}{\partial x_j} \right) + E_j \left(\frac{\partial E_i}{\partial x_j} \right) - E_j \left(\frac{\partial E_j}{\partial x_i} \right)$$

$$= \frac{\partial}{\partial x_j}(E_j E_i) - \frac{1}{2} \frac{\partial}{\partial x_i}(E_j E_j) = \frac{\partial}{\partial x_j} \left(E_j E_i - \frac{1}{2} \delta_{ij}(E_k E_k) \right)$$

A completely analogous result is obtained for the part of the integrand of equation (13.27) that depends only on $\mathbf{B}$:

$$[(\nabla \cdot \mathbf{B})\mathbf{B} + (\nabla \times \mathbf{B}) \times \mathbf{B}]_i = \frac{\partial}{\partial x_j} \left(B_j B_i - \frac{1}{2} \delta_{ij}(B_k B_k) \right) \tag{13.28}$$

With these relations, the integrand of equation (13.27) may be expressed as the divergence of a tensor

$$\frac{d}{dt}(\mathbf{P}_{\text{mech}} + \mathbf{P}_{\text{field}})_i = \frac{1}{4\pi} \int_V \frac{\partial}{\partial x_j} \left(E_i E_j + B_i B_j - \frac{1}{2} \delta_{ij}(E_k E_k + B_k B_k) \, dV \right) \tag{13.29}$$

This is the divergence $(\partial / \partial x_j)$ of the *Maxwellian stress tensor* with the components:

$$T_{ij} = \frac{1}{4\pi} \left(E_i E_j + B_i B_j - \frac{1}{2} \delta_{ij}(E_k E_k + B_k B_k) \right) \qquad (i, j, k = 1, 2, 3) \tag{13.30}$$

Obviously, the Maxwellian stress tensor is a symmetric tensor, that is $T_{ij} = T_{ji}$. We write it in matrix form using the components x, y, and z as indices:

$$4\pi (T_{ij}) = 4\pi \begin{pmatrix} T_{xx} & T_{xy} & T_{xz} \\ T_{yx} & T_{yy} & T_{yz} \\ T_{zx} & T_{zy} & T_{zz} \end{pmatrix}$$

$$= \begin{pmatrix} E_x^2 + B_x^2 - W & E_x E_y + B_x B_y & E_x E_z + B_x B_z \\ E_x E_y + B_x B_y & E_y^2 + B_y^2 - W & E_y E_z + B_y B_z \\ E_x E_z + B_x B_z & E_y E_z + B_y B_z & E_z^2 + B_z^2 - W \end{pmatrix} \tag{13.31}$$

with

$$W = \frac{1}{2}((\mathbf{E} \cdot \mathbf{E}) + (\mathbf{B} \cdot \mathbf{B})) = \frac{1}{2}((E_x^2 + E_y^2 + E_z^2) + (B_x^2 + B_y^2 + B_z^2)) \tag{13.32}$$

Considering the electromagnetic field in space filled with matter, we may obtain the components of the Maxwellian stress tensor in an analogous manner from equation (13.26a) instead of equation (13.27):

$$4\pi(T'_{ij}) = \begin{pmatrix} H_x B_x + D_x E_x - W' & H_x B_y + D_x E_y & H_x B_z + D_x E_z \\ H_x B_y + D_x E_y & H_y B_y + D_y E_y - W' & H_y B_z + D_y D_z \\ H_x B_z + D_x E_z & H_y B_z + D_y D_z & H_z B_z + D_z E_z - W' \end{pmatrix}$$

(13.33)

$$W' = \frac{1}{2}((\mathbf{E} \cdot \mathbf{D}) + (\mathbf{B} \cdot \mathbf{H}))$$

$$= \frac{1}{2}((E_x D_x + E_y D_y + E_z D_z) + (B_x H_x + B_y H_y + B_z H_z))$$

(13.34)

The trace of the Maxwellian stress tensor has the value

$$\mathrm{Tr}(T'_{ij}) = T'_{ii} = -\frac{1}{4\pi} W' = -\frac{1}{8\pi}(\mathbf{E} \cdot \mathbf{D} + \mathbf{B} \cdot \mathbf{H}) = -(w_{\mathrm{el}} + w_{\mathrm{m}})$$

(13.35)

With the help of the Maxwellian stress tensor, the momentum equation (13.22) can be written as

$$\frac{d}{dt}(\mathbf{P}_{\mathrm{mech}} + \mathbf{P}_{\mathrm{field}})_i = \int_V \frac{\partial}{\partial x_j} T_{ij}\, dV$$

(13.36)

Using Gauss' theorem, the volume integral is rewritten formally into a surface integral, so that

$$\int_V \frac{\partial}{\partial x_j} T_{ij}\, dV = \oint_{\mathrm{surface}} T_{ij} n_j\, da$$

(13.37)

where n_j means the jth component of the unit normal vector (direction cosine) perpendicular to the area da. The validity of (13.37) can be seen at once by the remark that $\{T_{i1}, T_{i2}, T_{i3}\}$ represent the components of the ith row vector of the Maxwellian tensor; so equation (13.37) is just the usual Gauss' theorem for the ith row vector. With this transformation, we obtain the *conservation law of the linear momentum*:

$$\frac{d}{dt}(\mathbf{P}_{\mathrm{mech}} + \mathbf{P}_{\mathrm{field}})_i - \oint_{\mathrm{area}} T_{ij} n_j\, da = 0$$

(13.38)

(for the ith component). If the area lies in a region shielded from the field or at infinity, then the surface integral vanishes. From (13.38) we learn that only the *total momentum = mechanical momentum plus field momentum* is conserved. It is clear that the quantity $-T_{ij} n_j$ does not vanish in general. Obviously, it represents the ith component of the momentum *flow* through the unit of area. We stress the term momentum *flow*, that is, momentum per second through unit area, thus a rate of change of the momentum per unit of area. Then, the expression $-T_{ij} n_j \Delta a$ may be understood as the ith component of a force

$$\Delta F_i = -T_{ij} n_j \Delta a$$

(13.39)

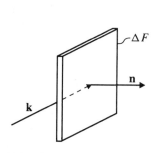

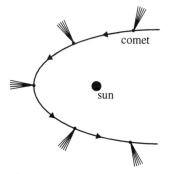

Figure 13.3. The force exerted by the radiation on the area element **da** = da**n** is given by **F** = $\{F_i\}$, with $F_i = -T_{ij}n_j\Delta a$.

Figure 13.4. The tail of the comet is always directed away from the sun.

on the area element. If the radiation is, e.g., absorbed by a black area element $\Delta\mathbf{a} = \Delta a\mathbf{n}$, then the force

$$\Delta\mathbf{F}\cdot\mathbf{n} = F_i n_i = -T_{ij}n_i n_j\Delta a$$

is exerted on the area Δa by the radiation. According to equation (13.38), the total system consisting of mechanical forces, field forces, and forces exerted by the field on its boundary is at equilibrium, as it has to be. Taking the force per area unit, then we are dealing obviously with a pressure, the *radiation pressure* (Figure 13.3):

$$p_{\text{radiation}} = \frac{\Delta\mathbf{K}\cdot\mathbf{n}}{\Delta a} = -T_{ij}n_i n_j \tag{13.40}$$

This means that, besides energy, momentum also is transferred from the radiation source to an absorber by electromagnetic radiation. The rate of change of this momentum per unit area can be measured on the absorber.

To detect the radiation pressure, Lebedev and Hull used a torsion balance. The metal plates attached to the end of a bar are exposed alternating to radiation with the frequency of the normal mode, and well observable deflections are obtained in resonance.

The electromagnetic radiation pressure reaches extremely high values in the interior of the stars. For the temperatures of many millions of degrees prevailing there, the momentum flow is rather remarkable and plays an important role for the stability of the stars but also for explosions (e.g., supernova). One of the most striking effects on the radiation pressure is found in the observation of comets: the tails of the comets are always directed away from the sun (Figure 13.4). Because the gas of the comets has a very low density and because the radiation acts on the comet for a long time, the radiation pressure originating from the sun is sufficient to create the tails and to orient them.[1]

---

[1] Using Halley's comet as an example, the structure of comets is very vividly discussed in H. Balsiger, H. Fechtig, and J. Geiss, *Scientific American*, Sept. 1988, p. 62.

Example 13.2: Exercise: Energy transport in a conducting wire

The importance of the Poynting vector may be illustrated by the example of energy transport in a wire of radius a carrying a direct current I. Calculate the Joule heat and the Poynting energy flux.

Solution   In a wire, the Joule heat $IV = I^2R$ arises. The Joule heat $I \cdot V/l = I \cdot E$ per unit length is spent. Here, V is the voltage, R is the resistance, and k is the length of the wire. But how did the energy come to the position l? The answer is given by the Poynting vector $\mathbf{S}$. In this case, $\mathbf{E}$ lies along the axis of the wire, and $\mathbf{H} \perp \mathbf{E}$ winds around the wire (see Figure 13.5). Therefore, at the surface of the wire $\mathbf{E} \times \mathbf{H}$ and thus $\mathbf{S} = (c/4\pi)\mathbf{E} \times \mathbf{H}$ are directed into the wire . Further, $|\mathbf{S}| = S = (c/4\pi)E \cdot H = (c/4\pi)E2I/(a \cdot c) = EI/(2\pi a)$. The energy flow into the wire per unit length is

$$S \cdot 2\pi a \cdot 1 = EI$$

and, hence, quantitatively equal to the heat released, $E \cdot I$ per unit length of the wire. At the positions where the energy is spent, the energy does not flow due to electron transport but from the field into the wire, where the density of the energy flow is given by the Poynting vector $\mathbf{S}$. Really, it seems to be rather unlikely from the very beginning that this energy is transported there by the conduction electrons in the wire, because due to their large number they move only with a relatively low average velocity (some cm/sec) throughout the wire even for the largest currents. Correspondingly, for example, for a transmitting antenna the energy flows from the wire to the radiated field.

$$\mathbf{I} = \mathbf{J} \cdot F = \mathbf{J} \cdot \pi a^2$$

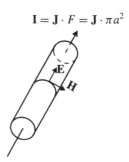

Figure 13.5. Conducting wire. The current I and the field intensities $\mathbf{E}$ and $\mathbf{H}$ are explained.

Example 13.3: Magnetic field energy of a coil

From Maxwell's equations, one obtains the magnetic field energy

$$U_{\text{mag}} = \frac{1}{4\pi} \int \mathbf{H} \cdot d\mathbf{B} \qquad \text{or} \qquad dU_{\text{mag}} = \frac{1}{4\pi}\mathbf{H} \cdot d\mathbf{B}$$

This result may be derived in another way and thus checked and explained: let the wire be wound uniformly on a rod with n turns per cm, cross sectional area q, and length l. A battery of voltage $V^{(e)}$ supplies the current I. (See Figure 13.6). In the time dt, the work

$$dA = I \cdot V^{(e)}dt$$

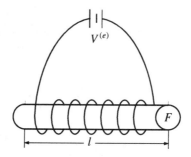

Figure 13.6. On the field energy of a solenoid.

is done by the battery. The power is $dA/dt = I V^{(e)}$. According to the induction law we have

$$I R = V^{(e)} - \frac{1}{c}\frac{d\phi}{dt} = V^{(e)} - \frac{1}{c} n \cdot l \cdot q \frac{dB}{dt}$$

The magnetic field inside the coil is given by

$$H = \frac{4\pi}{c} n I$$

The work done is

$$dA = I \cdot V^{(e)} dt = I\left(I \cdot R + \frac{1}{c} n l q \frac{dB}{dt} \right) dt = I^2 R dt + \frac{q l}{4\pi} H dB$$

Since $q \cdot l$ is equal to the volume of the solenoid, besides the Joule heat $I^2 R$, we find the energy $H dB/4\pi$ delivered to the unit volume of the field. The battery supplies the ohmic resistance as well as the magnetic field of the coil.

Example 13.4:   Exercise: Continuity equation and Maxwell's equations

Derive the continuity equation $\operatorname{div} \mathbf{j} = -\partial\rho/\partial t$ from Maxwell's equations.

Solution   The inhomogenous macroscopic Maxwell equations are

$$\nabla \cdot \mathbf{D} = 4\pi\rho, \qquad \nabla \times \mathbf{H} - \frac{1}{c}\frac{\partial \mathbf{D}}{\partial t} = \frac{4\pi}{c}\mathbf{j}$$

Differentiate the first equation with respect to t and take the divergence of the second one, then

$$\frac{\partial}{\partial t}\nabla \cdot \mathbf{D} = 4\pi\frac{\partial}{\partial t}\rho, \qquad c\nabla \cdot (\nabla \times \mathbf{H}) - \nabla \cdot \frac{\partial}{\partial t}\mathbf{D} = 4\pi\nabla \cdot \mathbf{j}$$

Due to the vector identity $\nabla \cdot (\nabla \times \mathbf{A}) = 0$ adding these equations together yields the assertion

$$\frac{\partial\rho}{\partial t} + \nabla \cdot \mathbf{j} = 0$$

Example 13.5: **Exercise: Magnetic field energy of steady currents**

(a) Show that the energy of the magnetic field of a stationary current distribution $\mathbf{j}(\mathbf{x})$ in vacuum is determined by

$$W = \frac{1}{2c^2} \int d^3x \int d^3x' \frac{\mathbf{j}(\mathbf{x}) \cdot \mathbf{j}(\mathbf{x}')}{|\mathbf{x} - \mathbf{x}'|}$$

(b) The energy of a system of n current-carrying conductors (current I_i) can be described by the quadratic form

$$W = \frac{1}{2} \sum_{i=1}^{n} L_i I^2_i + \sum_{i=1}^{n} \sum_{j>i}^{n} L_{ij} I_i I_j$$

Find the expression for the self-induction coefficients L_i and the mutual induction coefficients L_{ij}.

(c) Calculate the self-induction coefficient defined in (b) for a long current-carrying coaxial line (Figure 13.7). Let the inner conductor (radius a) have the permeability μ_0. The space between the conductors should be filled with a material of permeability μ.

Solution (a) The energy content of a volume containing the electric fields $\mathbf{E}$ and $\mathbf{D}$ is

$$W_{\text{el}} = \frac{1}{8\pi} \int d^3x \, \mathbf{E}(\mathbf{x}) \cdot \mathbf{D}(\mathbf{x})$$

As shown above for the magnetic fields $\mathbf{H}$ and $\mathbf{B}$, we have

$$W_{\text{mag}} = \frac{1}{8\pi} \int d^3x \, \mathbf{H}(\mathbf{x}) \cdot \mathbf{B}(\mathbf{x})$$

Using $\mathbf{B} = \nabla \times A$ and the identity

$$\nabla \cdot (\mathbf{H} \times \mathbf{A}) = \mathbf{A} \cdot (\nabla \times \mathbf{H}) - \mathbf{H} \cdot (\nabla \times \mathbf{A})$$

the integral may be transformed (W_{mag} is simply called W)

$$W = \frac{1}{8\pi} \int d^3x \mathbf{H} \cdot \mathbf{B} = \frac{1}{8\pi} \int d^3x \mathbf{H} \cdot (\nabla \times \mathbf{A})$$

$$= -\frac{1}{8\pi} \int d^3x \nabla \cdot (\mathbf{H} \times \mathbf{A}) + \frac{1}{8\pi} \int d^3x \mathbf{A} \cdot (\nabla \times \mathbf{H})$$

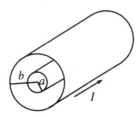

Figure 13.7. Coaxial line.

The first integral contains a divergence, so that according to the Gauss theorem, it can be transformed into a surface integral $\int d\mathbf{a} \cdot (\mathbf{H} \times \mathbf{A})$. If the volume is chosen large enough, this contribution vanishes. With $\nabla \times \mathbf{H} = (4\pi/c)\mathbf{j}$, we have

$$W = \frac{1}{2c} \int d^3x \mathbf{A}(\mathbf{x}) \cdot \mathbf{j}(\mathbf{x})$$

But on the other hand, one obtains the vector potential $\mathbf{A}$ from the current-density distribution

$$\mathbf{A} = \frac{1}{c} \int d^3x' \frac{\mathbf{j}(\mathbf{x}')}{|\mathbf{x} - \mathbf{x}'|}$$

So the relation assumed for the magnetic field energy W has been derived:

$$W = \frac{1}{2c^2} \int d^3x \int d^3x' \frac{\mathbf{j}(\mathbf{x}) \cdot \mathbf{j}(\mathbf{x}')}{|\mathbf{x} - \mathbf{x}'|}$$

(b) For a system of n current-carrying conductors,

$$\mathbf{j}(\mathbf{x}) = \sum_{i=1}^{n} \mathbf{j}_i(\mathbf{x})$$

This expression is substituted into the expression just derived:

$$W = \frac{1}{2} \sum_{i=1}^{n} \frac{1}{c^2} \int d^3x \int d^3x' \frac{\mathbf{j}_i(\mathbf{x}) \cdot \mathbf{j}_i(\mathbf{x}')}{|\mathbf{x} - \mathbf{x}'|}$$
$$+ \sum_{i=1}^{n} \sum_{j=i+1}^{n} \frac{1}{c^2} \int d^3x \int d^3x' \frac{\mathbf{j}_i(\mathbf{x}) \cdot \mathbf{j}_j(\mathbf{x}')}{|\mathbf{x} - \mathbf{x}'|}$$

Now we assume that the current density $\mathbf{j}_i(\mathbf{x})$ is proportional to the current I_i so that the spatial distribution $1/I_i \cdot \mathbf{j}_i(\mathbf{x})$ remains constant for all $\mathbf{x}$ when the current changes. This is surely correct for thin conductors, since the integral $\int d\mathbf{a}_i(1/I_i)\mathbf{j}_i(\mathbf{x}) = 1$ over a cross sectional area of the conductor is constant in any case. Hence, the magnetic energy may be written in the form

$$W = \frac{1}{2} \sum_{i=1}^{n} L_i I_i^2 + \sum_{i=1}^{n} \sum_{j=i+1}^{n} L_{ij} I_i I_j$$

where

$$L_i = \frac{1}{c^2 I_i^2} \int d^3x \, d^3x' \frac{\mathbf{j}_i(\mathbf{x}) \cdot \mathbf{j}_i(\mathbf{x}')}{|\mathbf{x} - \mathbf{x}'|}$$

and

$$L_{ij} = \frac{1}{c^2 I_i I_j} \int d^3x \, d^3x' \frac{\mathbf{j}_i(\mathbf{x}) \cdot \mathbf{j}_j(\mathbf{x}')}{|\mathbf{x} - \mathbf{x}'|}$$

are independent of the currents I_i.

For line-shaped conductors the double integral can be simplified. We may replace $d^3x\mathbf{j}_j$ by $I_i d\mathbf{l}_i$, using an element of the length of the conductor $d\mathbf{l}_i$ as for a current-carrying cylinder. Thus,

$$L_{ij} = \frac{1}{c^2} \oint \oint \frac{d\mathbf{l}_i \cdot d\mathbf{l}_j}{|\mathbf{x} - \mathbf{x}'|}$$

The integration is extended over the entire conducting loop. A similar representation of L_i is not possible for *line-shaped* conductors since their self-inducting coefficients are divergent. For small distances ρ from the conductor, B and $H \sim 1/\rho$, and the magnetic energy becomes infinite (compare Chapter 14).

(c) To calculate the self-inductance L of the coaxial line (a conductor), we have to evaluate the expression for the magnetic energy given above. The magnetic field H is restricted to the interior region $r \leq b$ since the total current through the line (inner conductor and outer conductor) vanishes. From $\oint \mathbf{H} \cdot d\mathbf{s} = \int (4\pi/c)\mathbf{j} \cdot d\mathbf{a}$, we have

$$\text{for } r \leq a: \qquad 2\pi r H = \frac{4\pi}{c} I \left(\frac{r}{a}\right)^2 \rightarrow H = \frac{2Ir}{ca^2}$$

$$\text{for } a \leq r \leq b: \quad 2\pi r H = \frac{4\pi}{c} I \rightarrow H = \frac{2I}{c}\frac{1}{r}$$

Then, the energy is

$$W = \frac{1}{8\pi} \int d^3x \mathbf{B} \cdot \mathbf{H} = \frac{1}{8\pi} \int_0^{2\pi} d\varphi \int_0^h dz \int_o^b B(r) H(r) r\, dr$$

$$= \frac{1}{8\pi} 2\pi h \left[\int_0^a \mu_0 \left(\frac{2I}{c}\frac{r}{a^2}\right)^2 r\, dr + \int_a^b \mu \left(\frac{2I}{c}\frac{1}{r}\right)^2 r\, dr \right]$$

$$= \frac{h}{4}\frac{4I^2}{c^2} \left[\frac{\mu_0}{a^4} \int_0^a r^3 dr + \mu \int_a^b \frac{1}{r} dr \right]$$

$$= \frac{I^2}{c^2} h \left[\frac{1}{4}\mu_0 + \mu \log \frac{b}{a} \right]$$

According to (b), L has been defined by $W = (1/2)LI^2$, so the self-inductance for a coaxial line of length h is

$$L = \frac{1}{c^2} h \left[\frac{1}{2}\mu_0 + 2\mu \log \frac{b}{a} \right]$$

As we can see, L is divergent for $a \rightarrow 0$.

Example 13.6: Exercise: Oscillating circuit

A circuit consists of a coil of inductance L, a capacitor of capacitance C, an ohmic resistor R, and an impressed voltage source of voltage $U^e(t)$. See Figure 13.8. Derive the oscillation equation

$$L\ddot{I} + R\dot{I} + \frac{1}{C}I = \dot{U}^e$$

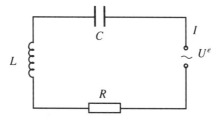

Figure 13.8. Oscillating circuit with the impressed circuit U^e.

with the help of the energy conservation law of electrodynamics (Poynting theorem), where $I = dq/dt$ is the current.

Solution   The Poynting theorem tells us that the change of the mechanical energy and of the field energy in a certain volume V is just given by the power flowing through the surface a:

$$\dot{W}_{\text{mech}}(V) + \dot{W}_{\text{mag}}(V) + \dot{W}_{\text{el}}(V) = L(a)$$

The volume is placed as indicated in Figure 13.9. Electric energy is stored in the capacitor, $W_{\text{el}} = (1/2)q^2/C$, where q is the charge on one capacitor plate. With $\dot{q} = I$, we have $\dot{W}_{\text{el}} = 1/C\dot{q}q = 1/CI \int_{t_0}^{t} I(t')\,dt'$. On the other hand, the energy of the coil is $W_{\text{mag}} = (1/2)LI^2$, or $\dot{W}_{\text{mag}} = LI\dot{I}$. In the resistor the power $\dot{W}_{\text{mech}} = RI^2$ is spent. We imagine the source (generator) to be far away so that the volume may be chosen to be large, too. Then, the power flowing through a consists only of the radiation loss of the switching circuit, which can be neglected for low frequencies, and of the power of the source, $L(a) = U^e I$. Thus, one obtains

$$\frac{1}{C}I \int_{t_0}^{t} I\,dt' + LI\dot{I} + RI^2 = U^e I$$

and after dividing by I and differentiating with respect to the time, we have

$$L\ddot{I} + R\dot{I} + \frac{1}{C}I = \dot{U}^e$$

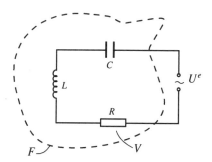

Figure 13.9. On the application of the Poynting theorem to the oscillating circuit.

Example 13.7: Exercise: Radiation pressure and comets

(a) The radiant power of the sun is $3.8 \times 10^{26}\,W$. What radiation pressure acts on a completely absorbing area in the vicinity of the earth ($R = 1.5 \times 10^{11}$m is the mean radius of the earth's orbit)?

(b) What is the value of the pressure exerted by the solar wind on the absorbing area? In the vicinity of the earth, the solar wind consists of about 5 ionized hydrogen atoms per cm^3 moving with a velocity of $v \approx 400$km/s.

(c) How can the appearance of the tail of the comet be explained? Why are there sometimes two tails? (See Figure 13.10.)

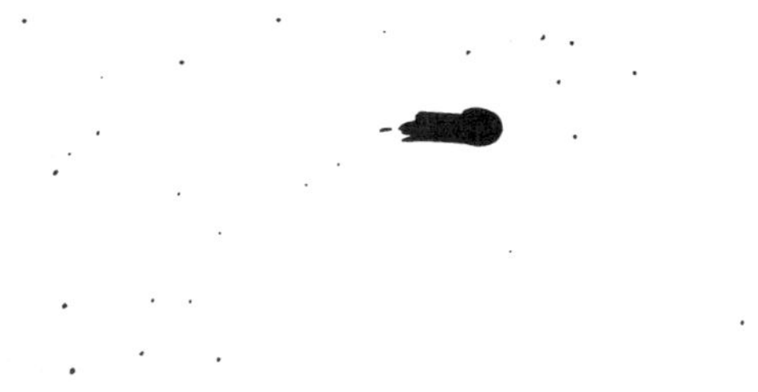

Figure 13.10. Comet Morehouse (1908). The straight "ion tail" goes far to the back. Furthermore, the "dust tail," bent something to the right is visible. Both tails can be seen clearly. Photograph courtesy of the Landessternwarte Heidelberg.

Solution (a) Assuming that the radiation of the sun is isotropic, the radiant power s in the vicinity of the earth is

$$s = \frac{S}{4\pi R^2} = \frac{3.8 \times 10^{26}\,W}{2.83 \times 10^{23}\,m^2} = 0.134\,W\,cm^{-2}$$

where S is the total power of the sun and R is the radius of the earth's orbit. Between the energy E and the momentum k of the electromagnetic radiation there is the relation $E = ck$,

$$p = \text{pressure} = \frac{\text{momentum transfer}}{\text{area} \cdot \text{time}} = \frac{\text{power}}{\text{area} \cdot c}$$

$$= \frac{s}{c} = 4.47 \times 10^{-10}\,N\,cm^{-2}$$

(b) The hydrogen mass is $m_H = 1.67 \times 10^{-27}$ kg. During the time T the momentum transfer to the area q corresponds to the momentum possessed by all the particles with the velocity $\mathbf{v}$ perpendicular to q in a cylinder of length $l = vT$ and the cross sectional area $q = Q$; the momentum transfer $= \rho m_H v l q$, where ρ is the particle density. Thus,

$$p = \text{pressure} = \frac{\text{momentum transfer}}{\text{area} \cdot \text{time}}$$

$$= \frac{Q}{qT} = \rho m_H v^2 = 1.34 \times 10^{-13}\,N\,cm^{-2}$$

that is, the pressure of the solar wind is much smaller than the pressure of the radiation.

(c) According to accepted theories, a comet consists of a small nucleus (diameter at most some kilometers) which may be compared to a dirty snowball. It may consist of ice crystals of various molecules (H_2O, CO_2, NH_3, C_4, ...) mixed with dust. Approaching the sun, the crystals evaporate, and gas and dust escape. The nucleus itself remains invisible, but the emerging gas-dust atmosphere (coma) with a diameter up to 100,000 km is visible. Now, the gas molecules and the dust particles are accelerated by the radiation pressure and by the solar wind, and form the tail of the comet. The acceleration depends on the density, absorptance, reflectance, shape, etc., of the particles and may be evaluated only very inaccurately (at the moment).

Assuming that all particles experience the same acceleration, their velocity is given by $\mathbf{v}(t) = \mathbf{v}_0 + \mathbf{a}t$ where $\mathbf{v}_0$ is the velocity of the comet and $\mathbf{a} = a\mathbf{n}$ is the acceleration of a particle in the direction away from the sun (the attractive force of the sun in the direction $(-\mathbf{n})$ is much smaller).

If the trajectory is curved, the initial velocity $\mathbf{v}_0$ is not conserved and differs for all points of the trajectory giving rise to the curvature of the tail. The greater the acceleration, the straighter is the tail. The dust absorbs or reflects the light. Since the pressure of the radiation is much larger than the pressure of the solar wind, the latter may be neglected here. Although the dust particles are rather heavy, they are accelerated appreciably; the associated fraction of the type II tail is homogeneous and curved more heavily.

By contrast, the gas is transparent, thus interacting only with the solar wind. On the other hand, one observes high velocities in the straight type I tail that cannot be explained simply by the pressure of the solar wind and the small mass of the molecules. Instead, one assumes that the gas molecules are ionized first by the solar wind, and the resulting plasma interacts with the plasma of the solar wind in a very complicated manner, which is far from understood. Also, magnetic effects (interplanetary magnetic field $\geq 10^5$ gauss, "magnetic storms," etc.) certainly play a great role in the "structure" of the type I-tails (unfortunately not clearly seen in the picture) caused to a large extend by the inhomogeneities of the nucleus of the comet itself.

Example 13.8: The question of magnetic monopoles

At the moment, there is no experimental evidence for the existence of magnetic charges or monopoles. Nevertheless, there is a great interest in magnetic monopoles, because on the one hand, electrodynamics (Maxwell's equations) would become very symmetric in its structure (e.g., $\operatorname{div} \mathbf{B} = 4\pi\rho_{\mathrm{mag}}$ in analogy to $\operatorname{div} \mathbf{D} = 4\pi\rho_{\mathrm{el}}$) and on the other hand (which would be much more serious), the mere existence of a simple monopole would lead necessarily to the quantization of the electric charge. The quantization of the electric charge in units of the elementary charge e (the charge of the electron) is a great mystery of physics, and its relationship to the existence of magnetic monopoles was pointed out for the first time by Dirac.[2] Let us postulate the existence of *magnetic charge and current densities* ρ_m and $\mathbf{j}_m$ in addition to the known *electric charge and current densities* ρ_e and $\mathbf{j}_e$. Then, the corresponding Maxwell equations are

$$\nabla \cdot \mathbf{D} = 4\pi\rho_e, \qquad \nabla \times \mathbf{H} = \frac{1}{c}\frac{\partial \mathbf{D}}{\partial t} + \frac{4\pi}{c}\mathbf{j}_e$$
$$\nabla \cdot \mathbf{B} = 4\pi\rho_m, \qquad -\nabla \times \mathbf{E} = \frac{1}{c}\frac{\partial \mathbf{B}}{\partial t} + \frac{4\pi}{c}\mathbf{j}_m \tag{13.41}$$

where besides $\partial\rho_e/\partial t + \operatorname{div}\mathbf{j}_e = 0$ also $\partial\rho_m/\partial t + \operatorname{div}\mathbf{j}_m = 0$; that is, the continuity equation, and thus charge conservation, holds for both kinds of charges. Now, one might expect these new Maxwell equations to lead to measurable physical effects distinct from those of the known, ordinary Maxwell equations. But this is not the case. One can see this in the following way: *quadratic forms* like

$$\mathbf{E} = \mathbf{E}'\cos\alpha + \mathbf{H}'\sin\alpha, \qquad \mathbf{D} = \mathbf{D}'\cos\alpha + \mathbf{B}'\sin\alpha$$
$$\mathbf{H} = -\mathbf{E}'\sin\alpha + \mathbf{H}'\cos\alpha, \qquad \mathbf{B} = -\mathbf{D}'\sin\alpha + \mathbf{B}'\cos\alpha \tag{13.42}$$

or $\mathbf{E} \times \mathbf{H}$, $\mathbf{E} \cdot \mathbf{D} + \mathbf{B} \cdot \mathbf{H}$, and also the components of the Maxwellian stress tensor $T_{\mu\nu}$ are invariant under the dual transformations (13.42), that is, $\mathbf{E} \times \mathbf{H} = \mathbf{E}' \times \mathbf{H}'$.

---

[2] P.A.M. Dirac, *Proc. Roy. Soc.* **A113** (1931) 60; *Phys. Rev.* **74** (1948) 817. See also the review of E. Amaldi, "On the Dirac Magnetic Poles" in *Old and New Problems in Elementary Particles*, ed. G. Puppi, Academic Press, New York (1968).

Furthermore, one can find by an easy computation that the Maxwell equations (13.41) are valid unchanged also for $\mathbf{E}'$, $\mathbf{H}'$, $\mathbf{D}'$, and $\mathbf{B}'$ if the sources are transformed analogously to (13.42), namely,

$$
\begin{aligned}
\rho_e &= \rho_e' \cos\alpha + \rho_m' \sin\alpha, & \mathbf{j}_e &= \mathbf{j}_e' \cos\alpha + \mathbf{j}_m' \sin\alpha \\
\rho_m &= -\rho_e' \sin\alpha + \rho_m' \cos\alpha, & \mathbf{j}_m &= -\mathbf{j}_e' \sin\alpha + \mathbf{j}_m' \cos\alpha
\end{aligned}
\tag{13.43}
$$

By substituting (13.42) and (13.43) into equation (13.41), one verifies easily that

$$
\begin{aligned}
\nabla \cdot \mathbf{D}' &= 4\pi \rho_e', & \nabla \times \mathbf{H}' &= \frac{1}{c}\frac{\partial \mathbf{D}'}{\partial t} + \frac{4\pi}{c}\mathbf{j}_e' \\
\nabla \cdot \mathbf{B}' &= 4\pi \rho_m', & -\nabla \times \mathbf{E}' &= \frac{1}{c}\frac{\partial \mathbf{B}'}{\partial t} + \frac{4\pi}{c}\mathbf{j}_m'
\end{aligned}
\tag{13.44}
$$

This demonstrates the invariance of the new Maxwell equations (13.41) under dual transformations (13.42) and (13.43). But this also demonstrates that to a great extent it is a question of convention what one calls the magnetic charge of a particle. If all particles had the same ratio between the electric and magnetic charge, the angle α would be chosen universally such that

$$
\rho_m = \rho_e'\left(-\sin\alpha + \frac{\rho_m'}{\rho_e'}\cos\alpha\right) = 0
$$

and, hence, also

$$
j_m = j_e\left(-\sin\alpha + \frac{j_m'}{j_e'}\cos\alpha\right) = j_e\left(-\sin\alpha + \frac{\rho_m'}{\rho_e'}\cos\alpha\right) = 0
$$

For this special angle α, equation (13.44) turns into the old, well-known Maxwell equations. Then, one would fix the charge of the electron by convention:

$$
q_e = -e, \qquad q_m = 0
$$

The charges of all other elementary particles would then be measured using these units. For example, for the proton one finds with a precision of $\sim 10^{-20}$:

$$
q_e^{(p)} = +e, \qquad q_m^{(p)} = 0
$$

This high precision for the vanishing magnetic charge is based on the fact that, for example, the magnetic field intensity of the earth at its surface amounts to 1 gauss only. Of course, it would have to be much larger if $q_m^{(p)} \neq 0$.

From these considerations, we learn the following: the question is not whether all particles in nature have the same ratio of the electric to the magnetic charges. In that case, we would choose the units appropriately such that the usual Maxwell equations are valid. Rather, the question is whether there are different particles with a distinct ratio of the electric and magnetic charges. If this were the case, then magnetic charges would have to be admitted generally, and the Maxwell equations would have to be discussed in the generalized form (13.41).

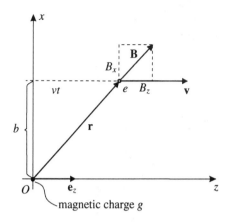

Figure 13.11. An electric charge e flying by a magnetic monopole g placed at the origin.

Now, we want to explain Dirac's basic idea on the relation between charge quantization and the existence of magnetic monopoles in an easily understandable manner:[3] let a magnetic charge g at the origin 0 generate the magnetic induction

$$\mathbf{B}(\mathbf{r}) = \frac{g}{r^2}\frac{\mathbf{r}}{r} \tag{13.45}$$

at the point $\mathbf{r}$. The $\mathbf{B}$-field (13.45) follows from $\operatorname{div}\mathbf{B} = 4\pi\rho(\mathbf{r})$ if $\rho(\mathbf{r}) = g\delta(\mathbf{r}')$ is a point-like density. This is completely analogous to the electromagnetic problem that we have discussed in detail, starting from equation (1.7). Then, let a particle with the charge e and the velocity $\mathbf{v} = v\mathbf{e}_z$ pass by along a straight line at the distance b (this is an approximation for large distances). See Figure 13.11. This particle experiences the Lorentz force

$$\mathbf{F} = F_y \cdot \mathbf{e}_y = \frac{ev}{c}B_x\mathbf{e}_y = \frac{eg}{c}\frac{vb}{(b^2 + v^2 t^2)^{3/2}}\mathbf{e}_y \tag{13.46}$$

In a collision, the particle experiences the momentum transfer

$$\Delta P_y = \int_{-\infty}^{\infty} F_y\, dt = \frac{egvb}{c}\int_{-\infty}^{\infty}\frac{dt}{(b^2 + v^2 t^2)^{3/2}} = \frac{2eg}{cb} \tag{13.47}$$

This momentum transfer leads to a change in the angular momentum

$$\Delta L_z = b\Delta P_y = \frac{2eg}{c} \tag{13.48}$$

Now, according to quantum mechanics, the orbital angular momentum is always quantized:

$$L_z = n\hbar, \qquad n = 0, \pm 1, \pm 2, \ldots \tag{13.49}$$

$\hbar$ is related to Planck's constant h: $\hbar = h/(2\pi)$. Hence, from (13.48) and (13.49),

$$\frac{eg}{\hbar c} = \frac{n}{2}, \qquad n \in N \tag{13.50}$$

---

[3] See A.S. Goldhaber, *Phys. Rev.* **140** (1965) B 1407.

But, this means that $e = (n/2)\left((\hbar c)/g\right)$, $n = 0, \pm 1, \pm 2, \ldots$ has to be quantized. The electric charge e may occur only in positive and negative multiples of the unit $(\hbar c)/(2g)$. As a consequence of the quantization of the angular momentum, the charge is also quantized. One should note: we get this result if anywhere in the world there exists a single magnetic monopole that would be chosen as the starting point (13.45) of our considerations. The coupling strength of magnetic fields is described by the fine structure constant $e^2/(\hbar c) = 1/137$. According to (13.50), the magnetic fine structure constant is $g^2/(\hbar c) = n^2/4\left((\hbar c)/e^2\right) = (137/4)n^2$, so it is much larger than $\left(e^2/(\hbar c)\right)$. We may conclude, that the forces exerted by the magnetic monopoles of the unit magnetic charge g are of the order of those exerted by the unit electric charge e

$$\sqrt{\frac{\frac{g^2}{\hbar c}}{\frac{e^2}{\hbar c}}} = \sqrt{\left(\frac{137}{2}\right)^2 n^2} = \frac{137}{2} n$$

In other words, the unit of magnetic charge exerts about the same force as an electrically charged nucleus with $Z = 137n/2$ protons.[4]

Example 13.9: Exercise: Conservation of angular momentum in the electromagnetic field

A solenoid of radius R with n turns per unit of length carries a stationary current I. Two hollow cylinders of length l are fixed coaxially and freely rotating. One cylinder of radius a is inside the coil ($a < R$) and carries the uniformly distributed charge Q. The outer cylinder of radius b ($b > R$) carries the charge $-Q$. If the current is switched off the cylinders start to rotate.

What is the value of the angular momentum? Where does the angular momentum come from?

Solution

The shut-down of the current causes a change of the magnetic field inside the coil. According to Faraday's induction law this change of the magnetic field induces an electric field

$$\oint \mathbf{E} \cdot d\mathbf{r} = -\frac{1}{c}\frac{d}{dt}\int \mathbf{B} \cdot d\mathbf{a} \tag{13.51}$$

The magnetic field in the interior of a solenoid has been calculated previously

$$\mathbf{B} = \frac{4\pi}{c} I n \mathbf{e}_z \tag{13.52}$$

Taking into account the present geometry we obtain for the cyclic closed field

$$\mathbf{E}(r) = -\frac{2\pi}{c^2} n \frac{dI}{dt}\frac{R^2}{r}\mathbf{e}_\varphi \qquad (r > R)$$

$$\mathbf{E}(r) = -\frac{2\pi}{c^2} n \frac{dI}{dt} r \mathbf{e}_\varphi \qquad (r < R) \tag{13.53}$$

---

[4]For further discussion of this problem, see J. D. Jackson, *Classical Electrodynamics*, John Wiley and Sons, New York (1975), second ed., and the cited article of Amaldi.

The torque acting on the outer cylinder is

$$\mathbf{D}_a = \mathbf{b} \times (-Q\mathbf{E}(b)) = \frac{2\pi}{c^2} Qn\frac{dI}{dt} R^2 \mathbf{e}_z$$

Thus, in the course of the shut-down time the cylinder receives an angular momentum $\int \mathbf{D} dt = \mathbf{L}$:

$$\mathbf{L}_a = \frac{2\pi}{c^2} QnR^2 \int_I^0 \frac{dI}{dt} dt \mathbf{e}_z = -\frac{2\pi}{c^2} QnR^2 I \mathbf{e}_z$$

The torque acting on the inner cylinder may be determined analogously

$$\mathbf{D}_i = -\frac{2\pi}{c^2} Qna^2 \frac{dI}{dt} \mathbf{e}_z \qquad (13.54)$$

After the current has been shut-down the angular momentum of the inner cylinder is

$$\mathbf{L}_i = \frac{2\pi}{c^2} Qna^2 I \mathbf{e}_z \qquad (13.55)$$

Finally, the total angular momentum of the system is

$$\mathbf{L} = \mathbf{L}_a + \mathbf{L}_i = \frac{2\pi}{c^2} QnI \left(a^2 - R^2 \right) \mathbf{e}_z$$

This nonzero angular momentum must have been present initially in the fields. To check the conservation of angular momentum explicitly, we calculate the angular momentum of the fields for the time in which no stationary current has been flowing. Before the current was shut-down we had the electric field of a cylindrical capacitor

$$\mathbf{E}(r) = \frac{2Q}{rl} \mathbf{e}_r \qquad (13.56)$$

and the magnetic field of the solenoid

$$\mathbf{B} = \frac{4\pi}{c} In\mathbf{e}_z \qquad (13.57)$$

The angular momentum of the fields is determined by the Poynting vector $\mathbf{S}$:

$$\mathbf{L}_{\text{field}} = \frac{1}{c^2} \int \mathbf{r} \times \mathbf{S} dV = \frac{1}{4\pi c} \int \mathbf{r} \times (\mathbf{E} \times \mathbf{B}) \, dV \qquad (13.58)$$

With

$$\mathbf{E} \times \mathbf{B} = -\frac{8\pi}{c} \frac{InQ}{rl} \mathbf{e}_\varphi \qquad (13.59)$$

and $V = (R^2 - a^2)\pi l$ The angular momentum of the electromagnetic field is connected with this momentum

$$\mathbf{L}_{\text{field}} = \frac{2\pi}{c^2} QnI \left(a^2 - R^2 \right) \mathbf{e}_z = \mathbf{L}_a + \mathbf{L}_i$$

This is the result just expected. The angular momentum the field looses by the shut-down of the current is transferred to the mechanical angular momentum of the rotating cylinder.

Biographical notes

James Clerk Maxwell, b. June 13, 1831, Edinburgh–d. Nov. 5, 1879, Cambridge. This physicist received a chair at the Marischal College, Aberdeen, in 1856, and at King's College, London, in 1860. From 1865 on he lived in private in his manor house in Glenair, until in 1871 when he was offered the Cavendish professorship for experimental physics in Cambridge. Maxwell renewed Th. Young's theory of trichromatic vision. In 1856–1857 he developed a theory of the stability of the rings of Saturn, and especially improved the kinetic theory of gases. He established an exponential law (known as Maxwell's distribution function) that enabled scientists, to calculate the distribution of the velocity of individual gaseous molecules much more exactly and closer to reality than using the hypotheses of A. Krönig and R. Clausius. His most important achievement was the mathematization of the notion of "lines of force," introduced into physics by M. Faraday. In the treatises "Faraday lines of force" (1855–1856), "On physical lines of force" (1861–1862), and "Dynamical theory of the electromagnetic field" (1865), Maxwell developed the foundation of a law of proximity theory (Maxwellian theory). Maxwell gave a summary of his theory in the two volumes of "Treatise on electricity and magnetism" (1873). He made the notion of "field" prominent in all physical theories. In general, Maxwell's basic representations have influenced the growth of physical knowledge for decades to come. (see also p. 528)

John Henry Poynting, b. Sept. 9, 1852, Monton (n. Manchester)–d. March 30, 1914, Birmingham. From 1880 Poynting was a professor at Mason College and Birmingham University. His experiments with electrodynamic energy flow were of fundamental importance for electrodynamics.

14 Quasi-Stationary Currents and Current Circuits

In engineering, the currents that play an important role are those whose strength is the same at any instant in any section of the conducting wire. In other words, the current is equal at any position of the wire but it may be time-varying. Such currents are called *quasi-stationary* if, additionally, the condition is fulfilled that the current filaments are lines unchanging in time. So, the current density $\mathbf{j}(\mathbf{r}, t)$ should be of the type

$$\mathbf{j}(\mathbf{r}, t) = \mathbf{j}(\mathbf{r})g(t) \tag{14.1}$$

Obviously, one can talk about current tubes constant in time and about the current in the individual tubes. For any segment of the tube one has

$$\int_{\text{surface}} \mathbf{j} \cdot \mathbf{n}\, da = 0 \quad \text{and therefore} \quad \operatorname{div} \mathbf{j} = 0 \tag{14.2}$$

Considering the Maxwell equation

$$\operatorname{curl} \mathbf{H} = \frac{4\pi}{c}\mathbf{j} + \frac{1}{c}\frac{\partial \mathbf{D}}{\partial t} \tag{14.3}$$

it becomes clear that equation (14.2) is valid always if the displacement current density $1/(4\pi)\partial\mathbf{D}/\partial t$ may be neglected compared to the conduction current density $\mathbf{j}$, because then

$$\operatorname{curl} \mathbf{H} = \frac{4\pi}{c}\mathbf{j} \tag{14.4}$$

and therefore $\operatorname{div} \mathbf{j} = c/(4\pi) \operatorname{div} \operatorname{curl} \mathbf{H} \equiv 0$.

Ohm's law for stationary currents reads $IR = V_e$, where V_e is the externally impressed voltage. If the induction flux ϕ through the surface enclosed by the conductor changes, then according to Faraday an additional voltage is induced in the conductor

$$V_i = -\frac{1}{c}\frac{\partial \phi}{\partial t} \tag{14.5}$$

Therefore, the generalized Ohm's law is

$$IR = V_e + V_i = V_e - \frac{1}{c}\frac{\partial \phi}{\partial t}$$ (14.6)

If there are several current circuits, for the kth circuit

$$I_k R_k = (V_e)_k - \frac{1}{c}\frac{\partial \phi_k}{\partial t}$$ (14.7)

where ϕ_k is the induction flux through the kth conductor circuit.

Induction coefficients

We consider n linear circuits. Let the kth circuit have resistance R_k and the applied, external voltage V_k.[1] Let I_k and ϕ_k be the corresponding current and the magnetic flux, respectively, in the kth circuit. There are no permanent magnets. Let the permeability of the embedded material be μ. For the flux ϕ_k we have

$$\phi_k = \int_{a_k} \mathbf{B} \cdot \mathbf{n} \, da$$ (14.8)

where a_k is the area enclosed by the kth circuit. Equation (14.8) may be rewritten further

$$\phi_k = \int_{a_k} (\text{curl}\,\mathbf{A}) \cdot \mathbf{n} \, da = \oint_k \mathbf{A} \cdot d\mathbf{r} = \oint \mathbf{A} \cdot d\mathbf{r}_k$$ (14.9)

According to Stoke's theorem, the line integral is extended along the kth conductor circuit. From equation (3.14)

$$\mathbf{A}(\mathbf{r}) = \frac{\mu}{c} \int \frac{\mathbf{j}(\mathbf{r}')}{|\mathbf{r} - \mathbf{r}'|} dV'$$ (14.10)

The volume integration dV' is restricted to the kth conductor because $\mathbf{j}(\mathbf{r}')dV' = I\,d\mathbf{r}'$, where $d\mathbf{r}'$ is the element of length along the conductor. For several conductor circuits equation (14.10) may be generalized,

$$\mathbf{A}(\mathbf{r}) = \frac{\mu}{c} \sum_{i=1}^{n} I_i \int \frac{d\mathbf{r}_i}{|\mathbf{r} - \mathbf{r}_i|}$$ (14.11)

and substituted into (14.9)

$$\frac{1}{c}\phi_k = \frac{\mu}{c^2} \sum_{i=1}^{n} I_i \int_i \int_k \frac{d\mathbf{r}_i \cdot d\mathbf{r}_k}{|\mathbf{r}_i - \mathbf{r}_k|} = \sum_i L_{ki} I_i$$ (14.12)

---

[1] For the sake of simplicity, we drop the index for the external (impressed) voltage.

Here we have introduced the *induction coefficients*

$$L_{ki} = \frac{\mu}{c^2} \int_i \int_k \frac{d\mathbf{r}_i \cdot d\mathbf{r}_k}{|\mathbf{r}_i - \mathbf{r}_k|} \tag{14.13}$$

which depend on the geometry (relative position, shape etc.) of the conductor circuits. Obviously, the symmetry relation

$$L_{ki} = L_{ik} \tag{14.14}$$

is valid in general. We state: *The fraction of the induction flux through the kth circuit originating from the ith one is* $\phi_k = cL_{ki}I_i$. This summarizes in words what is expressed precisely by equation (14.14).

In equation (14.14), one should pay particular attention to the term $L_{kk}I_k$. This fraction of the induction flux comes from the kth conductor itself: therefore, L_{kk} is also called the *self-induction coefficient* of the kth conductor,

$$L_{kk} = \frac{\mu}{c^2} \int_k \int \frac{d\mathbf{r}_k \cdot d\mathbf{r}_k'}{|\mathbf{r}_k - \mathbf{r}_k'|} \tag{14.15}$$

In calculating this self-induction coefficient, the cross sectional area of the conductor has to be taken into account; the conductor can no longer be viewed to be one-dimensional since the zero of $|\mathbf{r}_k - \mathbf{r}_k'|$ can be treated properly only in this way. The finite section of the conductor corresponds to reality.

Substituting equation (14.14) into (14.7),

$$I_k R_k + L_{k1}\dot{I}_1 + L_{k2}\dot{I}_2 + \cdots + L_{kn}\dot{I}_n = V_k \qquad (k = 1, 2, \ldots, n) \tag{14.16}$$

Here, we have assumed that the induction coefficients are constant in time, meaning that the wires are not allowed to bend as a function of time. Equation (14.16) represents a system of n ordinary differential equations of the first order for the unknown currents $I_1(t), I_2(t), \ldots, I_n(t)$. The applied voltages V_k, resistances R_k, and induction coefficients L_{ki} have to be regarded as given. The voltages V_k may be, time-periodic functions. We will illustrate this soon in some examples and exercises.

Magnetic energy of current circuits

Excluding ferromagnetic materials, the magnetic field energy is

$$W_m = \frac{1}{8\pi} \int \mathbf{H} \cdot \mathbf{B} \, dV \tag{14.17}$$

The magnetic induction $\mathbf{B}$ may be replaced again by curl $\mathbf{A}$. Then

$$\begin{aligned}
\mathbf{H} \cdot \mathbf{B} &= \mathbf{H} \cdot \text{curl}\,\mathbf{A} = \mathbf{H} \cdot \nabla \times \mathbf{A} = \mathbf{A} \cdot \nabla \times \mathbf{H} + \nabla \cdot \mathbf{A} \times \mathbf{H} \\
&= \mathbf{A} \cdot \text{curl}\,\mathbf{H} + \text{div}\,(\mathbf{A} \times \mathbf{H})
\end{aligned} \tag{14.18}$$

Substituting this into (14.17)

$$W_m = \frac{1}{8\pi} \int \mathbf{A} \cdot \operatorname{curl} \mathbf{H} \, dV + \frac{1}{8\pi} \int \operatorname{div} (\mathbf{A} \times \mathbf{H}) \, dV \tag{14.19}$$

The second integral vanishes according to Gauss' theorem; with equation (14.4) the first integral may be transformed to

$$W_m = \frac{1}{2c} \int \mathbf{A} \cdot \mathbf{j} \, dV \tag{14.20}$$

Considering linear conductors, then $\mathbf{j} \, dV = I \, d\mathbf{r}$ and

$$W_m = \frac{1}{2c} \sum_{k=1}^{n} I_k \int \mathbf{A} \cdot d\mathbf{r}_k \tag{14.21}$$

where we have summed over all conductors k. According to equation (14.9) the integral is the induction flux ϕ_k going through the kth conductor so that (14.21) also may be written as

$$W_m = \frac{1}{2c} \sum_{k=1}^{n} I_k \phi_k \tag{14.22}$$

Finally, we take expression (14.14) into account and obtain

$$W_m = \frac{1}{2} \sum_{k=1}^{n} \sum_{i=1}^{n} L_{ik} I_i I_k \tag{14.23}$$

Hence, the magnetic energy of the currents is a quadratic form of the current strengths. One should note the formal relation

$$\frac{\partial W_m}{\partial I_k} = \sum_{i=1}^{n} L_{ik} I_i = \frac{1}{c} \phi_k \tag{14.24}$$

The energy of a system of currents

The energy of a system of currents may change for two reasons: either the current strengths change or the relative positions of the conductors change (the network of the conductors bends itself). The first case occurs if, e.g., the applied voltage $V_k(t)$ changes as a function of time. Due to the change of the relative position, the coefficients of mutual inductance change; if the individual conductors are deformed then also the self-induction coefficients may be changed. Now, we will consider the two cases separately.

(a) **The energy equation for the change of the current strengths.** For the whole system, the work done per unit time by the current sources amounts to

$$\sum_{k=1}^{n} V_k I_k \tag{14.25}$$

This power (work per second) covers the Joule heat $\sum_{k=1}^{n} I_k^2 R_k$ occurring per second, as well as the increase of the magnetic energy $\partial W_m / \partial t$. Therefore, the conservation of energy holds in the following way:

$$\sum_k V_k I_k = \sum_k I_k^2 R_k + \frac{\partial W_m}{\partial t} \tag{14.26}$$

From equation (14.24) we get

$$\frac{\partial W_m}{\partial t} = \sum_i \sum_k L_{ik} \dot{I}_i I_k \tag{14.27}$$

One easily realizes that equation (14.26) may be derived directly from equation (14.16) (multiply of (14.16) with I_k and sum over k).

(b) The energy equation for the change of the geometric properties. Now, we assume that the current circuits are moved relative to each other. Then, on the right-hand side of equation (14.26), the work done per second dA/dt by the electromagnetic forces must be added:

$$\sum_k V_k I_k = \sum_k I_k^2 R_k + \frac{\partial W_m}{\partial t} + \frac{\partial A}{\partial t} \tag{14.28}$$

The rate of change of the magnetic energy is now caused by the change of the induction coefficients:

$$\frac{\partial W_m}{\partial t} = \frac{1}{2} \sum_i \sum_k \frac{dL_{ik}}{dt} I_i I_k \tag{14.29}$$

To calculate a concrete expression for dA we assume that the conductors are imbedded in a medium of permeability μ. Then, the Lorentz force

$$d\mathbf{F} = \frac{1}{c} \mathbf{j} \times \mathbf{B} \, dV = \frac{I}{c} d\mathbf{r} \times \mathbf{B} \tag{14.30}$$

acts on the volume element dV. If the conductor element is shifted by $d\mathbf{s}$ then the work done is

$$dA = d\mathbf{F} \cdot d\mathbf{s} = \frac{I}{c} d\mathbf{s} \cdot (d\mathbf{r} \times \mathbf{B}) = \frac{I}{c} \mathbf{B} \cdot (d\mathbf{s} \times d\mathbf{r}) = \frac{I}{c} d\phi \tag{14.31}$$

where $d\phi = \mathbf{B} \cdot (d\mathbf{s} \times d\mathbf{r})$ is the induction flux threading the area element $d\mathbf{s} \times d\mathbf{r}$ (Figure 14.1).

Now, the quantities in equation (14.31) are labeled by the index k to characterize the kth conductor:

$$dA_k = \frac{I_k}{c} d\phi_k \tag{14.32}$$

According to (14.14),

$$\frac{I}{c} d\phi_k = \sum_i (dL_{ki})_k I_i \qquad (i \neq k) \tag{14.33}$$

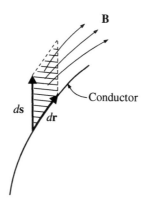

Figure 14.1. The conductor element *dr* is displaced by *ds*.

where $(dL_{ki})_k$ is the change of the induction coefficient L_{ki} due to the motion of the kth conductor. Without loss of generality we drop the condition $i \neq k$ because the change dL_{kk} due to the motion (not the deformation) of the kth conductor vanishes: $(dL_{kk})_k = 0$. The work done by the motion of the kth conductor is obtained by substituting (14.33) into (14.32)

$$dA_k = \sum_i (dL_{ik})_k I_i I_k \tag{14.34}$$

If the ith conductor is moved, one has similarly

$$dA_i = \sum_k (dL_{ik})_i I_i I_k \tag{14.35}$$

Thus, taking into account the motion of all conductors the total work is

$$dA = \sum_k dA_k = \frac{1}{2} \sum_i \sum_k (dA_i + dA_k)$$

$$= \frac{1}{2} \sum_i \sum_k [(dL_{ik})_i + (dL_{ik})_k] I_i I_k = \frac{1}{2} \sum_i \sum_k dL_{ik} I_i I_k \tag{14.36}$$

If the conductors are also deformed, the induction coefficients also change because of that. The formula (14.36) is correct in this form also in this case, because dL_{ik} contains the total change of the induction coefficients due to the motion and deformation of the conductors.

Now, we write for (14.36)

$$\frac{dA}{dt} = \frac{1}{2} \sum_i \sum_k \frac{dL_{ik}}{dt} I_i I_k \tag{14.37}$$

and recognize in comparison with (14.29) that the work done by the electromagnetic forces due to the motion and deformation of the conductors is equal to the work arising from the increase of the magnetic field energy (14.29). The work is performed by no means at the expense of the field energy; to the contrary, the field energy increases by the same amount.

The work is primarily done by the current sources of the individual circuits. For example, if during the motion of a conductor an electromagnetic force is generated by induction, then the voltage on the circuit has to be increased by means of a potentiometer to maintain the current intensity. (The constant current strength is a condition for the validity of the equations (14.29) and (14.36).) The increased voltage corresponds to an increased work by the current.

The natural motion of the conductors is connected with a positive work of the electromagnetic forces. According to our statement, the magnetic energy is increased this way. This circumstance may be used to set up a principle: *for constant currents the magnetic field energy of the conductors tends to a maximum. At equilibrium the maximum is reached.* Let us consider, for example, two rigid conductors. Then, the variable part of the energy is

$$W_{12} = L_{12} I_1 I_2 \tag{14.38}$$

This part of the magnetic energy arises from the common flux of both conductor currents. This flux is represented by the induction lines surrounding jointly both circuits. The two circuits attempt to orient themselves in such a way that the common flux is as large as possible. For plane conductors this will occur if the planes are parallel, the currents are equidirectional, and the regions of lightest density of lines are in a congruent position. The qualitative rule is that parallel, equidirectional currents attract each other. By the way, Coulomb's law leads to the same result if the two current circuits are replaced by magnetic double layers.

Example 14.1:   Exercise: The self-induction coefficient of a coil

Calculate the self-induction coefficient of a coil.

Solution   We assume that the radius of the wire is small compared to the radius R of the coil, and that the latter one is small compared to the length of the coil l. Then, according to Exercise 8.4, in the interior of the coil there are the magnetic field intensity

$$|\mathbf{B}| = \frac{4\pi}{c} \frac{NI}{l} = \frac{4\pi}{c} nI$$

and the energy

$$W_m = \frac{\mu}{8\pi} \int B^2 \, d\tau = \frac{\mu}{8\pi} B^2 R^2 \pi l = \frac{2\pi^2 \mu R^2 n^2}{c^2} l I^2$$

$d\tau$ is the volume element. On the other hand,

$$W_m = \frac{1}{2} L I^2$$

Equating both expressions we obtain

$$L = \frac{4\pi \mu n^2}{c^2} R^2 \pi l$$

This calculation assumes that the total energy is concentrated in the interior of the coil. For long and narrow coils this assumption is permissible, for toroidal coils even more so. If the coil has an iron core, the self-induction coefficient increases appreciably because $\mu \approx 500$ in the core.

Example 14.2: Exercise: The induction coefficient L_{12} of two loops

Calculate the induction coefficient L_{12} of two concentric circular conductors of the radii R_1 and R_2 and the distance b (Figure 14.2).

Solution   Let the radii of the two circles be R_1 and R_2, and the distance of their planes be b. We select two conductor elements $d\mathbf{r}_1$ and $d\mathbf{r}_2$, rotated from each other by the angle ϑ. From the cosine law their distance is

$$r_{12}^2 = b^2 + R_1^2 + R_2^2 - 2R_1 R_2 \cos \vartheta \tag{14.39}$$

and their scalar product is

$$d\mathbf{r}_1 \cdot d\mathbf{r}_2 = dr_1 \, dr_2 \cos \vartheta$$

We start from the general expression for the coefficients of mutual inductance:

$$L_{12} = \frac{\mu}{c^2} \oint \frac{d\mathbf{r}_1 \cdot d\mathbf{r}_2}{r_{12}} \tag{14.40}$$

and, at first, integrate with respect to $dr_2 = R d\vartheta$ for fixed dr_1

$$L_{12} = \frac{\mu}{c^2} \oint dr_1 \oint \frac{R_2 \cos \vartheta \, d\vartheta}{\sqrt{b^2 + R_1^2 + R_2^2 - 2R_1 R_2 \cos \vartheta}} \tag{14.41}$$

The first integral yields $\oint dr_1 = 2\pi R_1$; therefore,

$$L_{12} = \frac{2\pi\mu}{c^2} \int_0^{2\pi} \frac{R_1 \cdot R_2 \cos \vartheta \, d\vartheta}{\sqrt{b^2 + R_1^2 + R_2^2 - 2R_1 R_2 \cos \vartheta}} \tag{14.42}$$

Now, we introduce the following notations:

$$k^2 = \frac{4R_1 R_2}{(R_1 + R_2)^2 + b^2}, \qquad \vartheta = \pi - 2\varphi \tag{14.43}$$

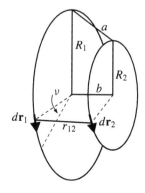

Figure 14.2. Two concentric circular inductors at the distance b.

Then,

$$\cos \vartheta = -\cos 2\varphi = 2\sin^2 \varphi - 1$$

$$r_{12} = \sqrt{b^2 + R_1^2 + R_2^2 - 4R_1 R_2 \sin^2 \varphi + 2R_1 R_2}$$

$$= \sqrt{b^2 + (R_1 + R_2)^2}\sqrt{1 - k^2 \sin^2 \varphi} \tag{14.44}$$

One obtains

$$L_{12} = \frac{4\pi \mu}{c^2}k\sqrt{R_1 R_2} \int_0^{\pi/2} \frac{2\sin^2 \varphi - 1}{\sqrt{1 - k^2 \sin^2 \varphi}} d\varphi \tag{14.45}$$

The integrand may be decomposed into two summands:

$$\frac{2\sin^2 \varphi - 1}{\sqrt{1 - k^2 \sin^2 \varphi}} = \frac{1}{k^2}\left\{ \frac{2 - k^2}{\sqrt{1 - k^2 \sin^2 \varphi}} - 2\sqrt{1 - k^2 \sin^2 \varphi} \right\} \tag{14.46}$$

So, we have succeeded in representing L_{12} as a sum of two monogenic elliptic integrals of the first and the second kind. We use the usual notations

$$K = \int_0^{\pi/2} \frac{d\varphi}{\sqrt{1 - k^2 \sin^2 \varphi}},$$

$$E = \int_0^{\pi/2} \sqrt{1 - k^2 \sin^2 \varphi}\, d\varphi \tag{14.47}$$

and obtain

$$L_{12} = \frac{4\pi \mu}{c^2}\sqrt{R_1 R_2}\left\{ \left(\frac{2}{k} - k\right)K - \frac{2E}{k} \right\} \tag{14.48}$$

For known k the values of the integrals E and K may be taken from corresponding tables.

Until now, the calculations have been performed without simplifications. But now, for sake of simplicity, we want to apply our results to the special case in which the radii of the circular conductors differ only very slightly, and the distance b is small compared to the radii:

$$|R_1 - R_2| \ll R_1, \qquad b \ll R_1$$

Then also the distance

$$a = \sqrt{(R_1 - R_2)^2 - b^2} \tag{14.49}$$

of the peripheries becomes small in comparison with the radii. From the equation

$$\epsilon \equiv 1 - k^2 = \frac{(R_1 - R_2)^2 + b^2}{(R_1 + R_2)^2 + b^2} \approx \left(\frac{a}{2R}\right)^2 \tag{14.50}$$

it is evident that k is about 1, and therefore to a good approximation,

$$E = \int_0^{\pi/2} \cos \varphi\, d\varphi = 1 \tag{14.51}$$

The calculation of K is somewhat more complicated because for $k = 1$ the integrand tends to infinity in the vicinity of $\pi/2$. By the transformation $\psi = \pi/2 - \varphi$ we obtain

$$K = \int_0^{\psi_0} \frac{d\psi}{\sqrt{1 - k^2 \cos^2 \psi}} + \int_{\psi_0}^{\pi/2} \frac{d\psi}{\sqrt{1 - k^2 \cos^2 \psi}} \tag{14.52}$$

Let ψ_0 be a small angle of the order $1 - k^2 \ll \psi_0 \ll 1$. In the second integral the value $k = 1$ may be substituted freely. One obtains

$$\int_{\psi_0}^{\pi/2} \frac{d\psi}{\sin \psi} = \ln \text{tg} \frac{\psi_0}{2} \approx \ln \frac{\psi_0}{2} \tag{14.53}$$

In the first integral ψ takes only small values and one writes $\cos^2 \psi = 1 - \psi^2$. Hence, we obtain for the integral

$$\int_0^{\psi_0} \frac{d\psi}{\sqrt{1 - k^2 + k^2 \psi^2}} = \int_0^{\psi_0} \frac{d\psi}{\sqrt{\varepsilon^2 + k^2 \psi^2}} \tag{14.54}$$

and

$$K = \ln \frac{1}{\varepsilon} \left(\psi_0 + \sqrt{\varepsilon^2 + \psi_0^2} \right) - \ln \frac{\psi_0}{2}$$

Here, under the root sign ε^2 may be neglected compared to ψ_0^2. Thus,

$$K = \ln \frac{2\psi_0}{\varepsilon} - \ln \frac{\psi_0}{2} = \ln \frac{4}{\varepsilon} = \ln \frac{8R}{a}$$

so that

$$L_{12} = \frac{4\pi\mu}{c^2} R \left(\ln \frac{8R}{a} - 2 \right) \tag{14.55}$$

The formula obtained enables us to calculate the mutual induction coefficient for each pair of turns of a coil short compared to its radius. The strict formula (14.48) allows the same for an arbitrary coil. But it is clear that the sum of these coefficients over all pairs does not yield yet the self-induction coefficient of the entire coil. For that it is still necessary to assess the self-induction coefficients of the individual turns. Therefore, our next exercise concerns the determination of the self-induction coefficient of one turn.

Example 14.3:   Exercise: Self-induction coefficient of a circular inductor

Determine the self-induction coefficient of a circular conductor.

Solution   Let the radius of the conductor be R, and that of the wire be r. We assume that $r \ll R$. Let the current be I.

The induction flux generated by the current consists of two parts. The first part proceeds entirely in the outer medium with the permeability μ_a, and it is bounded by the circle of radius $R - r$. The second part is restricted to the conductor with the permeability μ_i. Corresponding to these two parts we decompose the total self-induction coefficient L also into two parts:

$$L = L_a + L_i \tag{14.56}$$

To calculate L_a we introduce an approximation. We imagine the current flow in the middle line of the wire, that is, in a circle of radius R. Then, L_a is defined by the equation (14.14):

$$\frac{1}{c}\phi_a = L_a I \tag{14.57}$$

where ϕ_a is that flux created by the current through a circle of radius R. Hence, L_a is the mutual induction coefficient of the circles of the radii R and $R - r$, respectively, and it follows from equation (14.2) by replacing the distance of the peripheries a by r:

$$L_a = \frac{4\pi\mu_a}{c^2} R \left(\ln \frac{8R}{r} - 2 \right) \tag{14.58}$$

For L_i the definition implied by equation (14.24) turns out to be convenient. Applied to our case equation (14.24) reads

$$W_m = \frac{1}{2} L_i I^2 \tag{14.59}$$

where W_m is the field energy contained in the region of the wire only. This energy may be calculated easily if the field intensity $\mathbf{B}$ is taken from Exercise 8.1 where it was derived for a solenoid:

$$|\mathbf{B}| = \frac{2I}{cr^2}\rho \tag{14.60}$$

ρ denotes the distance from the axis of the wire. The energy referring to the element of length of the wire is

$$\frac{\mu_i}{8\pi} \int_0^r B^2 2\pi\rho \, d\rho = \frac{\mu_i I^2}{4c^2} \tag{14.61}$$

Since the length of the wire is $2\pi R$, the total interior magnetic energy is

$$W_m = \frac{\mu_i \pi I^2}{2c^2} R \tag{14.62}$$

A comparison with equation (14.59) yields

$$L_i = \mu_i \frac{\pi}{c^2} R \tag{14.63}$$

Due to equations (14.58) and (14.63), the full self-induction coefficient of the circular conductor is

$$L = \frac{4\pi R}{c^2} \left\{ \frac{\mu_i}{4} + \mu_a \left(\ln \frac{8R}{r} - 2 \right) \right\} \tag{14.64}$$

The relative magnitude of the two summands contained in the parantheses depends essentially on whether the wire consists of a ferromagnetic material. If it does, then $\mu_i \approx 500$, $\mu_a = 1$. If R/r does not exceed the value 100, then the first term is many times greater than the second one. The predominant fraction of the energy is located in the wire. But, if it is a wire of copper ($\mu_i = 1$), the greatest part of the energy lies in the exterior fields.

Now, by means of equation (14.55) of the preceding exercise and (14.64) the self-induction coefficient of a cylindrical coil can be calculated. For larger numbers of turns the calculation becomes much more difficult to perform. In textbooks on technology, one can find approximation formulas adapted to the particular type of the coil that allow for a more rapid calculation.

Example 14.4: Exercise: Mutual induction of two concentric coils

Calculate the mutual induction of two coils of the length l and with the number of turns N_1 and N_2, respectively, as shown in Figure 14.3.

Solution The inner coil (1) carries a current I_1 generating the magnetic induction

$$|\mathbf{B}_1| = \frac{4\pi}{c} \frac{N_1}{l} I_1$$

Hence, in the outer coil there arises an induction flux

$$\phi_1 = \pi R_1^2 B_1 = \frac{4\pi^2 R_1^2}{c} \frac{N_1}{l} I_1 \tag{14.65}$$

if the strong fields outside the inner coil are neglected. If the current $I_1(t)$ varies in time, according to Faraday's induction law the voltage

$$V_2 = -\frac{N_2}{c} \frac{d\phi_1}{dt} = -\frac{4\pi^2 R_1^2}{c^2} \frac{N_1 N_2}{l} \frac{dI_1}{dt}$$

$$\equiv -L_{12} \frac{dI_1}{dt} \tag{14.66}$$

is induced in the outer coil, and the mutual induction coefficient may be read off directly:

$$L_{12} = \frac{4\pi^2 R_1^2}{c^2} \frac{N_1 N_2}{l} \tag{14.67}$$

If a current $I_2(t)$ flows in the outer coil, generating the magnetic induction field

$$|\mathbf{B}_2| = \frac{4\pi}{c} \frac{N_2}{l} I_2 \tag{14.68}$$

and in the inner coil the induction flux is

$$\phi_2 = \pi R_1^2 B_2 = \frac{4\pi^2 R_1^2}{c} \frac{N_2}{l} I_2 \tag{14.69}$$

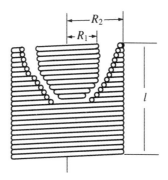

Figure 14.3. Mutual induction of two concentric coils.

then in the inner coil the voltage

$$V_1 = -\frac{N_1}{c}\frac{d\phi_2}{dt} = -\frac{4\pi^2 R_1^2}{c^2}\frac{N_1 N_2}{l}\frac{dI_2}{dt} \equiv -L_{12}\frac{dI_2}{dt} \tag{14.70}$$

is induced. We realize that by this consideration we are let to the same mutual induction coefficient L_{12} as in equation (14.67).

If we are dealing with a single coil only (radius R, number of turns N, length l) then the induction field in the interior is

$$|\mathbf{B}| = \frac{4\pi}{c}\frac{N}{l}I$$

if I is the current. If the current changes, then a countervoltage is induced in the coil itself due to Faraday's induction law. This voltage amounts to

$$V = -\frac{1}{c}\frac{d\phi}{dt} = -\frac{4\pi^2 R^2}{c^2}\frac{N^2}{l}\frac{dI}{dt} \equiv -L\frac{dI}{dt}$$

This gives the self-induction coefficient of the coil

$$L = \frac{4\pi^2 R^2}{c^2}\frac{N^2}{l}$$

If the coil is filled with a medium of permeability μ, the same consideration leads to

$$L = \frac{4\pi^2 R^2}{c^2}\mu\frac{N^2}{l}$$

For iron, with the permeability $\mu = 3000$, the self-induction and the mutual induction would be increased by a factor of 3000.

Example 14.5: Lenz's law

Let a closed conductor be moved in a magnetic field in such a way that the induction flux ϕ crossing it changes. Then, a current I is induced according to Faraday's induction law

$$IR = -\frac{1}{c}\frac{d\phi}{dt} \tag{14.71}$$

At the same time, electromagnetic forces act on the conductor which according to equation (14.31) do the work

$$dA = \frac{I}{c}d\phi = \frac{I}{c}\frac{d\phi}{dt}dt \tag{14.72}$$

If the expression $-1/(Rc)\,d\phi/dt$ following from the next to last equation is substituted for I, one obtains

$$dA = -\frac{1}{Rc^2}\left(\frac{d\phi}{dt}\right)^2 dt \tag{14.73}$$

The work is always negative. In moving the conductor one must overcome the electromagnetic forces. Therefore, the induction has always such a direction, that the occurring electromagnetic forces attempt to prevent the induced motion. This is the familiar *Lenz's law*.

Circuit with resistance and self-induction

We are concerned now with the application of the general equations to the phenomena of currents occurring in one or two conductors at rest.

As the first case, we consider a circuit possessing the *ohmic* resistance R and the self-induction coefficient L. This case is governed by the system of equations (14.16), which for one circuit reduces to

$$RI + L\frac{dI}{dt} = V \tag{14.74}$$

The simplest situation is realized if the voltage V vanishes. Practically, this means that at the moment $t = 0$ a current I_0 is generated in the conductor, e.g., by an induction shock and then is left to itself. The solution of the homogeneous differential equation

$$RI + L\frac{dI}{dt} = 0$$

is

$$I = I_0\, e^{-(R/L)t} \tag{14.75}$$

The current decreases exponentially in time. The decrease of the current proceeds more slowly, the higher the induction coefficient L is in comparison with R. A current of longer duration generates a greater *Joule* heat. But since for a large L also greater work is required to generate the current I_0 there is no contradiction.

The next simplest case is to have a constant voltage V ($V = V_0$). If, e.g., a battery is switched on at time $t = 0$, then the corresponding current is $I_0 = 0$. Now, the equation may be made homogeneous by the substitution $I = \bar{I} + V_0/R$:

$$R\bar{I} + L\frac{d\bar{I}}{dt} = 0$$

The solution is

$$I = \frac{V_0}{R}\left(1 - e^{-(R/L)t}\right) \tag{14.76}$$

The current does not immediately assume the magnitude required by Ohm's law after the voltage is turned on, but only after the decay of the exponential term. See Figure 14.4. We may acquire insight into the energetic relations if equation (14.74) is multiplied by I on both sides and written in the following order

$$VI = I^2 R + LI\frac{dI}{dt} = I^2 R + \frac{d}{dt}\left(\frac{1}{2}LI^2\right) \tag{14.77}$$

The last term is the change of the magnetic energy in time. Now, let the battery be switched off from the circuit. Then, the energy equation becomes

$$I^2 R = -\frac{d}{dt}\left(\frac{1}{2}LI^2\right) \tag{14.78}$$

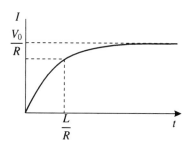

Figure 14.4. Current as a function of time during the switching on of a voltage V_0.

The Joule heat generated at the resistor R by the exponentially decaying current finds its source in the decreasing energy of the field.

Characteristic new insights are gained by the assumption that the voltage changes periodically in time:

$$V = V_0 \cos \omega t$$

V_0 is the amplitude, and ω is the angular frequency $2\pi \nu$ of the alternating voltage being switched on. Now, equation (14.74) becomes

$$RI + L\frac{dI}{dt} = V_0 \cos \omega t \tag{14.79}$$

Keeping in mind that $\cos \omega t$ is the real part of the function $e^{i\omega t}$, the equation becomes more practical if the exponential function is substituted. Then, one obtains a complex solution for I, but the real part satisfies the original equation (14.79). In this case, we write the equation in the form

$$RI + L\frac{dI}{dt} = V_0 e^{i\omega t} \tag{14.80}$$

As we know, the solution of an inhomogeneous differential equation consists of the sum of a particular solution and the complete solution of the homogeneous equation. The latter one is at our disposal already from equation (14.75):

$$I = J e^{-(R/L)t}$$

By variation of the constant J one obtains a particular solution. Therefore, we consider J as a function of t, thus $J = J(t)$. We obtain

$$\frac{dI}{dt} = \left(\frac{dJ}{dt} - \frac{R}{L}J\right) e^{-(R/L)t} \tag{14.81}$$

Substituting the expression for I and dI/dt into equation (14.80), then

$$\frac{dJ}{dt} = \frac{V_0}{L} e^{(R/L+i\omega)t} \tag{14.82}$$

and after integration

$$J(t) = \frac{V_0}{R + i\omega L} e^{(R/L + i\omega)t} + C \tag{14.83}$$

Thus

$$I = \frac{V_0}{R + i\omega L} e^{i\omega t} + Ce^{-(R/L)t} \tag{14.84}$$

The second term decaying in time does not contribute to the development of a steady-state situation and it is therefore neglected (it vanishes after the build-up of the oscillation). Hence, the remaining steady-state solution is

$$I = \frac{V_0}{R + i\omega L} e^{i\omega t} \tag{14.85}$$

Finally, we still have to perform the decomposition into the real and imaginary parts. For this, we write

$$\frac{1}{R + i\omega L} = \frac{R - i\omega L}{R^2 + \omega^2 L^2} = a - ib \tag{14.86}$$

with

$$a = \frac{R}{R^2 + \omega^2 L^2} \quad \text{and} \quad b = \frac{\omega L}{R^2 + \omega^2 L^2} \tag{14.87}$$

With the phase-angle δ defined by

$$tg\delta = \frac{\omega L}{R} \tag{14.88}$$

we obtain

$$I = \frac{V_0}{\sqrt{R^2 + \omega^2 L^2}} e^{i(\omega t - \delta)} \tag{14.89}$$

and the final form of the solution is

$$I = \frac{V_0}{\sqrt{R^2 + \omega^2 L^2}} \cos(\omega t - \delta) \tag{14.90}$$

It turns out that the current has the same frequency as the voltage. This has to be expressed for the steady state. But, the two quantities are not oscillating with the same phase. The current trails the corresponding phase of the electromotive force by the time interval δ/ω. As mentioned above, δ is called the phase angle.

A comparison with *Ohm's* law demonstrates that the expression $\sqrt{R^2 + \omega^2 L^2}$ has to be regarded as the resistance of the circuit. Its value depends not only on the data of the circuit, but also on the frequency of the current. This resistance is called the *impedance*. The term ωL is called the *inductance* or the *inductive resistance*.

Figure 14.5 illustrates the occurring quantities. The ohmic resistance R is drawn from the origin along the real axis. At its endpoint, the inductive resistance ωL is drawn vertically. Briefly formulated, in the complex plane the quantity $R + i\omega L$ is represented by the point P. The absolute value $\overline{OP}$ gives the impedance; the arc δ gives the phase angle.

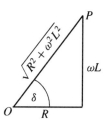

Figure 14.5. On understanding the complex resistance.

The amplitude of the current

$$\frac{V_0}{\sqrt{R^2 + \omega^2 L^2}}$$

is denoted by I_0 and we write

$$I = I_0 \cos(\omega t - \delta) \tag{14.91}$$

At first, we are interested in the power of the periodic electromotive force:

$$VI = V_0 I_0 \cos \omega t \cdot \cos(\omega t - \delta) \tag{14.92}$$

Due to the rapid variations, only the time average is of practical importance. We take the average of the time interval of a period $T = 2\pi/\omega$:

$$\overline{V(t) \cdot I(t)} = V_0 I_0 \frac{1}{T} \int_0^T \cos \omega t \cdot \cos(\omega t - \delta) \, dt$$

$$= V_0 I_0 \frac{1}{T} \int_0^T (\cos^2 \omega t \cdot \cos \delta + \cos \omega t \cdot \sin \omega t \cdot \sin \delta) \, dt$$

the average value of $\cos^2 \omega t$ and $\cos \omega t \cdot \sin \omega t$ are $1/2$ and zero, respectively. Therefore,

$$\overline{VI} = \frac{1}{2} V_0 I_0 \cos \delta = \frac{V_0}{\sqrt{2}} \frac{I_0}{\sqrt{2}} \cos \delta \tag{14.93}$$

According to equation (14.91), $I_0/\sqrt{2}$ is the root of the average value of I^2. Furthermore, $V_0/\sqrt{2} = \sqrt{\overline{V^2}}$. These quantities are called the *effective current* and the *effective electromotive force* (effective voltage, respectively). Hot-wire instruments measure these quantities. Hence, we have

$$\overline{VI} = V_{\text{eff}} I_{\text{eff}} \cos \delta \tag{14.94}$$

If the phase angle δ approaches the value $\pi/2$, the power of the current converges to zero. No Joule heat worth mentioning is generated. In this case one talks of a *wattless current*. Such a current flows, e.g., in an unloaded transformer. Its energy consumption is so small that switching the device off is unnecessary.

Example 14.6: For completion: The vector diagram

Up till now, the analytic solution of equation (14.80) has been discussed. But, this equation also may be solved graphically (see Figure 14.6). As we know, the complex number $a + ib$ is represented by a vector in the complex plane. If the absolute value of the number is denoted by r and its arc by φ, the representation is such that the angle φ is drawn with respect to the real axis, and the value r is measured along its free leg. In these polar coordinates,

$$a + ib = r(\cos \varphi + i \sin \varphi) = re^{i\varphi}$$

The given voltage $V_0 e^{i\omega t}$ may thus be represented by a vector of magnitude V_0 including the variable angle ωt with the real axis. Hence, the "vector V" is rotating at the angular velocity ω about the origin in the direction from the real axis to the imaginary axis. The sum of two complex numbers corresponds to the resultant of the vectors.

Concerning the product of the numbers $r_1 e^{i\varphi_1}$ and $r_2 e^{i\varphi_2}$, the absolute value becomes $r_1 \cdot r_2$, and the arc is $\varphi_1 + \varphi_2$. Since $i = e^{i\pi/2}$, multiplication by i implies the positive rotation of the vector by 90°. The differentiation of a quantity of the form $A e^{i\omega t}$ with respect to time amounts to multiplying with ω.

Hence, assuming that equation (14.80) has a solution of the type $I = I_0 e^{i\omega t}$, it can be written in the following way:

$$RI + i\omega LI = V$$

It means that the vector V has two orthogonal components, RI and ωLI. The given quantity V may be constructed easily with the help of the absolute value V_0 and the arc ωt. During one period the endpoint of the corresponding vector describes a circle about the origin. At any instant the direction of RI is given by the arc $\omega t - \delta$. So V must be decomposed into two perpendicular components. After solving this elementary task, the real components of the vectors V, RI, and ωLI also may be constructed by projection onto the real axis.

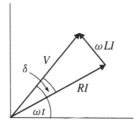

Figure 14.6. The vector diagram.

A circuit with self-induction, capacitance, and resistance

We now complete our previous considerations and imagine a capacitor with the capacitance C included in the circuit (Figure 14.7). The conduction current ends at the plates, the continuation between them is left for the displacement current. The surface currents over the plates, starting from the terminals of the capacitor, and the displacement current through

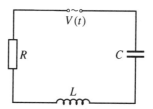

Figure 14.7. Circuit with resistance R, self-inductance L, and capacitance C.

the dielectric remove the linearity of the circuit. Strictly speaking, the closed current must be decomposed into elementary tubes, and the induction current must be calculated for each individual tube. But, if the distance of the plates is small then the induction flux of the individual tubes does not deviate much from what the circuit possesses for a shorted capacitor. Therefore, in this approximation one may talk of a linear current and the electromotive force of the induction may be set equal to $-L\,dI/dt$. But in this way, we neglect the magnetic field in the dielectric of the gap arising from the displacement current. However, this is very weak and restricted to a very small region.

Now, the induction current is not closed. Therefore, the *generalized Ohm's law* must be written in the sense of equation (14.16) in the following form:

$$RI + L\frac{dI}{dt} = V + (V_1 - V_2) \tag{14.95}$$

Here, $V_1 - V_2$ is the voltage (potential difference) at the open element (capacitor). If the direction of the current is referred to that of the potential gradient, then positive I means that the current flows in the direction of the gradient, that is, from the positive plate to the negative one. Therefore, if the instantaneous charge of the positive plate is q, then

$$I = -\frac{dq}{dt} \quad \text{and} \quad q = -\int_{t_0}^{t} I\,dt \tag{14.96}$$

where t_0 is the moment in which the plate has been discharged. The potential difference is

$$V_1 - V_2 = \frac{q}{C} = -\frac{1}{C}\int_{t_0}^{t} I\,dt \tag{14.97}$$

Substituting this expression the current equation becomes

$$RI = V - \frac{1}{C}\int_{t_0}^{t} I\,dt - L\frac{dI}{dt} \tag{14.98}$$

If we eliminate the integral by means of integration with respect to time we obtain the following equation of second order

$$L\ddot{I} + R\dot{I} + \frac{1}{C}I = \dot{V} \tag{14.99}$$

We integrate this equation at first for $V = 0$ and subsequently for $V = V_0 e^{i\omega t}$.

We consider the circuit containing no electromotive force (voltage) to be excited by a single induction shock acting, e.g., on the coil; then the circuit is left to itself. These processes are described by the equation

$$L\ddot{I} + R\dot{I} + \frac{1}{C}I = 0 \tag{14.100}$$

This is analogous to the equation in mechanics that describes damped oscillations (compare Volume 1, Chapter 23 of the series). The corresponding characteristic equation $Lk^2 + Rk + 1/C = 0$ has the roots

$$k_{1,2} = -\frac{R}{2L} \pm \sqrt{\left(\frac{R}{2L}\right)^2 - \frac{1}{CL}} \tag{14.101}$$

Therefore, the solution is

$$I = c_1 e^{k_1 t} + c_2 e^{k_2 t}$$

Now, we have to distinguish two cases, according to the value of the discriminant. If

$$\left(\frac{R}{2L}\right)^2 - \frac{1}{CL} \geq 0 \tag{14.102}$$

then the two roots k_1, k_2 are real and negative. I is not a periodic function of time. We choose the initial condition that the capacitor is discharged, and therefore also current starts developing at the moment $t = 0$. Hence, $c_2 = -c_1$ and

$$I = c_1(e^{k_1 t} - e^{k_2 t})$$

The current increases from zero up to a certain maximum in a time period given by the equation

$$k_1 e^{k_1 t} - k_2 e^{k_2 t} = 0$$

thus,

$$t = \frac{\ln k_2 - \ln k_1}{k_1 - k_2}$$

After having reached the maximum the current decreases rapidly to zero. Physically, the case of a negative discriminant is more important. We utilize the relation

$$\sqrt{\left(\frac{R}{2L}\right)^2 - \frac{1}{CL}} = i\omega$$

and with $c_1 = a/2i$, we write

$$I = \frac{a}{2i} e^{-[R/(2L)]t} \left(e^{i\omega t} - e^{-i\omega t}\right) = ae^{-[R/(2L)]t} \sin \omega t \tag{14.103}$$

The discharging of the capacitor proceeds now in the form of oscillating currents. The plates are charged alternately positively and negatively. If the circuit has a spark gap then not one spark but a sequence of such sparks arises (*Fedderson*, 1858) which may be observed in a rotating mirror; also, photographs may be taken. However, the amplitudes of the oscillation of the current are not constant. The factor $e^{-[R/(2L)]t}$ in equation (14.83) demonstrates that the oscillations are damped. If $(R/2L)^2$ may be neglected compared to $1/(CL)$ the frequency of the oscillation becomes

$$\omega_0 = \frac{1}{\sqrt{LC}} \tag{14.104}$$

This is the so-called *Thomson formula*. Here, ω_0 is the fundamental mode of the circuit. By a suitable dimensioning one may discern that the frequency increases correspondingly. Then, the induction flux going through the coil of the circuit changes very rapidly; $\partial/\partial t$ takes very large values. Tesla used a coil consisting of a few wide turns as the primary winding of a transformer. At the poles of a secondary winding consisting of many turns of a thin wire, very high alternating voltages have been obtained.

We discuss briefly the general case of a periodic external voltage switched on. The differential equation to be solved is

$$L\ddot{I} + R\dot{I} + \frac{1}{C}I = i\omega V_0 e^{i\omega t}$$

In it one observes the structure of the equation of forced oscillations. We restrict ourselves to stationary solutions, and with $I = I_0 e^{i\omega t}$, we obtain the algebraic equation

$$I_0\left(-\omega^2 L + i\omega R + \frac{1}{C}\right) = i\omega V_0$$

and from that

$$I_0 = \frac{V_0}{R + i\left(\omega L - \dfrac{1}{\omega C}\right)}$$

Therefore,

$$I = \frac{V_0 e^{i\omega t}}{R + i\left(\omega L - \dfrac{1}{\omega C}\right)}$$

The solution differs from equation (14.84) only in ωL being replaced by $\omega L - 1/\omega C$. The final real solution (14.85) is

$$I = \frac{V_0}{\sqrt{R^2 + \left(\omega L - \dfrac{1}{\omega C}\right)^2}} \cos(\omega t - \delta) \tag{14.105}$$

But now

$$tg\,\delta = \frac{\omega L - \dfrac{1}{\omega C}}{R}$$

Again, a phase shift exists between the current and the electromotive force. For a pure RC circuit, $L = 0$,

$$tg\,\delta = \frac{1}{\omega C R}$$

The phase of the current leads the phase of the electromotive force. Self-inductance causes a delay, but capacitance causes an acceleration of the phase of the current. In the case $\omega L = 1/\omega C$ the phase shift disappears and the impedance reduces to the ohmic resistance R. At the same time,

$$\omega = \frac{1}{\sqrt{LC}} = \omega_0$$

This has the following meaning: if the frequency of the switched-on electromotive force coincides with the fundamental circuit (resonance case) then the phase shift vanishes, and the amplitude of the current is V_0/R.

The expression $-1/\omega C$ is called the *capacitive resistance*. It becomes smaller for larger capacitance and higher frequency of oscillation. Figure 14.8 shows how the impedance and the phase shift may be represented graphically.

The power of the current is given now also by the expression $E_{\text{eff}} I_{\text{eff}} \cos\delta$. If the value of δ in a RL-circuit is infavourable from the viewpoint of the power then it is easy to see that it may be altered in the desired way by switching on a capacitance.

As we know, the current of an antenna goes through an oscillating circuit tuned to the frequency of the receiver by a variable capacitor (Figure 14.9). What is the resistance of the circuit? For simplicity, let the ohmic resistance be negligible. In the figure the inductive and the capacitive resistance are added. If the resulting resistance is denoted by R_{eff}, then

$$\frac{1}{R_{\text{eff}}} = \frac{1}{\omega L} - \omega C = \frac{1 - \omega^2 LC}{\omega L}$$

Due to the tuning of the circuit, ω becomes equal to the fundamental frequency:

$$\omega^2 = \frac{1}{LC}, \qquad \text{hence,} \qquad \frac{1}{R_{\text{eff}}} = 0, \qquad R_{\text{eff}} = \infty$$

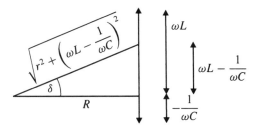

Figure 14.8. On the calculation of the impedance from the ohmic, inductive, and capacitive resistances.

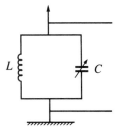

Figure 14.9. The current of an antenna goes through an oscillating circuit.

Taking into account also the ohmic resistance then R_{eff} does not become infinite, but it is still large. The tuned circuit represents an appreciable resistance. A high potential difference occurs at the supply points which is passed on to the first amplifier stage.

Example 14.7:   Exercise: Coupled circuits

Discuss the currents in two circuits that have no capacitance (RL circuits).

Solution   We consider two RL circuits (Figure 14.10). In the first, the voltage $V_0 e^{i\omega t}$ is present; the second one possesses no current source but obtains its excitation by the induction flux the first one sends through its coil. The processes are governed by the differential equations (compare equation (14.16)):

$$L_{11}\dot{I}_1 + L_{12}\dot{I}_2 + R_1 I_1 = V_0 e^{i\omega t}, \qquad L_{12}\dot{I}_1 + L_{22}\dot{I}_2 + R_2 I_2 = 0 \qquad (14.106)$$

If the induction coils of two circuits are arranged in such a way that only a small portion of the flux of the first coil goes to the second one, then one talks of *weak coupling*. Both coils may be installed also on a laminated iron ring as is the case for a transformer. Then the entire induction flux goes through the rings and thus through both coils. Then one talks about a *tight coupling*. However, one should keep in mind that, according to equations (14.13) and (14.15), the induction coefficients are no longer constants if an iron core is present since then μ depends essentially on the field intensity **H**.

Assuming that the external voltage generates currents of equal frequency we can make the following ansatz:

$$I_1 = i_1 e^{i\omega t}, \qquad I_2 = i_2 e^{i\omega t} \qquad (14.107)$$

i_1 and i_2 are complex numbers by means of which possible phase shifts may be described:

$$i_1 = i_1^0 e^{-i\delta_1}, \qquad i_2 = i_2^0 e^{-i\delta_2} \qquad (14.108)$$

Going back, we obtain

$$I_1 = i_1^0 e^{i(\omega t - \delta_1)}, \qquad I_2 = i_2^0 e^{i(\omega t - \delta_2)} \qquad (14.109)$$

δ_1 and δ_2 are the phase delays of the two currents against the electromotive force; i_1^0 and i_2^0 are the amplitudes of the currents. The expressions (14.107) are substituted into the equation (14.106), and we obtain

$$i\omega L_{11}i_1 + i\omega L_{12}i_2 + R_1 i_1 = V_0, \qquad i\omega L_{12}i_1 + i\omega L_{22}i_2 + R_2 i_2 = 0 \qquad (14.110)$$

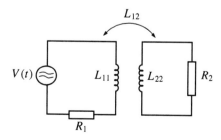

Figure 14.10. Two inductively coupled circuits.

Introducing the abbreviations

$$a_1 = R_1 + i\omega L_{11}, \qquad a_2 = R_2 + i\omega L_{22}, \qquad b = i\omega L_{12}$$

yields

$$a_1 i_1 + b i_2 = V_0, \qquad b i_1 + a_2 i_1 = 0$$

The solution is

$$i_1 = V_0 \frac{a_2}{a_1 a_2 - b^2}, \qquad i_2 = -V_0 \frac{b}{a_1 a_2 - b^2} \tag{14.111}$$

The next step would consist in the separation of the real and imaginary parts to determine the four unknowns i_1^0, i_2^0, δ_1, and δ_2. Let us be satisfied to calculate the ratio of the amplitudes i_2^0 / i_1^0 and the phase difference $\delta_2 - \delta_1$ of the two currents. We obtain

$$\frac{i_2^0}{i_1^0} e^{-i(\delta_2 - \delta_1)} = -\frac{b}{a_2} = -\frac{i\omega L_{12}}{R_2 + i\omega L_{22}}$$

Here the separation of real and imaginary parts may be performed easily. One obtains

$$\frac{i_2^0}{i_1^0} = \frac{\omega L_{12}}{\sqrt{R_2^2 + \omega^2 L_{22}^2}}, \qquad tg(\delta_2 - \delta_1) = -\frac{R_2}{\omega L_{22}}$$

It may be seen that the amplitude of the induced current becomes greater the tighter the coupling (the greater L_{12}) and the smaller the impedance of the second circuit. For $R_2 \ll \omega L_{22}$ the tangent of the phase difference of the two currents becomes a very small negative number, that is, $\delta_2 - \delta_1 = \pi$. The two currents have opposite directions resulting in mutual repulsion. Thus, the experiment of *Elihu Thomson* is explained for which the second conducting circuit was a light ring of aluminium with an extremely small ohmic resistance.

Example 14.8: The transformer

In this example we refer to Example 14.7.

Some relations between the induction coefficients follow from the tight coupling realized for the transformer (Figure 14.11). Let the number of turns of the primary and secondary coils be N_1 and N_2, respectively. Since there is a phase difference between I_1 and I_2 we choose a moment for which the secondary current just vanishes: $I_1 \neq 0$, $I_2 = 0$. Let the magnetic induction flux in the iron core be ϕ at this moment. Because this flux goes through all N_1 turns of the primary circuit, the total flux

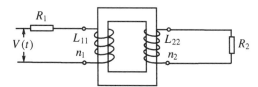

Figure 14.11. A transformer.

through this circuit is $\phi_1 = N_1\phi$. Thus, the secondary flux through the secondary circuit is $\phi_2 = N_2\phi$. On the other hand, from equation (14.12)

$$\phi_1 = cL_{11}I_1, \qquad \phi_2 = cL_{12}I_1$$

Therefore,

$$\frac{\phi_1}{\phi_2} = \frac{L_{11}}{L_{12}} = \frac{N_1}{N_2} \tag{14.112}$$

By similar considerations one finds for the case that at the moment the current is flowing only in the secondary circuit,

$$\frac{L_{12}}{L_{22}} = \frac{N_1}{N_2}$$

From these results we obtain

$$L_{12} = \sqrt{L_{11}L_{22}}, \qquad \frac{L_{11}}{L_{22}} = \frac{N_1^2}{N_2^2} \tag{14.113}$$

The circuit 1 consists of the generator and the primary winding of the transformer. The conductors in this circuit are dimensioned such that the resistance R_1 is vanishingly small. Then, $a_1 = i\omega L_{11}$, and therefore,

$$i_2 = -V_0 \frac{i\omega L_{12}}{i\omega L_{11}(R_2 + i\omega L_{22}) - i^2\omega^2 L_{12}^2} = V_0 \frac{L_{12}}{L_{11}R_2 + i\omega(L_{11}L_{22} - L_{12}^2)}$$

and taking further into account equations (14.112), (14.113),

$$i_2 = \frac{V_0}{R_2}\frac{N_2}{N_1} \tag{14.114}$$

According to the negative sign, the secondary current is opposite to the primary electromotive force. Its amplitude is

$$i_2^0 = \frac{V_0}{R_2}\frac{N_2}{N_1}$$

According to Ohm's law this may be explained such that the voltage in the secondary circuit is N_2/N_1 times the primary voltage. N_2/N_1 is called the step-up ratio of the transformer. If R_2 decreases for increasing load (parallel connection of consumers) then i_2^0 increases correspondingly.

In our considerations L_{22} has been the self-induction coefficient of the secondary winding of the transformer but not that of the entire secondary circuit. Hence, our results are valid only in cases in which the consumer circuit has only ohmic resistance (lamps) but not self-inductance (motors) switched on.

For an open secondary circuit, $R_2 = \infty$. For i_1 we obtain

$$i_1 = \frac{V_0}{a_1} = \frac{V_0}{i\omega L_{11}}i^{-i\pi/2} \tag{14.115}$$

Because now $\delta_1 = \pi/2$, the power of the primary circuit vanishes.

Biographical notes

Nikola Tesla, b. July 9, 1856, Smiljan (Croatia)–d. Jan. 7, 1943, New York. Tesla's career as an inventor began in Hungary at Graz university. In 1844 he emigrated to the United States where he worked with Edison for some time. After a quarrel about the payment for a discovery, he broke off relations. He had a fundamental part in the development of the technical use of alternating current. It was known that electric energy could be transported much more efficiently with high voltage. Tesla constructed adequate converters to raise the voltage of a current before transport, and to reduce it again before consumption. He also worked on the development of motors that run on alternating current. In 1912 he and Edison were awarded the Nobel Prize, the acceptance of which he declined because of his personal quarrels with Edison.

15 Electromagnetic Waves in Vacuum

Starting from Maxwell's equations, we want to treat electromagnetic fields that are rapidly alternating in time and space. At first, we will restrict ourselves to fields in vacuum. In this case, $\mathbf{D} = \mathbf{E}$ and $\mathbf{B} = \mathbf{H}$; furthermore, $\mathbf{j} = 0$ and $\rho = 0$. So, we have Maxwell's equations in the form

$$\nabla \times \mathbf{B} = \frac{1}{c}\frac{\partial \mathbf{E}}{\partial t} \tag{15.1}$$

$$\nabla \times \mathbf{E} = -\frac{1}{c}\frac{\partial \mathbf{B}}{\partial t} \tag{15.2}$$

$$\nabla \cdot \mathbf{E} = 0 \tag{15.3}$$

$$\nabla \cdot \mathbf{B} = 0 \tag{15.4}$$

We may eliminate the vectors $\mathbf{E}$ and $\mathbf{B}$ from the system of equations. Taking the curl of the equation (15.1):

$$\nabla \times (\nabla \times \mathbf{B}) = \frac{1}{c}\frac{\partial}{\partial t}(\nabla \times \mathbf{E})$$

and substituting the equation (15.2), then

$$\nabla \times (\nabla \times \mathbf{B}) = -\frac{1}{c^2}\frac{\partial^2 \mathbf{B}}{\partial t^2} \tag{15.5}$$

Using the identity $\nabla \times (\nabla \times \mathbf{B}) = \nabla(\nabla \cdot \mathbf{B}) - \Delta\mathbf{B}$ due to $\nabla \cdot \mathbf{B} = 0$ we obtain from equation (15.5)

$$\Delta\mathbf{B} = \frac{1}{c^2}\frac{\partial^2 \mathbf{B}}{\partial t^2} \tag{15.6a}$$

Correspondingly, we may eliminate $\mathbf{B}$ and obtain

$$\Delta\mathbf{E} = \frac{1}{c^2}\frac{\partial^2 \mathbf{E}}{\partial t^2} \tag{15.6b}$$

Hence, both vectors $\mathbf{E}$ and $\mathbf{B}$ fulfill the same *wave equation*. We have encountered wave equations already in mechanics in treating oscillation processes. There are various types of solutions for these equations. At first, we notice that the form of the wave equation

$$\left(\Delta - \frac{1}{c^2}\frac{\partial^2}{\partial t^2}\right)u(\mathbf{r}, t) = 0 \tag{15.7}$$

is relativistically invariant because the operator

$$\Delta - \frac{1}{c^2}\frac{\partial^2}{\partial t^2} = \sum_{\mu=1}^{4}\frac{\partial}{\partial x_\mu}\frac{\partial}{\partial x_\mu}$$

is equal to the scalar product of the four-gradient

$$\frac{\partial}{\partial x_\mu} = \left\{\frac{\partial}{\partial x_1}, \frac{\partial}{\partial x_2}, \frac{\partial}{\partial x_3}, \frac{\partial}{\partial ict}\right\}$$

with itself. An equation of the form (15.7) is solved by any function

$$u(\mathbf{r}, t) = u(k_\mu x_\mu) = u(k_1 x_1 + k_2 x_2 + k_3 x_3 + k_4 x_4) \tag{15.8}$$

if $k_\mu k_\mu \equiv \sum_\mu k_\mu^2 = k_1^2 + k_2^2 + k_3^2 + k_4^2 = 0$,[1] that is, k_μ is a four-null-vector (light vector). We verify easily that

$$\frac{\partial}{\partial x_\mu}\frac{\partial}{\partial x_\mu}u(k_\nu x_\nu) = k_\mu k_\mu \frac{d^2 u(z)}{dz^2}, \qquad z = k_\nu x_\nu$$

and therefore,

$$0 = \left(\Delta - \frac{1}{c^2}\frac{\partial^2}{\partial t^2}\right)u(k_\nu x_\nu) = \frac{\partial}{\partial x_\mu}\frac{\partial}{\partial x_\mu}u(k_\nu x_\nu) = k_\mu k_\mu \frac{d^2}{dz^2}u(z)$$

$$= (k_1^2 + k_2^2 + k_3^2 + k_4^2)\frac{d^2}{dz^2}u(z) \Rightarrow k_\mu k_\mu = 0$$

Writing $\hat{k} = \{k_\mu\} = \{k_1, k_2, k_3, i\omega/c\} = \{\mathbf{k}, i\omega/c\}$, the dispersion relation is $k_\mu k_\mu = 0 = \mathbf{k}^2 - \omega^2/c^2$ or $k^2 = \omega^2/c^2$, and thus $k = \omega/c$. Then, the solutions of the wave equations are: $u(z) = u(k_\mu x_\mu) = u(\mathbf{k} \cdot \mathbf{r} - \omega t)$. This may be understood in the framework of a one-dimensional wave equation,

$$\left(\frac{\partial^2}{\partial x^2} - \frac{1}{c^2}\frac{\partial^2}{\partial t^2}\right)u(x \pm ct) = 0$$

Here, $u(x \pm ct)$ are again arbitrary functions possessing always the special combination

$$x + ct = \{x, ict\} \cdot \{1, -i\} = \{x, ict\} \cdot \{k_1', k_4'\}$$

---

[1] We use Einstein's sum convention. This means to sum up automatically over identical indices $k_\mu k_\mu = \sum_{\mu=1}^{4} k_\mu^2$.

or

$$x - ct = \{x, ict\} \cdot \{1, i\} = \{x, ict\} \cdot \{k_1, k_4\}$$

as its argument ($k_\mu x_\mu$ in the general case). In these cases $k_1'^2 + k_4'^2 = 0$ or $k_1^2 + k_4^2 = 0$; hence, these functions are of the type (15.8). Some of them are illustrated in Figure 15.1. At time $t = 0$, the function $u(x \pm ct)$ describes the "mountain" $u(x)$. In the course of time this "mountain" does not change its shape but merely shifts to the position $x \pm ct = 0$.

More precisely, $u(x + ct)$ travels to the left to the position $x = -ct$, and $u(x - ct)$ travels to the right to the position $x = ct$. Obviously, the propagation of the wave train proceeds with the velocity $|v| = |x/t| = c$.

Now, we investigate *plane waves* representing space-time functions for which the geometric position of a certain phase of oscillation at a certain moment t is a plane. The equation of the surface of constant phase (Figure 15.2) is

$$k_\mu x_\mu = \mathbf{k} \cdot \mathbf{r} - \omega t = \text{const} \qquad \text{or} \qquad \mathbf{k} \cdot \mathbf{r} = \text{const} + \omega t \qquad \textbf{(15.9)}$$

Obviously it is the representation of a plane in the so-called *Hessian normal form*.

The wave vector $\mathbf{k}$ is given by

$$\mathbf{k} = \text{grad}(\mathbf{k} \cdot \mathbf{r} - \omega t)$$

It is normal to the plane of the wave. From (15.9), $\mathbf{k}/k \cdot \mathbf{r} = (\text{const} + \omega t)1/k$. The surfaces of constant phase of the solutions $u(k_\mu x_\mu)$ of the wave equation turn out to be planes. Therefore we talk about plane waves. Since $(\mathbf{k}/k) \cdot \mathbf{r} = r_0$ is the distance of the plane from the origin we may conclude from (15.9) that this distance increases according to $r_0 = \text{const} + \omega t/k$. Then, the velocity of propagation of the wave front is $v = dr_0/dt = \omega/k$. As already mentioned, the vector $\mathbf{k}$ is perpendicular to the surface of constant phase and points in the direction of propagation of the wave. Hence, we may

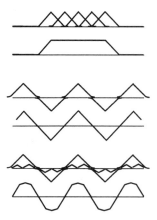

Figure 15.1. Various shapes of wave trains.

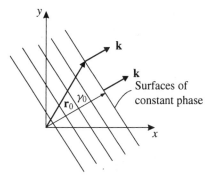

Figure 15.2. Surfaces of constant phase (planes) for plane waves. r_0 is the distance of the phase plane from the origin.

conclude $\mathbf{v} = (\omega/k)\,(\mathbf{k}/k)$. Therefore, an arbitrary plane wave $\mathbf{A}$, being a solution of the wave equation, has the form

$$\mathbf{A}\,(\mathbf{r}, t) = \mathbf{A}_k\,f(\mathbf{k}\cdot\mathbf{r} - \omega t) \tag{15.10}$$

where f is an arbitrary, twice continously diffferentiable function, and $\mathbf{A}_k$ is a vector constant in space and time. From (15.6a), (15.6b), or (15.7), $k^2 = \omega^2/c^2$ or $k = \omega/c$.

This is the *dispersion relation for electromagnetic waves in vacuum*. The relation $k^2 - \omega^2/c^2 = 0$ is the same as $k_\mu k_\mu = 0$ (15.8). The latter is written merely in a 4-dimensional manner. A comparison of both yields

$$\hat{k} = \{k_1, k_2, k_3, k_4\} = \left\{k_1, k_2, k_3, i\frac{\omega}{c}\right\} = \left\{\mathbf{k}, i\frac{\omega}{c}\right\}$$

Taking temporarily the direction of the normal vector $\mathbf{k}/|\mathbf{k}|$ as the z-axis, then

$$A(z, t) = \mathbf{A}_k f(kz - \omega t) = \mathbf{A}_k F\left(z - \frac{\omega t}{k}\right) \tag{15.11}$$

The most general solution for plane waves has the structure

$$\mathbf{A}'(z, t) = \mathbf{A}'_{k1} F\left(z - t\frac{\omega}{k}\right) + \mathbf{A}'_{k2} G\left(z + t\frac{\omega}{k}\right) \tag{15.12}$$

where F and G are functions representing waves propagating in the positive and negative z-direction, respectively. The (generally) complex valued $\mathbf{A}_k$ (amplitude function) gives the direction of oscillation of the plane wave. For $\mathbf{A}_k$ constant in time the wave is *linearly polarized*.

In the following, we will consider *monochromatic plane waves* and the propagation of harmonic oscillations (harmonic time-function f for fixed $\mathbf{r}$). The most general solution is the function

$$\mathbf{E}(\mathbf{r}, t) = \mathbf{E}_{01}\,e^{i(\mathbf{k}\cdot\mathbf{r} - \omega t)} + \mathbf{E}_{02}\,e^{i(\mathbf{k}\cdot\mathbf{r} + \omega t)} \tag{15.13}$$

satisfying the wave equation. Here, we have written $\mathbf{E}(\mathbf{r}, t)$ instead of $\mathbf{A}(\mathbf{r}, t)$ to remind us of the electric field intensity. According to equation (15.6b) it fulfills the wave equation. A corresponding statement holds for $\mathbf{B}$. We restrict ourselves to the explanation of the complex notation of the partial solutions:

$$\mathbf{E} = \mathbf{E}_0\,e^{i(\mathbf{k}\cdot\mathbf{r} - \omega t)} \quad \text{and} \quad \mathbf{B} = \mathbf{B}_0\,e^{i(\mathbf{k}\cdot\mathbf{r} - \omega t)} \tag{15.14}$$

By this notation we mean

$$\mathbf{E} = \text{Re}\left(\mathbf{E}_0\,e^{i(\mathbf{k}\cdot\mathbf{r} - \omega t)}\right) = \mathbf{E}_0\,\cos(\mathbf{k}\cdot\mathbf{r} - \omega t)$$

$$\mathbf{B} = \text{Re}\left(\mathbf{B}_0\,e^{i(\mathbf{k}\cdot\mathbf{r} - \omega t)}\right) = \mathbf{B}_0\,\cos(\mathbf{k}\cdot\mathbf{r} - \omega t) \tag{15.15}$$

since only the real part of the fields has a physical meaning. We make use of the complex notation because many formulas are thereby simplified and it often allows for a more lucid representation. But if a quantity written in a complex manner has to be taken to the second power, then the real part has to be calculated explicitly, and only that part has to be squared!

The restriction of our considerations to monochromatic waves is meaningful insofar as for nonharmonic waves it is always possible to perform a decomposition into monochromatic components by means of a Fourier analysis. A harmonic plane wave represents a wave field spread out periodically everywhere in space and time. Taking the divergence of the partial wave

$$\mathbf{E} = \mathbf{E}_0 \, e^{\, i \, (\mathbf{k} \cdot \mathbf{r} - \omega t)} \tag{15.16}$$

then, assuming $\mathbf{E}_0$ to be constant (in space and time), we obtain

$$\nabla \cdot \mathbf{E} = \mathbf{E}_0 \cdot \nabla \left(e^{i(\mathbf{k} \cdot \mathbf{r} - \omega t)} \right) = \mathbf{E}_0 \cdot (i\mathbf{k}) \, e^{\, i \, (\mathbf{k} \cdot \mathbf{r} - \omega t)} \tag{15.17}$$

Differentiating with respect to time, we obtain

$$\frac{\partial}{\partial t} \mathbf{E} = \mathbf{E}_0 \frac{\partial}{\partial t} \left(e^{\, i \, (\mathbf{k} \cdot \mathbf{r} - \omega t)} \right) = \mathbf{E}_0 (-i\omega) \, e^{i(\mathbf{k} \cdot \mathbf{r} - \omega t)}$$

that is, using the complex notation (compare equations (15.14) and (15.15)), the operators ∇ and $\partial / \partial t$ may be replaced by

$$\nabla \longrightarrow i\mathbf{k}, \qquad \frac{\partial}{\partial t} \longrightarrow -i\omega \tag{15.18}$$

A similar procedure holds for the complex conjugate solutions, for which the operators must be replaced by

$$\nabla \longrightarrow -i\mathbf{k}, \qquad \frac{\partial}{\partial t} \longrightarrow i\omega \tag{15.19}$$

Taking this fact into account by substituting the functions (15.14) into Maxwell's equations (15.1)–(15.4), we obtain the set of equations

$$\mathbf{k} \times \mathbf{B} = -\frac{\omega}{c} \mathbf{E} \tag{15.20}$$

$$\mathbf{k} \times \mathbf{E} = +\frac{\omega}{c} \mathbf{B} \tag{15.21}$$

$$\mathbf{k} \cdot \mathbf{E} = 0 \tag{15.22}$$

$$\mathbf{k} \cdot \mathbf{B} = 0 \tag{15.23}$$

and via the wave equation (15.6a) and (15.6b), the dispersion relation

$$\mathbf{k} \cdot \mathbf{k} = \frac{\omega^2}{c^2} \tag{15.24}$$

Here, $\mathbf{k}$ is the *wave vector* or *wave number vector*:

$$\mathbf{k} = \frac{\omega}{c} \mathbf{n} = \frac{2\pi}{\lambda} \mathbf{n} = \text{const} \tag{15.25}$$

where $\mathbf{n}$ is a unit vector in the direction of propagation of the wave. The magnitude of $\mathbf{k}$ is a measure for the number of wavelengths per unit length. As already mentioned, we get the relation $|\mathbf{k}| = k = \omega/c$ if the ansatz (15.14) is substituted into the wave equation:

$$\left(\Delta - \frac{1}{c^2} \frac{\partial^2}{\partial t^2} \right) \mathbf{E}_0 \, e^{i(\mathbf{k} \cdot \mathbf{r} - \omega t)} = -\left(k^2 - \frac{\omega^2}{c^2} \right) \mathbf{E}_0 \, e^{i(\mathbf{k} \cdot \mathbf{r} - \omega t)} = 0$$

and, therefore, $k^2 = (\omega^2/c^2)$, so in a rather elegant way the dispersion relation (15.24) is obtained.

If the z-axis points in the direction of propagation for a fixed phase $\varphi_0 = kz - \omega t$,

$$kz = \omega t + \varphi_0 \qquad \text{or} \qquad z = \frac{\omega t}{k} + \frac{\varphi_0}{k} = ct + \frac{\varphi_0}{k} \tag{15.26}$$

Thus, the phase velocity of the electromagnetic wave is the velocity of light

$$\frac{dz}{dt} = \frac{\omega}{k} = c \tag{15.27}$$

From equations (15.22) and (15.23) we may infer that $\mathbf{E}$ and $\mathbf{B}$ are oscillating in a plane perpendicular to the wave vector $\mathbf{k}$ (determining the direction of propagation of the wave), and hence, to the direction of motion. So, from Maxwell's equations we find *that the electromagnetic waves are transverse waves.* See Figure 15.3.

From equations (15.20)–(15.23) we can see that $\mathbf{k}$, $\mathbf{E}$, and $\mathbf{B}$, in this order, form an orthogonal right-handed system (Figure 15.4). Due to this orthogonality, equation (15.20) may be written in the form

$$kB = \frac{\omega}{c} E$$

With the relation $k = \omega/c$,

$$B_0 = E_0 \tag{15.28}$$

since $|\exp(i\varphi)| = 1$ for all real values φ. The energy density of the electromagnetic field follows from the time average of the sum of the squares of the field intensity $\mathbf{E}$ and the magnetic induction $\mathbf{B}$. In the calculation of these squares we have to note that only the real parts of $\mathbf{E}$ and $\mathbf{B}$ are allowed to be squared.

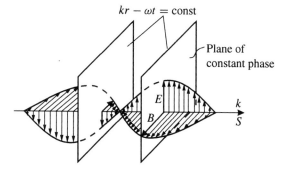

Figure 15.3. Linearly polarized electromagnetic wave.

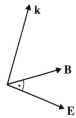

Figure 15.4. The field intensities $\mathbf{E}$ and $\mathbf{B}$, and the wave number vector $\mathbf{k}$ form a right-handed system.

Hence, for the energy density of a monochromatic plane wave we obtain (with $\overline{\cos^2\alpha} = 1/2$):

$$
\begin{aligned}
w &= \frac{1}{8\pi}\left(\overline{\mathbf{E}^2} + \overline{\mathbf{B}^2}\right) \\
&= \frac{1}{8\pi}\left(E_0^2\,\overline{\cos^2(\mathbf{k}\cdot\mathbf{r} - \omega t)} + B_0^2\,\overline{\cos^2(\mathbf{k}\cdot\mathbf{r} - \omega t)}\right) \\
&= \frac{1}{16\pi}\left(E_0^2 + B_0^2\right) = \frac{1}{8\pi}E_0^2
\end{aligned}
\tag{15.29}
$$

due to (15.28). For the Poynting vector of electromagnetic waves describing the energy flux we obtain

$$
\begin{aligned}
\mathbf{S} &= \frac{c}{4\pi}\,\overline{(\mathbf{E}\times\mathbf{B})} = \frac{c}{4\pi}\,\overline{(EB)}\,\frac{\mathbf{k}}{k} \\
&= \frac{c}{4\pi}\,\overline{E^2}\,\frac{\mathbf{k}}{k} = \frac{c}{4\pi}\,E_0^2\,\overline{\cos^2(\mathbf{k}\cdot\mathbf{r} - \omega t)}\,\frac{\mathbf{k}}{k} \\
&= \frac{c}{8\pi}\,E_0^2\,\frac{\mathbf{k}}{k} = cw\,\frac{\mathbf{k}}{k}
\end{aligned}
\tag{15.30}
$$

where w is the energy density (15.29) just calculated. So, the energy flux of the electromagnetic wave per unit of area proceeds in the direction of the wave number vector

$$
\mathbf{k} = \frac{\omega}{c|\mathbf{E}|^2}\,\mathbf{E}\times\mathbf{B}
$$

To explain the photoelectric effect Einstein postulated that electromagnetic waves consist of quanta (photons). These light particles should have the energy $\hbar\omega$. Here, $\hbar$ is Planck's constant divided by 2π (compare Volume 4 of this series: *Quantum Mechanics*). The energy density of a photon in the volume V is

$$
w = \frac{\hbar\omega}{V}
$$

If now $\mathbf{p}$ is the momentum of the photon, then obviously the momentum density has to be $\mathbf{p}/V$. With the relation (15.30)

$$
\frac{\mathbf{p}}{V} = \frac{\mathbf{S}}{c^2} = \frac{w}{c}\,\frac{\mathbf{k}}{k} = \frac{\hbar\omega/c}{V}\,\frac{\mathbf{k}}{k} = \frac{\hbar\mathbf{k}}{V}
$$

thus, $\mathbf{p} = \hbar\mathbf{k}$. That is, if a plane wave is contained in a volume V, and if according to Einstein, one requires that this plane wave represents an energy quantum with the energy $\hbar\omega$, then electrodynamics yields the *de Broglie relation* $\mathbf{p} = \hbar\mathbf{k}$, where $\mathbf{p}$ is the momentum of the contained photon.

Polarization of plane waves

Up till now, we have considered only monochromatic plane waves, for which the amplitudes $\mathbf{E}_0$ and $\mathbf{B}_0$ have been constant with respect to magnitude and direction. Such waves are called *linearly polarized*.

Therefore, it is suitable to introduce the constant unit vectors $\boldsymbol{\epsilon}_1$ and $\boldsymbol{\epsilon}_2$, perpendicular to each other, in directions of $\mathbf{B}_0$ and $\mathbf{E}_0$. They span the plane normal to $\mathbf{k}$, that is, the *polarization plane* (Figure 15.5). Then both solutions

$$\mathbf{E}_1(\mathbf{r}, t) = \boldsymbol{\epsilon}_1 E_1 e^{i(\mathbf{k} \cdot \mathbf{r} - \omega t)}$$

$$\mathbf{E}_2(\mathbf{r}, t) = \boldsymbol{\epsilon}_2 E_2 e^{i(\mathbf{k} \cdot \mathbf{r} - \omega t)}$$

fulfill the wave equation. $\mathbf{B}_1(\mathbf{r}, t)$ belongs to $\mathbf{E}_1(\mathbf{r}, t)$, and $\mathbf{E}_2(\mathbf{r}, t)$, correspondingly to $\mathbf{B}_2(\mathbf{r}, t)$. The amplitudes E_1 and E_2 may be complex-valued to take into account possibly occurring phase differences between E_1 and E_2. A general solution for a plane wave propagating in direction of the wave number vector $\mathbf{k}$ is a linear combination of the partial waves $\mathbf{E}_1$ and $\mathbf{E}_2$

$$\mathbf{E}(\mathbf{r}, t) = (\boldsymbol{\epsilon}_1 E_1 + \boldsymbol{\epsilon}_2 E_2)e^{i(\mathbf{k} \cdot \mathbf{r} - \omega t)} \tag{15.31}$$

Now one agrees upon the following: the vector of the electric field $\mathbf{E}$ is understood to be representative of the entire plane wave, and thus determines also the polarization properties. If E_1 and E_2 have the same phase then we have a linearly polarized wave (Figure 15.6) whose polarization vector includes the angle

$$\Theta = \arctan \frac{E_2}{E_1}$$

with the unit vector $\boldsymbol{\epsilon}_1$. The magnitude of the amplitude of the field intensity is given by

$$E = \sqrt{E_1^2 + E_2^2}$$

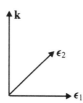

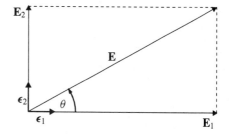

Figure 15.5. The polarization plane is spanned by ϵ_1 and ϵ_2, which are perpendicular to $\mathbf{k}$.

Figure 15.6. The composition of the field intensity vector $\mathbf{E}$ out of the linearly polarized waves $\mathbf{E}_1$ and $\mathbf{E}_2$.

If a phase shift occurs between E_1 and E_2, then we talk about an *elliptically polarized wave*.

$$\mathbf{E}(\mathbf{r}, t) = \boldsymbol{\epsilon}_1 E_1 e^{i(\mathbf{k}\cdot\mathbf{r}-\omega t)} + \boldsymbol{\epsilon}_2 E_2 e^{i(\mathbf{k}\cdot\mathbf{r}-\omega t+\varphi)} \tag{15.32}$$

A special case of the elliptically polarized wave is the *circularly polarized wave*, for which E_1 and E_2 are of equal magnitude but exhibit a phase difference of $\varphi = \pm\pi/2$ (Figure 15.7). Then, equation (15.31) becomes

$$\mathbf{E}(\mathbf{r}, t) = E_0(\boldsymbol{\epsilon}_1 \pm i\boldsymbol{\epsilon}_2)e^{i(\mathbf{k}\cdot\mathbf{r}-\omega t)} \tag{15.33}$$

where E_0 is the real amplitude. Now, we want to understand this notation in more detail. We choose the coordinate system in such a way that the wave propagates in the z-direction. Then, $\boldsymbol{\epsilon}_1$ and $\boldsymbol{\epsilon}_2$ point in the x- and y-directions, respectively. We consider the components of the instantaneous electric field by passing to the corresponding real part in equation (15.33):

$$\begin{aligned}
E_x(\mathbf{r}, t) &= E_0 \cos(kz - \omega t) = +E_0 \cos(\omega t - kz) \\
E_y(\mathbf{r}, t) &= \mp E_0 \sin(kz - \omega t) = \pm E_0 \sin(\omega t - kz)
\end{aligned} \tag{15.34}$$

For fixed points in space, $z = z_0$, these equations are the parametric representation of a circle; that is, the **E**-vector has a constant magnitude and rotates at constant angular velocity.

Depending on the choice of the sign in equation (15.33) we talk about left-handed circular polarization or right-handed circular polarization: Viewing toward the oncoming wave, that is, in direction of the negative z-axis, the **E**-vector in (15.34) rotates to the left for the upper sign and to the right for the lower sign. In the case of the elliptically polarized wave the equation (15.34) is changed in the following way if the phase is still given by $\varphi = \pm\pi/2$ (Figure 15.8):

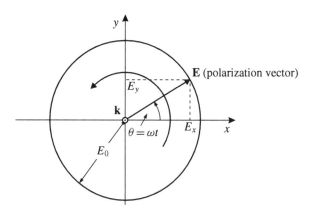

Figure 15.7. Circular polarization: The **k**-vector points out of the plane of the page. We have $E_x^2 + E_y^2 = E_0^2$ (equation of a circle).

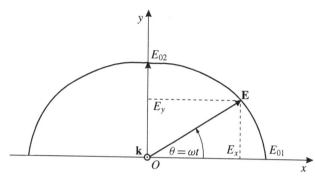

Figure 15.8. Elliptic polarization: The **k**-vector points out of the plane of the page. We have $(E_x/E_{01})^2 + (E_y/E_{02})^2 = 1$ (equation of an ellipse).

$$E_x(\mathbf{r}, t) = E_{01} \cos(kz - \omega t)$$

$$E_y(\mathbf{r}, t) = \mp E_{02} \sin(kz - \omega t), \qquad E_{01} \neq E_{02} \tag{15.35}$$

that is, the amplitude of the **E**-vector is no longer constant.

For fixed points in space this describes an elliptic trajectory with the semiaxis E_{01} and E_{02}.

Here, the axes E_{01} and E_{02} coincide with the x- and y-directions, respectively, of our coordinate system. In the general case (15.32), that is, for an arbitrary phase φ and for $E_{01} \neq E_{02}$ the ellipse is still rotated against the xy-coordinate system. This is realized by the superposition of two circularly polarized plane waves.

Example 15.1: Exercise: Linear and circular polarization

Discuss the notions of linear and circular polarization.

Two monochromatic waves of equal frequency are circularly polarized in opposite directions. They are in phase and both travel in z-direction. The amplitudes are A and B. Determine the kind of polarization for various ratios of the amplitudes.

Solution   A plane monochromatic wave with the wave vector **k** can be written as

$$\mathbf{E}(\mathbf{r}, t) = \mathrm{Re}\, \boldsymbol{\epsilon}(\mathbf{r}, t) \qquad \boldsymbol{\epsilon}(\mathbf{r}, t) = \boldsymbol{\epsilon}_0 e^{i(\mathbf{k} \cdot \mathbf{r} - \omega t)}$$

$$\mathbf{B}(\mathbf{r}, t) = \mathrm{Re}\, \boldsymbol{\beta}(\mathbf{r}, t) \qquad \boldsymbol{\beta}(\mathbf{r}, t) = \boldsymbol{\beta}_0 e^{i(\mathbf{k} \cdot \mathbf{r} - \omega t)}$$

$$\mathbf{k}^2 = \frac{\omega^2}{c^2}$$

with fixed complex vectors $\boldsymbol{\epsilon}_0$ and $\boldsymbol{\beta}_0$.

From Maxwell's equations we get

$$\mathbf{k} \cdot \mathbf{E} = \mathbf{k} \cdot \mathbf{B} = \mathbf{E} \cdot \mathbf{B} = 0$$

Thus, electromagnetic waves are transverse waves. Therefore, it is meaningful to introduce an orthogonal coordinate system consisting of **u**, $\boldsymbol{\epsilon}_1$, and $\boldsymbol{\epsilon}_2$. Here, **u** points in the direction of propagation while ϵ_1 and ϵ_2 are two real, fixed vectors spanning the oscillation plane of **E** and **B** (Figure 15.9). Thus,

$$\epsilon(\mathbf{r}, t) = (E_1\epsilon_1 + E_2\epsilon_2)e^{i(\mathbf{k}\cdot\mathbf{r}-\omega t)}, \qquad \beta(\mathbf{r}, t) = \mathbf{u} \times \epsilon(\mathbf{r}, t)$$

where E_1 and E_2 are two arbitrary complex numbers.

(a) If E_1 and E_2 have equal phases ($E_1 = e^{i\varphi}|E_1|$, $E_2 = e^{i\varphi}|E_2|$) the wave is linearly polarized because

$$\epsilon(\mathbf{r}, t) = (|E_1|\epsilon_1 + |E_2|\epsilon_2)e^{i(\mathbf{k}\cdot\mathbf{r}-\omega t+\varphi)}$$
$$\mathbf{E}(\mathbf{r}, t) = (|E_1|\epsilon_1 + |E_2|\epsilon_2)\cos(\mathbf{k}\cdot\mathbf{r} - \omega t + \varphi)$$

So, **E** oscillates along a fixed direction tilted against ϵ_1 by $\vartheta = \arctan(|E_2|/|E_1|)$. The amplitude is $\sqrt{|E_1|^2 + |E_2|^2}$. Of course, also in the case of a phase shift of π ($E_1 = e^{i\varphi}|E_1|$, $E_2 = -e^{i\varphi}|E_2|$) a linearly polarized wave arises.

If a phase shift ($E_1 = e^{i\varphi}|E_1|$, $E_2 = e^{i\varphi+\Delta\varphi}|E_2|$) occurs between E_1 and E_2, then one talks about an *elliptically polarized* wave.

The special case $|E_1| = |E_2| = E_0$, $\Delta\varphi = \pm\pi/2$ is called *circular polarization*. In this case,

$$\epsilon(\mathbf{r}, t) = (\epsilon_1 \pm i\epsilon_2)E_0 e^{i(\mathbf{k}\cdot\mathbf{r}-\omega t+\varphi)}$$
$$\mathbf{E}(\mathbf{r}, t) = E_0\cos(\mathbf{k}\cdot\mathbf{r} - \omega t + \varphi)\epsilon_1 \mp \sin(\mathbf{k}\cdot\mathbf{r} - \omega t + \varphi)\epsilon_2$$

So, **E** has the magnitude E_0, constant in space and time, while for $\mathbf{r} = \text{const}$ the direction rotates at the angular velocity ω (Figure 15.10):

$$\Delta\varphi = +\frac{\pi}{2}, \qquad \epsilon \sim \epsilon_1 + i\epsilon_2 \text{ left-handed (positive helicity)}$$

$$\Delta\varphi = -\frac{\pi}{2}, \qquad \epsilon \sim \epsilon_1 - i\epsilon_2 \text{ right-handed (negative helicity)}$$

(b) Since we are considering waves circularly polarized in opposite directions, we obtain ($\mathbf{k}\cdot\mathbf{r} + \varphi = 0$)

$$\left.\begin{array}{l} \mathbf{E}^{(1)} = A(\cos\omega t)\epsilon_1 + A(\sin\omega t)\epsilon_2 \\ \mathbf{E}^{(2)} = B(\cos\omega t)\epsilon_1 - B(\sin\omega t)\epsilon_2 \end{array}\right\} \qquad A, B \in R$$

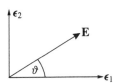

Figure 15.9. The electric field vector and the polarization plane.

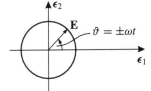

Figure 15.10. The field vector **E** rotates anti-clockwise (positive helicity) or clockwise (negative helicity) as seen against the oncoming wave.

Hence, the sum is

$$\mathbf{E} = \mathbf{E}^{(1)} + \mathbf{E}^{(2)} = (A + B)(\cos \omega t)\epsilon_1 + (A - B)(\sin \omega t)\epsilon_2$$

If A and B are allowed to vary, one obtains

$A = B$	linearly polarized in ϵ_1-direction
$A = -B$	linearly polarized in ϵ_2-direction
$A = 0$	right-handed circularly polarized
$B = 0$	left-handed circularly polarized
$|A| > |B|$	left-handed elliptically polarized
$|A| < |B|$	right-handed elliptically polarized

Example 15.2: On the speed of light (historical note)

The speed of light was first determined by Olaf Roemer[2] in 1675 by precise observation of the period of the moons of the Jupiter. Figure 15.11 illustrates the geometry underlying the measurement. An exact measurement of the period of a moon of the Jupiter was achieved by determining the entry into and the escape of the moon from the shadow of Jupiter. Roemer found distinct values for the period in distinct times of the year. Let t_1 be the time of the disappearance of the moon into the shadow of the Jupiter and t_2 the time of its reentry into the shadow after a full period. Then, the period is $T = t_2 - t_1$. The figure shows the relative positions of the Earth and the Jupiter during the time interval of observation.

Now consider Figure 15.12. Let v be the component of the velocity of the earth in direction of the Jupiter; then, the first moonset is observed on the Earth at the time $t_1 + L/c$. L is the initial Earth–Jupiter distance. (The Earth does not move far during the time of flight L/c of the light.)

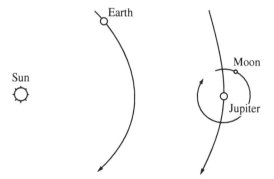

Figure 15.11. Positions of the planets Earth and Jupiter during the observation of a period of a moon of the Jupiter (not to scale).

---

[2]For biographical notes on O. Roemer, see page 545.

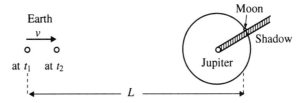

Figure 15.12. The motion of the Earth during the time interval of the observation of the moon of the Jupiter.

Between the two moonsets, the Earth travels the path $v(t_2 - t_1)$, relative to Jupiter. Therefore, on the Earth the second moonset occurs at the time

$$t_2 + \frac{[L - v(t_2 - t_1)]}{c}$$

and, hence, an *apparent period T'* of

$$T' = t_2 + \frac{[L - v(t_2 - t_1)]}{c} - \left(t_1 + \frac{L}{c}\right) = T\left(1 - \frac{v}{c}\right)$$

is ascertained. Drawing the apparent period determined in this way as a function of time, the graph in Figure 15.13 is obtained.

The apparent period is a function of time recuring each solar year, that is, it is periodic with a period of 1 year. The variation amounts to $2Tv_0/c$, where v_0 is the rotation speed of the Earth about the sun. (v_0 is also the maximal velocity of the Earth relative to the Jupiter.) Then the measurement of T', T, $2v_0T/c$, and v_0 allows us to determine c. Roemer's method yields the correct speed of the light to within 30%.

At the time of Maxwell, the speed of light was known from the Fizeau experiment to within a precision of 5%. In this experiment, a light ray passes between two teeth of a rotating gear to be reflected from a mirror standing behind it and to pass back the same way. If the speed of rotation is calculated such that the ray after passing the distance gear–mirror and back just hits again a gap, the light gets through, and the speed of light can be determined from the known gear–mirror distance.

Weber and Kohlrausch also determined the constant c in Maxwell's equations by comparing the electric and magnetic forces on a certain unit charge. Using Coulomb's force law a well defined

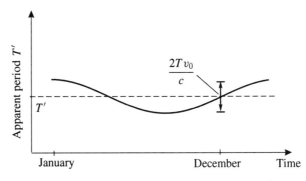

Figure 15.13. The apparent period of a moon of the Jupiter as a function of time.

magnitude of charge may be determined. The measurement of the magnetic forces between two currents in a known geometric arrangement allows us to determine the constant c based on the laws of Biot-Savart and Ampère (Chapter 8). Weber and Kohlrausch first performed the experiment in 1856, with the remarkable result

$$c = 3.1 \cdot 10^{10} \text{cm/sec}$$

That the constant c has to have the dimension of a velocity one may see directly from Coulomb's law, $F = q^2/r^2$, and the force law between currents, yielding $F = I^2/c^2$. Modern experiments fix the constant up to 6 significant digits after the decimal point. The numerical agreement with the speed of light is excellent, confirming that light is an electromagnetic wave.

16 Electromagnetic Waves in Matter

Now we consider the propagation of electromagnetic waves in infinitely extended, homogeneous media at rest characterized by the permittivity ϵ, the permeability μ, and the conductivity σ. We restrict ourselves to isotropic materials which may be polarized and magnetized in a normal manner. Then,

$$\mathbf{D} = \epsilon \mathbf{E}, \qquad \mathbf{B} = \mu \mathbf{H}, \qquad \mathbf{j} = \sigma \mathbf{E} \tag{16.1}$$

Hence, also Ohm's law should be valid everywhere in the medium. Furthermore, we assume $\rho = 0$ (so, the medium is neutral), and we obtain for Maxwell's equations:

$$\operatorname{curl} \mathbf{H} = \frac{\epsilon}{c} \frac{\partial \mathbf{E}}{\partial t} + \frac{4\pi \sigma}{c} \mathbf{E} \tag{16.2}$$

$$\operatorname{curl} \mathbf{E} = -\frac{\mu}{c} \frac{\partial \mathbf{H}}{\partial t} \tag{16.3}$$

$$\operatorname{div} \mathbf{H} = 0 \tag{16.4}$$

$$\operatorname{div} \mathbf{E} = 0 \tag{16.5}$$

First, we take the curl of equation (16.2) and substitute it subsequently into (16.3); due to $\operatorname{div} \mathbf{H} = 0$ we obtain the following differential equation:

$$\Delta \mathbf{H} = \frac{\epsilon \mu}{c^2} \frac{\partial^2}{\partial t^2} \mathbf{H} + \frac{4\pi \mu \sigma}{c^2} \frac{\partial}{\partial t} \mathbf{H} \tag{16.6a}$$

Analogously, we proceed with equation (16.3) and obtain

$$\Delta \mathbf{E} = \frac{\epsilon \mu}{c^2} \frac{\partial^2}{\partial t^2} \mathbf{E} + \frac{4\pi \mu \sigma}{c^2} \frac{\partial}{\partial t} \mathbf{E} \tag{16.6b}$$

Since these equations play a role also in the propagation of waves along wires, they are called *telegraph equations*.

In media for which, in particular, $\sigma = 0$ holds (insulators) the telegraph equations reduce to the ordinary wave equations with the velocity of propagation c':

$$c' = \frac{c}{\sqrt{\epsilon \mu}}$$

The index of refraction known from optics n is defined by

$$n = \frac{c}{c'} = \sqrt{\epsilon \mu}$$

For most media, $\mu \approx 1$, thus to a good approximation, $n \approx \sqrt{\epsilon}$.

This relation is confirmed in practice for long waves, and for some materials (for example O_2, N_2, CO_2, H_2) it is still valid for visible light. On the other hand, for water $\epsilon = 81$ and $n = 1.33$.

The failure of the relation $n = \sqrt{\epsilon}$ is based, obviously, on the fact that for high-frequency electromagnetic radiation the previous constants ϵ, μ, and σ become frequency dependent. In the derivation of these quantities we assumed steady fields, so that the eventually occurring dipols had time enough to align themselves in the fields (Chapter 5). But, if the material is exposed to rapidly alternating fields, these assumptions are no longer fulfilled, and the phenomena of forced oscillations occurs.

While we shall assume further that $\mu(\omega) \approx 1$, the dependence of ϵ and σ on the frequency will be investigated in greater detail in this chapter.

Since we assume $\epsilon = \epsilon(\omega)$ and $\sigma = \sigma(\omega)$, equation (16.1) can be given for special Fourier components only. Therefore, we restrict ourselves to one Fourier component, hence, to monochromatic (unifrequency) waves, and we check whether already the plane, harmonic waves

$$\mathbf{E} = \mathbf{E}_0 e^{i(\mathbf{k} \cdot \mathbf{r} - \omega t)}$$
$$\mathbf{H} = \mathbf{H}_0 e^{i(\mathbf{k} \cdot \mathbf{r} - \omega t)}$$

are solutions to Maxwell's equations (16.2) to (16.5). Applying the differentiation formulas for plane waves already derived, we obtain for equation (16.2):

$$\text{curl } \mathbf{H} = -\frac{\epsilon}{c} i \omega \mathbf{E} + \frac{4\pi \sigma}{c} \mathbf{E} = -i \frac{\omega}{c} \eta \mathbf{E}$$

with

$$\eta = \epsilon + \frac{4\pi i \sigma}{\omega}$$

We recognize that for the case $\sigma \neq 0$ the permittivity ϵ has to be replaced by the complex quantity η, while for $\sigma = 0$ this equation becomes the previous expression

$$\text{curl } \mathbf{H} = -i \frac{\omega}{c} \epsilon \mathbf{E}$$

This entitles us to denote η as the *generalized (complex) permittivity*.

Rewriting equations (16.2)–(16.6a) corresponding to the preceding chapter, we obtain the relations:

$$\mathbf{k} \times \mathbf{H} = -\frac{\omega}{c}\eta\mathbf{E} \tag{16.7}$$

$$\mathbf{k} \times \mathbf{E} = \frac{\omega}{c}\mu\mathbf{H} \tag{16.8}$$

$$\mathbf{k} \cdot \mathbf{E} = 0 \tag{16.9}$$

$$\mathbf{k} \cdot \mathbf{H} = 0 \tag{16.10}$$

that is, in a conducting medium the electromagnetic waves also are transverse waves, and also here $\mathbf{k}$, $\mathbf{E}$, and $\mathbf{H}$ form a right-handed orthogonal system in this order. However, the magnitudes of the electric and magnetic field intensities are no longer equal to each other; that is, we have *no* longer $|\mathbf{E}| = |\mathbf{H}|$.

First, we determine $\mathbf{k}$; for this purpose we multiply equation (16.7) vectorially from the left by $\mathbf{k}$, and solving the triple vector product we obtain the relation:

$$\mathbf{k} \times (\mathbf{k} \times \mathbf{H}) = (\mathbf{k} \cdot \mathbf{H})\mathbf{k} - k^2\mathbf{H} = -\frac{\omega}{c}\eta\mathbf{k} \times \mathbf{E}$$

and because of the equations (16.8) and (16.10):

$$k^2\mathbf{H} = \frac{\omega}{c}\eta\frac{\omega}{c}\mu\mathbf{H}$$

that is,

$$k^2 = \frac{\omega^2}{c^2}\eta\mu = \mu\epsilon\frac{\omega^2}{c^2}\left(1 + i\frac{4\pi\sigma}{\omega\epsilon}\right) \tag{16.11}$$

This is the *dispersion relation in a conducting medium.*

$$\eta = \epsilon\left(1 + i\frac{4\pi\sigma}{\omega\epsilon}\right) \tag{16.12}$$

is the *generalized permittivity.* For $\sigma = 0$, this equation reduces to

$$k = \frac{\omega}{c}\sqrt{\epsilon\mu} = \frac{\omega}{c}n \tag{16.13}$$

In order to calculate the wave number vector $\mathbf{k}$ for the general case, we set

$$k = \alpha + i\beta \tag{16.14}$$

Due to $k^2 = \alpha^2 + 2i\alpha\beta - \beta^2$, by comparison with equation (16.11) we obtain

$$\alpha^2 - \beta^2 = \mu\epsilon\frac{\omega^2}{c^2}, \qquad 2\alpha\beta = +\frac{4\pi\sigma}{\omega\epsilon}\mu\epsilon\frac{\omega^2}{c^2} \tag{16.15}$$

From this equation we determine α and β. After rewriting we obtain a biquadratic equation whose solution is

$$\beta^2 = -\frac{1}{2}\mu\epsilon\frac{\omega^2}{c^2}\left(1 \pm \sqrt{1 + \left(\frac{4\pi\sigma}{\omega\epsilon}\right)^2}\right)$$

Since we have assumed β to be real, and σ, ω, and ϵ are positive quantities, we obtain for the sign of the root

$$\beta^2 = \mu\epsilon\frac{\omega^2}{c^2}\frac{\sqrt{1 + \left(\frac{4\pi\sigma}{\omega\epsilon}\right)^2} - 1}{2}$$

and hence for β:

$$\beta = \pm\sqrt{\mu\epsilon}\frac{\omega}{c}\left(\frac{\sqrt{1 + \left(\frac{4\pi\sigma}{\omega\epsilon}\right)^2} - 1}{2}\right)^{1/2}$$

Analogously, we find

$$\alpha = \pm\sqrt{\mu\epsilon}\frac{\omega}{c}\left(\frac{\sqrt{1 + \left(\frac{4\pi\sigma}{\omega\epsilon}\right)^2} + 1}{2}\right)^{1/2}$$

Due to (16.15) the sign of α is fixed by the sign of β. (If β is positive α has to be positive and vice versa.) For physical reasons we have to choose the positive sign (see equation (16.18) ff.). Hence, for $k = \alpha + i\beta$ we obtain

$$k = \sqrt{\mu\epsilon}\,\frac{\omega}{c}\left[\left(\frac{\sqrt{1 + \left(\frac{4\pi\sigma}{\omega\epsilon}\right)^2} + 1}{2}\right)^{1/2} + i\left(\frac{\sqrt{1 + \left(\frac{4\pi\sigma}{\omega\epsilon}\right)^2} - 1}{2}\right)^{1/2}\right] \qquad \textbf{(16.16)}$$

Now, we determine the magnitude and the phase of the vector $\mathbf{k} = |\mathbf{k}|e^{i\varphi}$:

$$|\mathbf{k}| = \sqrt{\alpha^2 + \beta^2} = \sqrt{\epsilon\mu}\frac{\omega}{c}\left(1 + \left(\frac{4\pi\sigma}{\omega\epsilon}\right)^2\right)^{1/4}$$

$$\varphi = \arctan\frac{\beta}{\alpha} = \frac{1}{2}\left(2\arctan\frac{\beta}{\alpha}\right)$$

$$= \frac{1}{2} \arctan \frac{2\beta/\alpha}{1 - \beta^2/\alpha^2} = \frac{1}{2} \arctan \frac{2\alpha\beta}{\alpha^2 - \beta^2} = \frac{1}{2} \arctan \frac{4\pi\sigma}{\omega\epsilon}$$

(Here, the known relation $\arctan u + \arctan v = \arctan[(u + v)/(1 - uv)]$ has been used.) Now, we introduce the *generalized index of refraction* $p = p_1 + ip_2$:

$$p = \sqrt{\mu\eta} = p_1 + ip_2 \tag{16.17a}$$

Because $k = (\omega/c)\sqrt{\mu\eta} = (\omega/c)\, p = (\omega/c)\,(p_1 + ip_2) = \alpha + i\beta$,

$$p = \sqrt{\mu\epsilon} \left(1 + \left(\frac{4\pi\sigma}{\omega\epsilon} \right)^2 \right)^{1/4} e^{i\varphi} = p_1 + ip_2$$

with

$$p_{1/2} = \sqrt{\mu\epsilon} \left(\frac{\sqrt{1 + \left(\frac{4\pi\sigma}{\omega\epsilon} \right)^2} \pm 1}{2} \right)^{1/2}, \qquad \varphi = \frac{1}{2} \arctan \frac{4\pi\sigma}{\omega\epsilon} \tag{16.17b}$$

hence,

$$p_1 = \frac{c}{\omega}\alpha, \qquad p_2 = \frac{c}{\omega}\beta$$

Now, we consider a plane wave propagating in the x-direction and substitute our ansatz $k = \alpha + i\beta = (\omega/c)\, p = (\omega/c)\,(p_1 + ip_2)$ in the exponential part of the wavefunction:

$$\exp(i(kx - \omega t)) = \exp\left(i \left(\frac{\omega}{c} px - \omega t \right) \right)$$

$$= \exp\left(i \left(\frac{\omega}{c} p_1 x - \omega t \right) \right) \exp\left(-\frac{\omega}{c} p_2 x \right) \tag{16.18}$$

$$= \exp(i(\alpha x - \omega t)) \exp(-\beta x)$$

At once, we obtain a condition for the signs of α and β: To obtain a solution physically meaningful, α and β have to be positive, at the same time implying p_1 positive and p_2 positive. In this case, we obtain a wave that decays exponentially in a medium with a conductivity different from zero (Figure 16.1). The negative solution for β would lead to a wave $\exp(+|\beta|x)$, which is exponentially increasing, and physically ridiculous. Therefore, the negative β-value has to be excluded. Now, we introduce $d = |1/\beta| = ((\omega/c)\, p_2)^{-1}$.

This is the distance within which the wave decreases by $1/e$ of its original value. This quantity is denoted the *penetration depth* of the wave into the medium. Due to $\exp(i\varphi) = \exp(i(\varphi + 2\pi))$, for the wavelength the relation

$$\alpha\lambda = 2\pi$$

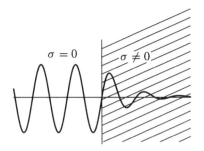

Figure 16.1. Decaying wave in a conducting medium.

has to be valid; that is,

$$\lambda = \frac{2\pi}{\alpha}$$

When the wave has traveled the distance $x = \lambda$, its amplitude has diminished by the factor

$$\exp\left(-\frac{2\pi\beta}{\alpha}\right) = \exp\left(-\frac{2\pi p_2}{p_1}\right) \tag{16.19}$$

the so-called *extinction coefficient*. This is a measure for the damping of the wave.

Example 16.1:   Exercise: Calculation of the index of refraction

A light wave has the frequency $\nu = 4 \times 10^{14}$ Hz and the wavelength $\lambda = 5 \times 10^{-7}$ m.

(a) What is its speed?

(b) What is the value of the index of refraction of the medium in which the wave propagates?

(c) What is the wavelength in vacuum?

Solution   In vacuum: $c = \lambda\nu$, but in a dispersive medium $v = c/n = \lambda\nu$.

(a) We obtain $v = \lambda\nu = 2 \cdot 10^8$ m/s.

(b) The index of refraction is $n = c/v = 3/2$.

(c) The wavelength in vacuum would be $\lambda_0 = c/\nu = n \cdot \lambda = 3/2\,\lambda = 7.5 \cdot 10^{-2}$ m.

Frequency dependence of the conductivity

In metals, the conductivity may be explained by the existence of free electrons. We assume that N such electrons are located in the volume V. Then, for the kth electron Newton's law is valid:

$$m \frac{d\mathbf{v}_k}{dt} = e\mathbf{E}_k - \beta \mathbf{v}_k$$

where β is the friction coefficient. Multiplying both sides of the equation with the elementary charge e and summing over all N electrons, because

$$\frac{1}{V} \sum_{k=1}^{N} e\mathbf{v}_k = \mathbf{j}$$

we obtain the following equation:

$$m \frac{d\mathbf{j}}{dt} = \frac{1}{V} \sum_{k=1}^{N} \mathbf{E}_k e^2 - \beta \mathbf{j}$$

In a first approximation, we set $\mathbf{E}_k$ equal to the average field intensity $\mathbf{E}$, and hence, if n gives the partial density ($n = N/V$),

$$m \frac{d\mathbf{j}}{dt} = n\mathbf{E}e^2 - \beta \mathbf{j} \qquad \text{or} \qquad \frac{m}{ne^2} \frac{d\mathbf{j}}{dt} = \mathbf{E} - \frac{\beta}{ne^2} \mathbf{j} \qquad \textbf{(16.20)}$$

To express the friction coefficient β by macroscopically measurable quantities we first start from the static case. For a constant current, $d\mathbf{j}/dt = 0$ and therefore,

$$\mathbf{E} = \frac{\beta}{ne^2} \mathbf{j}$$

With the *direct-current conductivity* σ_0 we may write $\mathbf{j} = \sigma_0 \mathbf{E}$ and conclude $\sigma_0 = (n/\beta)e^2$. From this relation we may determine the friction coefficient

$$\beta = \frac{ne^2}{\sigma_0} \qquad \textbf{(16.21)}$$

Substituting the friction coefficient obtained in this way into equation (16.20) we conclude

$$\frac{d\mathbf{j}}{dt} + \frac{ne^2}{\sigma_0 m} \mathbf{j} = \frac{ne^2}{m} \mathbf{E} \qquad \text{or} \qquad \frac{d\mathbf{j}}{dt} + \frac{1}{\tau} \mathbf{j} = \frac{ne^2}{m} \mathbf{E} \qquad \textbf{(16.22)}$$

To shorten the notation we have introduced

$$\tau = \frac{m}{\beta} = \frac{m\sigma_0}{ne^2}$$

The quantity $\tau = 1/\gamma$ is the *damping time* characterizing the decay (damping) of the oscillation, as (16.22) makes clear immediately.

If a harmonic, rapidly fluctuating electric field is applied to a conductor then it makes sense to expect that also the current density performs harmonic oscillations; that is, together with $E \sim \exp(-i\omega t)$ also $j \sim \exp(-i\omega t)$. In this case the differentiation with respect to time may be replaced by a multiplication by $-i\omega$, and

$$-i\omega \mathbf{j} + \frac{1}{\tau}\mathbf{j} = \frac{ne^2}{m}\mathbf{E} \qquad \text{or} \qquad \frac{\left(-i\omega + \frac{1}{\tau}\right)}{\frac{ne^2}{m}}\mathbf{j} = \mathbf{E}$$

This equation is written as $\mathbf{j} = \sigma(\omega)\mathbf{E}$ with

$$\sigma(\omega) = \frac{ne^2}{m\left(-i\omega + \frac{1}{\tau}\right)} = \frac{\sigma_0}{1 - i\omega\tau} \tag{16.23}$$

If $\omega\tau = 2\pi\tau/T \ll 1$, this term can be neglected; that is, for small frequencies the conductivity is constant and equal to σ_0. If $\omega\tau \gg 1$, a *purely imaginary conductivity* is obtained. This means that a phase shift of 90^0 between $\mathbf{E}$ and $\mathbf{j}$ occurs, since in this case $\mathbf{E}$ and $d\mathbf{j}/dt$ are of equal phase.

Substituting the found frequency dependence of the conductivity into equation (16.12)

$$\eta = \varepsilon + i\frac{4\pi\sigma}{\omega} \tag{16.24a}$$

we obtain

$$\eta = \varepsilon + \frac{4\pi}{\omega}i\left(\frac{\sigma_0}{1 - i\tau\omega}\right) = \varepsilon - \frac{4\pi\sigma_0}{i\omega(1 - i\tau\omega)} \tag{16.24b}$$

It is worthwhile to introduce the further abbreviation

$$\omega_p \equiv \sqrt{\frac{4\pi ne^2}{\varepsilon m}} = \sqrt{\frac{4\pi\sigma_0}{\varepsilon\tau}} \tag{16.25}$$

So we may write

$$\eta = \varepsilon\left(1 - \frac{\omega_p^2\tau}{i\omega(1 - i\omega\tau)}\right) = \varepsilon\left(1 + i\frac{\omega_p}{\omega}\frac{\omega_p\tau}{(1 - i\omega\tau)}\right) \tag{16.25a}$$

The quantity ω_p is denoted the *plasma frequency*. To illustrate this notion we want to investigate the decay of a charge density ρ (electron plasma, ion plasma) present in the interior of the conductor. For this purpose, we take the divergence of equation (16.22):

$$\text{div}\frac{d\mathbf{j}}{dt} + \text{div}\frac{1}{\tau}\mathbf{j} = \text{div}\frac{ne^2}{m}\mathbf{E}$$

or

$$\frac{d}{dt}(\text{div}\,\mathbf{j}) + \frac{1}{\tau}\text{div}\,\mathbf{j} = \frac{ne^2}{m}\text{div}\,\mathbf{E} \tag{16.26}$$

Due to $\text{div}\,\mathbf{E} = 4\pi\rho/\varepsilon$ and $d\rho/dt = -\,\text{div}\,\mathbf{j}$, we obtain the oscillation equation

$$0 = \frac{d^2\rho}{dt^2} + \frac{1}{\tau}\frac{d\rho}{dt} + \omega_p^2\rho \tag{16.27}$$

for the charge density (the plasma). This is the differential equation of a damped oscillation, where the expression introduced for abbreviation

$$\gamma = \frac{1}{\tau} = \frac{ne^2}{m\sigma_0} \tag{16.28}$$

has the physical meaning of a damping constant, like in equation (16.22). ω_p is the eigenfrequency of the oscillation of the charge density (electron plasma). Hence, a plasma may perform fundamental oscillations. The eigenfrequency ω_p increases with the density n of the particles, and decreases with increasing mass m of the particles. This is plausible.

Frequency dependence of the polarizability

Up till now, the complex permittivity became frequency dependent only by the conductivity $\sigma/\omega = \sigma(\omega)/\omega$. The question arises as to what extent also the dielectric susceptibility ε may be frequency dependent. In equation (5.20) the permittivity is connected with the dielectric susceptibility χ_e according to $\varepsilon = 1 + 4\pi\chi_e$. On the other hand, this χ_e characterizes the polarization via (5.19) $\mathbf{P} = \chi_e\mathbf{E}$. But the polarization is related to the average dipole moment of the atom (molecule) according to (5.11): $\mathbf{P} = N\langle\mathbf{P}_{\mathrm{Mol}}\rangle$. $\mathbf{P}$ gives the dipole density (dipoles per volume). Consequently, the polarizability of the atoms (molecules) must be investigated with respect to the frequency dependence in more detail, since the atomic dipoles are induced to a large extent by the electric field (Chapter 6). Up till now, we have considered only the polarizability of materials (more precisely, of atoms and molecules of materials) in *static and quasi-static electric fields.* Here we could assume that the charge carriers responsible for the polarization follow the fields inactively. This assumption is no longer fulfilled in high-frequency fields. Now, we will investigate a simple model which is able to explain the most important phenomena.

As the simplest atomic model, we imagine electrons elastically bound to the atomic nucleus and performing damped harmonic oscillations at the frequency ω_0. See Figure 16.2. In the force-free case (no external fields) the positive nuclear charge and the center-of-

Figure 16.2. Electron attached to a spring.

gravity of the negative electric charge should coincide. Our model is described by the differential equation of the harmonic oscillator.

$$m \left(\frac{d^2\mathbf{r}}{dt^2} + \gamma \frac{d\mathbf{r}}{dt} + \omega_0^2 \mathbf{r} \right) = \mathbf{F}(t)$$

where γ is a damping constant. For an electron in an electromagnetic field the Lorentz force has to be inserted on the right-hand side,

$$\mathbf{F} = e \left(\mathbf{E} + \frac{\mathbf{v}}{c} \times \mathbf{B} \right)$$

We may assume that $\mathbf{v} \ll \mathbf{c}$ and therefore neglect the influence of the magnetic field. The electric dipole moment $\mathbf{p}$ has been defined by the equation $\mathbf{p} = e\mathbf{r}$. Together with the simplified form of the Lorentz force we introduce this into the differential equation and obtain

$$m \left(\frac{d^2\mathbf{p}}{dt^2} + \gamma \frac{d\mathbf{p}}{dt} + \omega_0^2 \mathbf{p} \right) = e^2 \mathbf{E} \tag{16.29}$$

In solving this differential equation two cases may be distinguished again.

First, we consider again the *static case*; then $d\mathbf{p}/dt \equiv 0$, and we obtain

$$\omega_0^2 \mathbf{p} = \frac{e^2}{m} \mathbf{E} \qquad \text{or} \qquad \mathbf{p} = \frac{e^2}{\omega_0^2 m} \mathbf{E} \tag{16.30}$$

The prefactor $\alpha = e^2/(\omega_0^2 m)$ is the *static atomic polarizability*, $\mathbf{p} = \alpha \mathbf{E}$.

From the equations $\mathbf{P} = N \langle \mathbf{p} \rangle = \chi_e \mathbf{E}$, $\chi_e = N\alpha$, and $\varepsilon = 1 + 4\pi \chi_e$, we obtain for the permittivity:

$$\varepsilon = 1 + 4\pi N \frac{e^2}{m\omega_0^2} \tag{16.31}$$

Here, the density of the polarizable charge carries (electrons) have been denoted by N. In the *dynamic case* we assume that the atomic dipoles change with the same frequency at which the external field $\mathbf{E}$ oscillates; that is,

$$\mathbf{E} = \mathbf{E}_0 e^{-i\omega t} \qquad \text{and} \qquad \mathbf{p} = \mathbf{p}_0 e^{-i\omega t}$$

To grasp a possible phase shift between $\mathbf{E}$ and $\mathbf{p}$ we allow $\mathbf{p}_0$ (and thus α) to be complex. With the ansatz given above, from the differential equation (16.29) we obtain

$$\mathbf{p}_0 = \frac{e^2}{m} \frac{1}{\omega_0^2 - \omega^2 - i\gamma\omega} \mathbf{E}_0 = \alpha(\omega)\mathbf{E}_0 \tag{16.32}$$

The *atomic polarizability* $\alpha(\omega)$ *is now complex and frequency dependent*. For small ω the special case of the static approximation is obtained again. The real part of α is

$$\text{Re}\,\alpha = \frac{e^2}{m} \frac{\omega_0^2 - \omega^2}{(\omega_0^2 - \omega^2)^2 + \gamma^2\omega^2} \tag{16.33}$$

and the imaginary part is

$$\text{Im}\,\alpha = \frac{e^2}{m}\frac{\gamma\omega}{(\omega_0^2 - \omega^2)^2 + \gamma^2\omega^2} \tag{16.34}$$

In Figure 16.3 $\text{Re}\,\alpha$ and $\text{Im}\,\alpha$ are drawn. In the vicinity of the resonance $\omega = \omega_0$ both parts have extreme values, being very pronounced for small damping. The real part changes the sign (oscillating in opposition for $\omega > \omega_0$). The neighborhood of the resonance point is denoted the region of *anomalous dispersion*; in the region of *normal dispersion* α is largely real and increases with the frequency. We have derived the formulas for α and ε for an electron oscillating at the frequency ω_0. In general, several electrons of an atom are involved in the polarization. If N is the number of electrons per unit volume and f_k is the portion of electrons oscillating at ω_k, then from (16.32) we obtain similar to (16.31) *Drude's formula* for ε,

$$\varepsilon = 1 + \frac{4\pi N e^2}{m}\sum_k \frac{f_k}{\omega_k^2 - \omega^2 - i\gamma\omega} \tag{16.35}$$

With this formula the polarization may be described for materials with several regions of anomalous dispersion. The quantities f_k are called *oscillator strengths*. They obtain their real meaning in quantum mechanics, where they are characteristic quantities for the transition probabilities of the electron with respect to transitions with the energy $\hbar\omega_k$. For many nonconducting media we have for the index of refraction

$$n \cong \sqrt{\varepsilon} \approx 1 + \frac{2\pi N e^2}{m}\sum_k \frac{f_k}{\omega_k^2 - \omega^2 - i\gamma\omega} \tag{16.36}$$

where only the first term has been taken in the expansion of the root.

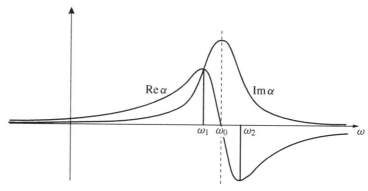

Figure 16.3. In the frequency domain $\omega_1 \le \omega \le \omega_2$ we talk about *anomalous dispersion* because light of shorter wavelength (higher frequency) is less refracted than light with a longer wavelength (lower frequency).

Frequency dependence of the index of refraction

The complex index of refraction of a material is given by (see equation (16.17a)):

$$p = \sqrt{\mu\eta} = \frac{ck}{\omega} = p_1 + ip_2$$

Furthermore, we have derived already (see equation (16.25a)):

$$\eta = \varepsilon \left(1 - \frac{\omega_p^2 \tau}{i\omega(1 - i\omega\tau)} \right)$$

We will discuss the equation given above for insulators and conductors. For insulators, $\sigma_0 = 0$ and hence also $\sigma(\omega) = 0$. Furthermore, $\tau = 0$; that is, $\eta = \varepsilon$ and $p = \sqrt{\varepsilon\mu}$. Disregarding anomalous dispersion, ε is largely real; hence, the index of refraction p becomes real, and $p = p_1 = (c/\omega)\alpha$ (see equation (16.17a)).

Because $\sigma_0 \neq 0$, damping of the electromagnetic waves has to be expected for metals. To get an overview of the damping, we set $\varepsilon = \mu = 1$ and so investigate only the influence of the frequency-dependent conductivity. With this simplification we obtain

$$\eta = 1 - \frac{\omega_p^2 \tau}{i\omega(1 - i\omega\tau)}$$

For long waves, $\omega \ll 1/\tau \ll \omega_p$; therefore, we can neglect $i\omega\tau$ in the denominator and assume η to be large compared to 1. That is,

$$\eta \approx 1 - \frac{\omega_p^2 \tau}{i\omega} \approx \frac{i\omega_p^2 \tau}{\omega}$$

Hence, we get for p

$$p = \sqrt{\eta} = \omega_p \sqrt{\frac{\tau}{\omega}} \sqrt{i}$$

The index of refraction is

$$p = p_1 + ip_2$$

Substituting $\sqrt{i} = e^{+i\pi/4} = [(1 + i)/2]\sqrt{2}$, then

$$p = \omega_p \sqrt{\frac{\tau}{\omega}} \frac{1 + i}{2} \sqrt{2} \tag{16.37}$$

Hence, for long waves, $p_1 = p_2$. According to equation (16.19) the extinction coefficient is given by $\exp(-2\pi p_2/p_1)$. Therefore, along a distance of one wavelength λ the amplitude of the wave is weakend by a factor $\exp(-2\pi) \approx 0.0019$. The penetration depth has been defined as the distance within which the amplitude of the wave decreases by $1/e$; according to (16.18), it is given by

$$d = \frac{1}{\frac{\omega}{c} p_2} = \frac{1}{\beta} \tag{16.38}$$

Hence, for long waves the penetration depth is

$$d = \frac{\sqrt{2}c}{\omega_p \sqrt{\dfrac{\tau}{\omega} - \omega}} = \frac{c}{\sqrt{2\pi \sigma_0 \omega}}$$

(16.39)

In the following table d is given for copper and various vacuum wavelengths ($\sigma_0 = 5.4 \cdot 10^{+17}$/sec):

λ (m)	10^{-2}	1	10^2	10^4
d (μm)	0.37	3.7	37	370

At this point, we should mention the *skin effect*. We know how the energy loss (Joule heat) in current-carrying conductors is compensated: the Poynting vector describing the energy flow through an area per unit time is directed into the wire; the fields penetrate into the wire (compare Exercise 13.2). High-frequency currents in the frequency region up to about 1 GHz have wavelengths for which our considerations on the penetration depth are correct. But, if high-frequency fields just penetrate into a conductor, then in the interior of the wire no current may flow. The current density is different from zero only in a thin surface layer. This phenomenon is called the skin effect.

Now we will consider the case of short wavelengths. Here, $\omega \gg 1/\tau$, and we obtain at once

$$\eta = 1 - \frac{\omega_p^2}{\omega^2}$$

The index of refraction is

$$p \approx \sqrt{\eta} = \sqrt{1 - \frac{\omega_p^2}{\omega^2}} = p_1 - ip_2$$

(16.40)

For $\omega \ll \omega_p$, p is imaginary, $p_1 \approx 0$, and $p_2 \approx \omega_p/\omega$.

We notice at once that the penetration depth of the wave is frequency-dependent (see equation (16.18)), namely,

$$d = \frac{c}{\omega_p}$$

(16.41)

The case $\omega > \omega_p$ is more interesting; the imaginary part of p (and, thus, the damping) vanishes, p becomes real, and

$$p = p_1 = \frac{c}{\omega}\alpha = \frac{1}{\omega}\sqrt{\omega^2 - \omega_p^2}$$

(16.42)

The inertia of the conduction electrons has such a strong influence, that, practically, no conduction current flows; at the highest frequencies metals are *transparent*. This effect may be demonstrated with hard X-rays; otherwise, transparent metals hardly may be realized.

Example 16.2: Exercise: Nonlocal generalization of the relation D(x, t) = εE(x, t)

In many dielectrics the local and linear relation $\mathbf{D}(\mathbf{x}, t) = \varepsilon \mathbf{E}(\mathbf{x}, t)$ is valid. For space- and time-dependent fields this relation may be generalized nonlocally

$$\mathbf{D}(\mathbf{x}, t) = \int d^3 x' \int dt' \, \varepsilon(\mathbf{x}', t') \mathbf{E}(\mathbf{x} - \mathbf{x}', t - t')$$

where $\varepsilon(\mathbf{x}', t')$ is different from zero only in the neighborhood $(\mathbf{x}', t') = (\mathbf{0}, 0)$.

(a) Discuss this relation and introduce the Fourier transforms of $\mathbf{D}$, $\mathbf{E}$, and ε.

(b) What is the value of the velocity at which a wave with wave vector $\mathbf{k}$ propagates in this medium? How is the index of refraction defined?

Solution   **(a)** The relation $\mathbf{D}(\mathbf{x}, t) = \varepsilon \mathbf{E}(\mathbf{x}, t)$ implies that the **D**-field, and thus, the polarization, that is, the response of the medium, depends only on the electric field **E** at the same position and at the same time. Certainly this is correct only for slowly varying fields. Slow variation means the period of the electromagnetic variation is large compared to the periods of the electrons and nuclei, and the characteristic wavelengths are large compared to the distance of the individual atoms, thus, e.g., for steady fields. Furthermore, the material has to be (macroscopically) uniform, since ε has assumed to be constant. If the fields are no longer varying, then the relation may be extended nonlocally as

$$\mathbf{D}(\mathbf{x}, t) = \int d^3 x' \int dt' \, \varepsilon(\mathbf{x}', t') \mathbf{E}(\mathbf{x} - \mathbf{x}', t - t') \tag{16.43}$$

that is, the polarization at the position $(\mathbf{x}, t)$ depends on the electric field within a small space-time neighborhood of the point $(\mathbf{x}, t)$. For reasons of causality we must have

$$\varepsilon(\mathbf{x}', t') = 0 \qquad \text{for } |\mathbf{x}'| > ct' \tag{16.44}$$

in particular, $\varepsilon(\mathbf{x}', t') = 0$ for $t' < 0$.

Often, for larger wavelengths the space dependence of ε may be neglected:

$$\mathbf{D}(\mathbf{x}, t) = \int dt' \, \varepsilon(t') \mathbf{E}(\mathbf{x}, t - t') \tag{16.45}$$

By means of

$$f(\mathbf{k}, \omega) = \int d^3 x \, dt \, f(\mathbf{x}, t) e^{-i(\mathbf{k} \cdot \mathbf{x} - \omega t)}$$

$$f(\mathbf{x}, t) = \frac{1}{(2\pi)^4} \int d^3 k \, d\omega \, f(\mathbf{k}, \omega) e^{+i(\mathbf{k} \cdot \mathbf{x} - \omega t)}$$

with $f = D_i$, E_i, ε, we use the Fourier transforms.

Multiplying (16.43) by $\exp[-i(\mathbf{k} \cdot \mathbf{x} - \omega t)]$ and integrating over **x** and t, we obtain

$$\int d^3 x \, dt \, \mathbf{D}(\mathbf{x}, t) e^{-i(\mathbf{k} \cdot \mathbf{x} - \omega t)} = \int d^3 x \, dt \int d^3 x' \, dt' \, \varepsilon(\mathbf{x}', t') \, \mathbf{E}(\mathbf{x} - \mathbf{x}', t - t') \, e^{-i(\mathbf{k} \cdot \mathbf{x} - \omega t)}$$

thus, with $\mathbf{x} - \mathbf{x}' = \mathbf{x}''$ and $t - t' = t''$

$$\mathbf{D}(\mathbf{k}, \omega) = \int d^3 x'' \, dt'' \int d^3 x' \, dt' \, \varepsilon(\mathbf{x}', t') \, \mathbf{E}(\mathbf{x}'', t'') \, e^{-i(\mathbf{k} \cdot \mathbf{x}'' - \omega t'')} \, e^{-i(\mathbf{k} \cdot \mathbf{x}' - \omega t')}$$

$$= \varepsilon(\mathbf{k}, \omega) \mathbf{E}(\mathbf{k}, \omega) \tag{16.46}$$

So, the integral becomes a product of the Fourier transforms (convolution). Again neglecting the space dependence, we obtain simply

$$\mathbf{D}(\mathbf{x}, \omega) = \varepsilon(\omega)\mathbf{E}(\mathbf{x}, \omega) \tag{16.47}$$

Thus, the permittivity depends on the frequency. We emphasize that the Fourier-transformed quantities $\mathbf{D}$ and $\mathbf{E}$ may be complex. But, since the space-dependent quantities are real,

$$f(\mathbf{k}, \omega) = f^*(-\mathbf{k}, -\omega), \qquad (f = D_i, E_i, \varepsilon)$$

In the case of the local relation $\mathbf{D}(\mathbf{x}, t) = \varepsilon_0 \mathbf{E}(\mathbf{x}, t)$, $\varepsilon(\mathbf{x}', t') = \varepsilon_0 \delta(\mathbf{x}')\delta(t')$, and therefore, $\varepsilon(\mathbf{k}, \omega) = \varepsilon_0$ is independent of wave number and frequency.

(b) Since we are interested in plane waves, we make the ansatz

$$\mathbf{E}(\mathbf{x}, t) = \mathrm{Re}\,\mathcal{E}(\mathbf{x}, t), \qquad \mathcal{E}(\mathbf{x}, t) = \mathcal{E}_0 e^{i(\mathbf{k}\cdot\mathbf{x} - \omega t)}$$
$$\mathbf{B}(\mathbf{x}, t) = \mathrm{Re}\,\mathcal{B}(\mathbf{x}, t), \qquad \mathcal{B}(\mathbf{x}, t) = \mathcal{B}_0 e^{i(\mathbf{k}\cdot\mathbf{x} - \omega t)} \tag{16.48}$$

where, at first, $\mathcal{E}_0$ and $\mathcal{B}_0$ are two constant vectors, and ω is not yet fixed. Substituting this ansatz together with the relation (16.43) and $\mathbf{H} = \mathbf{B}$ into Maxwell's equations for the source-free space

$$\nabla \cdot \mathcal{D}(\mathbf{x}, t) = 0, \qquad \nabla \times \mathcal{E}(\mathbf{x}, t) + \frac{1}{c}\frac{\partial \mathcal{B}}{\partial t}(\mathbf{x}, t) = 0$$

$$\nabla \cdot \mathcal{B}(\mathbf{x}, t) = 0, \qquad \nabla \times \mathcal{H}(\mathbf{x}, t) - \frac{1}{c}\frac{\partial \mathcal{D}}{\partial t}(\mathbf{x}, t) = 0 \tag{16.49}$$

then

$$\mathbf{k} \cdot \mathcal{E}_0 \varepsilon(\mathbf{k}, \omega) = 0, \qquad \mathbf{k} \times \mathcal{E}_0 - \frac{\omega}{c}\mathcal{B}_0 = 0$$

$$\mathbf{k} \cdot \mathcal{B}_0 = 0, \qquad \mathbf{k} \times \mathcal{B}_0 + \frac{\omega}{c}\mathcal{E}_0 \varepsilon(\mathbf{k}, \omega) = 0 \tag{16.50}$$

For $\varepsilon(\mathbf{k}, \omega) = 0$, there is a longitudinal solution

$$\mathcal{E}_0 \sim \mathbf{k}, \qquad \mathcal{B}_0 = \mathbf{0}$$

This solution is related to the plasma oscillations, but we will not elaborate on this here.

Let $\varepsilon(\mathbf{k}, \omega)$ be always different from 0. Then, we see directly that $\mathbf{k}$, $\mathcal{E}_0$, and $\mathcal{B}_0$ form an orthogonal system. Multiplying both of the right-hand sides with $\mathbf{k}$ and utilizing both of the left-hand sides, then we obtain the further relation

$$\left[k^2 - \frac{\omega^2}{c^2}\varepsilon(\mathbf{k}, \omega) \right] \mathcal{E}_0 = 0, \qquad \left[k^2 - \frac{\omega^2}{c^2}\varepsilon(\mathbf{k}, \omega) \right] \mathcal{B}_0 = 0 \tag{16.51}$$

This yields the relationship of $\mathbf{k}$ to ω: $\omega(\mathbf{k})$ is just the zero of $c^2 k^2 - \omega^2 \varepsilon(\mathbf{k}, \omega)$. In the case of the local relation $\mathbf{D}(\mathbf{x}, t) = \varepsilon_0 \mathbf{E}(\mathbf{x}, t)$, we have simply $\omega^2 = c^2 k^2/\varepsilon_0$.

In (a) we have seen that $\varepsilon(\mathbf{k}, \omega)$ may be also complex or negative; this leads to complex ω-values. Hence,

$$\begin{Bmatrix} \mathcal{E} \\ \mathcal{B} \end{Bmatrix} = \begin{Bmatrix} \mathcal{E}_0 \\ \mathcal{B}_0 \end{Bmatrix} e^{i(\mathbf{k}\cdot\mathbf{x} - \omega(k)t)}$$

$$\mathbf{k} \cdot \mathcal{E}_0 = \mathbf{k} \cdot \mathcal{B}_0 = 0$$

$$\mathcal{B}_0 = \frac{c}{\omega} \mathbf{k} \times \mathcal{E}_0 \qquad (16.52)$$

So, a negative imaginary part of ω corresponds to a damped oscillation, and there is no wavelike solution. Positive imaginary parts cannot occur.

If there is no damping, the velocity of propagation is $v = \omega/k = c/\sqrt{\varepsilon}$. When $\varepsilon > 1$ ($\varepsilon < 1$), one has $v < c$ ($v > c$).

Why is this result not a contradiction to the principle of causality?

Example 16.3: Exercise: Waves along a two-conductor line

A two-conductor line of length l (submarine cable) is connected to an alternating current source $U^e(t)$ and a resistor R_0 (Figure 16.4a). The wire has the self-inductance L, capacitance C_v, and resistance R per unit length.

(a) Imagine the two-conductor line to be decomposed into small pieces of length Δx, and regard it as a chain of oscillating circuits (Figure 16.4b). Think of the relation between the current I and the voltage U at the points x and $x + \Delta x$. By means of the limit $\Delta x \to 0$, derive the telegraph equations

$$L\frac{\partial I}{\partial t} + RI + \frac{\partial U}{\partial x} = 0, \qquad \frac{\partial I}{\partial x} + C\frac{\partial U}{\partial t} = 0$$

taking into account the capacitance coefficient C_i. What are the boundary conditions?

(b) Show that for $R = 0$ the quantities I and U obey a wave equation. What is the velocity v at which they are propagating?

(c) Show that for $R_0 = \sqrt{L/C}$ there are solutions of the type

$$U(x, t) = U_0 f(x - vt)$$

$$I(x, t) = \sqrt{\frac{C}{L}} U(x, t)$$

By which $U_0 f(x - vt)$ is determined. What is the physical meaning of the solutions?

Solution **(a)** According to Kirchhoff's rules

$$I_x = I_{C_v} + I_{C_i} + I_{x+\Delta x} \qquad (16.53)$$
$$U_x = U_R + U_L + U_{x+\Delta x} \qquad (16.54)$$

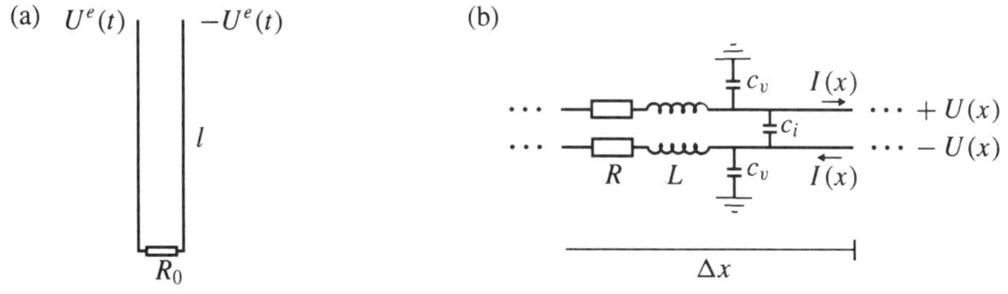

Figure 16.4. A two-conductor line connected to an alternating current source.

The relation $U = Q/C$ is known to exist between the voltage at a capacitor and the charge Q on one plate if C is the capacitance. Hence,

$$2U = \frac{Q_{C_i}}{C_i \Delta x}$$

$$U = \frac{Q_{C_V}}{C_V \Delta x} \Rightarrow I_{C_i} + I_{C_V} = \dot{Q}_{C_i} + \dot{Q}_{C_V} = \dot{U}(x)[2C_i + C_V]\Delta x$$

With $C = 2C_i + C_V$ one obtains from (16.53):

$$I(x + \Delta x, t) - I(x, t) + \frac{\partial}{\partial t} U(x, t) C \Delta x = 0$$

After dividing by Δx and taking the limit $\Delta x \to 0$,

$$\frac{\partial I}{\partial t} + C \frac{\partial U}{\partial x} = 0 \tag{16.55}$$

On the other hand, $U_R = (\Delta x R)I_x$, $U_L = (L\Delta x)\dot{I}_x$, so after a limiting procedure (16.53) becomes

$$\frac{\partial U}{\partial x} + RI + L\frac{\partial I}{\partial t} = 0 \tag{16.56}$$

(b) For $R = 0$,

$$\frac{\partial^2 I}{\partial x^2} = -C\frac{\partial}{\partial x}\frac{\partial}{\partial t}U = -C\frac{\partial}{\partial t}\frac{\partial}{\partial x}U = CL\frac{\partial^2 I}{\partial t^2}$$

$$\Rightarrow \left(\frac{\partial^2}{\partial x^2} - LC\frac{\partial^2}{\partial t^2}\right)I = 0$$

and analogously $\left(\partial^2/\partial x^2 - LC\partial^2/\partial t^2\right)U = 0$

From this equation one may read off the velocity of the phase, $v = (LC)^{-1/2}$, as can be realized immediately.

(c) From this equation one may read off directly the boundary condition

$$U(x = l, t) = R_0 I(x = l, t)$$

as can be realized immediately. $U_0 f(x - vt)$ is determined from the boundary condition $U(x = 0, t) = U^e(t)$. The general solutions of the oscillation equations contain waves traveling forward $(f(x - vt))$ as well as waves traveling backward $(f'(x + vt))$, or reflected waves. The reflection is avoided by a special choice of the resistance R_0.

17 Index of Reflection and Refraction

In this chapter we want to derive the reflection and refraction laws of optics from Maxwell's equations. For this purpose, we will consider the behavior of an electromagnetic wave at the interface between two distinct media. We will assume that the media are not ferromagnetic and have a constant polarizability, permeability, and conductivity.

Normal incidence

First, we will consider plane electromagnetic waves striking the interface in the normal direction. From the elementary laws of geometric optics we know that part of the wave is reflected and part of the wave is refracted.

We want to compute the reflection coefficient for a wave. For this purpose, we start from the conditions for continuity for the normal and tangential components of the fields. The distinct media are denoted by the indices 1 and 2.

From Maxwell's equations

$$\operatorname{div}\mathbf{D} = 4\pi\rho$$
$$\operatorname{div}\mathbf{B} = 0$$

neglecting the surface charge, earlier we found the boundary conditions

$$D_n^{(1)} = D_n^{(2)}, \qquad B_n^{(1)} = B_n^{(2)} \tag{17.1}$$

(The first equation follows from equation (6.6) with the assumption of vanishing surface charge, the second one from equation (11.16).)

Remark B_n is not the normal component of the $\mathbf{B}$-field (17.1) which is continuous between media (1) and (2), but is normal to da, see Figure 17.1. The Maxwell equations from which we had concluded the continuity of the tangential components E_t and H_t contain now further

333

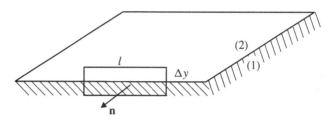

Figure 17.1. The boundary layer between two media, seen in perspective.

terms, so that the validity of the boundary conditions still must be verified. The equation $\text{curl}\,\mathbf{E} = -(1/c)(\partial/\partial t)\mathbf{B}$ yields the integral relation

$$\int_{\text{area}} \text{curl}\,\mathbf{E} \cdot \mathbf{n}\,da = -\frac{1}{c}\frac{\partial}{\partial t}\int_{\text{area}} \mathbf{B} \cdot \mathbf{n}\,da$$

The figure shows the position of the surface of integration a lying normal to the interface between the media. With help of Stokes' theorem we obtain from the left-hand side of the equation in the limit $\Delta y \to 0$

$$\int_{\text{area}} \text{curl}\,\mathbf{E} \cdot \mathbf{n}\,da = \oint_C \mathbf{E} \cdot d\mathbf{s} = \left(E_t^{(1)} - E_t^{(2)}\right)l$$

and for the right-hand side, as $\Delta y \to 0$,

$$\frac{\partial}{\partial t}\int \mathbf{B} \cdot \mathbf{n}\,da = \frac{\partial}{\partial t}B_n l\Delta y \to 0$$

Thus, we obtain the continuity of the tangential components

$$E_t^{(1)} = E_t^{(2)} \tag{17.2}$$

Correspondingly, from the Maxwell equation

$$\text{curl}\,\mathbf{H} = \frac{1}{c}\frac{\partial}{\partial t}\mathbf{D} + \frac{4\pi}{c}\mathbf{j}$$

assuming that no surface currents are present, in an analogous calculation we obtain the boundary condition

$$H_t^{(1)} = H_t^{(2)} \tag{17.3}$$

Having been convinced of the validity of the former boundary conditions we want to investigate now the following special case: let a wave be incident from the vacuum on a medium with the generalized index of refraction $p = p_1 + ip_2$ (16.16). Let the (y, z)-plane be the interface, as in Figure 17.2. With this simplification the wavevector is $\mathbf{k} = k\mathbf{e}_x$, and also $\mathbf{k} \cdot \mathbf{r} = kx$. We suppose the incident wave to be a plane wave:

$$E_y = ae^{i(kx-\omega t)}, \qquad H_z = be^{i(kx-\omega t)} \tag{17.4}$$

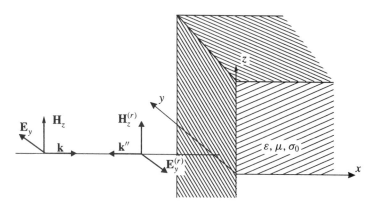

Figure 17.2. Normal to the interface incident wave and reflected wave.

where $x < 0$. The fact that $\mathbf{H}$ has only a z-component follows from equation (17.8) for $\mathbf{k} = (k, 0, 0)$ and $\mathbf{E} = (0, E_y, 0)$. $\mathbf{k}$, $\mathbf{E}$, and $\mathbf{H}$ form a right-handed system. Since the incident wave is traveling in vacuum, $p = 1$ and thus $k = \omega/c$.

Utilizing the relations for vacuum derived in Chapter 15 (see equations (15.20) and (15.21)), we obtain

$$\mathbf{E} = -\frac{\mathbf{k}}{k} \times \mathbf{H} \qquad \text{and} \qquad \mathbf{H} = \frac{\mathbf{k}}{k} \times \mathbf{E}$$

From these equations we find in (17.4) that $b = +a$. For the reflected wave we make the ansatz:

$$E_y^{(r)} = -a'' e^{i(k''x - \omega''t)}, \qquad H_z^{(r)} = b'' e^{i(k''x - \omega''t)} \tag{17.5}$$

where again $x < 0$. b'' is fixed by the preceeding equation involving a''. ($\mathbf{k}''$ points in the $-x$-direction.) Since the reflected wave, like the incident wave, propagates in vacuum, $k'' = k$.

In the medium, $p \neq 1$, and hence, the wave number vector of the penetrating wave $k' = (\omega/c)\sqrt{\mu\eta} = (\omega/c)p$ (see equation (16.16)). Furthermore, in a medium (see equation (16.8)),

$$\mathbf{k}' \times \mathbf{E}^{(e)} = \frac{\omega}{c}\mu\mathbf{H}^{(e)}$$

Dividing this equation by $k' = (\omega/c)p$ and solving for $\mathbf{H}$ we obtain

$$\mathbf{H}^{(e)} = \frac{p}{\mu}\frac{\mathbf{k}'}{k'} \times \mathbf{E}^{(e)} \tag{17.6}$$

The penetrating wave is described by the following ansatz:

$$E_y^{(e)} = a' e^{i(k'x - \omega' t)}$$

and taking into account the relation $H^{(e)} = (p/\mu) E^{(e)}$ for $\mathbf{k} \perp \mathbf{E}^{(e)}$, we note that for the magnetic field

$$H_z^{(e)} = \frac{p}{\mu} a' e^{i(k'x - \omega' t)}$$

Considerations of the interface

We are interested now in the ratios of the amplitudes a', a'', and a. From the continuity of the tangential components E_t and H_t at the interface and from the wave equations of the incident, reflected, and refracted waves, for $x = 0$ and $t = 0$ we infer

$$E_y\big|_{x=0,t=0} + E_y^{(r)}\big|_{x=0,t=0} = E_y^{(e)}\big|_{x=0,t=0} \qquad \Leftrightarrow \qquad a - a'' = a'$$

$$H_z\big|_{x=0,t=0} + H_z^{(r)}\big|_{x=0,t=0} = H_z^{(e)}\big|_{x=0,t=0} \qquad \Leftrightarrow \qquad a + a'' = \frac{p}{\mu} a' \qquad \textbf{(17.7)}$$

Furthermore, for $x = 0$ and arbitrary t there must be $\omega = \omega' = \omega''$, as demonstrated by the following calculation:

$$E_y\big|_{x=0,t} + E_y^{(r)}\big|_{x=0,t} = E_y^{(e)}\big|_{x=0,t} \qquad \Leftrightarrow$$

$$a e^{-i\omega t} - a'' e^{-i\omega'' t} = a' e^{-i\omega' t} \equiv (a - a'') e^{-i\omega' t}$$

$$H_z\big|_{x=0,t} + H_z^{(r)}\big|_{x=0,t} = H_z^{(e)}\big|_{x=0,t} \qquad \Leftrightarrow$$

$$a e^{-i\omega t} + a'' e^{-i\omega'' t} = \frac{p}{\mu} a' e^{-i\omega' t} \equiv (a + a'') e^{-i\omega' t} \qquad \textbf{(17.8)}$$

The addition of both equations yields at once $\omega = \omega'$, and from that immediately $\omega = \omega' = \omega''$. Equations (17.7) may be solved for a' and a'', with the result

$$a'' = a \frac{p - \mu}{p + \mu} \qquad \text{and} \qquad a' = a \frac{2\mu}{p + \mu} \qquad \textbf{(17.9)}$$

The *reflection coefficient* R is defined as the ratio of the intensities of the reflected and incident waves. The intensity is proportional to the square of the amplitudes

$$R = \frac{|a''|^2}{|a|^2} = \frac{a'' a''^*}{a a^*}$$

Substituting and taking into account[1] $p = p_1 + ip_2 \equiv n + i\kappa$ (κ corresponds to β in (17.18)), we obtain

$$R = \frac{(p-\mu)(p^*-\mu)}{(p+\mu)(p^*+\mu)} = \frac{(p_1-\mu)^2 + p_2^2}{(p_2+\mu)^2 + p_2^2} = \frac{(n-\mu)^2 + \kappa^2}{(n+\mu)^2 + \kappa^2}$$

With increasing absorption also the reflection increases. For long waves ($n \approx \kappa$) we obtain ($\mu \approx 1$):

$$R = \frac{2n^2 - 2n + 1}{2n^2 + 2n + 1} \approx \frac{n-1}{n+1} \approx 1 - \frac{2}{n}$$

This formula has been checked for various metals. It is satisfied up to about $\lambda = 25\mu m$ or $25 \cdot 10^{-6} m = 25 \cdot 10^{-3} mm$.

In the case of low-frequency electromagnetic waves we may set approximately

$$n = \kappa = \sqrt{\frac{2\pi\sigma_0}{\omega}}$$

and for the reflection coefficient we obtain

$$R = 1 - \sqrt{\frac{2\omega}{\pi\sigma_0}}$$

We know that for high-frequency fields the conductivity σ is smaller than the direct current conductivity σ_0 (compare equation (16.23)). So we have a qualitative result, at least, for R: the reflection coefficient should decrease with increasing frequency. This effect may be observed, e.g., for H_2SO_4: ammonium sulfate possesses—after being diluted with water—an excellent direct current conductivity, but it is completely transparent to light ($R \approx 0$). The heavy ions are not able to follow the high-frequency fields.

Example 17.1:   Exercise: Frequency-dependent index of refraction, reflection coefficient

Write the Fourier-transformed Maxwell equations. For low frequencies ω the assumptions $\mathbf{D}_0 = \varepsilon_0\mathbf{E}$, $\mathbf{B} = \mu\mathbf{H}$, $\mathbf{j}_0 = \sigma_0\mathbf{E}$, and $\rho_0 = 0$, with the static constants ε_0, μ, and $\sigma_0 \in \mathbb{R}$ are valid.

(a) Show that in Maxwell's equations the quantities $\mathbf{D}_0$ and $\mathbf{j}_0$ may be replaced by the quantities $\mathbf{D}$ and $\mathbf{j}$, so that for $\omega \neq 0$, $\mathbf{D} = \varepsilon\mathbf{E}$ and $\mathbf{j} \equiv \mathbf{0}$. What is the difference between the fields $\mathbf{D}_0$ and $\mathbf{D}$? Think about the concept of polarization and explain why in the static case ($\omega = 0$) the field $\mathbf{D}$ becomes useless. Describe ε.

(b) Calculate $p = \sqrt{\varepsilon}$.

(c) Give the reflection coefficient for very low frequencies.

---

[1] Now, we write the complex index of refraction as $p = n + i\kappa$ to assure the usual relation $p = n$ ($\kappa \to 0$) of the real index of refraction.

Solution   By the Fourier transformation Maxwell's equations

$$\nabla \cdot \mathbf{D_0} = \rho_0 4\pi, \qquad\qquad\qquad \nabla \cdot \mathbf{B} = 0$$

$$\nabla \times \mathbf{H} - \frac{1}{c}\frac{\partial}{\partial t}\mathbf{D_0} = \frac{4\pi}{c}\mathbf{j_0}, \qquad \nabla \times \mathbf{E} + \frac{1}{c}\frac{\partial}{\partial t}\mathbf{B} = 0$$

are transformed into

$$\mathbf{k} \cdot \mathbf{D_0}(\omega, \mathbf{k}) = \frac{4\pi}{i}\rho_0(\omega, \mathbf{k}), \qquad\qquad\qquad \mathbf{k} \cdot \mathbf{B}(\omega, \mathbf{k}) = 0$$

$$\mathbf{k} \times \mathbf{H}(\omega, \mathbf{k}) + \frac{1}{c}\omega\mathbf{D_0}(\omega, \mathbf{k}) = \frac{4\pi}{ic}\mathbf{j_0}(\omega, \mathbf{k}), \qquad \mathbf{k} \times \mathbf{E}(\omega, \mathbf{k}) - \frac{1}{c}\omega\mathbf{B}(\omega, \mathbf{k}) = 0$$

The constraints $\rho_0 = 0$, $\mathbf{j_0} = \sigma_0 \mathbf{E}$, $\mathbf{D_0} = \varepsilon_0 \mathbf{E}$, and $\mathbf{B} = \mu\mathbf{H}$ change the inhomogeneous Maxwell's equations:

$$\frac{1}{\mu}\mathbf{k} \times \mathbf{B}(\omega, \mathbf{k}) + \frac{1}{c}\omega\left[\varepsilon_0 + i\frac{4\pi\sigma_0}{\omega}\right]\mathbf{E}(\omega, \mathbf{k}) = 0$$

$$\mathbf{k} \cdot \mathbf{E} = 0$$

(a) Defining $\mathbf{D} = \varepsilon(\omega)\mathbf{E}$, with $\varepsilon(\omega) = \varepsilon_0 + i4\pi\sigma_0/\omega$, then

$$\mathbf{k} \times \mathbf{H}(\omega, \mathbf{k}) + \frac{1}{c}\omega\mathbf{D}(\omega, \mathbf{k}) = 0$$

$$\mathbf{k} \cdot \mathbf{D} = \varepsilon(\omega)\mathbf{k} \cdot \mathbf{E} = 0 \qquad \text{and} \qquad \mathbf{j} \equiv 0$$

The definition of $\mathbf{D}$ is $\mathbf{D} = \mathbf{E} + 4\pi\mathbf{P} = \varepsilon\mathbf{E}$, where $\mathbf{E}$ is the macroscopic field, and the definition of $\mathbf{P}$ depends on the definition of the "molecules" and the "free charges." In the case $\omega \neq 0$, an electron which only oscillates and does not move far from the ion core may be viewed as a constituent of the molecule. Then there are no conduction electrons, and consequently, the macroscopic current is $\mathbf{j} \equiv 0$. Of course, the polarization depends on ω, hence also $\varepsilon = \varepsilon(\omega)$. As well, the ion may be regarded "the molecule," then the macroscopic current is $j_0 \neq 0$ and $\mathbf{D_0} = \mathbf{E} + 4\pi\mathbf{P_0}$, where $\mathbf{P_0}$ describes only the polarization of the ion core.

(b) We suppose $\mu = 1$, then the index of refraction is $\sqrt{\varepsilon} = p$. Let $N = \left[\varepsilon_0^2 + (4\pi\sigma_0/\omega)^2\right]^{+1/2}$ and $\varphi = \arccos(\varepsilon_0/N)$, then

$$p^2(\omega) = \varepsilon(\omega) = Ne^{i\varphi} \quad\Rightarrow\quad p(\omega) = \pm\sqrt{N}\left[\cos\frac{\varphi}{2} + i\sin\frac{\varphi}{2}\right]$$

Because $\cos(\varphi/2) = \sqrt{\frac{1}{2}(1 + \cos\varphi)}$ and $\sin(\varphi/2) = \sqrt{\frac{1}{2}(1 - \cos\varphi)}$,

$$p(\omega) = \pm\sqrt{N}\left(\sqrt{\frac{1}{2}\left(1 + \frac{\varepsilon_0}{N}\right)} + i\sqrt{\frac{1}{2}\left(1 - \frac{\varepsilon_0}{N}\right)}\right)$$

$$= \pm\sqrt{\varepsilon_0}\left(\sqrt{\frac{\sqrt{1 + [4\pi\sigma_0/(\omega\varepsilon_0)]^2} + 1}{2}} + i\sqrt{\frac{\sqrt{1 + [4\pi\sigma_0/(\omega\varepsilon_0)]^2} - 1}{2}}\right)$$

One can make a choice of the sign so that for $\sigma_0 = 0$ the previous result $p = \sqrt{\varepsilon_0}$ for insulators is obtained; that is, the positive value is the correct one.

(c) As $\omega \to 0$, $p \approx \sqrt{2\pi\sigma_0/\omega}(1+i)$. With $p = \alpha + i\beta$, we obtain at once the reflection coefficient

$$R = \left|\frac{1-p}{1+p}\right|^2 = \frac{(1-\alpha)^2 + \beta^2}{(1+\alpha)^2 + \beta^2}$$

$$\approx \frac{1 - 2\alpha + 2\alpha^2}{1 + 2\alpha + 2\alpha^2} = \frac{1 - \dfrac{1}{\alpha} + \dfrac{1}{2\alpha^2}}{1 + \dfrac{1}{\alpha} + \dfrac{1}{2\alpha^2}}$$

$$\omega \to 0$$

$$= \left(1 - \frac{1}{\alpha} + \frac{1}{2\alpha^2}\right)\left[1 - \left(\frac{1}{\alpha} + \frac{1}{2\alpha^2}\right) + O\left(\frac{1}{\alpha^2}\right)\right]$$

$$= 1 - \frac{1}{\alpha} - \frac{1}{\alpha} + O\left(\frac{1}{\alpha^2}\right) \approx 1 - \frac{2}{\alpha} = 1 - \sqrt{\frac{2\omega}{\pi\sigma}}$$

General derivation of the laws of refraction and reflection (Fresnel's formulas)

Having discussed normal incidence, we consider the incidence of the electromagnetic wave at an angle α with respect to the interface of two distinct media. At first, we are concerned with the kinematical aspects of reflection and refraction; that is, we will calculate the geometrical relations, the angles between the various wavevectors.

A plane wave having the frequency ω and the wavevector $\mathbf{k}$ approaches from the isotropic medium 1 with the permittivity ε and permeability μ, and strikes the interface to medium 2 (ε', μ').[2] The angle of incidence to the origin of the coordinate system is α, and the interface is the plane $z = 0$. The incident wave is refracted into the isotropic, nonferromagnetic medium 2 (wavevector $\mathbf{k}'$) and reflected into the medium 1 at the angle γ (wavevector $\mathbf{k}''$). Figure 17.3 illustrates the notation.

For the electric field intensity of the incident wave we again make the ansatz

$$\mathbf{E} = \mathbf{E}_0 e^{i(\mathbf{k}\cdot\mathbf{r} - \omega t)}$$

and then, using the relations $\mathbf{k} \times \mathbf{E} = (\omega/c)\mu\mathbf{H} = (\omega/c)\mathbf{B}$ and $k = (\omega/c)\sqrt{\varepsilon\mu}$, we obtain for the magnetic induction (see equation (16.8))

$$\mathbf{B} = \sqrt{\varepsilon\mu}\frac{\mathbf{k} \times \mathbf{E}}{k} \tag{17.10}$$

Correspondingly, for the refracted wave

$$\mathbf{E}' = \mathbf{E}_0' e^{i(\mathbf{k}'\cdot\mathbf{r} - \omega't)}, \qquad \mathbf{B}' = \sqrt{\varepsilon'\mu'}\frac{\mathbf{k}' \times \mathbf{E}'}{k'} \tag{17.11}$$

---

[2]We suppose that in both media the conductivity is $\sigma = 0$; otherwise, the waves are damped and the media would not be transparent.

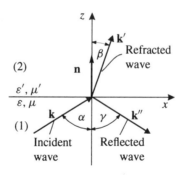

Figure 17.3. Incident, refracted, and reflected waves.

and for the reflected wave

$$\mathbf{E}'' = \mathbf{E}_0'' e^{i(\mathbf{k}''\cdot\mathbf{r}-\omega''t)}, \qquad \mathbf{B}'' = \sqrt{\varepsilon\mu}\frac{\mathbf{k}'' \times \mathbf{E}''}{k''} \tag{17.12}$$

As for normal incidence, in this case also the four conditions of continuity for

$$E_t \quad \text{and} \quad H_t \tag{17.13}$$

$$D_n \quad \text{and} \quad B_n \tag{17.14}$$

are valid at the interface, where we have excluded surface currents and surface charges.

To fulfill such boundary conditions at all times and at all points of the interface, the time-dependent and space-dependent factors in front of the incident, refracted, and reflected waves must be equal in the plane $z = 0$:

$$e^{i(\mathbf{k}\cdot\mathbf{r}-\omega t)} = e^{i(\mathbf{k}'\cdot\mathbf{r}-\omega't)} = e^{i(\mathbf{k}''\cdot\mathbf{r}-\omega''t)}$$

From the properties of the exponential function we obtain

$$\mathbf{k}\cdot\mathbf{r} - \omega t = \mathbf{k}'\cdot\mathbf{r} - \omega't + m_1 2\pi = \mathbf{k}''\cdot\mathbf{r} - \omega''t + m_2 2\pi$$

for all times t, where m_1 and m_2 are integers. This must hold for all $\mathbf{r}$ lying in the interface, and in particular, for $m_1 = m_2 = 0$ and for $\mathbf{r} = 0$. The wave strikes the interface at the origin. For $\mathbf{r} = 0$ and $t \neq 0$ we conclude in particular for $m_1 = m_2 = 0$,

$$\omega = \omega' = \omega'' \tag{17.15}$$

Hence, the frequency of the wave is not altered in refraction and reflection. Furthermore, this implies

$$k = \frac{\omega}{c}\sqrt{\varepsilon\mu} = \frac{\omega''}{c}\sqrt{\varepsilon\mu} = k''$$

that is, the wavevectors of the *incident* and *reflected* wave are equal to each other with respect to this magnitude.

Due to (17.15), and at a point $\mathbf{r}$ from the boundary layer (plane interface),

$$(\mathbf{k} \cdot \mathbf{r})_{z=0} = (\mathbf{k}' \cdot \mathbf{r})_{z=0} = (\mathbf{k}'' \cdot \mathbf{r})_{z=0} \qquad (17.16)$$

This relation implies that the projections of $\mathbf{k}$, $\mathbf{k}'$, and $\mathbf{k}''$ onto any $\mathbf{r}$ of the interface are all equal to each other. But, this is possible only if $\mathbf{k}$, $\mathbf{k}'$ and $\mathbf{k}''$ lie in one plane, the so-called *plane of incidence*.

Utilizing these facts the figure may be interpreted in the following way: Choosing a particular value of $\mathbf{r}$ on the x-axis,

$$\mathbf{r} = x\mathbf{e}_1$$

from equation (17.16)

$$kx \sin \alpha = k'x \sin \beta = k''x \sin \gamma \qquad (17.17)$$

Since $k = k''$,

$$\sin \alpha = \sin \gamma \qquad \text{or} \qquad \alpha = \gamma$$

that is, the angle of incidence equals the angle of reflection. Furthermore, we obtain *Snellius' law of refraction*

$$\frac{\sin \alpha}{\sin \beta} = \frac{k'}{k} = \sqrt{\frac{\varepsilon' \mu'}{\varepsilon \mu}} = \frac{n'}{n} = \frac{c_1}{c_2} = \frac{\lambda_1}{\lambda_2} \qquad (17.18)$$

where c_1, λ_1 and c_2, λ_2 are the speed of light and the wavelength in media 1 and 2, respectively. Furthermore,

$$n = \frac{c}{c_1} = \sqrt{\varepsilon \mu} \qquad (17.19)$$

is the index of refraction, from Maxwell's relation. The speed of light, the wavelength, and the angle are smaller in the medium of higher optical density, while the index of refraction, the permittivity, and the magnitude of the wave vector are larger ($\mu \approx 1$).

After deriving the relations known from geometric optics, now we must consider the dynamical aspect of the phenomenon, in particular to calculate the intensities of the reflected or refracted waves. While the kinematical statements follow merely from the existence of general boundary conditions at $z = 0$, we deduce the dynamical properties only from special boundary conditions on the continuity of the tangential and normal components, relations (17.13) and (17.14). With the normal unit vector we write these relations in the following form: For the *tangential component of* $\mathbf{E}$, we have at $z = 0$

$$(\mathbf{E}_0 + \mathbf{E}_0'' - \mathbf{E}_0') \times \mathbf{n} = 0 \qquad (17.20a)$$

Using equation (17.10) the *normal component of* $\mathbf{B}$ is expressed by the electric field intensity:

$$(\mathbf{k} \times \mathbf{E}_0 + \mathbf{k}'' \times \mathbf{E}_0'' - \mathbf{k}' \times \mathbf{E}_0') \cdot \mathbf{n} = 0 \qquad (17.20b)$$

because $\sqrt{\varepsilon'\mu'}/k' = \sqrt{\varepsilon\mu}/k$, according to equation (17.18). With equation (17.3) we obtain the *tangential component of* $\mathbf{H} = [c/(\omega\mu)]\mathbf{k} \times \mathbf{E}$:

$$\left(\frac{1}{\mu} \left(\mathbf{k} \times \mathbf{E}_0 + \mathbf{k}'' \times \mathbf{E}_0'' \right) - \frac{1}{\mu'} \left(\mathbf{k}' \times \mathbf{E}_0' \right) \right) \times \mathbf{n} = 0 \qquad \textbf{(17.20c)}$$

and for the *normal component of* $\mathbf{D}$

$$\left(\varepsilon \left(\mathbf{E}_0 + \mathbf{E}_0'' \right) - \varepsilon' \mathbf{E}_0' \right) \cdot \mathbf{n} = 0 \qquad \textbf{(17.20d)}$$

In the derivation we have merely to note that

$$\frac{\sqrt{\varepsilon\mu}}{k} = \frac{c}{\omega} = \frac{c}{\omega'} = \frac{\sqrt{\varepsilon'\mu'}}{k'} \qquad \text{and} \qquad \mathbf{D} = \varepsilon\mathbf{E}$$

because the media have been assumed to be isotropic.

In the following, we want to investigate two special cases. We will consider a linearly polarized wave with the electric field intensity, first, normal to the plane of incidence, and second, in the plane of incidence. The general case of an elliptically polarized wave results from a superposition of these two cases.

If $\mathbf{E}_0$ is perpendicular to the plane of incidence (Figure 17.4) then the boundary condition (17.20a) yields the relation

$$(\mathbf{E}_0 + \mathbf{E}_0'' - \mathbf{E}_0') \times \mathbf{n} = +(E_0 + E_0'' - E_0')\mathbf{e}_x = 0$$

which leads to

$$E_0 + E_0'' - E_0' = 0 \qquad \textbf{(17.21)}$$

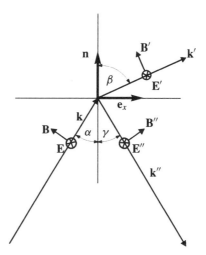

Figure 17.4. Refraction and reflection of a linearly polarized wave if $\mathbf{E}_0$ lies perpendicular to the plane of incidence.

Equation (17.21) follows also from the equation (17.20b) if the vector product is rewritten and the refraction law is taken into account:

$$(\mathbf{n} \times \mathbf{k}) \cdot \mathbf{E}_0 + (\mathbf{n} \times \mathbf{k}'') \cdot \mathbf{E}_0'' - (\mathbf{n} \times \mathbf{k}') \cdot \mathbf{E}_0'$$
$$= (\sin \alpha) k \mathbf{e}_y \cdot \mathbf{E}_0 + (\sin \gamma) k \mathbf{e}_y \cdot \mathbf{E}_0'' - (\sin \beta) k' \mathbf{e}_y \cdot \mathbf{E}_0'$$
$$= \mathbf{e}_y \cdot (\mathbf{E}_0 + \mathbf{E}_0'' - \mathbf{E}_0') = 0$$

since $\sin(\pi - \gamma) = \sin \gamma$. On the other hand, by rewriting the cross product equation (17.20c) yields

$$\frac{1}{\mu} (\mathbf{k} \times \mathbf{E}_0) \times \mathbf{n} = -\frac{1}{\mu} (\mathbf{k}(\mathbf{n} \cdot \mathbf{E}_0) - \mathbf{E}_0(\mathbf{n} \cdot \mathbf{k}))$$

for the incident wave ($\mathbf{n} \cdot \mathbf{E}_0 = 0$)

$$\frac{1}{\mu} (\mathbf{k} \times \mathbf{E}_0) \times \mathbf{n} = \frac{1}{\mu} k \mathbf{E}_0 \cos \alpha = \frac{\omega}{c} \sqrt{\frac{\varepsilon}{\mu}} \mathbf{E}_0 \cos \alpha$$

Corresponding relations result for the reflected and refracted waves ($\cos(\pi - \gamma) = -\cos \gamma$):

$$\frac{1}{\mu} (\mathbf{k}'' \times \mathbf{E}_0'') \times \mathbf{n} = -\frac{\omega''}{c} \sqrt{\frac{\varepsilon}{\mu}} \mathbf{E}_0'' \cos \gamma$$

$$\frac{1}{\mu'} (\mathbf{k}' \times \mathbf{E}_0') \times \mathbf{n} = \frac{\omega'}{c} \sqrt{\frac{\varepsilon'}{\mu'}} \mathbf{E}_0' \cos \beta$$

Taking into account the fact that the frequencies are equal to each other, as are the angles of incidence and reflection, then the equation (17.20c) becomes

$$(E_0 - E_0'') \sqrt{\frac{\varepsilon}{\mu}} \cos \alpha - E_0' \sqrt{\frac{\varepsilon'}{\mu'}} \cos \beta = 0 \tag{17.22}$$

Condition (17.20d) yields nothing new, since $\mathbf{E}_0 \cdot \mathbf{n} \equiv 0$. Using equation (17.21) we may eliminate E_0' or E_0'', correspondingly. Thus, we obtain the amplitudes of the reflected and refracted wave relative to the incident one. On one hand,

$$\frac{E_0''}{E_0} = \frac{1 - \sqrt{\dfrac{\varepsilon' \mu \cos \beta}{\varepsilon \mu' \cos \alpha}}}{1 + \sqrt{\dfrac{\varepsilon' \mu \cos \beta}{\varepsilon \mu' \cos \alpha}}} = \frac{1 - \dfrac{\mu \tan \alpha}{\mu' \tan \beta}}{1 + \dfrac{\mu \tan \alpha}{\mu' \tan \beta}} \tag{17.23}$$

and on the other hand,

$$\frac{E_0'}{E_0} = \frac{2}{1 + \sqrt{\dfrac{\varepsilon' \mu \cos \beta}{\varepsilon \mu' \cos \alpha}}} = \frac{2}{1 + \dfrac{\mu \tan \alpha}{\mu' \tan \beta}} \tag{17.24}$$

If we set $\mu = \mu'$, which is, in general, the case for the frequencies of visible radiation, then we obtain Fresnel's formulas for light polarized perpendicular to the plane of incidence

$$\frac{E_0''}{E_0} = \frac{\sin(\beta - \alpha)}{\sin(\alpha + \beta)} \quad \text{and} \quad \frac{E_0'}{E_0} = \frac{2 \sin \beta \cos \alpha}{\sin(\alpha + \beta)} \tag{17.25}$$

If $\mathbf{E}_0$ lies in the plane of incidence (Figure 17.5), according to the conditions of continuity (17.20a–d) the following equations are obtained:

$$(17.20a) \to (E_0 - E_0'') \cos \alpha - E_0' \cos \beta = 0 \tag{17.26}$$

$$(17.20d) \to (E_0 + E_0'') \sqrt{\frac{\varepsilon}{\mu}} - E_0' \sqrt{\frac{\varepsilon'}{\mu'}} = 0 \tag{17.27}$$

In (17.26) $\sin(\pi/2 - \alpha) = \cos(\alpha)$, and correspondingly, $\sin(\pi/2 - \beta) = \cos(\beta)$ are used. The sign of E_0'' results from its opposite orientation in comparison to E_0 when building the vector product. Equations (17.20b) and (17.20d) do not contain any new information. Equation (17.27) follows by use of (17.18). Therefore, in this case Fresnel's formulas read

$$\frac{E_0''}{E_0} = \frac{1 - \sqrt{\dfrac{\varepsilon \mu'}{\varepsilon' \mu}} \dfrac{\cos \beta}{\cos \alpha}}{1 + \sqrt{\dfrac{\varepsilon \mu'}{\varepsilon' \mu}} \dfrac{\cos \beta}{\cos \alpha}} = \frac{1 - \dfrac{\varepsilon}{\varepsilon'} \dfrac{\tan \alpha}{\tan \beta}}{1 + \dfrac{\varepsilon}{\varepsilon'} \dfrac{\tan \alpha}{\tan \beta}} \tag{17.28}$$

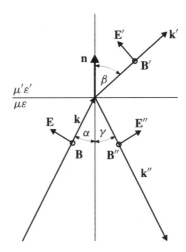

Figure 17.5. Refraction and reflection of a linarly polarized wave if $\mathbf{E}_0$ lies in the plane of incidence.

$$\frac{E'_0}{E_0} = \frac{2\sqrt{\dfrac{\varepsilon\mu'}{\varepsilon'\mu}}}{1 + \sqrt{\dfrac{\varepsilon\mu'}{\varepsilon'\mu}}\dfrac{\cos\beta}{\cos\alpha}} = \frac{2\dfrac{n'}{n}\dfrac{\varepsilon}{\varepsilon'}}{1 + \dfrac{\varepsilon}{\varepsilon'}\dfrac{\tan\alpha}{\tan\beta}} \tag{17.29}$$

Setting again $\mu = \mu'$, the equations simplify according to

$$\frac{E''_0}{E_0} = \frac{\tan(\alpha - \beta)}{\tan(\alpha + \beta)} \tag{17.30}$$

and

$$\frac{E'_0}{E_0} = \frac{2\sin\beta\cos\alpha}{\sin(\alpha + \beta)\cos(\alpha - \beta)} \tag{17.31}$$

In the special case of normal incidence, both angles vanish, $\alpha = 0$ and $\beta = 0$. Therefore, from the cosine representation in equations (17.23), (17.24), (17.28), and (17.29), we obtain the following relations:

$$\frac{E''_0}{E_0} = \frac{1 - \sqrt{\varepsilon'\mu/(\varepsilon\mu')}}{1 + \sqrt{\varepsilon'\mu/(\varepsilon\mu')}} \rightarrow \frac{n - n'}{n + n'}, \qquad \mu = \mu'$$

$$\frac{E'_0}{E_0} = \frac{2}{1 + \sqrt{\varepsilon'\mu/(\varepsilon\mu')}} \rightarrow \frac{2}{1 + n'/n}, \qquad \mu = \mu' \tag{17.32}$$

$$\frac{E''_0}{E_0} = \frac{1 - \sqrt{\varepsilon\mu'/(\varepsilon'\mu)}}{1 + \sqrt{\varepsilon\mu'/(\varepsilon'\mu)}} \rightarrow \frac{n' - n}{n' + n}, \qquad \mu = \mu'$$

$$\frac{E'_0}{E_0} = \frac{2\sqrt{\varepsilon\mu'/(\varepsilon'\mu)}}{1 + \sqrt{\varepsilon\mu'/(\varepsilon'\mu)}} \rightarrow \frac{2}{1 + n'/n}, \qquad \mu = \mu' \tag{17.33}$$

If both indices of refraction are equal to each other, we see at once that nothing is reflected:

$$E''_0 = 0, \qquad E'_0 = E_0$$

as we expect. If the index of refraction becomes very large ($n' \rightarrow \infty$), then the whole wave is reflected:

$$|E''_0| = E_0, \qquad E'_0 = 0$$

The inversion of the sign of E'_0/E_0 from (17.24), compared to E''_0/E_0 from (17.28), is caused by the distinct definition of $\mathbf{E}''_0$: In the first case, this vector points per definition in the same direction as $\mathbf{E}_0$; in the second case it changes direction (see the figures). However, it is interesting that for $n' > n$ a real phase jump of $\pi/2$ appears. Then, in the first case $E''_0/E_0 < 0$, or corresponding to the definition of the vectors, $\mathbf{E}''_0$ changes direction compared to $\mathbf{E}_0$, and in the second case, $E''_0/E_0 > 0$ and according to the figures, this means also a change of direction.

Total reflection

If the electromagnetic wave passes from a medium of higher optical density into a medium of lower optical density, the angle of refraction of the refracted wave is greater than the angle of incidence of the incident wave. Hence, there is a certain angle of incidence for which the angle of refraction reaches 90°. Then, the wave no longer penetrates into the medium of lower optical density. (See Figure 17.6.) We will consider this case of total reflection in somewhat more detail. A wave propagates in a medium having an index of refraction n and hits the interface to a medium having an index of refraction n'. The relation

$$n' < n$$

is valid. Snell's law of refraction yields the relation between the angle and the indices of refraction:

$$\frac{\sin \alpha}{\sin \beta} = \frac{n'}{n}, \qquad \text{hence,} \qquad \sin \alpha = \frac{n'}{n} \sin \beta \leq \sin \beta$$

For an angle of refraction of $\beta_0 = \pi/2$ there is an angle of incidence α_0 such that

$$\sin \alpha_0 = \frac{n'}{n}$$

It will turn out that in refraction this angle α_0 has the meaning of a critical angle of incidence. For example, by measuring this critical angle α_0, the optical density of an unknown medium may be determined experimentally with respect to a standard medium.

Replacing the ratio of the indices of refraction by the critical angle, we obtain the angle of refraction for an arbitrary angle of incidence α:

$$\cos \beta = \sqrt{1 - \sin^2 \beta} = \sqrt{1 - \left(\frac{n^2}{n'^2}\right) \sin^2 \alpha} = \sqrt{1 - \left(\frac{\sin \alpha}{\sin \alpha_0}\right)^2}$$

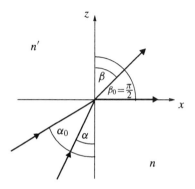

Figure 17.6. Total reflection.

For an angle of incidence $\alpha \leq \alpha_0$ we still get a real value of $\cos \beta$, and hence, for β, but $\cos \beta$ becomes purely imaginary for angles of incidence $\alpha > \alpha_0$; that is, the angle of refraction becomes complex:

$$\cos \beta = i\sqrt{\left(\frac{\sin \alpha}{\sin \alpha_0}\right)^2 - 1}, \qquad \alpha > \alpha_0$$

If the (x, y)-plane is the plane of incidence (see Figure 17.6), we obtain for the phase of refracted wave:

$$e^{i\mathbf{k}' \cdot \mathbf{r}} = e^{i(k'_x x + k'_z z)} = e^{i(k' x \sin \beta + k' z \cos \beta)}$$
$$= e^{ik' x (n/n') \sin \alpha} e^{-k' z \sqrt{(\sin \alpha / \sin \alpha_0)^2 - 1}}$$

We see that along the interface we are dealing with a standing wave of the phase $\exp\left(ik' x (n/n') \sin \alpha\right)$; however, the second exponential factor with a real, negative exponent shows that perpendicular to the plane of incidence (z-direction) the wave decays exponentially (damping factor). The penetration depth of the wave into the medium of lower optical density is on the order of magnitude of wavelengths. Therefore, in general, there is no refracted wave if $\alpha > \alpha_0$. If the layer of the medium of lower optical density is smaller than one wavelength and a layer of higher optical density lies directly over it, the wave may jump over to some extent. (See Figure 17.7.) In *Theoretical Physics*, volume 4, *Quantum Mechanics*, we shall recognize this phenomenon in the energetically forbidden tunneling through a potential barrier. We further want to calculate the energy flow through the boundary surface in the case of total reflection, that is, for $\alpha > \alpha_0$. In the time average indicated by $\langle \rangle$, this is given by

$$\langle \mathbf{S} \cdot \mathbf{n} \rangle = \frac{c}{8\pi} \left\langle \text{Re}[\mathbf{n} \cdot (\mathbf{E}' \times \mathbf{H}'^*)] \right\rangle$$

If we note that

$$\mathbf{H}' = \frac{c}{\mu' \omega} \mathbf{k}' \times \mathbf{E}'$$

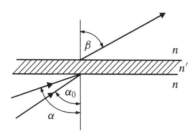

Figure 17.7. For $\alpha > \alpha_0$, a light wave "jumps" over a thin layer of lower optical density.

then, with $\mathbf{n} \cdot (\mathbf{E}' \times (\mathbf{k}' \times \mathbf{E}'^*)) = \mathbf{n} \cdot (|\mathbf{E}'|^2 \mathbf{k}' - (\mathbf{k}' \cdot \mathbf{E}')\mathbf{E}'^*) = |\mathbf{E}'|^2 \mathbf{n} \cdot \mathbf{k}'$, since $\mathbf{k}' \perp \mathbf{E}'$, we obtain

$$\mathbf{S} \cdot \mathbf{n} = \frac{c^2}{8\pi \mu' \omega} \operatorname{Re} \left[(\mathbf{n} \cdot \mathbf{k}') \left| E_0' \right|^2 \right] = \frac{c^2}{8\pi \mu' \omega} \operatorname{Re}(k' \cos \beta) \left| E_0' \right|^2 = 0$$

because $\cos \beta$ is purely imaginary as we have proved above. As expressed, no energy crosses the interface; all energy flows into the reflected wave (total reflection). At first glance, this result seems to be in contrast to our previous proposition for the limiting case of thin separation layers for which a certain portion of the wave should jump over (tunnel through). Without transport of energy this would be hardly possible. But, to treat this problem in a consistent way we would have to formulate our derivation more generally; that is, in this case we could not neglect the reflection of the refracted wave at the second interface ($n' \rightarrow n$). We could do this up till now since we started out with two semi-infinite media (with the indices of refraction n and n'). However, in the limiting case of thin separating layers the necessary consideration of the reflected refracted wave requires a not-too-small flow of energy through the boundary layer.

Brewster's angle

We assume that $\mu' = \mu$ in both media. Then, if in the incident wave the electric vector $\mathbf{E}$ is oscillating parallel to the plane of incidence, the amplitude of the reflected wave is

$$E_0'' = E_0 \frac{\tan(\alpha - \beta)}{\tan(\alpha + \beta)}$$

If for the angle

$$\tan(\alpha + \beta) = \infty, \qquad \text{that is,} \qquad \alpha + \beta = \frac{\pi}{2}$$

the amplitude of the reflected wave is zero, $E_0'' = 0$. This implies simply that $\sin \beta = \cos \alpha$. At once, Snell's law of refraction yields:

$$\frac{n'}{n} = \frac{\sin \alpha}{\sin \beta} = \tan \alpha_B$$

The angle α_B is called *Brewster's angle*.

For Brewster's angle there is no refracted ray whose electric vector oscillates parallel to the plane of incidence. If arbitrary light is incident under Brewster's angle then the reflected ray is linearly polarized perpendicular to the plane of incidence. See Figure 17.8.

This process may be understood if the process of reflection is considered microscopically: the incident wave excites dipole oscillations of the electrons (Figure 17.9). These electrons themselves emit dipole radiation. As we shall see later (Chapter 20) this radiation is strongest in the direction perpendicular to the direction of oscillation of the electron. It vanishes completely in direction of the oscillation. At the boundary surface two directions of effective radiation are formed by interference of the radiation of all electrons; one of them forms the refracted wave. The other one forms the reflected wave. If the electrons are

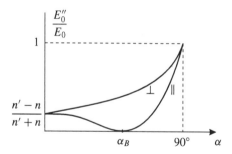

Figure 17.8. Amplitude of the reflected wave for light polarized parallel and perpendicular to the plane of incidence.

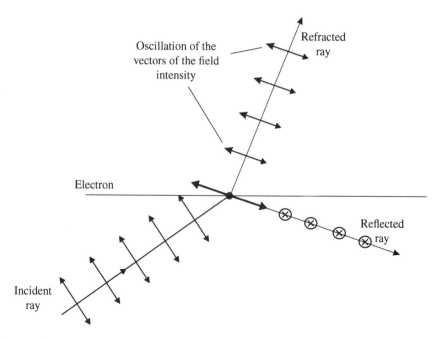

Figure 17.9. Refraction and reflection of light at the Brewster angle. The reflected ray is perpendicular to the refracted ray. Its electric field vectors oscillate perpendicular to the plane of incidence (indicated by ⊗).

oscillating on average just in direction of the reflected wave, the intensity of the radiation is equal to zero, so no wave is reflected. Then, the average radiation of these electrons obviously propagates only perpendicular to the direction of oscillation: the reflected and refracted waves are perpendicular.

Example 17.2:   Exercise: Reflection of a wave from a metallic surface

An electromagnetic wave is incident normally on a metal (Figure 17.10). The index of refraction and the coefficient of extinction are

$$n = \kappa = \left(\frac{2\pi\sigma\mu}{\omega}\right)^{1/2} \gg 1$$

(a) What are the amplitudes of the reflected and penetrating waves?

(b) The penetrating wave exerts a force

$$\mathbf{F} = \frac{1}{c}\int \mathbf{j} \times \mathbf{H}\,dV$$

on the metal. What is the time average of the pressure p that the wave exerts on the material?

Solution   **(a)** In medium 1 there is a vacuum, $\varepsilon = \mu = 1$. We choose the z-axis to be the direction of the incident wave; then the wave vector is

$$\mathbf{k}_A = k\mathbf{e}_z = \frac{\omega}{c}\mathbf{e}_z$$

For the **E**-field and the **H**-field we make the ansatz of plane waves:

$$\mathbf{E}_A = A\exp(ikz - i\omega t)\,\mathbf{e}_x, \qquad \mathbf{H}_A = A\exp(ikz - i\omega t)\,\mathbf{e}_y$$

Both fields are related by $\mathbf{k} \times \mathbf{E} = (\omega/c)\mathbf{H}$.
Similarly, for the reflected wave

$$\mathbf{k}_c = -k\mathbf{e}_z = -\frac{\omega}{c}\mathbf{e}_z$$

The fields of the reflected wave take the form

$$\mathbf{E}_C = C\exp(-ikz - i\omega t)\,\mathbf{e}_x, \qquad \mathbf{H}_C = -C\exp(-ikz - i\omega t)\,\mathbf{e}_y$$

In the medium 2, the wavevector of the penetrating wave is

$$\mathbf{k}_B = k_B\mathbf{e}_z = \frac{\omega}{c}(n + i\kappa)\mathbf{e}_z, \qquad \text{since} \qquad \frac{kc}{\omega} = n + i\kappa$$

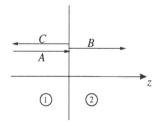

Figure 17.10. Incident, reflected, and transmitted waves.

Then, the field intensities of the penetrating wave are

$$\mathbf{E}_B = B \exp{(ik_B z - i\omega t)}\,\mathbf{e}_x = B \exp\left(i\frac{\omega}{c}nz - i\omega t\right)\exp\left(-\frac{\omega}{c}\kappa z\right)\mathbf{e}_x$$

$$\mathbf{H}_B = \frac{1}{\mu}(n + i\kappa)B\exp(ik_B z - i\omega t)\mathbf{e}_y$$

because

$$\mathbf{k} \times \mathbf{E} = \frac{\omega}{c}\mu\mathbf{H}$$

Since the tangential components of $\mathbf{E}$ and $\mathbf{H}$ have to be continuous, the amplitudes are related at $z = 0$ according to

$$A + C = B \tag{17.34}$$

$$A - C = \frac{1}{\mu}(n + i\kappa)B \tag{17.35}$$

From this we obtain the amplitude of the penetrating wave

$$B = \frac{2A}{1 + \frac{1}{\mu}(n + i\kappa)} \cong \frac{2A\mu}{n(1 + i)}$$

since $n = \kappa \gg 1$. With the relation

$$\frac{1}{1 + i} = \frac{1}{\sqrt{2}}\exp\left(-i\frac{\pi}{4}\right)$$

we get (compare Chapter 16)

$$B = \left(\frac{\mu\omega}{\sigma\pi}\right)^{1/2}\exp\left(-i\frac{\pi}{4}\right)A$$

According to equation (17.35), the amplitude of the incident wave is

$$A = C + \frac{1}{\mu}(n + i\kappa)B$$

Substituting the result for the amplitude $B \cong 2A\mu/[n(1 + i)]$, then

$$A \approx C + 2A, \qquad \text{hence} \qquad A \approx -C$$

Since the amplitude of the reflected wave is about the same as that of the incident wave and a jump in phase of 2π occurs, a standing wave arises in medium (1) with the field intensities

$$\mathbf{E} = A\,(\exp(ikz) - \exp(-ikz))\exp(-i\omega t)\mathbf{e}_x$$
$$\qquad = 2iA\sin(kz)\exp(-i\omega t)\mathbf{e}_x$$
$$\mathbf{H} = A\,(\exp(ikz) + \exp(-ikz))\exp(-i\omega t)\mathbf{e}_y$$
$$\qquad = 2A\cos(kz)\exp(-i\omega t)\mathbf{e}_y$$

The wave penetrating into medium (2) is damped:

$$\mathbf{E}_B = A \left(\frac{\mu\omega}{\sigma\pi}\right)^{1/2} \exp\left(-i\frac{\pi}{4}\right) \exp\left(i\left(\frac{\omega}{c}nz - \omega t\right)\right) \exp\left(-\frac{\omega}{c}\kappa z\right) \mathbf{e}_x$$

$$\mathbf{H}_B = 2A \exp\left(i\left(\frac{\omega}{c}nz - \omega t\right)\right) \exp\left(-\frac{\omega}{c}\kappa z\right) \mathbf{e}_y$$

(b) The integral is rewritten

$$\mathbf{F} = \frac{1}{c}\int \mathbf{j} \times \mathbf{H}\, dV = \frac{1}{c}\int \sigma \mathbf{E}_B \times \mathbf{H}_B\, dV$$

$$= \frac{\sigma}{c}\int_0^\infty E_{B_x}(z) H_{B_y}(z)\, dz\, \Delta F \mathbf{e}_z$$

where ΔF is the surface of the metal. For the fields one has to take a penetrating wave since only inside the metal are the current and the conductivity different from zero. Then the pressure exerted by the wave is

$$p = \frac{\sigma}{c}\int_0^\infty E_{B_x} H_{B_y}\, dz$$

We consider the time average of $E_{B_x} H_{B_y}$:

$$E_{B_x} = \mathrm{Re}\left(A\left(\frac{\mu\omega}{\sigma\pi}\right)^{1/2}\exp\left(-i\frac{\pi}{4}\right)\exp\left(i\frac{\omega}{c}(n+i\kappa)z - i\omega t\right)\right)$$

$$H_{B_y} = \mathrm{Re}\left(2A \exp\left(i\frac{\omega}{c}(n+i\kappa)z - i\omega t\right)\right)$$

Then, the product of the real parts is

$$E_{B_x} H_{B_y} = 2A^2 \left(\frac{\mu\omega}{\sigma\pi}\right)^{1/2} \exp\left(-\frac{2\omega}{c}\kappa z\right) \cos\left(\frac{\omega}{c}nz - \omega t - \frac{\pi}{4}\right) \cos\left(\frac{\omega}{c}nz - \omega t\right)$$

In order to perform the time average we consider the time-dependent cosine-terms only:

$$\frac{1}{t}\int_0^t \cos\left(\frac{\omega}{c}nz - \omega t - \frac{\pi}{4}\right)\cos\left(\frac{\omega}{c}nz - \omega t\right) dt$$

$$= \frac{1}{t}\int_0^t \frac{1}{4}\left(\exp\left(i\frac{\omega}{c}nz - i\omega t - i\frac{\pi}{4}\right) + \exp\left(-i\frac{\omega}{c}nz + i\omega t + i\frac{\pi}{4}\right)\right)$$

$$\times \left(\exp\left(i\frac{\omega}{c}nz - i\omega t\right) + \exp\left(-i\frac{\omega}{c}nz + i\omega t\right)\right) dt$$

In averaging, the oscillating terms drop out (multiplying out)

$$\frac{1}{4}\left(\exp\left(-i\frac{\pi}{4}\right)\exp\left(+i\frac{\pi}{4}\right)\right) = \frac{1}{2}\cos\frac{\pi}{4} = \frac{1}{2\sqrt{2}}$$

Thus, the time-average of the pressure exerted by the wave is

$$\overline{p} = \frac{|\mathbf{F}|}{\Delta F} = \frac{\sigma}{c} 2A^2 \left(\frac{\mu\omega}{\sigma\pi}\right)^{1/2} \frac{1}{2\sqrt{2}} \int_0^\infty \exp\left(-\frac{2\omega}{c}\kappa z\right) dz$$

$$= \frac{\sigma}{c} A^2 \left(\frac{\mu\omega}{\sigma\pi}\right)^{1/2} \frac{1}{\sqrt{2}} \frac{c}{2\omega\kappa}$$

We have $\kappa = (2\pi\mu\sigma\omega^{-1})^{1/2}$. Thus, the pressure is $\overline{p} = A^2/(4\pi)$. Hence, the pressure is independent of the nature of the material and of the frequency of the wave.

Biographical notes

Willebrord Snellius (Snell von Rojen), b. 1580, Leyden–d. Oct. 30, 1626, Leyden. Dutch mathematician and physicist. He was the first to carry out a Gradmessung to determine the shape of the earth by ascertaining the length of the graduated arc with the help of triangulation. In the course of this, he found a solution of the resection as well. In the publication of his findings ("De terrae ambitus vera quantitate," 1617) he introduced himself as *Eratosthenes Batavus*; independent of Descartes, he found the law of refringency (about 1620).

Augustin Jean Fresnel, b. May 10, 1788, Broglie (n. Bernay/Eure)–d. July 14, 1827, Ville d'Avray (n. Paris). In 1815 this French physicist substantiated the wave theory of light in detail for the first time. His experimental and theoretical works dealt with diffraction, interference, polarization, double-refraction, and aberration of light. To demonstrate interference Fresnel among other things invented a double mirror and a biprism. On a suggestion by Young, Fresnel in 1821 developed a theory of transverse light waves. Furthermore, he constructed annular lenses still in use today in lighthouses.

David Brewster, b. Dec. 11, 1781, Jedburgh (Scotland)–d. Feb. 10, 1868, Allerby (Scotland). Initially Brewster studied theology. His interest for natural sciences, however, prevented him from taking holy orders. He became a lawyer first, then professor for physics at St. Andrews (Scotland). The Royal Society catalogue lists 299 works written by Brewster, the most important of them dealing with optics (reflection, absorption, and polarization of light). In 1815 he discovered the law of complete polarization of light which is reflected and refracted on glass surfaces. In addition, he was a writer of biographies and publisher of several scientific journals. He thus contributed fundamentally to the popularization of natural sciences. His works include memoirs of the life, writings, and discoveries of Sir Isaac Newton, repr. of the 1855 edition, New York (1967).

18 Wave Guides and Resonant Cavities

We investigate the propagation of electromagnetic waves in wave guides and resonant cavities, that is, in hollow, cylinder-like metal bodies. The metal body has the same cross-sectional area everywhere, and its surface is assumed to be an ideal conductor. See Figure 18.1. If the body is closed, one talks of a resonant cavity, if its ends are open it is denoted a wave guide. Let a uniform material of permittivity ε and permeability μ be inside the cylinder. For the time dependence $e^{-i\omega t}$ of the electromagnetic field in the interior of the cylinder, Maxwell's equations take the following form:

$$\nabla \times \mathbf{E} = i\frac{\omega}{c}\mathbf{B}, \qquad \nabla \cdot \mathbf{E} = 0$$

$$\nabla \times \mathbf{B} = -i\mu\varepsilon\frac{\omega}{c}\mathbf{E}, \qquad \nabla \cdot \mathbf{B} = 0$$

(18.1)

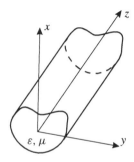

Figure 18.1. Hollow cylinder-like metal body with permittivity ε and permeability μ.

In these equations, $\mathbf{E}$ and $\mathbf{B}$ are still coupled. Again, we take the curl, and after rewriting we obtain the wave equations known from Chapter 16 (see equations (16.6a) and (16.6b))

$$\nabla^2 \mathbf{E} + \mu\varepsilon\frac{\omega^2}{c^2}\mathbf{E} = 0, \qquad \nabla^2 \mathbf{B} + \mu\varepsilon\frac{\omega^2}{c^2}\mathbf{B} = 0 \tag{18.2}$$

Due to the cylindrical symmetry of the problem we expect to find waves travelling in positive and negative z-directions or corresponding standing waves:

$$\left.\begin{aligned} \mathbf{E}(x, y, z, t) &= \mathbf{E}(x, y) \\ \mathbf{B}(x, y, z, t) &= \mathbf{B}(x, y) \end{aligned}\right\} e^{\pm ikz - i\omega t} \tag{18.3}$$

where the wave number k is a real or complex parameter still unknown.

With this ansatz the wave equation (18.2) is simplified if the differential operator ∇^2 is rewritten:

$$\nabla^2 = \left(\frac{\partial^2}{\partial x^2} + \frac{\partial^2}{\partial y^2}\right) + \frac{\partial^2}{\partial z^2} = \nabla_t^2 + \nabla_z^2$$

where ∇_t^2 is the transverse part of the operator. Then, we obtain for $\mathbf{E}$ (correspondingly also for $\mathbf{B}$):

$$\left(\nabla^2 + \mu\varepsilon\frac{\omega^2}{c^2}\right)\mathbf{E} = \left(\nabla_t^2 + \frac{\partial^2}{\partial z^2} + \mu\varepsilon\frac{\omega^2}{c^2}\right)\mathbf{E}(x, y)e^{\pm ikz - i\omega t}$$

$$= \left(\nabla_t^2 + \mu\varepsilon\frac{\omega^2}{c^2} - k^2\right)\mathbf{E}(x, y)e^{\pm ikz - i\omega t} = 0$$

This equation has to be valid for all t and z, so that

$$\left(\nabla_t^2 + \mu\varepsilon\frac{\omega^2}{c^2} - k^2\right)\mathbf{E}(x, y) = 0 \tag{18.4}$$

The field is decomposed into a component parallel to the z-axis and a component perpendicular to the z-axis:

$$\mathbf{E} = \mathbf{E}_z + \mathbf{E}_t, \qquad \mathbf{B} = \mathbf{B}_z + \mathbf{B}_t, \tag{18.5}$$

Now we shall show that it is sufficient to know the z-components of $\mathbf{E}$ and $\mathbf{B}$ since the transverse components may be represented by them. Namely, according to equation (18.1)

$$\nabla \times \mathbf{E} = \frac{i\omega}{c}\mathbf{B}$$

and using (18.5)

$$(\nabla_t + \nabla_z) \times (\mathbf{E}_t + \mathbf{E}_z) = \frac{i\omega}{c}(\mathbf{B}_t + \mathbf{B}_z)$$

Now,

$$\nabla_z \times \mathbf{E}_z = \frac{\partial E_z}{\partial z}\mathbf{e}_z \times \mathbf{e}_z = 0$$

The two terms $\nabla_t \times \mathbf{E}_t$ and $\mathbf{B}_z$ are the only vectors in z-direction so that we may split the components:

$$\nabla_t \times \mathbf{E}_t = \frac{i\omega}{c}\mathbf{B}_z, \qquad \nabla_z \times \mathbf{E}_t + \nabla_t \times \mathbf{E}_z = \frac{i\omega}{c}\mathbf{B}_t \tag{18.6}$$

This equation is multiplied vectorially from the left-hand side by ∇_z, leading to

$$\frac{i\omega}{c}\nabla_z \times \mathbf{B}_t = \nabla_z \times (\nabla_z \times \mathbf{E}_t) + \nabla_z \times (\nabla_t \times \mathbf{E}_z)$$

Solving the two double vector-products, then

$$\nabla_z \times (\nabla_z \times \mathbf{E}_t) = \nabla_z (\nabla_z \cdot \mathbf{E}_t) - (\nabla_z \cdot \nabla_z)\mathbf{E}_t$$
$$\nabla_z \times (\nabla_t \times \mathbf{E}_z) = \nabla_t (\nabla_z \cdot \mathbf{E}_z) - (\nabla_z \cdot \nabla_t)\mathbf{E}_z$$

Two of the parentheses are equal to zero because they represent the scalar product of orthogonal vectors. Thus, we obtain the equation

$$-\nabla_z^2 \mathbf{E}_t + \nabla_t \left(\frac{\partial E_z}{\partial z}\right) = \frac{\omega i}{c}\nabla_z \times \mathbf{B}_t$$

According to equation (18.3)

$$-\nabla_z^2 \mathbf{E}_t = -\frac{\partial^2}{\partial z^2}\left(\mathbf{E}_t(x, y)e^{\pm ikz - i\omega t}\right) = k^2 \mathbf{E}_t$$

and we obtain

$$k^2 \mathbf{E}_t + \nabla_t \left(\frac{\partial E_z}{\partial z}\right) = \frac{\omega i}{c}\nabla_z \times \mathbf{B}_t \tag{18.7}$$

Correspondingly, we proceed with the second Maxwell equation of (18.1):

$$\nabla \times \mathbf{B} = -\frac{i\omega}{c}\mu\varepsilon\mathbf{E}$$

which is split also into transverse and longitudinal parts:

$$\nabla_t \times \mathbf{B}_t + \nabla_t \times \mathbf{B}_z + \nabla_z \times \mathbf{B}_t + \nabla_z \times \mathbf{B}_z = -\frac{i\omega}{c}\mu\varepsilon\mathbf{E}_t - \frac{i\omega}{c}\mu\varepsilon\mathbf{E}_z$$

According to the same considerations, the equation is reduced to

$$\nabla_t \times \mathbf{B}_z = -\nabla_z \times \mathbf{B}_t - \frac{i\omega}{c}\mu\varepsilon\mathbf{E}_t \tag{18.8}$$

In equation (18.7) the right-hand side is replaced by equation (18.8):

$$k^2 \mathbf{E}_t + \nabla_t \left(\frac{\partial E_z}{\partial z}\right) = -\frac{i\omega}{c}\nabla_t \times \mathbf{B}_z + \frac{\omega^2}{c^2}\mu\varepsilon\mathbf{E}_t$$

Solving for $\mathbf{E}_t$,

$$\left(\frac{\omega^2}{c^2}\mu\varepsilon - k^2\right)\mathbf{E}_t = \nabla_t \left(\frac{\partial}{\partial z}E_z\right) + \frac{i\omega}{c}(\nabla_t \times \mathbf{B}_z)$$

Supposing $(\omega^2/c^2)\mu\varepsilon - k^2 \neq 0$, after a division we obtain the equation

$$\mathbf{E}_t = \frac{1}{\mu\varepsilon\dfrac{\omega^2}{c^2} - k^2}\left(\nabla_t\frac{\partial E_z}{\partial z} - \frac{i\omega}{c}(\mathbf{e}_z \times \nabla_t)B_z\right) \tag{18.9a}$$

and completely analogously,

$$\mathbf{B}_t = \frac{1}{\mu\varepsilon\dfrac{\omega^2}{c^2} - k^2}\left(\nabla_t\frac{\partial B_z}{\partial z} + \frac{i\omega}{c}\mu\varepsilon(\mathbf{e}_z \times \nabla_t)E_z\right) \tag{18.9b}$$

The transverse components are entirely determined by the longitudinal components. So, we have to consider only the z-component of equation (18.4) (or the corresponding equation for **B**). In the following, we will see that the case $(\omega^2/c^2)\mu\varepsilon - k^2 = 0$, first excluded, may occur. Due to the dispersion relation for electromagnetic waves, it corresponds to the case of a wave propagating in the z-direction only. Such waves are not allowed to have a z-component of the electric or magnetic field, $E_z = B_z = 0$ (transverse electromagnetic waves), and therefore the right-hand sides of equations (18.9a) and (18.9b) do not become infinity but are simply not defined mathematically.

Boundary conditions

Since we have assumed the surface S of the cylinder to be an ideal conductor, the boundary conditions (Chapter 17)

$$\mathbf{n} \cdot \mathbf{B} = 0, \qquad \mathbf{n} \times \mathbf{E} = 0$$

have to be fulfilled there, where **n** is the unit vector normal to the surface. (Since surface charges and surface currents are allowed to occur, we are not able to make direct statements on D_n and B_t.) This condition is equivalent to the requirement

$$E_z|_S = 0, \qquad (\mathbf{n} \cdot \mathbf{B}_t)|_S = 0$$

Substituting $\mathbf{B}_t$ from equation (18.9b) into the second equation, then

$$(\mathbf{n} \cdot \mathbf{B}_t)_S = \mathbf{n} \cdot \frac{1}{\mu\varepsilon\dfrac{\omega^2}{c^2} - k^2}\left(\nabla_t\frac{\partial B_z}{\partial z} + i\mu\varepsilon\frac{\omega}{c}(\mathbf{e}_z \times \nabla_t)E_z\right)\Bigg|_S$$

$$= \frac{1}{\mu\varepsilon\dfrac{\omega^2}{c^2} - k^2}\mathbf{n} \cdot \nabla_t\frac{\partial B_z}{\partial z}\Bigg|_S = 0$$

because $(\mathbf{e}_z \times \nabla_t E_z)$ is tangential to the surface S and therefore $\mathbf{n} \cdot (\mathbf{e}_z \times \nabla_t E_z = 0)$. Hence, for B_z we have the condition

$$\mathbf{n} \cdot \nabla_t \left. \frac{\partial B_z}{\partial z} \right|_S = 0 \Rightarrow \left. \frac{\partial}{\partial n} \frac{\partial B_z}{\partial z} \right|_S = \left. \frac{\partial}{\partial z} \frac{\partial B_z}{\partial n} \right|_S = 0, \qquad \left. \frac{\partial B_z}{\partial n} \right|_S = 0$$

The last step is understandable due to the fact that the entire z-dependence of the wave has to be of the form $B_z = B_z(x, y)e^{i(kz - \omega t)}$, and therefore, $\partial B_z/\partial z = ik B_z$.

Classification of fields in wave guides: TM, TE, and TEM waves

The two-dimensional wave equation (18.4) for E_z and B_z, together with the boundary conditions for E_z and B_z at the cylindrical surface, form an eigenvalue problem (compare the problem of the oscillating membrane in Mechanics II).

For a given frequency ω, only definite axial wave numbers k obey the differential equation and boundary conditions (wave guides); or for a given wave number k, only definite frequencies are allowed (resonant cavity). In general, both boundary conditions cannot be fulfilled at the same time since the boundary conditions are distinct, although the eigenvalue equations coincide formally. According to the boundary conditions fulfilled, one distinguishes the fields:

- Transverse magnetic (TM): $B_z = 0$ everywhere, $E_z|_S = 0$

- Transverse electric (TE): $E_z = 0$ everywhere, $\left. \dfrac{\partial B_z}{\partial z} \right|_S = 0$

For TM waves, the boundary condition $B_z = 0$ everywhere follows from the following argument: One has always $\partial B_z/\partial z|_S = 0$. But, one has to require also $B_z|_S = 0$, to match the boundary conditions formally to those of $\mathbf{E}$. But, if $B_z|_S = 0$ and $\partial B_z/\partial z|_S = 0$, then B_z has to vanish everywhere. One may argue correspondingly for the TE waves.

In the special case $B_z = E_z = 0$ everywhere, one talks of transverse electromagnetic waves (TEM). Then, from the equations (18.9a) and (18.9b) for

$$\gamma^2 \equiv \mu\varepsilon \frac{\omega^2}{c^2} - k^2 \neq 0$$

one gets only the trivial solution $B_t = E_t = 0$. Hence, we have to consider $\mu\varepsilon\omega^2/c^2 - k^2 = 0$ to get nontrivial solutions; that is, for TEM waves there holds $k = \sqrt{\mu\varepsilon}\,\omega/c$. This corresponds precisely to the usual dispersion relation for electromagnetic waves. Since k has been defined as the component of the wave number vector representing propagation in z-direction (see equation (18.3)), in this case $\mathbf{k} = (0, 0, k)$; that is the wave propagates exclusively in the z-direction. Since $\mathbf{k}$, $\mathbf{E}$, and $\mathbf{B}$ form a right-handed system, it is clear further that $E_z = B_z = 0$. In equations (18.9a) and (18.9b), the undefined expression $0/0$ appears (see the remark to these equations). But we will show that for $k = \sqrt{\mu\varepsilon}\,\omega/c$ and $B_z = E_z = 0$ (the boundary conditions are fulfilled automatically) transverse solutions of Maxwell's equations for $\mathbf{E}$ and $\mathbf{B}$ exist. TEM waves satisfy Laplace's equation in *two* dimensions (18.4):

$$\Delta_t \mathbf{E}_{TEM} = 0, \qquad \Delta_t \mathbf{B}_{TEM} = 0 \tag{18.10a}$$

We will show that $\mathbf{E}_{TEM}$ is perpendicular to $\mathbf{B}_{TEM}$. From Maxwell's equations we get at once

$$i\frac{\omega}{c}\mathbf{B}_{TEM} = \nabla \times \mathbf{E}_{TEM} = (\nabla_t + \nabla_z) \times \mathbf{E}_{TEM}$$

$$= \nabla_t \times \mathbf{E}_{TEM} + \frac{\partial}{\partial z}(\mathbf{e}_z \times \mathbf{E}_{TEM})$$

Due to $B_z = 0$, $\mathbf{B}_{TEM}$ has to be a vector in the (x, y)-plane; the same holds for $\mathbf{E}_{TEM}$. But then $\nabla_t \times \mathbf{E}_{TEM}$ is a vector in the z-direction. But, since on the left-hand side of the equation only one vector lies in the (x, y)-plane,

$$\nabla_t \times \mathbf{E}_{TEM} = 0 \quad \text{and} \quad \frac{\partial}{\partial z}(\mathbf{e}_z \times \mathbf{E}_{TEM}) = \frac{i\omega}{c}\mathbf{B}_{TEM} \tag{18.10b}$$

By the way, these two equations are obtained from equation (18.6), with $\mathbf{E}_z = \mathbf{B}_z = 0$. From this,

$$\mathbf{B}_{TEM} = \frac{c}{i\omega}\frac{\partial}{\partial z}(\mathbf{e}_z \times \mathbf{E}_{TEM}) \tag{18.11}$$

If $\mathbf{E}_{TEM}$ may be represented by

$$\mathbf{E}_{TEM} = \mathbf{E}_{0\,TEM}e^{i(kz-\omega t)}$$

then we obtain

$$\mathbf{B}_{TEM} = \frac{c}{i\omega}ik(\mathbf{e}_z \times \mathbf{E}_{TEM}) \tag{18.12}$$

and with $k = \sqrt{\mu\varepsilon}\omega/c$ we get the relation

$$\mathbf{B}_{TEM} = \sqrt{\mu\varepsilon}(\mathbf{e}_z \times \mathbf{E}_{TEM}) \tag{18.13}$$

Thus, the same relation holds between the electric field and the magnetic field as for the propagation of a wave in an unbounded medium.

Equation (18.10a) demonstrates that both $\mathbf{E}_{TEM}$ and $\mathbf{B}_{TEM}$ satisfy Laplace's equation. Moreover, $\mathbf{E}_{TEM}$ and $\mathbf{B}_{TEM}$ may be derived from scalar potentials also obeying Laplace's equation. Since the surface of the conductor is an equipotential surface (because the conductivity is assumed to be infinetely high), the two-dimensional wave equation can be fulfilled only trivially, since $\mathbf{E}$ vanishes inside the surface. Thus, TEM fields cannot propagate inside a single conductor; it is necessary to have, for example, two concentric, circular metal cylinders (coaxial cable), as in Figure 18.2.

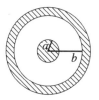

Figure 18.2. TEM waves can propagate only in coaxial cables.

Example 18.1: The coaxial cable

A coaxial cable (Figure 18.3) may transmit a pure TEM wave in addition to TE waves and TM waves, while in a tubular wave guide only TE waves and TM waves may be transported. For the TEM waves the following relations are satisfied

$$k = \sqrt{\mu\varepsilon}\frac{\omega}{c}, \qquad \Delta_t \mathbf{E}_{\text{TEM}} = 0$$

as well as

$$\mathbf{B}_{\text{TEM}} = \sqrt{\mu\varepsilon}(\mathbf{e}_z \times \mathbf{E}_{\text{TEM}})$$

Instead of solving the wave equation for $\mathbf{E}$, we derive the electric field from a potential ϕ. This is possible because $\nabla_t \times \mathbf{E}_t = 0$, see equation (18.10b),

$$\mathbf{E}_{\text{TEM}} = (\nabla_t \phi)e^{-i(\omega t - kz)}$$

Now we solve the wave equation for ϕ:

$$\Delta_t \phi = 0$$

Rewritten in polar coordinates Laplace's equation reads

$$\left(\frac{\partial^2}{\partial \rho^2} + \frac{1}{\rho}\frac{\partial}{\partial \rho} + \frac{1}{\rho^2}\frac{\partial^2}{\partial \varphi^2}\right)\phi(\rho, \varphi) = 0$$

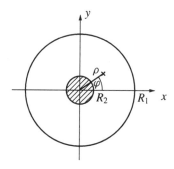

Figure 18.3. Cross section of a coaxial cable.

The potential is constant on the inner and outer conductors: $\phi(R_1, \varphi) = \phi_1$ and $\phi(R_2, \varphi) = \phi_2$. Since the boundary conditions are independent of the azimuthal angle φ, ϕ itself depends merely on ρ. Therefore, the wave equation is simplified,

$$\frac{1}{\rho}\frac{\partial}{\partial\rho}\left(\rho\frac{\partial\phi}{\partial\rho}\right) = 0$$

Its solution is

$$\phi(\rho) = A\ln\rho + B$$

with

$$A = \frac{\phi_1 - \phi_2}{\ln R_1 - \ln R_2}, \qquad B = \frac{\phi_2\ln R_1 - \phi_1\ln R_2}{\ln R_1 - \ln R_2}$$

as can be proved easily by substitution.

The field components of the TEM waves are:

(1) **Cylindrical coordinates**

$$\mathbf{E}_{\text{TEM}}(\rho, \varphi, z, t) = \frac{A}{\rho}\mathbf{e}_\rho e^{-i(\omega t - kz)}$$

$$\mathbf{B}_{\text{TEM}}(\rho, \varphi, z, t) = \sqrt{\mu\varepsilon}\frac{A}{\rho}\mathbf{e}_\varphi e^{-i(\omega t - kz)}$$

(2) **Cartesian coordinates**

$$\mathbf{E}_{\text{TEM}}(x, y, z, t) = \frac{A}{x^2 + y^2}(x\mathbf{e}_x + y\mathbf{e}_y)e^{-i(\omega t - kz)}$$

$$\mathbf{B}_{\text{TEM}}(x, y, z, t) = \frac{\sqrt{\varepsilon\mu}A}{x^2 + y^2}(-y\mathbf{e}_x + x\mathbf{e}_y)e^{-i(\omega t - kz)}$$

The coefficient A may be expressed also with the help of the maximal value of one field, $\mathbf{E}$ or $\mathbf{B}$, on the inner or outer conductor so that the knowledge of ϕ_1 and ϕ_2 becomes superfluous. The coaxial line transmits TEM waves of any frequency with the velocity of light in the dielectric and is therefore suitable as a *wide-band cable for the transmission of wide frequency bands*.

If the inner conductor is removed ($R_2 \to 0$), B may be determined from the boundary condition $\phi(R_1) = \phi_1$. But the potential $\phi(\rho) = A\log(\rho/R_1) + \phi_1$ obtained in this way is divergent for $\rho = 0$. Therefore, A has to be zero. The potential becomes a constant value, that is, in hollow conductors there are no TEM waves.

The two equations (18.9a) and (18.9b) are simplified considerably if only TE waves or TM waves are considered. If the z-dependence is given by the exponential e^{ikz}, then the two equations for *TM waves* are given by

$$\mathbf{B}_t = \frac{1}{\gamma^2}\left(i\mu\varepsilon\frac{\omega}{c}\right)(\mathbf{e}_z \times \nabla_t)E_z, \qquad \mathbf{E}_t = \frac{1}{\gamma^2}ik\nabla_t E_z \tag{18.14}$$

since B_z vanishes everywhere. For brevity, we have set $\gamma^2 = \mu\varepsilon\omega^2/c^2 - k^2$. Substituting $\nabla_t E_z$ from the second equation into the first one, then $\mathbf{B}_t$ is simplified further,

$$\mathbf{B}_t = \frac{1}{\gamma^2}\left(i\mu\varepsilon\frac{\omega}{c}\mathbf{e}_z \times \frac{\gamma^2}{ik}\mathbf{E}_t\right) = \frac{\mu\varepsilon\omega}{ck}(\mathbf{e}_z \times \mathbf{E}_t)$$

Correspondingly, the equations (18.9a) and (18.9b) yield for *TE waves*:

$$\mathbf{E}_t = -\frac{\omega}{ck}(\mathbf{e}_z \times \mathbf{B}_t), \qquad \mathbf{B}_t = \frac{ik}{\gamma^2}\nabla_t B_z \tag{18.15}$$

since the component E_z is zero everywhere. The important result of these considerations is that in the interior of a conducting tube (wave guide) pure transverse waves (TM, TE) are no longer possible, except for the special TEM waves. *The TM waves and the TE waves have also longitudinal components.*

The two-dimensional wave equation (18.4) holds for each of the three scalar vector components, in particular for E_z and B_z. Together with the boundary conditions they form an eigenvalue problem, namely,

$$\text{TM waves}: \quad (\nabla_t^2 + \gamma^2)E_z(x,y) = 0 \qquad E_z|_S = 0$$

$$\text{TE waves}: \quad (\nabla_t^2 + \gamma^2)B_z(x,y) = 0 \qquad \left.\frac{\partial B_z}{\partial n}\right|_S = 0$$

The constant γ^2 is never allowed to be negative because E_z and B_z have to be solutions of the wave equation, and they have to satisfy also the boundary conditions. If γ^2 were negative the solution would be an exponential function with a real exponent. This solution is not periodic, and it could satisfy the boundary condition for positive and negative xy-values only in a trivial manner. For the positive eigenvalues γ_λ^2 we obtain a sequence of $B_{z\lambda}$ or $E_{z\lambda}$, $\lambda = 1, 2, \ldots$. (The explicit solution yields γ as a function of the dimension of the cable, see the Example 18.3.) Then, we may calculate the TM field from (18.14) and (18.15), respectively. For a definite frequency ω, from the eigenvalues γ_λ^2 we get a wave number k dependent on λ. From $\gamma_\lambda^2 = \mu\varepsilon\omega^2/c^2 - k^2$,

$$k_\lambda^2 = \frac{\mu\varepsilon}{c^2}\left(\omega^2 - \frac{c^2\gamma_\lambda^2}{\mu\varepsilon}\right), \qquad k_\lambda = \frac{\sqrt{\mu\varepsilon}}{c}\sqrt{\omega^2 - \frac{c^2\gamma_\lambda^2}{\mu\varepsilon}}$$

We set $c^2\gamma_\lambda^2/(\mu\varepsilon) = \omega_\lambda^2$, since this expression has the dimension of a frequency. This yields

$$k_\lambda = \frac{\sqrt{\mu\varepsilon}}{c}\sqrt{\omega^2 - \omega_\lambda^2}$$

The wave number k_λ is real for $\omega \geq \omega_\lambda$ and imaginary for $\omega < \omega_\lambda$. For imaginary wave numbers k_λ we obtain a real exponential function for the z-dependence (Figure 18.4); that is, the wave decays exponentially: $e^{i(kz-\omega t)} = e^{-i\omega t}e^{-|k|z}$. Thus, only waves of the frequency $\omega > \omega_\lambda$ may propagate in the wave guide, ω_λ represents a *critical frequency*. The dependence of the wave number k_λ on the frequency is represented in Figure 18.5.

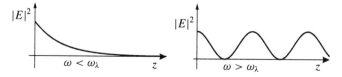

Figure 18.4. Decaying and travelling TM and TE waves.

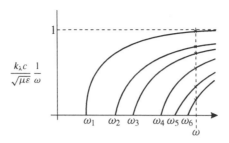

Figure 18.5. Dependence of the critical frequencies ω_λ on the frequency ω and the wave number k.

Along the abscissa, the critical frequencies ω_λ as well as the frequency ω are drawn; along the ordinate

$$\frac{k_\lambda c}{\sqrt{\mu \varepsilon}} \frac{1}{\omega}$$

is drawn. We have the relation

$$\frac{k_\lambda c}{\sqrt{\mu \varepsilon}} \frac{1}{\omega} = \frac{\sqrt{\omega^2 - \omega_\lambda^2}}{\omega} = \sqrt{1 - \left(\frac{\omega_\lambda}{\omega}\right)^2}$$

so that the left-hand side $(k_\lambda c/\sqrt{\mu \varepsilon})(1/\omega)$ tends to 1 for $\omega \gg \omega_\lambda$.

From Figure 18.5 we infer that for a given frequency ω with $\omega > \omega_1$ only a finite number of oscillation modes may propagate in a wave guide. The existence of a bound for the frequency is the reason for the fact that the phenomenon appears only in the high-frequency region. The associated wavelength has to be smaller than $\lambda_1 = 2\pi \varepsilon/\omega_1 \approx a$, the dimension of the system, to be able to travel along the table (compare the Example 18.4). The main application of wave guides as transmitters of microwaves (telephone, etc.) is based on this property.

Phase velocity and group velocity

If we consider the instantaneous picture of a sine-like travelling wave, the wavelength λ is the distance between two points possessing an equal phase at any time, and the phase velocity is just the velocity at which the oscillation phase propagates. If T is the oscillation period, then the phase velocity is

$$v_p = \frac{\lambda}{T} = \frac{\lambda \omega}{2\pi} = \frac{\omega}{k} \tag{18.16}$$

But in nature, we never deal with monochromatic waves of definite frequency and wave number, since even for apparently monochromatic light sources, frequency spectra and regions of wavelength occur, though they are very small.

The more general motion of a wave may be described in terms of a superposition of harmonic waves. As a group of waves, we define a wave packet (Figure 18.6) of finite length (each emitter emits only for a finite time; the wave packet emitted is finite). Groups of waves may be repeated in a periodic sequence or may be shaped periodically. Since

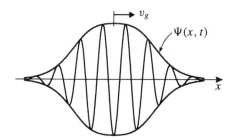

Figure 18.6. A wave packet results from overlaying many waves $\psi(x, t) = \int c(k) e^{i(kx - \omega t)} dk$ (see volume 1: *Mechanics*).

in a medium with dispersion the phase velocity depends on the frequency ω (because the permittivity is a function of the frequency, $\varepsilon = \varepsilon(\omega)$), the phase velocities of the distinct waves are different from each other. In the propagation of the wave the phase differences change and also the shape of the group of waves changes.

Due to the dispersion, some mark of the resulting picture of the waves, e.g., the highest peak of the wave, does not move at the average phase velocity of the various components of the wave, but rather at the group velocity

$$v_g = \frac{d\omega}{dk} \tag{18.17}$$

The phase velocity v_p equals the group velocity only in a dispersive-free medium. But before studying the general case, we will explain the difference between phase velocity and group velocity for a simple example.

We consider the superposition of two waves of equal amplitude and distinct, but neighboring frequencies ω_1, k_1 and ω_2, k_2:

$$
\begin{aligned}
U(x, t) &= A \left(e^{i(k_1 x - \omega_1 t)} + e^{i(k_2 x - \omega_2 t)} \right) \\
&= A e^{i[(k_1 + k_2)/2]x - i[(\omega_1 + \omega_2)/2]t} \\
&\quad \times \left(e^{i[(k_1 - k_2)/2]x - i[(\omega_1 - \omega_2)/2]t} + e^{i[(k_2 - k_1)/2]x - i[(\omega_2 - \omega_1)/2]t} \right) \\
&= 2A \cos \left(\frac{k_1 - k_2}{2} x - \frac{\omega_1 - \omega_2}{2} t \right) e^{i[(k_1 + k_2)/2]x - i[(\omega_1 + \omega_2)/2]t}
\end{aligned}
$$

By this transformation we have split the wave into an amplitude factor slowly oscillating at $\omega_1 - \omega_2$ and a phase factor rapidly oscillating at $\omega_1 + \omega_2$ The phase moves at the velocity:

$$v_p = \frac{\omega_1 + \omega_2}{k_1 + k_2} \approx \frac{\omega}{k} \qquad \text{for} \qquad \omega_1 \approx \omega_2 \approx \omega$$

the amplitude (group of waves) moves at the velocity (Figure 18.7)

$$v_g = \frac{\omega_1 - \omega_2}{k_1 - k_2} \Rightarrow \frac{d\omega}{dk}$$

The energy of the wave is determined by its amplitude, so, in general, the group velocity gives also the velocity of the energy transport. However, the domain of anomalous dispersion

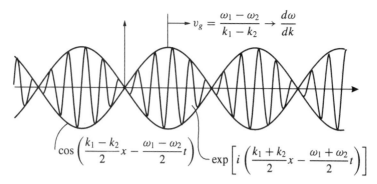

Figure 18.7. Part of a periodic sequence of peaks of the wave. The individual peak moves with the group velocity v_g.

demands a particular consideration (see Chapter 19). We further note that group velocity, phase velocity, and the notion of the wave packet have been introduced and discussed in a more rigorous manner in volume 1, Chapter 32. The present considerations remind the reader of things already known and illustratively support them.

Example 18.2: Wave velocity in a wave guide

In a wave guide the following relation holds for the wave number of an oscillation mode λ,

$$k = k_\lambda = \frac{\sqrt{\mu \varepsilon}}{c} \sqrt{\omega^2 - \omega_\lambda^2}$$

If for very large frequencies $\omega \gg \omega_\lambda$ the second term under the root sign may be neglected, then

$$k = \frac{\sqrt{\mu \varepsilon}}{c} \omega$$

and we obtain equal values of the phase velocity and the group velocity

$$v_p = \frac{\omega}{\lambda} = \frac{c}{\sqrt{\mu \varepsilon}}, \qquad v_g = \frac{d\omega}{dk} = \frac{c}{\sqrt{\mu \varepsilon}}$$

Quite a different situation occurs if we are not allowed to neglect this term. Then, the phase velocity is

$$v_p = \frac{\omega}{k_\lambda} = \frac{c}{\sqrt{\mu \varepsilon}} \frac{1}{\sqrt{1 - \frac{\omega_\lambda^2}{\omega^2}}} > \frac{c}{\sqrt{\mu \varepsilon}}$$

In the wave guide the phase velocity is higher than in free space. The group velocity is

$$v_g = \frac{d\omega}{dk_\lambda} = \frac{c^2 k_\lambda}{\mu \varepsilon \sqrt{\frac{c^2}{\mu \varepsilon} k_\lambda^2 + \omega_\lambda^2}} = \frac{c}{\sqrt{\mu \varepsilon}} \frac{\sqrt{\omega^2 - \omega_\lambda^2}}{\omega}$$

If the frequency of an oscillation mode ω approaches the critical frequency ω_λ, then the corresponding phase velocity becomes infinite, and the group velocity tends to zero ($k_\lambda = 0$). The wave can no longer propagate in the wave guide. Obviously, always

$$v_p \cdot v_g = \frac{c^2}{\mu\varepsilon} = c'^2$$

Example 18.3:   The rectangular wave guide

We consider a wave guide of rectangular cross section (Figure 18.8). The position of the coordinate system and the dimensions are given in the sketch. The surface of the wave guide is assumed to be ideally conducting. Then, the boundary conditions are

$$\mathbf{n} \cdot \mathbf{B} = 0, \qquad \mathbf{n} \times \mathbf{E} = 0 \tag{18.18}$$

where $\mathbf{n}$ is the unit vector normal to the surface of the conductor. Writing down condition (18.18) explicitly this implies

$$E_y = E_z = B_x = 0, \qquad x = 0, a$$
$$E_x = E_z = B_y = 0, \qquad y = 0, b \tag{18.19}$$

For the field intensities we make the ansatz of a wave propagating in the z-direction

$$\mathbf{E}(x, y, z, t) = \mathbf{E}(x, y)e^{i(kz-\omega t)}, \qquad \mathbf{B}(x, y, z, t) = \mathbf{B}(x, y)e^{i(kz-\omega t)} \tag{18.20}$$

Substituting this ansatz into the wave equation

$$\left(\Delta - \frac{1}{c^2}\frac{\partial^2}{\partial t^2}\right)\mathbf{C} = 0$$

we obtain

$$\left(\frac{\partial^2}{\partial x^2} + \frac{\partial^2}{\partial y^2}\right)\mathbf{C} + \left(\frac{\omega^2}{c^2} - k^2\right)\mathbf{C} = 0 \tag{18.21}$$

where $\mathbf{C}$ stands substitutionally for $\mathbf{E}$ and $\mathbf{B}$.

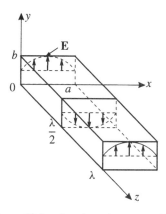

Figure 18.8.   A rectangular wave guide.

Besides the wave equation (18.21) the field intensities have also to obey Maxwell's equations

$$\text{div } \mathbf{E} = 0, \qquad \text{curl } \mathbf{B} = \frac{1}{c}\frac{\partial}{\partial t}\mathbf{E} = -\frac{i\omega}{c}\mathbf{E}$$

$$\text{div } \mathbf{B} = 0, \qquad \text{curl } \mathbf{E} = -\frac{1}{c}\frac{\partial}{\partial t}\mathbf{B} = \frac{i\omega}{c}\mathbf{B} \tag{18.22}$$

For the amplitudes $\mathbf{E}(x, y)$ and $\mathbf{B}(x, y)$ from (18.20), we make the following ansatz satisfying the boundary conditions

$$E_x = \alpha \cos\frac{m\pi}{a}x \sin\frac{n\pi}{b}y, \qquad B_x = \alpha' \sin\frac{m\pi}{a}x \cos\frac{n\pi}{b}y$$

$$E_y = \beta \sin\frac{m\pi}{a}x \cos\frac{n\pi}{b}y, \qquad B_y = \beta' \cos\frac{m\pi}{a}x \sin\frac{n\pi}{b}y \tag{18.23}$$

$$E_z = \gamma \sin\frac{m\pi}{a}x \sin\frac{n\pi}{b}y, \qquad B_z = \gamma' \cos\frac{m\pi}{a}x \cos\frac{n\pi}{b}y$$

The entire wave is obtained by multiplying with a phase factor $e^{i(kz-\omega t)}$. The special kind of ansatz becomes clear immediately by substituting into Maxwell's equations. Then we obtain relations between the constants of the ansatz.

The substitution into the wave equation (18.21) yields the relation

$$\left(\frac{m\pi}{a}\right)^2 + \left(\frac{n\pi}{b}\right)^2 + k^2 = \frac{\omega^2}{c^2} \tag{18.24}$$

Obviously, there is only one real value of k if the frequency ω is higher than the critical frequency ω_g, where

$$\omega_g = \omega_{mn} = c\sqrt{\left(\frac{m\pi}{a}\right)^2 + \left(\frac{n\pi}{b}\right)^2}$$

According to the definition, a TE wave occurs if $E_z = 0$; thus, $\gamma = 0$ in equation (18.23). TM waves ($B_z = 0$) are obtained for $\gamma' = 0$. From $\nabla \times \mathbf{E} = (i\omega/c)\mathbf{B}$ and $\nabla \times \mathbf{B} = -(i\omega/c)\mathbf{E}$ one finds for the relation between the prefactors in equation (18.23):

$$\frac{i\omega}{c}\alpha' = \gamma\frac{n\pi}{b} - ik\beta$$

$$\frac{i\omega}{c}\beta' = \alpha ik - \gamma\frac{m\pi}{a} \tag{18.25}$$

$$\frac{i\omega}{c}\gamma' = \beta\frac{m\pi}{a} - \alpha\frac{n\pi}{b}$$

and

$$-\frac{i\omega}{c}\alpha = \gamma'\frac{n\pi}{b} - \beta'ik$$

$$-\frac{i\omega}{c}\beta = \alpha'ik + \gamma'\frac{m\pi}{a} \tag{18.26}$$

$$-\frac{i\omega}{c}\gamma = -\beta'\frac{m\pi}{a} + \alpha'\frac{n\pi}{b}$$

For TE waves ($\gamma = 0$):

$$\beta' \frac{m\pi}{a} = \alpha' \frac{n\pi}{b} \Rightarrow \alpha' \sim \frac{m}{n} \quad \text{and} \quad \beta' \sim \frac{n}{m} \tag{18.27}$$

as well as

$$\left. \begin{array}{l} \dfrac{\omega}{c}\alpha' = -k\beta \\[2mm] \dfrac{\omega}{c}\beta' = \alpha k \end{array} \right\} \Rightarrow \quad \begin{array}{l} \beta \sim \dfrac{m}{n} \\[2mm] \alpha \sim \dfrac{n}{m} \end{array} \tag{18.28}$$

Analogously, for TM waves ($\gamma' = 0$):

$$\beta \frac{m\pi}{a} = \alpha \frac{n\pi}{b} \quad \Rightarrow \quad \begin{array}{l} \beta \sim \dfrac{n}{m} \\[2mm] \alpha \sim \dfrac{m}{n} \end{array} \tag{18.29}$$

as well as

$$\beta' \sim \frac{m}{n} \quad \text{and} \quad \alpha' \sim \frac{n}{m} \tag{18.30}$$

just the opposite of TE waves. According to the equations (18.23), (18.29), and (18.30), nontrivial TM waves arise only of $m \neq 0$ and $n \neq 0$. The TM wave of lowest frequency is

$$\omega_{11}^{\text{TM}} = c\sqrt{\frac{\pi^2}{a^2} + \frac{\pi^2}{b^2}} \tag{18.31}$$

The nonvanishhing TE wave of lowest frequency (without loss of generality $a > b$) is obtained analogously from equation (18.23), (18.27), and (18.28), if either $n = 0, m \neq 0$ or $m = 0, n \neq 0$:

$$\omega_{10} = \frac{c\pi}{a}$$

This is lower than the critical frequency of the TM wave. The corresponding wavelength is $\lambda_{10} = 2\pi c/\omega_{10} = 2a$. Further, we consider the fundamental wave of the TE-type, ω_{10}. For the electric field we have

$$E_x = E_z = 0 \quad \text{and} \quad E_y = \beta \sin\frac{\pi x}{a} e^{i(kz - \omega t)}$$

Representing the sine function by the exponential function, the y-component may be written as superposition of two waves:

$$E_y = \frac{\beta}{2i}\left(e^{i(kz + (\pi/a)x - \omega t)} - e^{i(kz - (\pi/a)x - \omega t)}\right)$$

The factor i gives a phase shift of $\pi/2$.

So, the fundamental wave appears as a superposition of two waves whose normal lies in the (x, y)-plane. The angle ε between the normal and the x-direction is given by

$$\cos\varepsilon = \mp\frac{\pi}{a} \cdot \frac{1}{\sqrt{\left(\dfrac{\pi}{a}\right)^2 + k^2}}$$

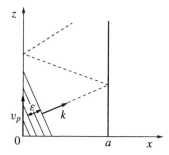

Figure 18.9. TE wave in a rectangular wave guide may be regarded as a plane wave repeatedly reflected from the walls.

Introducing the critical frequency ω_g so that $\omega^2/c^2 = k^2 + \omega_g^2/c^2$, where according to (6.24) $\omega_g^2 = c^2 \left[(m\pi/a)^2 + (n\pi/b)^2 \right]$, and therefore, $kc = \sqrt{\omega^2 - \omega_g^2}$ for $m = 1, n = 0$, then

$$\cos \varepsilon = -\frac{\omega_g}{\omega}$$

for the first, and

$$\cos \varepsilon = +\frac{\omega_g}{\omega}$$

for the second wave. The total **E**-field may be thought to arise by a repeated reflection of a plane wave, entering at the angle ε, at the planes $x = 0$ and $x = a$ (Figure 18.9). For the phase velocity, we get v_p

$$v_p = \frac{\omega}{k} = \frac{\omega}{\frac{1}{c}\sqrt{\omega^2 - \omega_g^2}} = \frac{c}{\sqrt{1 - \left(\frac{\omega_g}{\omega}\right)^2}} = \frac{c}{\sqrt{1 - \cos^2 \varepsilon}} = \frac{c}{\sin \varepsilon}$$

This is equal to the velocity of the intersect of the wave plane with the plane $x = 0$.

Resonant cavities

Basically, any closed, hollow body with a conducting surface may serve as a conducting cavity. Here, we restrict ourselves to cylindrical cavities. These can be thought to be constructed from cylindrical wave guides closed on both sides by conducting, plane surfaces perpendicular to the axis. The cavity is described by the two constants ε and μ. Since electromagnetic waves are reflected at both end surfaces, *standing waves* arise in direction of the axis. (All other waves disappear due to destructive interference.) Therefore, for TM waves, we make the following ansatz for the longitudinal part:

$$\mathbf{E}_z = \psi(x, y)(A \sin kz + B \cos kz)\mathbf{e}_z, \qquad \mathbf{B}_z = 0$$

Remark　　The ansatz $E_z = \psi(x, y)(A e^{ikz} + B e^{-ikz}) e_z$ would also be possible, but this would lead to a redefinition of the constants A and B. The transverse part follows from the equations (18.9a) and (18.9b),

$$\mathbf{E}_t = \frac{k}{\gamma^2} \nabla_t \psi(x, y)(A \cos kz - B \sin kz)$$

$$\mathbf{B}_t = \frac{1}{\gamma^2} i \mu \varepsilon \frac{\omega}{c} \mathbf{e}_z \times \nabla_t \psi(x, y)(A \sin kz + B \cos kz)$$

Let the cylinder have the height d; the z-axis is chosen such that the end faces lie at $z = 0$ and $z = d$, then the following boundary conditions are valid:

$$\mathbf{E}_t(z = 0) = 0, \qquad A = 0$$

For nontrivial solutions, from $\mathbf{E}_t(z = d) = 0$ we get the relation

$$\sin kd = 0$$

yielding a condition for the wave number k:

$$k = \frac{l\pi}{d}, \quad l = 0, 1, 2, \ldots$$

With these relations, for the amplitudes we obtain the equations

$$\mathbf{E}_z = B\psi(x, y) \cos\left(\frac{l\pi}{d} z\right) \mathbf{e}_z, \qquad \mathbf{E}_t = -B\frac{l\pi}{d\gamma^2} \sin\left(\frac{l\pi}{d} z\right) \nabla_t \psi(x, y)$$

$$\mathbf{B}_z = 0, \qquad \mathbf{B}_t = B\frac{i\mu\varepsilon\omega}{\gamma^2 c} \cos\left(\frac{l\pi}{d} z\right) \mathbf{e}_z \times \nabla_t \psi(x, y) \quad \textbf{(18.32)}$$

For the TE waves we make the analogous ansatz of a standing wave in the direction of the cylindrical axis for the longitudinal part of the magnetic field. Since the z-component of $\mathbf{B}$ crosses the end face without slope, and the field vanishes in the exterior

$$\mathbf{B}_z(z = 0) = \mathbf{B}_z(z = d) = 0$$

Corresponding to the considerations above we obtain the amplitudes of the TE waves:

$$\mathbf{E}_z = 0, \qquad \mathbf{E}_t = -A\frac{i\omega}{c\gamma^2} \sin\left(\frac{l\pi}{d} z\right) \mathbf{e}_z \times \nabla_t \psi(x, y)$$

$$\mathbf{B}_z = A\psi(x, y) \sin\left(\frac{l\pi}{d} z\right) \cdot \mathbf{e}_z, \qquad \mathbf{B}_t = A\frac{l\pi}{d\gamma^2} \cos\left(\frac{l\pi}{d} z\right) \nabla_t \psi(x, y) \quad \textbf{(18.33)}$$

For both oscillation modes the scalar function $\psi(x, y)$ depending on x and y follows from the wave equation (18.4)

$$(\Delta_t + \gamma^2)\psi(x, y) = 0$$

subject to the boundary condition that ψ (for $\mathbf{E}$) or $\partial\psi/\partial n$ (for $\mathbf{B}$) vanish on the surface of the cavity.

Expressing the constant γ^2 by frequency and wave number, we obtain

$$\gamma^2 = \varepsilon\mu\frac{\omega^2}{c^2} - \left(\frac{l\pi}{d}\right)^2$$

By the boundary conditions for the solution of the ψ-part we obtain equations of the type

$$\sin\gamma x' = 0$$

if x' lies on the surface of the wave guide. This yields a dependence of γ^2 on a parameter λ,

$$\gamma_\lambda^2 = \varepsilon\mu\frac{\omega_\lambda^2}{c^2} - \left(\frac{l\pi}{d}\right)^2 \qquad (18.34)$$

Solving for the eigenfrequency (resonance frequency), we get

$$\omega_\lambda = \frac{c}{\sqrt{\mu\varepsilon}}\sqrt{\gamma_\lambda^2 + \left(\frac{l\pi}{d}\right)^2} \qquad (18.35)$$

So, the resonant frequencies of the resonant cavity may be changed by displacing the end faces (distance d).

Example 18.4: The cylindrical resonant cavity

We consider a resonant cavity as given in Figure 18.10. Due to the cylindrical symmetry of the problem we introduce cylindrical coordinates so that the required function $\psi(x, y)$ becomes a function of ρ and φ. Due to the rotational symmetry the differential equation may be separated, and for $\psi(\rho, \varphi)$ we obtain

$$\psi(\rho, \varphi) = \psi(\rho)e^{im\varphi}, \qquad m = 0, 1, \ldots$$

So, in cylindrical coordinates the two-dimensional wave equation for ψ

$$\left(\nabla_t^2 + \gamma^2\right)\psi = 0 \qquad (18.36)$$

takes the form

$$\left(\frac{\partial^2}{\partial\rho^2} + \frac{1}{\rho}\frac{\partial}{\partial\rho} + \gamma^2 - \frac{m^2}{\rho^2}\right)\psi(\rho) = 0$$

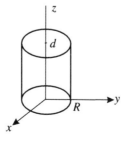

Figure 18.10. Cylindrical resonant cavity.

This is Bessel's differential equation (compare the lectures on Mechanics, Chapter 10). Thus, the radial solutions are Bessel functions. Then, the total wave function reads

$$\psi(\rho, \varphi) = J_m(\gamma_m \rho) e^{im\varphi}$$

We became familiar with the Bessel functions in mechanics, in the treatment of the oscillations of a membrane.

In the case of TM waves, from equations (18.16) we obtain for the electric field

$$E_z = B \cdot J_m(\gamma_m \rho) e^{im\varphi} \cos \frac{l\pi z}{d}$$

The boundary condition $E_z(\rho = R)$ yields a further condition for γ_m:

$$J_m(\gamma_m R) = 0$$

So, the γ_m values contain a further index n giving the zero of the Bessel function. If x_{mn} is the nth zero of the Bessel function J_m, then

$$\gamma_m \Rightarrow \gamma_{mn} = \frac{x_{mn}}{R}$$

By this E_z is now determined completely. Substituted into equation (18.35) we obtain for the resonance frequencies

$$\omega_{mnl} = \frac{c}{\sqrt{\mu\varepsilon}} \sqrt{\gamma_{mn}^2 + \left(\frac{\pi l}{d}\right)^2} = \frac{c}{\sqrt{\mu\varepsilon}} \sqrt{\frac{x_{mn}^2}{R^2} + \frac{\pi^2 l^2}{d^2}}$$

The lowest frequency is obtained for $m = 0, n = 1, l = 0$ (with $x_{01} = 2.4$). It is

$$\omega_{010} = \frac{2.4}{\sqrt{\mu\varepsilon}} \frac{c}{R}$$

This is independent of the height of the cylinder; therefore, tuning by changing d is not possible. With the time-dependence $e^{i\omega t}$, the associated TM$_{010}$ wave becomes

$$\mathbf{E}_z = B J_0\left(\frac{2.4\rho}{R}\right) e^{i\omega t} \mathbf{e}_z, \qquad \mathbf{B}_t = \frac{i\mu\varepsilon\omega_{010}}{c\gamma_{010}^2} B \mathbf{e}_z \times \nabla_t J_0\left(\frac{2.4\rho}{R}\right) e^{i\omega t}$$

In cylindrical coordinates the transverse gradient is

$$\nabla_t = \frac{\partial}{\partial\rho}\mathbf{e}_\rho + \frac{\partial}{\partial\varphi}\mathbf{e}_\varphi$$

Since $J_0(\gamma_{010}\rho)$ depends only on ρ, $\nabla_t J_0(\gamma_{010}\rho)$ has a component in $\mathbf{e}_\varphi$-direction only; thus, it is perpendicular to the z-axis; that is, $\mathbf{e}_z \times J_0(\gamma_{010} \cdot \rho)$ points in $\mathbf{e}_\varphi$-direction,

$$\mathbf{B}_t = \mathbf{B}_\varphi = \frac{i\mu\varepsilon\omega_{010}}{c\gamma_{010}^2} B \frac{\partial}{\partial\rho} J_0(\gamma_{010}\rho) e^{i\omega t} \mathbf{e}_\varphi$$

Without proof we give the following interrelations between the Bessel functions

$$J_{m-1}(x) - J_{m+1}(x) = 2\frac{dJ_m(x)}{dx}, \qquad J_{-m}(x) = (-1)^m J_m(x)$$

For $m = 0$,

$$\frac{d}{dx} J_0(x) = -J_1(x)$$

Employing this relation for the derivative of the Bessel function in $\mathbf{B}_\varphi$, for the magnetic field we obtain

$$\mathbf{B}_\varphi = -i\sqrt{\mu\varepsilon}\,B\,J_1(\gamma_{010}\rho)e^{i\omega t}\mathbf{e}_\varphi$$

Starting from equation (18.17) we obtain correspondingly for the TE waves

$$B_z = A\,J_m(\gamma_{mn}\rho)e^{im\varphi}\sin\frac{l\pi z}{d} \tag{18.37}$$

From $(\partial B_z/\partial n)(\rho = R) = (\partial B_z/\partial\rho)(\rho = R) = 0$ we have

$$\left.\frac{\partial}{\partial\rho}J_m(\gamma_{mn}\rho)\right|_{\rho=R} = 0$$

equivalent to

$$\left.\frac{\partial}{\partial(\gamma_{mn}\rho)}J_m(\gamma_{mn}\rho)\right|_{\rho=R} = 0$$

So, the γ_{mn} values are fixed now by the zeros of the derivatives of the Bessel functions. For the resonance frequencies we have

$$\omega_{mnl} = \frac{c}{\sqrt{\mu\varepsilon}}\sqrt{\frac{x_{mn}^2}{R^2} + \frac{l^2\pi^2}{d^2}}$$

According to the assumption, let $B_z \neq 0$, so that we obtain for the lowest frequency (with $J_1'(x_{11}) = 0$, $x_{11} = 1.8$)

$$\omega_{111} = \frac{1.8}{\sqrt{\mu\varepsilon}}\frac{c}{R}\sqrt{1 + 2.9\frac{R^2}{d^2}}$$

Taking into account the time dependence $e^{i\omega t}$ of $\mathbf{B}$, we obtain for the TE_{111} wave

$$B_z = A\,J_1(\gamma_{111}\rho)\sin\left(\frac{\pi z}{d}\right)e^{i(\omega t + \varphi)}$$

In contrast to the fundamental frequency of the TM mode, the TE_{111} wave may be tuned by changing the height of the cylinder d. For large values of d, ω_{111} lies below the frequency of the TM mode, and thus represents the fundamental frequency of the resonant cavity.

Example 18.5: The free-electron laser

Laser is an abbreviation for *Light Amplification by Stimulated Emission of Radiation*. Conventional light sources (thermal radiators and gas-discharge lamps) emit a broad, random mixture of frequencies; there is no phase correlation between distinct space points of the electromagnetic radiation field. This radiation is called polychromatic and incoherent. With the invention of lasers it became possible to generate a radiation field of high monochromacy and coherency. Actually, the mechanism of the laser is based on purely quantum mechanical processes of the interaction of the radiation field with matter. For most lasers, one uses the fact that atoms, ions, and molecules may exist in different energy states. The associated energy levels are E_1 and E_2 with $E_1 < E_2$. If a transition proceeds from level 1 to level 2 then radiation whose wavelength corresponds to the energy difference is absorbed; if the transition proceeds from level 2 to level 1 emission occurs. In general, the transition from the higher lying level to the lower lying one proceeds spontaneously. This radiation process is of statistical nature, correspondingly the light emitted is incoherent. But, if a transition from the higher level to the lower level is forced one talks of induced or stimulated emission. Then, the radiation is monochromatic and coherent. This is the process which is of fundamental importance for the laser. To understand the mechanism of a typical laser the knowledge of quantum mechanics is essential.

This is different from the free-electron laser. Its mechanism can be understood in the framework of classical electrodynamics, as will be shown in this example. Its mechanism differs in many respects from the mechanism of other lasers. Like other lasers, it is based on the principle of light amplification by stimulated emission, but the emission does not proceed as the transition between bound electron states of definite energy. The principle scheme of the free-electron laser is shown in Figure 18.11. An electromagnetic plane wave hits, together with a beam of free electrons, a magnetic field. The interaction af an appropriate magnetic field with the electrons may cause the electrons to transfer their energy in a coherent manner to the electromagnetic wave and thus leads to the amplification of the latter. An appropriate field is the *wiggler* field. The wiggler field drawn in Figure 18.11 is a steady magnetic field that varies periodically in space. The spatial period is denoted by λ_w. The arrows shown denote the direction of the linear transverse magnetic field

$$\mathbf{B}_w = B_w \cos \frac{2\pi z}{\lambda_w} \mathbf{e}_x \tag{18.38}$$

at half wavelength along the z-direction. Instead of such a field we may use a more complicated one which varies in x- and y-direction:

$$\mathbf{B}_w = B_w \left(\cos \frac{2\pi z}{\lambda_w} \mathbf{e}_x + \sin \frac{2\pi z}{\lambda_w} \mathbf{e}_y \right) \tag{18.39}$$

The $\mathbf{B}_w$-field exerts a force on an electron,

$$\mathbf{F}_w = \frac{e}{c} \mathbf{v} \times \mathbf{B}_w \tag{18.40}$$

Because $\mathbf{F}_w \cdot \mathbf{v} = 0$ is identically zero, the $\mathbf{B}_w$-field does no work on the electron. But it induces an oscillation of the electron motion transverse to the z-direction. Such an oscillating electron generates synchrotron radiation. Now it is of importance that the electric field of the electromagnetic wave applied externally may exchange energy with the oscillating electron. Here, $\mathbf{F}_E \cdot \mathbf{v} = e\mathbf{E} \cdot \mathbf{v} \neq 0$. This energy transfer is used to amplify the electromagnetic wave moving along the electron beam.

Now, we want to investigate in more detail the forces acting on an electron of velocity

$$\mathbf{v} = v_x \mathbf{e}_x + v_y \mathbf{e}_y + v_z \mathbf{e}_z \tag{18.41}$$

Here, the electron moves at the longitudinal velocity $v_z \approx c$, the speed of light. Substituting this into equation (18.40) and using the $\mathbf{B}_w$-field (18.39) then we obtain the following Newtonian equations of motion

$$\dot{v}_x = -\frac{e}{m_0} B_w \sin \frac{2\pi ct}{\lambda_w}$$

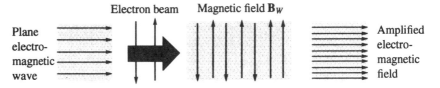

Figure 18.11. Free-electron laser.

$$\dot{v}_y = \frac{e}{m_0} B_w \cos \frac{2\pi c t}{\lambda_w} \tag{18.42}$$

Here, we have used $v_z \approx c$, and we have written $v_z t \approx ct$ for the z-coordinate of the electron. Integrating this equation we obtain

$$v_x = Kc \cos \frac{2\pi z}{\lambda_w}$$

$$v_y = Kc \sin \frac{2\pi z}{\lambda_w} \tag{18.43}$$

where we have introduced the wiggler parameter K

$$K = \frac{eB_w \lambda_w}{2\pi m_0 c^2} \tag{18.44}$$

Now, the classical equations of motion are not correct for relativistic electrons. We must take into account the velocity dependence of the electron mass. For a particle of speed v the mass is given by

$$m = \frac{m_0}{\sqrt{1 - v^2/c^2}} = \gamma m_0 \tag{18.45}$$

m_0 is the rest mass of the particle. This is substituted in (18.42), leading to a replacement of K in equation (18.43) by K/γ:

$$v_x = \frac{Kc}{\gamma} \cos \frac{2\pi z}{\lambda_w}$$

$$v_y = \frac{Kc}{\gamma} \sin \frac{2\pi z}{\lambda_w} \tag{18.46}$$

Now, using the definition of γ in equation (18.45) we obtain

$$\begin{aligned}
v_z^2 &= \mathbf{v}^2 - v_x^2 - v_y^2 \\
&= c^2 \frac{\gamma^2 - 1}{\gamma^2} - v_x^2 - v_y^2 \\
&= c^2 \frac{\gamma^2 - 1}{\gamma^2} - \frac{K^2 c^2}{\gamma^2} \\
&= c^2 \left(1 - \frac{1 + K^2}{\gamma^2} \right)
\end{aligned} \tag{18.47}$$

employing equation (18.46). For $v \approx c$, γ is large, and we may write

$$\begin{aligned}
v_z &= c \left(1 - \frac{1 + K^2}{\gamma^2} \right)^{1/2} \\
&\approx c \left(1 - \frac{1 + K^2}{2\gamma^2} \right)
\end{aligned} \tag{18.48}$$

if K is not too large. This is fulfilled due to proposition that v_x and v_y should be small compared to $v_z \approx c$. Now, $\mathbf{B}_w$ is usually chosen strong enough to determine the trajectory of the electron, contrary to the electromagnetic field of the incoming plane wave. $\mathbf{B}_w$ does not exchange energy with the electron, but by determining its trajectory it also determines the energy exchange between the electron and the electromagnetic wave.

The electric field of the wave is written in the form

$$\mathbf{E} = E_0 \left[\sin\left(\frac{2\pi z}{\lambda} - \omega t + \phi_0\right) \mathbf{e}_x + \cos\left(\frac{2\pi z}{\lambda} - \omega t + \phi_0\right) \mathbf{e}_y \right] \tag{18.49}$$

This representation is most convenient for combination with $\mathbf{B}_w$. The form of $\mathbf{B}_w$ determines the polarization of the emitted radiation. Linear transverse $\mathbf{B}_w$-fields lead to linear polarization; helical $\mathbf{B}_w$-fields like (18.39) lead to circular polarization. Then, the rate of work done by the plane wave on an electron of velocity $\mathbf{v}$ is

$$
\begin{aligned}
\dot{W} &= e\mathbf{E} \cdot \mathbf{v} \\
&= eE_0 \frac{Kc}{\gamma} \left[\cos\frac{2\pi z}{\lambda_w} \sin\left(\frac{2\pi z}{\lambda} - \omega t + \phi_0\right) + \sin\frac{2\pi z}{\lambda_w} \cos\left(\frac{2\pi z}{\lambda} - \omega t + \phi_0\right) \right] \\
&= eE_0 \frac{Kc}{\gamma} \sin\left[2\pi\left(\frac{1}{\lambda} + \frac{1}{\lambda_w}\right) z - \omega t + \phi_0 \right] \\
&= eE_0 \frac{Kc}{\gamma} \sin\phi
\end{aligned}
\tag{18.50}
$$

with

$$\phi = 2\pi\left(\frac{1}{\lambda} + \frac{1}{\lambda_w}\right) z - \omega t + \phi_0 \tag{18.51}$$

According to Einstein's formula $E = mc^2 = \gamma m_0 c^2$ (18.50) can be written also as

$$\dot{W} = \dot{\gamma} m_0 c^2 = eE_0 \frac{Kc}{\gamma} \sin\phi \tag{18.52}$$

or

$$\dot{\gamma} = eE_0 \frac{K}{\gamma m_0 c} \sin\phi \tag{18.53}$$

Using $\dot{z} = v_z$, where v_z is given by (18.48), we may also give an equation for the change in phase seen by the moving electron

$$
\begin{aligned}
\dot{\phi} &= 2\pi\left(\frac{1}{\lambda} + \frac{1}{\lambda_w}\right) v_z - \omega \\
&= 2\pi c\left(\frac{1}{\lambda} + \frac{1}{\lambda_w}\right)\left(1 - \frac{1 + K^2}{2\gamma^2}\right) - \frac{2\pi c}{\lambda} \\
&= \frac{2\pi c}{\lambda_w}\left[1 - \left(1 + \frac{\lambda_w}{\lambda}\right)\frac{1 + K^2}{2\gamma^2} \right]
\end{aligned}
\tag{18.54}
$$

In the free-electron laser the period of excitation λ_w is of the order of centimeters, while the wavelength of the electromagnetic wave λ is much smaller. Therefore, $\lambda_w/\lambda \gg 1$ and hence

$$\dot{\phi} \approx \frac{2\pi c}{\lambda_w}\left(1 - \frac{\lambda_w}{\lambda}\frac{1 + K^2}{2\gamma^2}\right) \tag{18.55}$$

For the value of the electric energy $\gamma m_0 c^2$, ϕ is constant ($\dot{\phi} = 0$) so that

$$\gamma^2 = \gamma_R^2 = \frac{\lambda_w}{2\lambda}(1 + K^2) \tag{18.56}$$

γ_R defines the electron resonance energy. This can be understood in the following way. If the electron in direction of the z-axis passes through the distance Δz in the time $\Delta t = \Delta z/v_z$ then it observes a change in the phase of the field (18.49). This change in phase is given by

$$\Delta\Theta = \omega\left(\Delta t - \frac{\Delta z}{c}\right) = \frac{2\pi c}{\lambda}\Delta z\left(\frac{1}{v_z} - \frac{1}{c}\right)$$

$$= \frac{2\pi}{\lambda}\Delta z\left(\frac{c}{v_z} - 1\right) \tag{18.57}$$

$$\approx \frac{2\pi}{\lambda}\Delta z\frac{1 + K^2}{2\gamma^2}$$

Here, we have used (18.48) with $v_z \approx c$. In particular, if $\Delta z = \lambda_w$ and $\gamma^2 = \gamma_R^2$, then $\Delta\Theta = 2\pi$. That is, for an electron with the resonance energy the period of the electromagnetic field equals the period of the exciting $\mathbf{B}_w$-field. According to equation (18.55), the resonance energy depends on the period λ_w and the intensity $\mathbf{B}_w$ of the magnetic field. Now, we have two coupled equations (18.53) and (18.55) for γ and ϕ:

$$\dot{\gamma} = eE_0\frac{Kc}{\gamma m_0 c}\sin\phi$$

$$\dot{\phi} = \frac{2\pi c}{\lambda_w}\left(1 - \frac{\gamma_R^2}{\gamma^2}\right) \tag{18.58}$$

These equations describe a single electron in the field of the exciting magnet and in the field of the monochromatic, plane electromagnetic wave. These equations point out that an electron may receive energy from the electromagnetic wave ($\dot{\gamma} > 0$), or it may deliver energy to the latter ($\dot{\gamma} < 0$). An energy gain of the electron corresponds to the absorption of the electromagnetic wave, the energy release of the electron corresponds to the stimulated emission. One could assume that a pulse of electrons injected into the zone of excitation exhibits a uniform distribution of ϕ-values so that the average value of $\sin\phi$ is about zero, and on the average absorption and emission balance each other. But, this is not the case. From equations (18.48) and (18.55) we obtain

$$v_z \approx c\left(1 - \frac{\lambda\gamma_R^2}{\lambda_w\gamma^2}\right) \tag{18.59}$$

and therefore

$$\dot{v}_z \approx \frac{2c\lambda\gamma_R^2}{\lambda_w\gamma^3}\dot{\gamma} \tag{18.60}$$

That is, an electron is accelerated or decelerated in the longitudinal direction depending on whether it receives energy ($\dot{\gamma} > 0$) or releases energy ($\dot{\gamma} < 0$). So, the exchange of energy with the electromagnetic wave leads not only to a change in the energy of the electrons but also to a reordering of the spatial electron distribution along the z-axis, whereby the faster electrons overtake the slower ones. On a macroscopic scale the electrons remain equally distributed, but the process of bundeling on the microscopic scale may lead to stimulated emission instead of absorption. Which process occurs depends on the distribution of the electron energies. A detailed analysis leads to the following three conclusions:

(a) Emission occurs if the electron energy is chosen, such that $\gamma > \gamma_R$, absorption occurs if $\gamma < \gamma_R$.

(b) The maximum emission occurs if for the electron energy $\gamma \approx (1 + 0.2/N_w)\gamma_R$. N_w is number of periods of the exciting $\mathbf{B}_w$-field.

(c) The emission is very small unless the electron pulse has a very sharp energy distribution.

The discussed process of stimulated emission in the free-electron laser is a classical one. But, it is different, in principle, from the stimulated emission in common lasers in which fixed energy levels in atoms or molecules are populated or depopulated. Therefore, the type of laser discussed here is not bounded to certain frequencies but it may be tuned over a large frequency range. Equation (18.56) defines for any wavelength λ a resonant electron energy in the vicinity of which emission may occur. Because the resonance energy depends also on the characteristics of $\mathbf{B}_w$, that is, on the magnitude of the field intensity and the period λ_w, tuning to a certain wavelength may be achieved either by variation of the $\mathbf{B}_w$-field parameter or of the electron energy. So, it is possible that the free-electron laser may be used in the range of wavelengths from 1 mm to 10^{-8} cm.

19 Light Waves

For light waves

$$\omega(k) = \frac{ck}{n(k)} \qquad \text{or} \qquad k(\omega) = n(\omega)\frac{\omega}{c} \tag{19.1a}$$

where c is the speed of light in vacuum and $n(k)$ is the index of refraction of the corresponding medium. According to equation (18.3) we obtain the phase velocity

$$v_p = \frac{\omega(k)}{k} = \frac{c}{n(k)} \qquad \text{or} \qquad v_p = \frac{\omega}{k(\omega)} = \frac{c}{n(\omega)}$$

To calculate the group velocity, we write formally

$$1 = \frac{d\omega}{d\omega} \qquad \text{or} \qquad 1 = \frac{dk}{dk}$$

and with equation (19.1a) we obtain

$$1 = \frac{c}{n^2}\left(n\frac{dk}{d\omega} - k\frac{dn}{d\omega}\right) \qquad \text{or} \qquad 1 = \frac{n}{c}\frac{d\omega}{dk} + \frac{\omega}{c}\frac{dn}{d\omega}\frac{d\omega}{dk} \tag{19.1b}$$

if the functions $\omega(k)$ and $n(\omega)$ are differentiable so that $dn/dk = (dn/d\omega)(d\omega/dk)$.

Taking into account (18.14) and if $dk/d\omega$ may be inverted the solution of the first of the equations (19.1b) yields

$$\frac{d\omega}{dk} = v_g = \frac{c}{n(k) + \omega\dfrac{dn}{d\omega}}$$

The second of the equations (19.1a) yields the same result. However, in both cases the functions $\omega(k)$, $n(k)$ or $k(\omega)$ are subject to certain conditions (differentiability, existence of the inverse function). For the most optical wavelengths, $n(k)$ is greater than 1 for many media.

In the case of *normal dispersion*, $dn/d\omega > 0$. Then, the velocity of the energy flow is smaller than the phase velocity and thus also smaller than the speed of light in vacuum c. But, in the case of *anomalous dispersion* $dn/d\omega$ may become negative and of large magnitude. The possible cases are represented in Figure 19.1. If $dn/d\omega$ becomes too small, v_g may even have poles or become negative (see the figure, dashed curve). This is just the case in which dn/dk cannot be inverted everywhere. But in this case, the equation for v_g is no

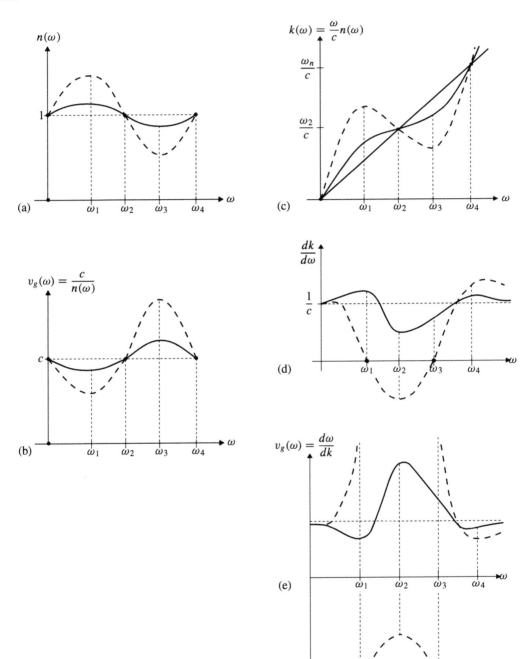

Figure 19.1. (a) The index of refraction $n(\omega)$, (b) phase velocity v_p, (c) wave number $k(\omega)$, (d) $dk/d\omega$, and (e) group velocity v_g in the domain of normal and anomalous dispersion.

longer valid, since the assumptions for its derivation are violated. However, also in the case that v_g remains well-defined (solid curve) there is no contradiction to the theory of relativity since for $d\omega/dk = v_g > 1$ the series expansion $\omega(k) = \omega(k_0) + (d\omega/dk)|_{k_0}(k - k_0) + \cdots$ which we have performed previously in deriving the group velocity (see Exercise 19.1) is no longer valid. In other cases of anomalous dispersion, absorption effects occure; the frequencies with the highest group velocity are absorbed first, so that again energy may not be transported with a velocity greater than c. In the following, these statements will be justified.

The propagation of signals in dispersive media

In this section we will investigate the question just raised about the propagation of signals (wave packets) within dispersive media; in particular, with respect to the agreement of this process with the principles of causality and the theory of relativity.[1] At first, we will consider a plane wave-packet $u_i(x, t)$ which, coming from vacuum ($x < 0$), penetrates perpendicularly into a semi-infinite medium ($x > 0$) with the index of refraction $n(\omega)$. See Figure 19.2. At the interface $x = 0$, let the incident wave packet be given by $u_i(0, t)$. We decompose $u_i(x, t)$ into its Fourier frequencies

$$u_i(x, t) = \int_{-\infty}^{+\infty} A(\omega)e^{i(kx - \omega t)}\, d\omega = \int_{-\infty}^{+\infty} A(\omega)e^{i(\omega/c)(x - ct)}\, d\omega, \qquad k = \frac{\omega}{c} \qquad (19.2)$$

where, in particular, at the interface $x = 0$,

$$A(\omega) = \frac{1}{2\pi} \int_{-\infty}^{+\infty} u_i(0, t)e^{i\omega t}\, dt \tag{19.3}$$

In a dispersive medium, each individual frequency propagates with the wave number vector (19.1a):

$$k(\omega) = \frac{\omega}{c}n(\omega)$$

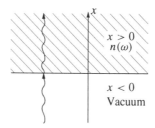

Figure 19.2. A wave entering from vacuum ($n = 1$) into a dispersive medium ($n = n(\omega)$).

[1]In 1914, A. Sommerfeld and L. Brillouin first studied these questions in the *Annals* of Physics.

According to equation (17.32), at the interface the amplitude $A(\omega)$ of the incident wave transforms as ($n = 1$, in vacuum)

$$\frac{A'(\omega)}{A(\omega)} = \frac{2n}{n + n'} \rightarrow \frac{2}{1 + n(\omega)}, \qquad A'(\omega) \rightarrow \frac{2}{1 + n(\omega)} A(\omega) \tag{19.4}$$

and for the *wave in the dispersive medium* we obtain

$$u(x, t) = \int_{-\infty}^{+\infty} \left(\frac{2}{1 + n(\omega)} \right) A(\omega) e^{i((\omega/c)n(\omega)x - \omega t)} \, d\omega \tag{19.5}$$

Further evaluation of this integral requires a knowledge of the analytic properties of the functions $A(\omega)$ and $n(\omega)$. First, we investigate $A(\omega)$, afterward $n(\omega)$. According to (19.2), the properties of $A(\omega)$ may be deduced from the known wave $u_i(x, t)$ incident in vacuum and its property to move at the speed of light. For $n(\omega)$ we will study a simple but realistic model below.

Analytic properties of $A(\omega)$

To proceed, we must know $A(\omega)$ as a function of ω, also for complex ω because we want to solve the Fourier integral (19.5) by integration in the complex plane. If the incident wave packet has a somewhat sharp edge (Figure 19.3) reaching the plane $x = 0$ at the time $t = 0$ we can say that

$$u_i(0, t) = 0 \qquad \text{for} \qquad t < 0 \tag{19.6}$$

If the wave packet is bell-shaped, then at the position $x = 0$, the wavefront may be approximated by

$$u_i(0, t) \approx \frac{at^m}{m!} \theta(t) e^{-\varepsilon t} \tag{19.7}$$

It vanishes for $t < 0$, then increases proportionally to t^m, and for long times it decays exponentially. The function (m is arbitrary) may be represented graphically by allowing the wave to travel across the origin (compare Figure 19.4). For this wavefront, one may give

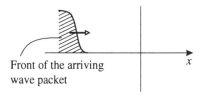

Front of the arriving
wave packet

Figure 19.3. Model for the front of a wave packet entering a new medium at $x = 0$.

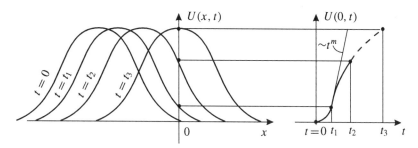

Figure 19.4. For short times, $U(0, t)$ is a power of t. t^m represents the shape of the travelling wave front.

also the monochromatic amplitudes $A(\omega)$ out of which the wave packet is composed. With equation (19.3) we obtain

$$A(\omega) = \frac{1}{2\pi} \int_{-\infty}^{+\infty} \frac{a}{m!} t^m \theta(t) e^{-\varepsilon t} e^{i\omega t} \, dt$$

$$= \frac{1}{2\pi} \frac{a}{m!} \int_0^\infty t^m e^{-(\varepsilon - i\omega)t} \, dt$$

$$= \frac{1}{2\pi} \frac{a}{m!} m! \frac{1}{(\varepsilon - i\omega)^{m+1}} = \frac{1}{2\pi} a(i)^{m+1} \left[\frac{1}{(\omega + i\varepsilon)} \right]^{m+1} \tag{19.8}$$

Hence, we have the result, important for the following, that the Fourier amplitude is

$$u_i(0, t) = \frac{at^m}{m!} \theta(t) e^{-\varepsilon t}, \qquad A(\omega) \sim \left(\frac{1}{\omega} \right)^{m+1} \tag{19.9}$$

for wave packets of the type (19.7).

In general, it is known that for real indices of refraction the Fourier amplitude $A(\omega)$ is an analytic function. There are no resonance effects in the propagation within transparent media. For complex or even purely imaginary indices of refraction one has no information on $A(\omega)$. In general, it may be a complicated function. In the following, we need further the important property that $A(\omega)$ is a regular function in the upper half-plane of the complex ω-plane and that it has no poles. See Figure 19.5. For the special wave packet (19.7) this may be seen directly from the result (19.8) since the zero of the denominator lies at $\omega = -i\varepsilon$ in the lower half-plane. *But quite generally, the poles of $A(\omega)$ may lie only in the lower half plane.* This can be seen in the following way: According to (19.2), in vacuum

$$u_i(x, t) = \int_{-\infty}^{+\infty} A(\omega) e^{i(kx - \omega t)} \, d\omega = \int_{-\infty}^{+\infty} A(\omega) e^{i(\omega/c)(x - ct)} \, d\omega, \qquad k = \frac{\omega}{c} \tag{19.10}$$

As we know from the wave equation, in vacuum any partial wave, hence, also the wave packet $u_i(x, t)$, propagates at the velocity of light. Consequently, the wave packet has to vanish in the space-like domain of the light cone, $u_i(x, t) \equiv 0$ for $x - ct > 0$. The left-hand side of (19.10) vanishes for $x - ct > 0$. Now, the right-hand side is considered further.

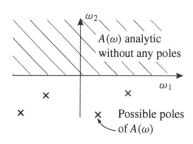

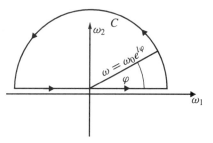

Figure 19.5. Complex ω-plane $\omega =$ $\omega_1 + i\omega_2$.

Figure 19.6. Integration path C; the sense of revolution is given.

We think of $A(\omega)$ as given and transform the integral over ω along the entire real ω-axis in (19.10) into a (closed) Cauchy integral over the infinitely distant semicircle (compare Figure 19.6):

$$0 = \int_{-\infty}^{+\infty} A(\omega)e^{i(\omega/c)(x-ct)}\,d\omega = \int_{\frown} A(\omega)e^{i(\omega/c)(x-ct)}\,d\omega - \int_{\frown} A(\omega)e^{i(\omega/c)(x-ct)}\,d\omega$$

$$(19.11)$$

Now we want to show that the integral along the ∞-distant semicircle in the upper complex half-plane yields zero. For this aim we substitute

$$\omega = \omega_0 e^{i\varphi} = \omega_0(\cos\varphi + i\sin\varphi)$$

According to equation (4.13)

$$\left|\int_C f(z)dz\right| \le Ml$$

we want to estimate the maximal function value of the integrand on the semicircle.

$$\left|A(\omega)e^{i(\omega/c)(x-ct)}\right| = \left|A(\omega)e^{i(\omega_0/c)(\cos\varphi + i\sin\varphi)(x-ct)}\right|$$
$$= |A(\omega)|\left|e^{i(\omega_0/c)\cos\varphi(x-ct)}\right|\left|e^{-(\omega_0/c)\sin\varphi(x-ct)}\right|$$
$$= |A(\omega)|\,e^{-(\omega_0/c)\sin\varphi(x-ct)} \qquad (19.12)$$

In equation (19.9) we derived $A(\omega)$ for a special assumption (equation (19.7)). $A(\omega)$ has been found to decay proportionally to $(1/\omega_0)^{m+1}$. In the following, we want to assume merely that $A(\omega)$ does not increase with ω more rapidly than the exponential function decreases.

Thus, in the upper complex half-plane ($\varphi > 0$) the integrand tends to zero as $\omega_0 \to \infty$ because we have assumed $(x - ct) > 0$! (Employing the property $A(\omega) \sim (1/\omega_0)^{m+1}$ from equation (19.9), we may also allow $(x - ct) \ge 0$.) So, in this case, the integral over the upper semicircle gives no contribution. The equation (19.11) simplifies itself according to

$$0 = \int_{-\infty}^{+\infty} A(\omega)e^{i(\omega/c)(x-ct)}\,d\omega = \int_{\frown} A(\omega)e^{i(\omega/c)(x-ct)}\,d\omega \qquad (19.11a)$$

Necessarily, the sum of the residues (compare the residue theorem in Part I, Chapter 4) in the (encircled) upper half-plane has to be zero:

$$0 \equiv \int_{\frown} A(\omega)e^{i(\omega/c)(x-ct)}\,d\omega$$
$$= 2\pi i \sum_k \mathrm{Res}\left\{A(\omega_k)e^{i(\omega_k/c)(x-ct)}\right\} = 2\pi i \sum_k e^{i(\omega_k/c)(x-ct)}\mathrm{Res}\,A(\omega_k) \tag{19.13}$$

because the exponential function has no poles in the complex domain.

Since this relation has to be valid for any position coordinate x and time t, it is fulfilled only if $A(\omega)$ possesses no poles in the upper half-plane. (Of course, for one special choice of x and t an accidental mutual cancellation of two or several residues of $A(\omega)$ cannot be excluded, correspondingly $A(\omega)$ could then not be analytical.)

In other words, because in vacuum the (incident) wave is forbidden to propagate at a velocity greater than the velocity of light, $A(\omega)$ cannot have poles in the upper half-plane. (For this proof we have assumed the condition $(x - ct) > 0$. Of course, the analytic properties of $A(\omega)$ in the upper half-plane cannot be different for $(x - ct) < 0$.)

Now, we want to derive the analytic properties of $A(\omega)$ in the lower complex half-plane. For this aim we assume $x - ct < 0$ and integrate along the integration path C' (compare Figure 19.7):

$$u_i(x, t) = \int_{-\infty}^{+\infty} A(\omega)e^{i(\omega/c)(x-ct)}\,d\omega$$
$$= \int_{\frown} A(\omega)e^{i(\omega/c)(x-ct)}\,d\omega - \int_{\smile} A(\omega)e^{i(\omega/c)(x-ct)}\,d\omega \tag{19.14}$$

The contribution of the second integral

$$\int_{\varphi=0}^{-\pi} A(\omega)e^{i(\omega_0/c)\cos\varphi(x-ct)}e^{-(\omega_0/c)\sin\varphi(x-ct)}i\omega_0 e^{i\varphi}\,d\varphi$$
$$= -i\omega_0 \int_{\varphi=0}^{\pi} A(\omega)e^{i(\omega_0/c)\cos\varphi(x-ct)}e^{(\omega_0/c)\sin\varphi(x-ct)}e^{-i\varphi}\,d\varphi$$

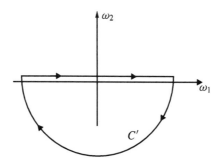

Figure 19.7. Integration path C': The sense of revolution is given.

vanishes analogously to the discussion above, since we assume $(x - ct) < 0$ (for $A(\omega) \sim (1/\omega_0)^{m+1}$ one may also allow $x - ct = 0$), and the integrand vanishes exponentially as $\omega_0 \to \infty$. In summary,

$$u_i(x, t)|_{x \leq ct} = -2\pi i \sum \text{residues of the integrand of (19.14)}$$

The minus sign comes from the sense of the integration in the integral in (19.14). Now, this equation is different from zero $(u_i(x, t)|_{x \leq ct} \neq 0)$ precisely if $A(\omega)$ has poles in the lower half-plane. Just then a wave packet exists propagating in vacuum. As we know, this propagates at a velocity smaller than the speed of light, $v_s < c$. But this may be seen also from the preceding consideration, since we consider the time-like light cone $x - ct < 0$.

Conclusion

From the propagation of waves (wavefronts) in vacuum and in media of low optical density at velocities smaller than the velocity of light, one may conclude that the poles of $A(\omega)$ lie in the lower half-plane of the complex ω-plane.

On the other hand, the position of the poles determines the group velocity (better: the signal velocity which contains the group velocity as a special case but also the propagation of a wavefront). For wavefronts in vacuum, one has always $v_{\text{signal}} < c$. This important information yields the statement that $A(\omega)$ has to be analytic in the upper ω-half-plane and must have poles in the lower ω-half-plane. Equation (19.8) confirms this result explicitly for the special ansatz (19.7) which contains all essentials.

From simple dispersion theory it follows that in the domains of anomalous dispersion one may have $v_g > c$. But, the result derived at the beginning of this chapter,

$$v_g = \frac{c}{n(\omega) + \omega \dfrac{dn}{d\omega}} \tag{19.15}$$

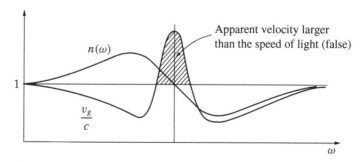

Figure 19.8. On the illustration of the region of normal and anomalous dispersion. The position of the poles of the amplitude function $A(\omega)$ is determined from the behavior for normal dispersion. The position of the poles is invariable: the poles cannot slide into the upper half-plane since $\omega = \omega'$ in the transition between the media. Then we will show that, due to the analytic behavior of $A(\omega)$, the signal velocity must be always smaller than the speed of light.

is not always applicable, concerning the propagation of signals. We have seen that the function $n(\omega)$ has to satisfy certain properties to obtain this equation. But even if these properties are fulfilled, we had obtained $v_g > c$ in regions of anomalous dispersion. But we want to show in the following that this formula fails generally in these regions. See Figure 19.8.

Analytic properties of the index of refraction

For a dielectric (with only one resonance energy) the index of refraction has the form (Chapter 16)

$$n^2(\omega) = 1 + \frac{\omega_p^2}{\omega_0^2 - \omega^2 - i\omega\gamma}, \qquad \omega_p^2 = \frac{4\pi N e^2}{m} f \tag{19.16}$$

$\gamma > 0$ is the damping factor, and, therefore, $-i\omega\gamma$ is always negative imaginary, which will be important later on. We want to calculate $n(\omega)$ because we need it in (19.5). For this aim we rewrite $n^2(\omega)$ and bring both terms in (19.16) to a common denominator:

$$n^2(\omega) = \frac{\omega_0^2 - \omega^2 - i\omega\gamma + \omega_p^2}{\omega_0^2 - \omega^2 - i\omega\gamma} = \frac{\text{polynomial of the second order}}{\text{polynomial of the second order}}$$

$$n^2(\omega) = \frac{(\omega - \omega_a)(\omega - \omega_b)}{(\omega - \omega_c)(\omega - \omega_d)}$$

where ω_a and ω_b are the zeros of the nominator, and ω_c and ω_d are the zeros of the denominator, namely:

$$\omega_a = -\frac{i\gamma}{2} + \sqrt{\omega_0^2 + \omega_p^2 - \frac{\gamma^2}{4}}, \qquad \omega_b = -\frac{i\gamma}{2} - \sqrt{\omega_0^2 + \omega_p^2 - \frac{\gamma^2}{4}}$$

$$\omega_c = -\frac{i\gamma}{2} + \sqrt{\omega_0^2 - \frac{\gamma^2}{4}}, \qquad \omega_d = -\frac{i\gamma}{2} - \sqrt{\omega_0^2 - \frac{\gamma^2}{4}} \tag{19.17}$$

From this equation we obtain

$$n(\omega) = \left[\frac{(\omega - \omega_a)(\omega - \omega_b)}{(\omega - \omega_c)(\omega - \omega_d)}\right]^{1/2}$$

Conclusion

From the formula for $n(\omega)$ and the values of ω_c, ω_d, one can conclude (see Figure 19.9):[2]

---

[2] We suggest that the reader review Chapter 4 (mathematical supplement).

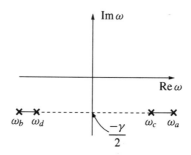

Figure 19.9. The poles ω_c, ω_d and the zeros ω_a, ω_b of $n(\omega)$.

(1) The poles of $n(\omega)$ (being identical with the poles of $n^2(\omega)$) lie in the lower half-plane (negative imaginary part) for a fixed value ω_0.

(2) The limiting value $\lim\limits_{\omega \to \infty} n(\omega) = 1$.

Summarizing, one may state: $n(\omega)$ as well as $A(\omega)$ can have poles in the lower half-plane only. The following proof for the propagation of signals with velocities lower than the speed of light in media with an index of refraction of the form (19.16) is based on this important statement.

Proof of the finite speed of light (signal velocity or group velocity) in matter

We note again the expression (19.5) for the wave packet in a dispersive medium:

$$u(x,t) = \int_{-\infty}^{\infty} \frac{2}{1 + n(\omega)} A(\omega) e^{i(k(\omega)x - \omega t)} \, d\omega$$

We assume $x - ct > 0$, that is, we consider the space-like light cone. If $u(x, t)$ should not vanish here, then in fact, wave propagation with a velocity higher than the speed of light would exist. We close the ω-integral in the upper half-plane (see Figure 19.10):

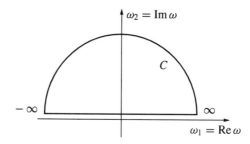

Figure 19.10. This integration path leads to wave packets with a velocity higher than the speed of light. Such wave packets vanish.

$$u(x,t) = \int_{-\infty}^{\infty} \frac{2}{1+n(\omega)} A(\omega) e^{i(k(\omega)x - \omega t)} \, d\omega$$

$$= \int_{\cap} - \int_C \frac{2}{1+n(\omega)} A(\omega) e^{i(k(\omega)x - \omega t)} \, d\omega \tag{19.18}$$

Now, we know that on the infinitely distant semicircle, $|\omega| \to \infty$; that is, $n(\omega) = 1$ and hence, $k = \omega/c$. Therefore, the integral over the infinitely distant semicircle becomes simply

$$\int_C \Rightarrow \int_C A(\omega) e^{i\omega(x/c - t)} \, d\omega$$

$A(\omega)$ is an analytic function in the upper half-plane. With $\omega = \omega_1 + i\omega_2$ the integrand of the curvilinear integral over C is

$$e^{i(\omega_1 + i\omega_2)(x/c - t)} = e^{i\omega_1(x/c - t)} e^{-\omega_2(x/c - t)}$$

Obviously, this vanishes exponentially, according to the assumption $x/c - t > 0$ and $\omega_2 > 0$ in the upper half-plane. Hence, $u(x,t) = \int_{\cap}$. But, from the residue theorem, $\int_{\cap} = 0$ because the integrand has no poles in the upper half-plane. But this means only that there are no wave packets (wavefronts) propagating at a velocity higher than the speed of light $v_s > c$. In other words: in the space-like light cone there is no propagation of waves.

Now we consider the case $x - ct < 0$ and close the integral in the lower half-plane (see Figure 19.11):

$$u(x,t) = \int_{-\infty}^{\infty} \frac{2}{1+n(\omega)} A(\omega) e^{i(k(\omega)x - \omega t)} \, d\omega$$

$$= \int_{\cup} - \int_{C'} \frac{2}{1+n(\omega)} A(\omega) e^{i(k(\omega)x - \omega t)} \, d\omega$$

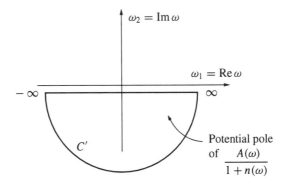

Figure 19.11. This integration path leads to wave packets with a velocity lower than the speed of light. Such wave packets do exist.

Again, we treat $\int_{C'}$. Here also $n(\omega) = 1$ as $|\omega| \to \infty$ and $k = \omega/c$. The integral is

$$\int_{C'} \Rightarrow \int_{C'} A(\omega) e^{i(\omega_1 - i\omega_2)(x/c - t)} \, d\omega$$

But now, in contrast to the preceding case, the imaginary part of the frequency is negative: $\omega = \omega_1 - i\omega_2$, $\omega_2 > 0$. $A(\omega)$ is analytic on C' (this follows from our previous consideration in which we found $A(\omega) \sim 1/\omega^{m+1}$).

Hence,

$$\int_{C'} \Rightarrow \int_{C'} A(\omega) e^{i\omega_1(x/c - t)} e^{\omega_2(x/c - t)} \, d\omega, \qquad \omega_2 > 0$$

The integral vanishes since according to our assumption $x/c - t < 0$. Now, we may calculate $\int_{\ominus}$: due to the poles of $A(\omega)$ and $n(\omega)$ in the lower half-plane, one obtains a definite value of $\int_{\ominus}$. The wave packet has a solution:

$$u(x, t) = \int_{-\infty}^{\infty} \frac{2}{1 + n(\omega)} A(\omega) e^{i(k(\omega)x - \omega t)} \, d\omega$$

$$= \int_{\ominus} = -2\pi i \sum_{\text{residues}} \cdots \neq 0 \tag{19.19}$$

So, in the time-like light cone $x - ct < 0$ a propagation of waves is possible, that is, there is a propagation of waves at a velocity lower than the velocity of light $v_s < c$.

Conclusion

The existence of wave packets with $v_s < c$ follows from the analytic properties of $n(\omega)$ (correct sign of the damping factor) and $A(\omega)$. However, in this discussion it is impossible to give the formula for v_s, but one may conclude only on a bound: *all signals propagate at velocities lower than the speed of light. This also holds for those which are emitted in the frequency domain of anomalous dispersion.*

Example 19.1: Exercise: Nearly monochromatic waves

A nearly monochromatic one-dimensional wave may be written as

$$u_{k_0}(x, t) = \frac{1}{(2\pi)^{1/2}} \int_{-\infty}^{\infty} A_{k_0}(k) e^{i(kx - \omega(k)t)} \, dk$$

where $A_{k_0}(k)$ is significantly different from zero only in a small neighborhood of k_0. Show that approximately

$$u(x, t) = u\left(x - t\frac{d\omega}{dk}\bigg|_{k=k_0}, 0\right) e^{i(k_0(d\omega/dk)|_{k=k_0} - \omega(k_0))t}$$

What can one conclude from that?

Solution   A one-dimensional, nearly monochromatic wave possessing the average wave number k_0 may be written as

$$u(x, t) = \frac{1}{\sqrt{2\pi}} \int_{-\infty}^{\infty} A(k) e^{i(kx - \omega(k)t)} \, dk \tag{19.20}$$

where $A(k)$ is peaked about k_0. (For a monochromatic wave, $A(k) \sim \delta(k - k_0)$.)

For $t = 0$,

$$u(x, 0) = \frac{1}{\sqrt{2\pi}} \int_{-\infty}^{\infty} A(k) e^{ikx} \, dk$$

Thus, $A(k)$ is the Fourier transform of $u(x, 0)$:

$$A(k) = \frac{1}{\sqrt{2\pi}} \int_{-\infty}^{\infty} u(x, 0) e^{-ikx} \, dx \tag{19.21}$$

Let, e.g., $u(x, 0)$ be a plane wave restricted to a spatial region of length $2L$:

$$u(x, 0) = e^{ik_0 x} \cdot \frac{1}{\sqrt[4]{2\pi}} \cdot \frac{1}{\sqrt{L}} e^{-x^2/(4L^2)}$$

The average position and the uncertainty of the position are defined by

$$\langle x \rangle = \frac{\displaystyle\int_{-\infty}^{\infty} x |u(x, 0)|^2 \, dx}{\displaystyle\int_{-\infty}^{\infty} |u(x, 0)|^2 \, dx} = \int_{-\infty}^{\infty} x |u(x, 0)|^2 \, dx \tag{19.22}$$

$$(\Delta x)^2 = \int_{-\infty}^{\infty} (x - \langle x \rangle)^2 |u(x, 0)|^2 \, dx$$

where we have utilized the normalization

$$\int_{-\infty}^{\infty} |u(x, 0)|^2 \, dx = 1$$

In our example, $\langle x \rangle = 0$, $\Delta x = L$.

For the Fourier transform, we get

$$A(k) = \frac{1}{\sqrt{2\pi}} \frac{1}{\sqrt[4]{2\pi}} \frac{1}{\sqrt{L}} \int_{-\infty}^{\infty} e^{-x^2/(4L^2) + i(k_0 - k)x} \, dx = \frac{\sqrt{2L}}{\sqrt[4]{2\pi}} e^{-L^2(k - k_0)^2}$$

If $\langle k \rangle$ and Δk are defined analogously, then

$$\langle k \rangle = k_0, \ \Delta k = \frac{1}{2L}$$

Hence,

$$\Delta x \cdot \Delta k = \frac{1}{2}$$

The longer the wave train the more monochromatic it is; conversely, many wave numbers contribute to spatially short waves. Generally, $\Delta k \cdot \Delta x \geq 1/2$.

Since $A(k)$ is concentrated about k_0, $\omega(k)$ may be expanded about k_0:

$$\omega(k) = \omega_0 + \frac{d\omega}{dk}\bigg|_0 \cdot (k - k_0) + \cdots; \qquad \omega_0 = \omega(k_0), \qquad \frac{d\omega}{dk}\bigg|_0 = \frac{d\omega}{dk}\bigg|_{k=k_0}$$

Thus,

$$u(x, t) = \frac{1}{\sqrt{2\pi}} \exp\left(i\left[k_0 \frac{d\omega}{dk}\bigg|_0 - \omega_0\right]t\right) \int_{-\infty}^{\infty} A(k) e^{i[x - (d\omega/dk)|_0 t]k} \, dk$$

Here, we substitute (19.21):

$$u(x, t) = \exp\left(i\left[k_0 \frac{d\omega}{dk}\bigg|_0 - \omega_0\right]t\right)$$

$$\times \frac{1}{2\pi} \int_{-\infty}^{\infty} dk \int_{-\infty}^{\infty} dx' \, u(x', 0) \exp\left(i\left[x - \frac{d\omega}{dk}\bigg|_0 t - x'\right]k\right)$$

The k-integration yields a δ-function which simplifies the x'-integration:

$$u(x, t) = u\left(x - \frac{d\omega}{dk}\bigg|_0 t, 0\right) \exp\left(i\left[k_0 \frac{d\omega}{dk}\bigg|_0 - \omega_0\right]t\right)$$

While the phase velocity, that is, the velocity of the surfaces of equal phase, is $v_p = \omega(k)/k$, the envelope, that is, the wave group, moves at the group velocity $v_g = (d\omega/dk)|_0$ which is different from v_p in general. In the case of electromagnetic waves, $\omega(k) = ck/n(k)$ and therefore

$$v_p = \frac{c}{n(k)}$$

The group velocity is

$$v_g = \frac{c}{\left[n(\omega) + \omega \dfrac{dn}{d\omega}\right]}$$

where we have expressed the derivation with respect to k in terms of that with respect to ω. We note without further explanation that in the case of anomalous dispersion also the group velocity so defined may become higher than the speed of light.

As a second example, we will consider the quantum mechanical treatment of particles of mass m without spin. These particles are described by the wave function $\psi(\mathbf{x}, t)$ obeying the Klein-Gordon equation

$$\left[\frac{1}{c^2} \frac{\partial^2}{\partial t^2} - \nabla^2 + \left(\frac{mc}{\hbar}\right)^2\right] \psi(\mathbf{x}, t) = 0$$

Here, the solutions are of the form

$$\psi(\mathbf{x}, t) = \frac{1}{(2\pi)^{3/2}} \int d^3k \, A(\mathbf{k}) e^{i(\mathbf{k}\cdot\mathbf{x} - \omega t)}$$

with $1/c^2 \omega^2 = k^2 + (mc/\hbar)^2$. The solution $\psi_k(\mathbf{x}, t) = e^{i(\mathbf{k}\cdot\mathbf{x} - \omega t)}$ describes a particle of momentum $\mathbf{p} = \hbar\mathbf{k}$ and energy $E = \hbar\omega$. Expressed in terms of these quantities the dispersion relation reads

$E^2 = p^2c^2 + m^2c^4$, this is simply the relativistic relation between energy and momentum. For such a wave,

$$v_p = \frac{\omega}{k} = \frac{E}{p} = \sqrt{1 + \frac{m^2c^2}{p^2}} \, c > c \qquad \text{and} \qquad v_g = \frac{d\omega}{dk} = \frac{k}{\omega}c^2 = \frac{p}{E}c^2$$

is the velocity of the particle.

Example 19.2: Causality and dispersion relations

Dispersion relations are relations between physical quantities which hold, in general, for every linear, time-independent, causal physical system. Therefore, we shall formulate them in a somewhat more abstract manner for a *cause $U(t)$* and an *effect $E(t)$*, and later on, explain them for some examples of interest to us. The quantities $U(t)$ and $E(t)$ are assumed to be *real physical quantities* which are connected by some physical process, not necessarily an electromagnetic one. But this *process of connection has to be linear, time-independent, and causal*. Examples for such $U(t)$ and $E(t)$ are the electric field $\mathbf{E}(t)$ or the polarization $\mathbf{P}(t)$, respectively, at a certain point in a piece of matter.

We start our general considerations by noting the Fourier transforms

$$U(t) = \frac{1}{\sqrt{2\pi}} \int_{-\infty}^{\infty} d\omega \, e^{-i\omega t} u(\omega)$$

$$E(t) = \frac{1}{\sqrt{2\pi}} \int_{-\infty}^{\infty} d\omega \, e^{-i\omega t} e(\omega) \tag{19.23}$$

Since U(t) and E(t) are real, one may derive conditions for $u(\omega)$ and $e(\omega)$. From

$$U(t) = U^*(t) \tag{19.24}$$

we get

$$\int_{-\infty}^{\infty} d\omega \, e^{-i\omega t} u(\omega) = \int_{-\infty}^{\infty} d\omega \, e^{+i\omega t} u^*(\omega) = \int_{-\infty}^{\infty} d\omega' \, e^{-i\omega' t} u^*(-\omega')$$

where we have transformed the variable, $\omega' = -\omega$. Because it is of no significance how the integration variable is denoted, in the last integral we may write ω instead of ω', and we obtain

$$u(\omega) = u^*(-\omega) \qquad \text{and} \qquad e(\omega) = e^*(-\omega) \tag{19.25}$$

These relations are called *crossing relations*. With this result the integrals (19.23) can be taken over positive frequencies only. This can be seen in the following way:

$$U(t) = \frac{1}{\sqrt{2\pi}} \int_{-\infty}^{0} d\omega \, e^{-i\omega t} u(\omega) + \frac{1}{\sqrt{2\pi}} \int_{0}^{\infty} d\omega \, e^{-i\omega t} u(\omega) \tag{19.26}$$

$$= \frac{1}{\sqrt{2\pi}} \int_{0}^{\infty} d\omega' \, e^{+i\omega' t} u(-\omega') + \frac{1}{\sqrt{2\pi}} \int_{0}^{\infty} d\omega \, e^{-i\omega t} u(\omega)$$

$$= \frac{1}{\sqrt{2\pi}} \int_{0}^{\infty} d\omega \left[u(\omega)e^{-i\omega t} + u^*(\omega)e^{+i\omega t} \right]$$

and similarly for $E(t)$. If we write $u(\omega)$ in the form

$$u(\omega) = \frac{1}{2} r(\omega) e^{i\theta(\omega)} \tag{19.27}$$

equation (19.26) becomes

$$U(t) = \frac{1}{\sqrt{2\pi}} \int_0^\infty d\omega \, \frac{r(\omega)}{2} \left[e^{-i(\omega t - \theta(\omega))} + e^{+i(\omega t - \theta(\omega))} \right] \tag{19.28}$$

$$= \frac{1}{\sqrt{2\pi}} \int_0^\infty d\omega \, r(\omega) \cos[\omega t - \theta(\omega)]$$

Consequently, the negative frequencies are superfluous, and the *real function U(t) contains exclusively positive frequencies*.

Next, we use the assumption that the relation between the cause $U(t)$ and the effect $E(t)$ is linear. This implies the following: If the cause U_1 gives rise to the effect E_1, and U_2 gives rise to the effect E_2 then $\alpha U_1 + \beta U_2$ causes the effect $\alpha E_1 + \beta E_2$. This property of linearity may be summarized in the following form

$$E(t) = \int_{-\infty}^\infty dt' \, G(t, t')U(t') \tag{19.29}$$

where $G(t, t')$ is a weight function that weights the distinct causes at time t' contributing to the effect at time t. Hence, studying the physical system with respect to cause and effect is equivalent to the investigation of the Green function $G(t, t')$.

The next property of the physical process to be taken into account is that this physical process, connecting cause and effect, should itself be independent of the time at which it proceeds. This implies that $G(t, t')$ can be a function of the time-difference $t - t'$ only; then,

$$E(t) = \int_{-\infty}^\infty dt' \, G(t' - t)U(t') \tag{19.30}$$

Let us make this clear: for example, if a switch in a current circuit is closed then the currents flowing in this circuit (or the fields or the voltages) depend only on the time after closing the switch, supposing that all elements in the circuit are time independent. If, e.g., the circuit contains time-dependent resistors, the currents flowing would depend on the resistance present.

Now, it is suitable to introduce the Fourier transform of $G(t)$

$$G(t) = \frac{1}{2\pi} \int_{-\infty}^\infty d\omega \, e^{-i\omega t} g(\omega) \tag{19.31}$$

The choice of the prefactor $1/(2\pi)$ instead of $1/(\sqrt{2\pi})$, will turn out to be useful when we derive the relation between $e(\omega)$, $u(\omega)$, and $g(\omega)$:

$$e(\omega) = \frac{1}{\sqrt{2\pi}} \int_{-\infty}^\infty dt \, e^{i\omega t} E(t) \tag{19.32}$$

$$= \frac{1}{\sqrt{2\pi}} \int_{-\infty}^\infty dt \, e^{i\omega t} \int_{-\infty}^\infty dt' \, G(t - t')U(t')$$

$$= \frac{1}{\sqrt{2\pi}^3} \int_{-\infty}^\infty dt \, e^{i\omega t} \int_{-\infty}^\infty dt' \, U(t') \int_{-\infty}^\infty d\omega' \, e^{-i\omega'(t-t')} g(\omega')$$

$$= \frac{1}{\sqrt{2\pi}} \int_{-\infty}^\infty dt' \, U(t') \int_{-\infty}^\infty d\omega' \, g(\omega')e^{i\omega't'} \frac{1}{2\pi} \int_{-\infty}^\infty dt \, e^{i(\omega-\omega')t}$$

$$= \frac{1}{\sqrt{2\pi}} \int_{-\infty}^\infty dt' \, U(t')g(\omega)e^{i\omega t'} = g(\omega)u(\omega)$$

Mathematically, this result is called the *convolution theorem*. Since

$$g(\omega) = \frac{e(\omega)}{u(\omega)}$$

together with (19.25) also $g(\omega)$ obeys the crossing relation:

$$g(\omega) = g^*(-\omega) \tag{19.33}$$

This is clear because $G(t)$ must be real. The properties used up till now are sufficient to justify that for a unifrequent cause U also the effect E is unifrequent. For this proof, in the intermediate calculation we use the complex notation, allowed due to the linearity of the relation (19.30). Thus, we write

$$U(t) = U_0 e^{-i\omega t} \tag{19.34}$$

and agree to use only the real part of this and the following relations. Then, according to (19.30),

$$
\begin{aligned}
E(t) &= \int_{-\infty}^{\infty} dt' \, G(t'-t) U_0 e^{-i\omega t'} \\
&= U_0 e^{-i\omega t} \int_{-\infty}^{\infty} dt' \, G(t'-t) e^{i\omega(t-t')} \\
&= U_0 e^{-i\omega t} g(\omega) \equiv E_0 e^{-i\omega t}
\end{aligned}
\tag{19.35}
$$

Obviously, $g(\omega) = E_0/U_0$. Really, the Green function may be probed by unifrequent causes, that is, one can perform a Fourier analysis. Often, one writes simply

$$E = gU \tag{19.36}$$

although this is satisfied only for the stationary operation, that is, in the unifrequent case. As an example this situation is explained within the classical model for a dielectric for which the relation (compare equation (5.19))

$$\mathbf{P} = \chi_e \mathbf{E} \tag{19.37}$$

is valid, and at a definite frequency, the dielectric susceptibility is given by

$$\chi_e(\omega) = \frac{Ne^2}{m} \frac{1}{\omega_a^2 - \omega^2 - i\gamma\omega} \tag{19.38}$$

(see equation (16.35)). Here, $\chi_e(\omega)$ is the function $g(\omega)$ discussed above in general.

Next, we further take into account the causality between cause $U(t)$ and effect $E(t)$. Precisely, we mean that $E(t)$ can depend only on causes $U(t')$ which happened at earlier times, that is, for $t' < t$. Therefore, one has to require

$$G(t - t') = 0 \qquad \text{for} \qquad t - t' < 0 \tag{19.39}$$

Again, some examples will support one's understanding: The current $\mathbf{I}$ is caused only by an electric field $\mathbf{E}$; similarly, the magnetization $\mathbf{M}$ of a material is generated by the magnetic induction $\mathbf{B}$.

Another example is the current $\mathbf{I}$ in a circuit caused by a generator of voltage V (compare Figure 19.12). The differential equation for the current is well known. It reads

$$L\frac{dI}{dt} + RI + \frac{Q}{C} = V \tag{19.40}$$

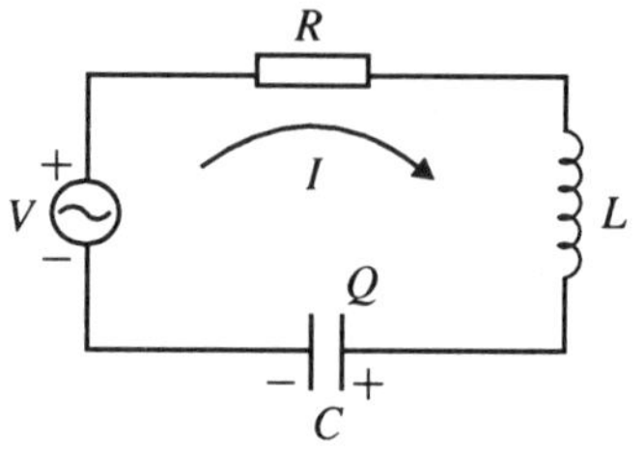

Figure 19.12. Linear circuit with resistor R, inductor L, and capacitor C.

Since $I = dQ/dt$, $I(\omega) = -i\omega Q(\omega)$ and one finds

$$g(\omega) = \frac{e(\omega)}{u(\omega)} = \frac{I(\omega)}{V(\omega)} = \frac{1}{Z}$$

(19.41)

where

$$Z(\omega) = -i\omega L + R + \frac{i}{\omega C}$$

(19.42)

is the complex resistance or the impedance. In this case the *transfer function* $g(\omega)$ is $Y(\omega) = 1/Z(\omega)$. It is also called the admittance of the circuit.

A further example is the scattering of an electromagnetic wave arising from the reflection of an incident wave from a conducting object. The incident wave is the cause, and the scattered wave is the effect.

The fact that the processes are proceeding in a causal manner leads to a constraint on the transfer function $g(\omega)$. According to (19.31),

$$g(\omega) = \int_{-\infty}^{\infty} dt\, e^{i\omega t} G(t) = \int_{0}^{\infty} dt\, e^{i\omega t} G(t)$$

(19.43)

The condition of causality (19.39) has been employed in the last step. This leads to additional information on the function $g(\omega)$. The same equation may be used to extend the definition of $g(\omega)$ to *complex frequencies* $\omega = \omega_1 + i\omega_2$. Then one can also expect that $g(\omega)$ possesses no singularities in the upper complex ω-half-plane if $g(\omega)$ behaves reasonably along the real ω-axis. To better understand the sense of the extension of the definition of $g(\omega)$ to complex ω-values in the upper half-plane we compare

$$g(\omega_1) = \int_{0}^{\infty} dt\, e^{i\omega_1 t} G(t)$$

(19.44)

$$g(\omega_1 + i\omega_2) = \int_{0}^{\infty} dt\, e^{i\omega_1 t} G(t) e^{-\omega_2 t}$$

For complex ω the integrand is smaller by the factor $e^{-\omega_2 t}$. If ω lies in the upper half-plane, then $\omega_2 > 0$ and therefore always $0 \leq e^{-\omega_2 t} \leq 1$. For the derivation of $g(\omega)$ we get

$$\frac{dg}{d\omega}\bigg|_{\omega = \omega_1 + i\omega_2} = \int_{0}^{\infty} dt\, e^{i\omega_1 t} G(t) i t e^{-\omega_2 t}$$

(19.45)

In general, this derivative exists in the upper half-plane, supposing $g(\omega)$ is finite on the real axis: although the integrand, compared to (19.44), is multiplied by (it) the factor $e^{-\omega_2 t}$ will suppress any power of t for long times. Now, if the derivative of a complex function exists in the neighborhood of a point in the complex plane, then the function is analytic at this point (see Chapter 4). Due to these considerations we may conclude that the derivative of $g(\omega)$ exists everywhere in the upper half-plane, and therefore, the function $g(\omega)$ is analytic in the entire upper half-plane. This implies that for a linear, time-independent, and causal system the Fourier transform of the Green function $G(t)$ (transfer function) $g(\omega)$ originally defined for real ω equals the limiting value of a function g of complex argument ω which is analytic in the entire upper half-plane.

As an example we consider again the dielectric susceptibility already noted in (19.38):

$$\chi_e = \frac{Ne^2}{m} \frac{1}{\omega_n^2 - \omega^2 - i\gamma\omega} \tag{19.46}$$

whose damping constant is $\gamma > 0$. This function is analytic except at the poles (the zeros of the denominator). The poles result from the equation

$$\omega^2 + i\gamma\omega - \omega_n^2 = 0 \qquad \text{as} \qquad \omega = -i\frac{\gamma}{2} \pm \sqrt{\omega_n^2 - \left(\frac{\gamma}{2}\right)^2} \tag{19.47}$$

The positions of these singular points are illustrated in Figure 19.13. They lie in the lower half-plane. The fact that $g(\omega)$ is analytic in the upper half-plane allows us to derive a relation between the real and imaginary parts of this function. This relation is called a dispersion relation.

Our present knowledge of the Fourier transform of $g(\omega)$ of the Green function is of quite general nature, independent of the details of the theory which connects cause and effect. If a differential equation describing the processes mentioned is available, a Green function with the properties described above can always be found with certainty. However, certain peculiarities of the system may be formulated directly via $g(\omega)$ without making use of the differential equation. The latter one is a luxury which is not absolutely necessary if the Green function is known.

The dispersion relations are derived from the line integral

$$\int d\omega' \frac{g(\omega')}{\omega' - \omega} \tag{19.48}$$

taken along the real axis (possibly shifted into the upper half-plane) and a semicircle of radius R lying in the upper half-plane (Figure 19.14).

According to our detailed discussion, $g(\omega)$ has no singularities in the interior and on the boundary of this closed curve. Therefore, with ω on the real axis, $g(\omega')/(\omega' - \omega)$ is also analytic in the domain mentioned; hence, the integral (19.48) vanishes. Now, it is further *assumed* that $g(\omega)$ behaves such (that is, that the physical system considered behaves in such a manner) that the part of the integral (19.48) along the upper semicircle vanishes as $R \to \infty$. We note that this is not always the

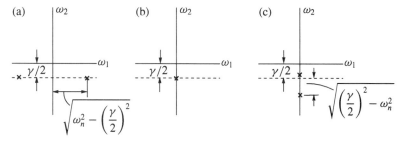

Figure 19.13. Position of the poles for the model of susceptibility: (a) Undercritical damping, $\gamma^2 < 4\omega_n^2$. (b) Critical damping, $\gamma^2 = 4\omega_n^2$. (c) Overcritical damping, $\gamma^2 > 4\omega_n^2$.

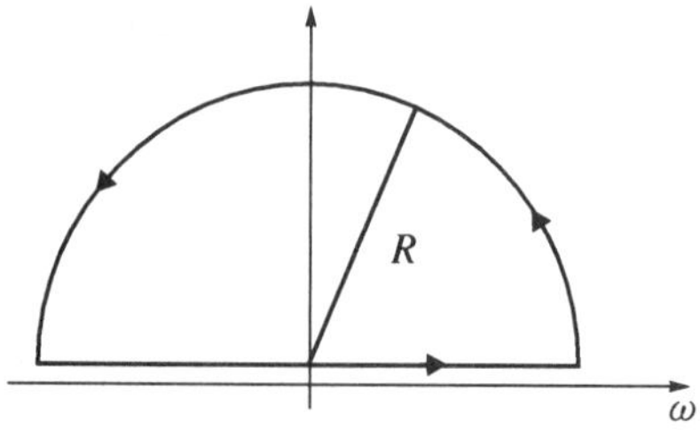

Figure 19.14. Illustration of the integration path used in equation (19.45).

case. But, soon we will become familiar with a so-called *subtraction method* which allows us to make the integral over the semicircle vanish.

If $g(\omega)$ is the dielectric susceptibility $\chi_e(\omega)$, all of these conditions are fulfilled: obviously, $\chi_e(\omega)$ has no singularity on the real axis, and for large ω it tends to

$$\lim_{\omega \to \infty} \chi_e(\omega) = -\frac{Ne^2}{m\omega^2}$$

so that the integral over the semicircle is of the order $2\pi R / R^3 \overset{R \to \infty}{\longrightarrow} 0$.

In general, we may write

$$0 = \int_\circ d\omega' \, \frac{g(\omega')}{\omega' - \omega} = \int_{-\infty}^\infty d\omega' \, \frac{g(\omega')}{\omega' - \omega} + \int_\cap d\omega' \, \frac{g(\omega')}{\omega' - \omega} \tag{19.49}$$

$$= \int_{-\infty}^\infty d\omega' \, \frac{g(\omega')}{\omega' - \omega} = P \int_{-\infty}^\infty d\omega' \, \frac{g(\omega')}{\omega' - \omega} + \int_\curvearrowright d\omega' \, \frac{g(\omega')}{\omega' - \omega}$$

where P means the principal value and $\int_\curvearrowright d\omega' \, g(\omega')/(\omega' - \omega)$ is the integral taken along an infinitesimally small semicircle about ω in the sense of revolution marked by the arrow. The latter integral may be calculated easily by substituting $\omega' - \omega = \rho e^{i\varphi}$, $d\omega' = i\rho e^{i\varphi} d\varphi$:

$$\int_\curvearrowright d\omega' \, \frac{g(\omega')}{\omega' - \omega} = \lim_{\rho \to 0} i g(\omega) \int_{-\pi}^0 \rho \frac{e^{i\varphi} \, d\varphi}{\rho e^{i\varphi}} = i g(\omega) \int_{-\pi}^0 d\varphi = -i\pi g(\omega) \tag{19.50}$$

From (19.49) we can conclude that

$$g(\omega) = \frac{1}{i\pi} P \int_{-\infty}^\infty d\omega' \, \frac{g(\omega')}{\omega' - \omega} \tag{19.51}$$

This is the expression for $g(\omega)$, which itself contains $g(\omega)$ again. But it is striking that on the right-hand side the factor i appears, so that the real part and the imaginary part of $g(\omega)$ are connected in a remarkable manner. By equating the real part or the imaginary part of (19.51), we obtain the relations

$$\text{Re}\,[g(\omega)] = \frac{1}{\pi} P \int_{-\infty}^\infty d\omega' \, \frac{\text{Im}\,[g(\omega')]}{\omega' - \omega} \tag{19.52}$$

$$\text{Im}\,[g(\omega)] = -\frac{1}{\pi} P \int_{-\infty}^\infty d\omega' \, \frac{\text{Re}\,[g(\omega')]}{\omega' - \omega}$$

These are the so-called *dispersion relations*. Their characteristic is to connect the real part and the imaginary part of $g(\omega)$. One may cast the dispersion relations (19.52) in a form more appropriate for comparison with experiment by writing them in such a way that only positive frequencies occur. So, e.g.,

$$\text{Re}\,[g(\omega)] = \frac{1}{\pi} P \int_{-\infty}^{0} d\omega' \frac{\text{Im}\,[g(\omega')]}{\omega' - \omega} + \frac{1}{\pi} P \int_{0}^{\infty} d\omega' \frac{\text{Im}\,[g(\omega')]}{\omega' - \omega}$$

$$= \frac{1}{\pi} P \int_{0}^{\infty} d\omega' \frac{\text{Im}\,[g(-\omega')]}{-\omega' - \omega} + \frac{1}{\pi} P \int_{0}^{\infty} d\omega' \frac{\text{Im}\,[g(\omega')]}{\omega' - \omega} \tag{19.53}$$

Now, for $g(\omega)$ a crossing relation of the kind (19.25) is valid; thus,

$$g(\omega) = g^*(-\omega) \tag{19.54}$$

and therefore,

$$\text{Re}\,[g(\omega)] = \text{Re}\,[g(-\omega)], \qquad \text{Im}\,[g(\omega)] = -\text{Im}\,[g(-\omega)] \tag{19.55}$$

So, (19.53) becomes

$$\text{Re}\,[g(\omega)] = \frac{1}{\pi} P \int_{0}^{\infty} d\omega' \frac{\text{Im}\,[g(\omega')]}{\omega' + \omega} + \frac{1}{\pi} P \int_{0}^{\infty} d\omega' \frac{\text{Im}\,[g(\omega')]}{\omega' - \omega}$$

$$= \frac{2}{\pi} P \int_{0}^{\infty} d\omega' \, \omega' \frac{\text{Im}\,[g(\omega')]}{\omega'^2 - \omega^2} \tag{19.56}$$

Similar transformations of the second equation in (19.52) yield

$$\text{Im}\,[g(\omega)] = -\frac{1}{\pi} P \int_{-\infty}^{0} d\omega' \frac{\text{Re}\,[g(\omega')]}{\omega' - \omega} - \frac{1}{\pi} P \int_{0}^{\infty} d\omega' \frac{\text{Re}\,[g(\omega')]}{\omega' - \omega}$$

$$= \frac{1}{\pi} P \int_{0}^{\infty} d\omega' \frac{\text{Re}\,[g(-\omega')]}{\omega' + \omega} - \frac{1}{\pi} P \int_{0}^{\infty} d\omega' \frac{\text{Re}\,[g(\omega')]}{\omega' - \omega}$$

$$= -\frac{2\omega}{\pi} P \int_{0}^{\infty} d\omega' \frac{\text{Re}\,[g(\omega')]}{\omega'^2 - \omega^2} \tag{19.57}$$

The dispersion relations were introduced by Kramers and Kronig for the permittivity. We may verify their result if we identify the dielectric susceptibility $\chi_e(\omega)$ with the function $g(\omega)$ and write

$$\chi_e(\omega) = \frac{\epsilon - 1}{4\pi} = \frac{\epsilon' - 1}{4\pi} + i\frac{\epsilon''}{4\pi} \tag{19.58}$$

where $\epsilon = \epsilon' + i\epsilon''$. Then equations (19.56) and (19.57) read

$$\epsilon'(\omega) - 1 = \frac{2}{\pi} P \int_{0}^{\infty} dx \, \frac{x\epsilon''(x)}{x^2 - \omega^2} \tag{19.59}$$

$$\epsilon''(\omega) = -\frac{2}{\pi} \omega P \int_{0}^{\infty} dx \, \frac{\epsilon'(x) - 1}{x^2 - \omega^2}$$

These are the famous *Kramers-Kronig dispersion relations*.

A medium is called *dispersive* if the phase velocity v_p depends on the frequency. The Kramers-Kromig relations imply that a medium is dispersive only if it is also absorptive. If absorption does not occur, that is, if the imaginary part of the permittivity vanishes for all frequencies ($\epsilon''(\omega) = 0$), then

$$\epsilon'(\omega) = 1, \qquad v_p = \frac{\omega}{k} = \frac{c}{\sqrt{\epsilon}} = c \tag{19.60}$$

Really, there is no dispersion. On the other hand, if dispersion occurs in a certain frequency region, there is also absorption in a (generally distinct) frequency region. This is the statement of the second equation in (19.59). Sometimes, there is absorption within a relatively small frequency band in the vicinity of $\omega = \omega_n$, thus, $\epsilon''(x) = K'\delta(x - x_n)$. Then, for ω different from ω_n, from the first equation in (19.59):

$$\epsilon'(\omega) - 1 \approx \frac{K}{\omega_n^2 - \omega^2} \tag{19.61}$$

Here, $K = (2/\pi)K'\omega_n$ is a constant independent of ω. The shape of the function $\epsilon'(\omega) - 1$ is represented in Figure 19.15.

If absorption is present within a broader but finite frequency region, then for high frequencies ω (more precise, for $\omega \gg \Omega$, where Ω characterizes the highest frequency of the absorption band) the first equation in (19.59) may be transformed into the following suitable approximation:

$$\epsilon'(\omega) - 1 \approx \frac{2}{\pi} \int_0^\infty dx \, \frac{x\epsilon''(x)}{-\omega^2} \tag{19.62}$$

Now, the width γ of the resonance curve is usually small compared to the resonance frequency: $\gamma < \omega_n$. In this case, (19.62) may be simplified further

$$\epsilon'(\omega) - 1 \approx -\frac{2}{\pi} \frac{\omega_n}{\omega^2} \int_0^\infty dx \, \epsilon''(x) \tag{19.63}$$

$$\int_0^\infty dx \, \epsilon''(x) \approx -\frac{\pi}{2\omega_n} \lim_{\omega \to \infty} \left[\omega^2(\epsilon'(\omega) - 1)\right]$$

This is a relation between the area under the curve of the imaginary part $\epsilon''(\omega)$ of the permittivity and the high-frequency limit of the real part of the permittivity. Such relations are called *sum rules*. For example, in the model (19.46) for the dielectric

$$\epsilon' - 1 = \frac{4\pi Ne^2}{m} \frac{\omega_n^2 - \omega^2}{(\omega_n^2 - \omega^2)^2 + \gamma^2\omega^2} \xrightarrow{\omega \to \infty} -\frac{4\pi Ne^2}{m} \frac{1}{\omega^2} \tag{19.64}$$

Hence, according to (19.63),

$$\int_0^\infty dx \, \epsilon''(x) \approx \frac{\pi}{2\omega_n} \frac{4\pi Ne^2}{m} \tag{19.65}$$

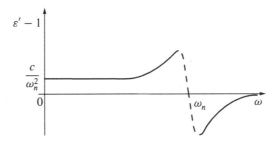

Figure 19.15. Typical behavior of the real part of the permittivity for resonance absorption.

It is interesting that the area under the absorption curve is obviously proportional to the number N, the number of the molecules per unit volume taking part actively in the absorption. Of course, this result may be derived also directly by integrating the imaginary part following from (19.46).

We further discuss the following case: let the absorption be restricted again to a finite frequency band; we want to know the low-frequency behavior of the real part of the permittivity. From (19.59) we obtain

$$\epsilon'(\omega) - 1 \approx \frac{2}{\pi} \int_0^\infty dx \, \frac{x\epsilon''(x)}{x^2} \approx \frac{2}{\pi} \frac{1}{\omega_n} \int_0^\infty dx \, \epsilon''(x) \tag{19.66}$$

This links the area under the curve of the imaginary permittivity to the low-frequency behavior of the real permittivity:

$$\int_0^\infty dx \, \epsilon''(x) \approx \frac{\pi \omega_n}{2} \lim_{\omega \to 0} \left[\epsilon'(\omega) - 1 \right] \tag{19.67}$$

Thus, it is possible to conclude the resonance frequency ω_n from the high-frequency and low-frequency behavior of the permittivity. The comparison of (19.63) and (19.67) yields

$$-\frac{\pi}{2\omega_n} \lim_{\omega \to \infty} \left[\omega^2(\epsilon'(\omega) - 1) \right] \approx \frac{\pi \omega_n}{2} \lim_{\omega \to 0} \left[\epsilon'(\omega) - 1 \right]$$

and therefore

$$\omega_n^2 \approx \frac{\lim_{\omega \to \infty} \left[\omega^2(\epsilon'(\omega) - 1) \right]}{\lim_{\omega \to 0} \left[\epsilon'(\omega) - 1 \right]} \tag{19.68}$$

In fact, this method has been used practically. Namely, the permittivity may be measured in the frequency region of microwaves and below, and on the other hand, also in the region of optical frequencies. In the intermediate range measurements of this kind hardly can be performed.

Frequently, it happens that the molecular absorption lines lie far in the infrared region. If in such a case there is only one resonance frequency, then it may be estimated with the help of the relation (19.68) although it is not directly measurable itself.

The dispersion relation and all relations and properties which may be derived from it may be obtained from the equation

$$P \int_{-\infty}^\infty d\omega' \, \frac{g(\omega')}{\omega' - \omega} = i\pi g(\omega) \tag{19.69}$$

This equation is valid for a linear, time-independent, and causal system for which additionally

$$\lim_{R \to \infty} \int_\cap d\omega' \, \frac{g(\omega')}{\omega' - \omega} = 0 \tag{19.70}$$

is valid. Now it may happen that the last relation (19.70) is not fulfilled. If, e.g., $g(\omega)$ tends to a constant (g_0) for $|\omega| \to \infty$, then the line integral over the infinitely distant semicircle yields

$$\lim_{R \to \infty} \int_\cap d\omega' \, \frac{g(\omega')}{\omega' - \omega} = g_0 \lim_{R \to \infty} \int_\cap \frac{d\omega'}{\omega'} = i\pi g_0 \tag{19.71}$$

Thus, it is different from zero and has to be taken into account in the considerations. Then we may help ourselves by developing dispersion relations for the function $g(\omega) - g_0$ according to the framework shown. The function $G(\omega) = g(\omega) - g_0$ has the same properties as $g(\omega)$ and vanishes as $|\omega| \to \infty$:

$\lim_{|\omega|\to\infty} G(\omega) = 0$. This strategy leads to the so-called *subtracted dispersion relations*. Sometimes, more than one subtraction is necessary. For example, if

$$g(\omega) \xrightarrow{|\omega|\to\infty} g_1\omega + g_0$$

dispersion theory may be applied to the function

$$g(\omega) - g_1\omega - g_0$$

In this case, one deals with two subtractions, namely, $-g_1\omega$ and $-g_0$. For each subtraction to be done a new constant is introduced into the considerations. The dispersion relations of Kramers and Kronig are of single-subtracted type because they are formulated for $\epsilon - 1$.

When one intends to make use of the dispersion relations in a new, developing branch of research, then the process of subtraction needs much effort because, at first, one does not know the number of subtractions to be done. One has to find out experimentally step by step how many subtractions are necessary in the particular complex of problems.

Example 19.3: Exercise: On the optical theorem

Let a material consisting of N scattering and absorption centers per unit volume be characterized by the complex index of refraction n. At large distances from a single scattering center, the electric field in the direction of the initial polarization is given by

$$\mathbf{E} = \mathbf{E}_0 e^{-i\omega t}\left[e^{ikz} + f(\omega, \theta)\frac{e^{ikr}}{r}\right]$$

The first term is the incident wave, and the second term describes the scattered component; $f(\omega, \theta)$ is the so-called *scattering amplitude*. For forward scattering and backward scattering only $f(\omega, 0)$ and $f(\omega, \pi)$ respectively are different from zero.

(a) Show that the index of refraction is connected with the scattering amplitude by

$$n = 1 + \frac{2\pi c^2}{\omega^2}Nf(\omega, 0)$$

supposing $|n - 1| \ll 1$. For that consider an incident wave e^{ikz} scattered from a thin plate of thickness δh in the (x, y)-plane.

Note: At the point $(0, 0, z)$ behind the plate the wave tends to

$$e^{ikz} + N\delta h\int_0^\infty f\left(\omega, \arctan\frac{l}{z}\right)\frac{e^{ik\sqrt{l^2+z^2}}}{\sqrt{l^2+z^2}}2\pi l\,dl$$

$$= e^{ikz}\left[1 + \frac{2\pi N\delta h}{k}\int_1^\infty f(\omega, \operatorname{arcsec}\mu)kze^{ikz(\mu-1)}\,d\mu\right]$$

For a distance of many wavelengths behind the plate one can find the solution by considering the limit for large kz. For this purpose one uses

$$\lim_{\alpha\to\infty}\alpha e^{i\alpha x} = i\lim_{\epsilon\to 0}[\delta(x - \epsilon) - \delta(x + \epsilon)]$$

and obtains

$$e^{ikz}\left[1 + \frac{2\pi N\delta h c}{\omega}f(\omega, 0)\right]$$

On the other hand, the macroscopic solution of the reflection-transmission problem for a plate is given by $e^{ikz}\left[1 + i\delta h\omega(n^2 - 1)/2c\right]$.

(b) Prove the optical theorem

$$\text{Im } f(\omega, 0) = \frac{\omega}{4\pi c}\sigma_t$$

where σ_t is the total scattering cross section for scattering and absorption. Note that the damping factor for a wave propagating in a medium with the complex index of refraction n is given by $|e^{i\omega nz/c}|^2$, which may be expressed also by $e^{-N\sigma_t z}$.

Solution   At the point z behind the thin plate not only is the unscattered part e^{ikz} arriving but also scattered parts from all the scattering centers, which have to be summed up. The plate is thin enough that it is sufficient to take into account only single scattering. Let a scattering center be at the distance l from the z-axis. The portion of the scattered light appearing under the scattering angle $\theta = \arctan(l/z)$ (Figure 19.16a) reaches the point $(0, 0, z)$ at the distance $r = \sqrt{l^2 + z^2}$. However, with respect to this point all scattering centers within an annulus of thickness δl are equal; there are $2\pi l\, dl\, N\delta h$ scattering centers of this kind (Figure 19.16b). Summing up all of these parts, one obtains

$$e^{ikz} + N\delta h \int_0^\infty 2\pi l\, f\left(\omega, \arctan\frac{l}{z}\right) \frac{e^{ik\sqrt{l^2+z^2}}}{\sqrt{l^2 + z^2}}\, dl$$

We substitute $\sqrt{l^2 + z^2} = z\mu$; that is, $\mu = \sqrt{l^2 + z^2}/z = \sec\theta$ or $\arctan(l/z) = \text{arcsec}\,\mu$ and $d\mu = l\, dl/z\sqrt{l^2 + z^2}$, and obtain

$$\int_0^\infty 2\pi l\, f\left(\omega, \arctan\frac{l}{z}\right) \frac{e^{ik\sqrt{l^2+z^2}}}{\sqrt{l^2 + z^2}}\, dl = \int_1^\infty 2\pi z\, f(\omega, \text{arcsec}\,\mu)e^{ik\mu z}\, d\mu$$

Hence, at the point $(0, 0, z)$ behind the plate the electric field in the direction of the initial polarization is given by

$$\mathbf{E} = \mathbf{E}_0\, e^{i(kz-\omega t)}\left[1 + \int_1^\infty \frac{2\pi N\delta h}{k} f(\omega, \text{arcsec}\,\mu)kze^{ikz(\mu-1)}\, d\mu\right]$$

The wavelength of the light is $\lambda = 2\pi/k$; thus, for $z \gg \lambda$, $z/\lambda = zk/2\pi \gg 1$. So, within a macroscopic distance from the plate only the limit $kz \to \infty$ has to be taken into account. Now,

$$\lim_{\alpha\to\infty} \alpha e^{i\alpha x} = i \lim_{\epsilon\to 0}\left[\delta(x - \epsilon) - \delta(x + \epsilon)\right]$$

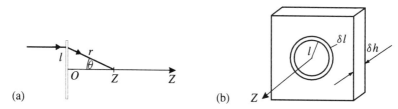

Figure 19.16. Plate with scattering centers: (a) In a cut. (b) In perspective.

(for the proof see below); that is,

$$\lim_{kz\to\infty}\int_1^\infty f(\omega,\arcsec\mu)kze^{ikz(\mu-1)}\,d\mu$$

$$= i\int_1^\infty f(\omega,\arcsec\mu)\left[\delta(\mu-1-0)-\delta(\mu-1+0)\right]d\mu$$

$$= if(\omega,\arcsec 1) = if(\omega,0)$$

Then, with $k=\omega/c$ one obtains

$$\mathbf{E} = \mathbf{E}_0 e^{i(kz-\omega t)}\left[1+\frac{i2\pi cN\delta h}{\omega}f(\omega,0)\right]$$

Only the forward scattering amplitude comes into play. From Chapter 16, it turns out that after passing the plate the amplitude of the wave is damped by the factor $e^{-\omega\mu\delta h/c}$, where μ is the imaginary part of the index of refraction. But, due to $|n-1|\ll 1$ with $n=\nu+i\mu$ we get

$$n^2 = \nu^2 - \mu^2 + 2i\nu\mu \approx 1+2i\mu \qquad \Rightarrow \qquad \mu = -i\frac{n^2-1}{2}$$

On the other hand, δh is very small, that is,

$$e^{-\omega\mu\delta h/c} \approx 1 - \frac{\delta h}{c}\omega\mu \approx 1+\frac{i\delta h\omega}{2c}(n^2-1)$$

Hence, on a macroscopic scale

$$\mathbf{E} = \mathbf{E}_0 e^{i(kz-\omega t)}\left[1+i\frac{\delta h\omega}{2c}(n^2-1)\right]$$

A comparison of both expressions yields

$$\frac{2\pi iNc\delta h}{\omega}f(\omega,0) = i\frac{\omega\delta h}{2c}(n^2-1)$$

$$\Rightarrow \quad n = \left[1+\frac{4\pi Nc^2}{\omega^2}f(\omega,0)\right]^{1/2} \approx 1+\frac{2\pi Nc^2}{\omega^2}f(\omega,0)$$

The total scattering cross section σ_t of a scattering center is the fictious area the incident wave has to pass perpendicularly to be deflected or absorbed at all. Let the incident intensity be I_0. In a part of the plate of thickness Δz the total area able to scatter or to absorb is given by $FN\sigma_t\Delta z$. After passing the segment between z and $z+\Delta z$ only $I(z)(1-FN\sigma_t\Delta z/F)$ of the incident intensity is left. See Figure 19.17. Hence,

$$I(z+\Delta z) = I(z)(1-N\sigma_t\Delta z) \quad \overset{\Delta z\to 0}{\Rightarrow} \quad \frac{dI}{dz}(z) = -N\sigma_t I(z)$$

$$\Rightarrow \quad I(z) = I_0 e^{-N\sigma_t z}$$

On the other hand, the macroscopic damping factor is $I(z) = I_0 e^{-2\omega\mu z/c}$. We have

$$-N\sigma_t z = -\frac{2\omega z}{c}\operatorname{Im}n \quad \Rightarrow \quad \sigma_t = \frac{2\omega}{Nc}\frac{2\pi Nc^2}{\omega^2}\operatorname{Im}f(\omega,0)$$

$$\Rightarrow \quad \operatorname{Im}f(\omega,0) = \frac{\omega}{4\pi c}\sigma_t$$

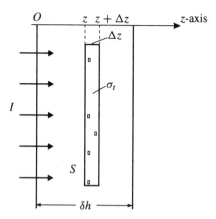

Figure 19.17. On the calculation of the total scattering cross section.

This is the so-called *optical theorem*.

Now, we further prove the relation used above, which is summarized in the following statement:

Proposition: If $\lim_{\alpha \to \infty} \alpha(e^{i\alpha x} - e^{-i\alpha x})$ exists, then

$$\lim_{\alpha \to \infty} \alpha(e^{i\alpha x} - e^{-i\alpha x}) = 2i \lim_{\epsilon \to 0} (\delta(x - \epsilon) - \delta(x + \epsilon))$$

Justification: Of course, we have to apply both sides to a test function $f(x)$. If we define the Fourier transform by

$$f(x) =: \frac{1}{\sqrt{2\pi}} \int_{-\infty}^{\infty} dk\, e^{-ikx} F(k)$$

$$F(k) = \frac{1}{\sqrt{2\pi}} \int_{-\infty}^{\infty} dx\, e^{ikx} f(x)$$

the proposition reads

$$\sqrt{2\pi} \lim_{\alpha \to \infty} \alpha\, (F(\alpha) - F(-\alpha)) = 2i \lim_{\epsilon \to 0} (f(\epsilon) - f(-\epsilon))$$

where the limit on the left-hand side should exist. Now, if

$$F(k) - F(-k) =: \frac{2\Delta}{\sqrt{2\pi}\, k} + O\left(\frac{1}{k^{1+\delta}}\right), \qquad (\delta > 0, k \to \infty)$$

then

$$f(\epsilon) - f(-\epsilon) = \frac{1}{\sqrt{2\pi}} \int_{-\infty}^{\infty} dk\, (e^{-ik\epsilon} - e^{ik\epsilon}) F(k)$$

$$= \frac{-2i}{\sqrt{2\pi}} \int_{-\infty}^{\infty} dk\, \sin(k\epsilon) F(k)$$

$$= \frac{-2i}{\sqrt{2\pi}} \int_{0}^{\infty} dk\, \sin(k\epsilon)(F(k) - F(-k))$$

We choose a $K(\epsilon)$ in such a way that $K(\epsilon) \to \infty$, $\epsilon K(\epsilon) \to 0$ as $\epsilon \to 0$:

$$f(\epsilon) - f(-\epsilon) = \frac{-2i}{\sqrt{2\pi}} \left\{ \int_0^K dk \, \sin(k\epsilon)(F(k) - F(-k)) \right.$$
$$\left. + \int_K^\infty dk \, \sin(k\epsilon)(F(k) - F(-k)) \right\}$$

Assuming that for all k, there is a constant C with

$$|F(k) - F(-k)| < \frac{C}{k}$$

the first integral is

$$\left| \int_0^K dk \ldots \right| < \int_0^K dk \, k\epsilon \frac{C}{k} = C\epsilon K$$

According to the definition we write for the second integral

$$\int_K^\infty dk \ldots = \int_K^\infty dk \, \sin(k\epsilon) \left(\frac{2\Delta}{\sqrt{2\pi}k} + O\left(\frac{1}{k^{1+\delta}}\right) \right)$$
$$= \frac{2\Delta}{\sqrt{2\pi}} \int_{K\epsilon}^\infty dz \, \frac{\sin z}{z} + \int_K^\infty dk \, O\left(\frac{1}{k^{1+\delta}}\right)$$
$$= \frac{2\Delta}{\sqrt{2\pi}} \int_0^\infty dz \, \frac{\sin z}{z} - \frac{2\Delta}{\sqrt{2\pi}} \int_0^{K\epsilon} dz \, \frac{\sin z}{z} + \int_K^\infty dk \, O\left(\frac{1}{k^{1+\delta}}\right)$$
$$= \sqrt{\frac{\pi}{2}} \Delta + O(\epsilon K) + O(K^{-\delta})$$

altogether

$$f(\epsilon) - f(-\epsilon) = -i \, \Delta + O(\epsilon K) + O(K^{-\delta})$$

Now, we perform the limiting procedure and obtain the proposition

$$f(\epsilon) - f(-\epsilon) = -i\Delta = -\frac{i}{2}\sqrt{2\pi} \lim_{k\to\infty} k(F(k) - F(-k))$$

Example 19.4: Exercise: The spin of the electromagnetic field

Show that the total angular momentum of an electromagnetic radiation field in vacuum may be decomposed into an orbital angular momentum depending on the reference frame and an angular momentum (spin) independent of the frame of reference.

Determine the spin of a right-handed circularly polarized or left-handed circularly polarized plane wave.

Note: Express the field intensities in terms of the potentials. Here, choose the three-dimensional transverse gauge $\phi = 0$, div $\mathbf{A} = 0$.

Solution   The total angular momentum of the electric field in empty space is given by

$$\mathbf{J} = \frac{1}{c^2} \int \mathbf{r} \times \mathbf{S} \, dV = \frac{1}{4\pi c} \int \mathbf{r} \times (\mathbf{E} \times \mathbf{B}) \, dV \tag{19.72}$$

The field intensities **E** and **B** are determined from the potentials

$$\mathbf{E} = -\nabla \Phi - \frac{1}{c} \frac{\partial \mathbf{A}}{\partial t}$$

$$\mathbf{B} = \nabla \times \mathbf{A} \tag{19.73}$$

The gauge is choosen such that $\Phi = 0$ and $\nabla \cdot \mathbf{A} = 0$. In this way the field intensities are determined by the vector potential alone. Then, written in components (19.73) reads

$$B_i = \varepsilon_{ijk} \frac{\partial A_k}{\partial x_j}, \qquad E_i = -\frac{1}{c} \frac{\partial A_i}{\partial t} \tag{19.74}$$

By expressing the vector product by means of the completely antisymmetric ε_{ijk} tensor we obtain for (19.72)

$$
\begin{aligned}
J_i &= \frac{1}{4\pi c} \varepsilon_{ijk} \int x_j \, (\mathbf{E} \times \mathbf{B})_k \, dV \\
&= \frac{1}{4\pi c} \varepsilon_{ijk} \varepsilon_{klm} \int x_j E_l \, (\nabla \times \mathbf{A})_m \, dV \\
&= -\frac{1}{4\pi c^2} \varepsilon_{ijk} \varepsilon_{klm} \varepsilon_{mnp} \int x_j \dot{A}_l \frac{\partial}{\partial x_n} A_p \, dV
\end{aligned}
\tag{19.75}
$$

The integrand is rewritten such that the Gaussian theorem may be used

$$x_j \dot{A}_l \frac{\partial}{\partial x_n} A_p = \frac{\partial}{\partial x_n} \left(x_j \dot{A}_l A_p \right) - \frac{\partial x_j}{\partial x_n} \dot{A}_l A_p - x_j \frac{\partial \dot{A}_l}{\partial x_n} A_p \tag{19.76}$$

After applying the generalized Gaussian theorem in the integration the first term on the right-hand side becomes a surface term and vanishes. Hence, (19.75) becomes

$$J_i = \frac{1}{4\pi c^2} \varepsilon_{ijk} \varepsilon_{klm} \varepsilon_{mnp} \int \left(\delta_{jn} \dot{A}_l A_p - x_j \frac{\partial \dot{A}_l}{\partial x_n} A_p \right) dV \tag{19.77}$$

For the contraction of two ε-tensors,

$$\varepsilon_{klm} \varepsilon_{mnp} = \delta_{kn} \delta_{lp} - \delta_{kp} \delta_{ln} \tag{19.78}$$

So, equation (19.77) becomes

$$
\begin{aligned}
J_i &= \frac{1}{4\pi c^2} \varepsilon_{ijk} \int \left[\left(\delta_{kn} \delta_{lp} - \delta_{kp} \delta_{ln} \right) \left(\delta_{jn} \dot{A}_l A_p - x_j \frac{\partial \dot{A}_l}{\partial x_n} A_p \right) \right] dV \\
&= \frac{1}{4\pi c^2} \varepsilon_{ijk} \int \left[\delta_{jk} \dot{A}_p A_p - x_j \frac{\partial \dot{A}_p}{\partial x_k} A_p - \dot{A}_j A_k + x_j \frac{\partial \dot{A}_n}{\partial x_n} A_k \right]
\end{aligned}
\tag{19.79}
$$

Two of the four terms vanish:

$$
\begin{aligned}
&\varepsilon_{ijk} \delta_{jk} = \varepsilon_{ikk} = 0 \\
&\frac{\partial \dot{A}_n}{\partial x_n} = \frac{\partial}{\partial t} \frac{\partial A_n}{\partial x_n} = \frac{\partial}{\partial t} (\nabla \cdot \mathbf{A}) = 0
\end{aligned}
\tag{19.80}
$$

Two terms of equation (19.79) remain:

$$J_i = \frac{1}{4\pi c^2}\varepsilon_{ijk}\int \left(A_j \dot{A}_k\right) dV + \frac{1}{4\pi c^2}\varepsilon_{ijk}\int \left(x_j \frac{\partial \dot{A}_p}{\partial x_k}\right) A_p \, dV \tag{19.81}$$

Obviously the first term is independent of the reference frame. Thus, we may interpret it as an intrinsic angular momentum of the field, the spin. Via x_j, the second term depends on the reference frame. It represents the angular momentum. Introducing the orbital angular momentum operator with which we will become familiar in Quantum Mechanics,

$$\hat{L}_i = \varepsilon_{ijk}x_j \frac{\partial}{\partial x_k} \tag{19.82}$$

this becomes obvious.

The vector potential of a right-handed or left-handed circularly polarized plane wave of frequency ω propagating in z-direction reads

$$\mathbf{A} = A_0 \left(\cos(kz - \omega t)\mathbf{e}_x \pm \sin(kz - \omega t)\mathbf{e}_y\right) \tag{19.83}$$

Then,

$$\dot{\mathbf{A}} = \omega A_0 \left(\sin(kz - \omega t)\mathbf{e}_x \mp \cos(kz - \omega t)\mathbf{e}_y\right) \tag{19.84}$$

and

$$\mathbf{A} \times \dot{\mathbf{A}} = \mp \omega A_0^2 \mathbf{e}_z \tag{19.85}$$

Then, the spin of a right-handed or left-handed circularly polarized wave per unit of volume is simply

$$\mathbf{S} = \mp \frac{1}{4\pi c^2}\omega A_0^2 \mathbf{e}_z \tag{19.86}$$

Thus, there are two spin orientations: in the direction of propagation and opposite to it.

Biographical notes

Hendrik Anthony Kramers, b. Dec. 17, 1894, Rotterdam–d. Apr. 24, 1952, Leiden. Kramers was a professor for theoretical physics at Utrecht university from 1926 to 1932, a professor at Delft technical college in 1931, and from 1934 to 1952 he lectured at Leiden university. He was president of the Atomic Energy Commission of the UN and of the "International Union for Pure and Applied Physics." Kramer's main field of research was the quantum theory of electrons and of radiation. Between 1926 and 1929, Kramer and Ralph Kronig developed the dispersion relation which was named for them.

Ralph Kronig, b. March 10, 1904, Dresden–d. Nov. 16, 1995. From 1931 to 1939, he lectured on mechanics and quantum mechanics at Groningen university, and from 1939 to 1969 he had a chair for theoretical physics at Delft technical college. Kronig intensively dealt with optics, electron optics, and absorption in metals. In the course of this research, he developed the dispersion relation together with H. Kramers. In 1962 he was awarded the Max-Planck-Medal of the Deutsche Physikalische Gesellschaft.

20 Moving Charges in Vacuum

In this chapter we want to treat the phenomena occurring for rapidly moving charges in vacuum. For this aim we start out from Maxwell's equations. Since we restrict ourselves to vacuum, $\mathbf{E} = \mathbf{D}$, $\mathbf{H} = \mathbf{B}$, and for Maxwell's equations we may write:

$$\nabla \cdot \mathbf{E} = 4\pi \rho \tag{20.1}$$

$$\nabla \cdot \mathbf{H} = 0 \tag{20.2}$$

$$\nabla \times \mathbf{E} = -\frac{1}{c}\frac{\partial \mathbf{H}}{\partial t} \tag{20.3}$$

$$\nabla \times \mathbf{H} = \frac{4\pi}{c}\mathbf{j} + \frac{1}{c}\frac{\partial \mathbf{E}}{\partial t} \tag{20.4}$$

where $\rho(\mathbf{r}, t)$ and $\mathbf{j}(\mathbf{r}, t)$ are general, time-dependent distributions of charge densities and current densities, respectively.

Introducing the vector potential $\mathbf{A}$ and the scalar potential ϕ,

$$\mathbf{H} = \nabla \times \mathbf{A} \tag{20.5}$$

and from equation (20.3)

$$\nabla \times \left(\mathbf{E} + \frac{1}{c}\frac{\partial \mathbf{A}}{\partial t}\right) = 0$$

Hence, $(\mathbf{E} + (1/c)\partial \mathbf{A}/\partial t)$ may be derived from a potential ϕ as the gradient, equivalent to the relation

$$\mathbf{E} = -\nabla \phi - \frac{1}{c}\frac{\partial \mathbf{A}}{\partial t} \tag{20.6}$$

This representation of fields by potentials (Coulomb potential ϕ and vector potential $\mathbf{A}$) follows directly from Maxwell's equations (20.2) and (20.3). Substituting this into equations (20.1) and (20.4), we obtain two coupled differential equations for the potentials,

$$\Delta \phi + \frac{1}{c}\frac{\partial (\nabla \cdot \mathbf{A})}{\partial t} = -4\pi \rho \tag{20.7}$$

$$\Delta\mathbf{A} - \frac{1}{c^2}\frac{\partial^2\mathbf{A}}{\partial^2 t} - \nabla\left(\nabla\cdot\mathbf{A} + \frac{1}{c}\frac{\partial\phi}{\partial t}\right) = \frac{4\pi}{c}\mathbf{j} \tag{20.8}$$

To decouple these equations one uses an appropriate gauge of the potentials.

Gauge transformation

The magnetic field (20.5) does not change in a transformation of the vector potential of the form

$$\mathbf{A}' = \mathbf{A} + \nabla\Lambda(\mathbf{r}, t)$$

From equation (20.6), we calculate for the scalar potential

$$\phi' = \phi - \frac{1}{c}\frac{\partial\Lambda}{\partial t}$$

if the electric field also must not be changed by the transformation. The potentials are at our disposal to the extent that the fields are not altered. The electromagnetic fields represent the measurable quantities; the potentials are auxiliary constructions. Such transformations are called *gauge transformations*. Maxwell's equations do not change under such transformations; they are *gauge invariant*.

The *Lorentz gauge* yields a condition between the two potentials, requiring

$$\frac{1}{c}\frac{\partial\phi'}{\partial t} + \nabla\cdot\mathbf{A}' = 0 \tag{20.9}$$

This condition does not result from chance. As can be seen from (20.8) and (20.7), a decoupling of the two coupled differential equations for the potentials is achieved by the additional condition. Then, equations (20.7) and (20.8) become

$$\Delta\phi(\mathbf{r}, t) - \frac{1}{c^2}\frac{\partial^2}{\partial t^2}\phi(\mathbf{r}, t) = -4\pi\rho(\mathbf{r}, t) \tag{20.10a}$$

$$\Delta\mathbf{A}(\mathbf{r}, t) - \frac{1}{c^2}\frac{\partial^2}{\partial t^2}\mathbf{A}(\mathbf{r}, t) = -\frac{4\pi}{c}\mathbf{j}(\mathbf{r}, t) \tag{20.10b}$$

which are two *decoupled differential equations* to determine the potentials.

Substituting the primed potentials into the Lorentz condition (20.9), we obtain the wave equation for the gauge function $\Lambda(\mathbf{r}, t)$:

$$\Delta\Lambda - \frac{1}{c^2}\frac{\partial^2}{\partial t^2}\Lambda = -\left(\nabla\cdot\mathbf{A} + \frac{1}{c}\frac{\partial\phi}{\partial t}\right)$$

So, if $\mathbf{A}(\mathbf{r}, t)$ and $\phi(\mathbf{r}, t)$ are given, the Lorentz gauge function $\Lambda(\mathbf{r}, t)$ may be determined by this equation. Another gauge that decouples equations (20.7) and (20.8) is the *Coulomb gauge*, which is also called the *transverse gauge*. It requires

$$\nabla\cdot\mathbf{A}' = 0 \tag{20.11}$$

The prime on the vector potential, which is to remind us of the Coulomb gauge, will be dropped in the following. Then, we obtain the two decoupled differential equations

$$\Delta\phi = -4\pi\rho \qquad\qquad\qquad \textbf{(20.12a)}$$

$$\Delta\mathbf{A} - \frac{1}{c^2}\frac{\partial^2}{\partial t^2}\mathbf{A} = -\frac{4\pi}{c}\mathbf{j} + \frac{1}{c}\nabla\frac{\partial\phi}{\partial t} \qquad\qquad \textbf{(20.12b)}$$

$$= -\frac{4\pi}{c}\left[\mathbf{j} - \frac{1}{4\pi}\nabla\frac{\partial\phi}{\partial t}\right] \equiv -\frac{4\pi}{c}\mathbf{j}_t$$

Equation (20.12a) demonstrates the origin of the name Coulomb gauge. Like the Coulomb potential in electrostatics, the electric potential is given by the instantaneous charge distribution.

In (20.12b) we have introduced the transverse component $\mathbf{j}_t$ of the current density $\mathbf{j}$. So, the longitudinal component $\mathbf{j}_l = 1/(4\pi)\nabla(\partial\phi/\partial t)$ is fixed. At the moment, neither $\mathbf{j}_l$ nor $\mathbf{j}_t$ exhibit why they are so named, or how they may be expressed completely by a given current density $\mathbf{j}$. But this will become clear soon. Hence, according to the definition, the right-hand side of equation (20.12b) represents the transverse component of the current density. The total current density $\mathbf{j}$ is the sum of the *transverse* component and the *longitudinal* component,

$$\mathbf{j} = \mathbf{j}_l + \mathbf{j}_t \qquad\qquad\qquad \textbf{(20.13a)}$$

where

$$\mathbf{j}_l = \frac{1}{4\pi}\nabla\frac{\partial\phi}{\partial t} \qquad \text{and} \qquad \mathbf{j}_t = \mathbf{j} - \mathbf{j}_l = \mathbf{j} - \frac{1}{4\pi}\nabla\frac{\partial\phi}{\partial t}$$

Now, we will express $\mathbf{j}_l$ and $\mathbf{j}_t$ exclusively by $\mathbf{j}$ and show that the individual components are given, respectively, by

$$\nabla \times \mathbf{j}_l = 0 \qquad \text{or} \qquad \nabla \cdot \mathbf{j}_t = 0 \qquad\qquad \textbf{(20.13b)}$$

The notations transverse and longitudinal will become clear at the end of this section, subsequent to equation (20.18).

Starting from the relation

$$\phi = \int \frac{\rho(\mathbf{r}', t)}{|\mathbf{r} - \mathbf{r}'|}\,dV'$$

we take the gradient of the time derivative

$$\nabla\frac{\partial}{\partial t}\phi = \nabla\int \frac{\frac{\partial}{\partial t}\rho(\mathbf{r}', t)}{|\mathbf{r} - \mathbf{r}'|}\,dV' \qquad\qquad \textbf{(20.14)}$$

With the continuity equation $\partial\rho(\mathbf{r}', t)/\partial t = -\nabla' \cdot \mathbf{j}(\mathbf{r}', t)$ we obtain

$$\nabla\frac{\partial}{\partial t}\phi = -\nabla\int \frac{\nabla' \cdot \mathbf{j}(\mathbf{r}', t)}{|\mathbf{r} - \mathbf{r}'|}\,dV' \qquad\qquad \textbf{(20.15)}$$

Now, we still have to show that the right-hand side of equation (20.15) corresponds to the longitudinal component $4\pi \mathbf{j}_l$ of the current density. For this aim, utilizing $\nabla \times (\nabla \times \mathbf{v}) = \nabla(\nabla \cdot \mathbf{v}) - \nabla^2 \mathbf{v}$, we write for the two components of the current density

$$\mathbf{j} = \mathbf{j}_t + \mathbf{j}_l \equiv \frac{1}{4\pi} \left\{ \underbrace{\nabla \times \nabla \times \int \frac{\mathbf{j}(\mathbf{r}', t)}{|\mathbf{r} - \mathbf{r}'|} dV'}_{4\pi \mathbf{j}_t} - \underbrace{\nabla \left(\nabla \cdot \int \frac{\mathbf{j}(\mathbf{r}', t)}{|\mathbf{r} - \mathbf{r}'|} dV' \right)}_{4\pi \mathbf{j}_l} \right\} \qquad \textbf{(20.16)}$$

corresponding to the decomposition (20.13a). Also relation (20.13b) may be verified easily. First, we check the validity of (20.16). Rewriting the double vector product, the second term cancels against the second term in the right-hand side of (20.16), and there remains

$$\mathbf{j} = \mathbf{j}_t + \mathbf{j}_l = -\frac{1}{4\pi} \Delta \int \frac{\mathbf{j}(\mathbf{r}', t)}{|\mathbf{r} - \mathbf{r}'|} dV'$$

Since the Δ-operator acts only on the unprimed coordinate, with

$$\Delta \frac{1}{|\mathbf{r} - \mathbf{r}'|} = -4\pi \delta(\mathbf{r} - \mathbf{r}')$$

we verify the correctness of our decomposition (20.16) at once:

$$-\frac{1}{4\pi} \Delta \int \frac{\mathbf{j}(\mathbf{r}', t)}{|\mathbf{r} - \mathbf{r}'|} dV' = \mathbf{j}(\mathbf{r}, t)$$

Thus, it has been shown that for equation (20.15)

$$\nabla \frac{\partial}{\partial t} \phi = 4\pi \mathbf{j}_l(\mathbf{r}, t) \qquad \textbf{(20.17)}$$

Really, from (20.16) we obtain

$$4\pi \mathbf{j}_l(\mathbf{r}, t) = -\nabla \nabla \cdot \int \frac{\mathbf{j}(\mathbf{r}', t)}{|\mathbf{r} - \mathbf{r}'|} dV'$$

$$= -\nabla \cdot \int \mathbf{j}(\mathbf{r}', t) \cdot \nabla \frac{1}{|\mathbf{r} - \mathbf{r}'|} dV'$$

$$= +\nabla \int \mathbf{j}(\mathbf{r}', t) \cdot \nabla' \frac{1}{|\mathbf{r} - \mathbf{r}'|} dV'$$

$$= \nabla \left[\int \nabla' \cdot \left(\frac{\mathbf{j}(\mathbf{r}', t)}{|\mathbf{r} - \mathbf{r}'|} \right) dV' - \int \frac{\nabla' \cdot \mathbf{j}(\mathbf{r}', t)}{|\mathbf{r} - \mathbf{r}'|} dV' \right]$$

Using Gauss' theorem, the integral over the divergence $\nabla \cdot (\ )$ vanishes, so that finally

$$4\pi \mathbf{j}_l(\mathbf{r}, t) = -\nabla \int \frac{\nabla' \cdot \mathbf{j}(\mathbf{r}', t)}{|\mathbf{r} - \mathbf{r}'|} dV'$$

This is the right-hand side of equation (20.15).

Now, equation (20.17) means for equation (20.12b)

$$\Delta \mathbf{A} - \frac{1}{c^2} \frac{\partial^2 \mathbf{A}}{\partial t^2} = -\frac{4\pi}{c} \mathbf{j}_t(\mathbf{r}, t) \tag{20.18}$$

In the case of the (transverse) Coulomb gauge, the vector potential is given by the transverse part of the current distribution $\mathbf{j}_t$. The notation transverse and longitudinal for the components of the current (20.16) can be justified by calculating (20.18). For plane waves $\mathbf{A}(\mathbf{r}, t) = \boldsymbol{\epsilon} \, e^{i(\mathbf{k} \cdot \mathbf{r} - \omega t)}$, condition (20.11) of the Coulomb gauge yields $\boldsymbol{\epsilon} \cdot \mathbf{k} = 0$; that is, the waves of the vector potential are polarized transversely. Then, in the vector equation (20.18) also the current density appearing at the right must have the same vector direction as the vector potential $\mathbf{A}(\mathbf{r}, t)$, that is, transverse with respect to $\mathbf{k}$. This also explains the name transverse gauge, used alternatively to Coulomb gauge.

The time-dependent Green function

To solve the wave equations (20.10a) and (20.10b), we consider equations of this type in general:

$$\nabla^2 \psi - \frac{1}{c^2} \frac{\partial^2}{\partial t^2} \psi = -4\pi f(\mathbf{r}, t) \tag{20.19}$$

As we have seen already in electrostatics, it is suitable to introduce a Green function to solve this equation, but now a time-dependent one, $G(\mathbf{r}, t; \mathbf{r}', t')$ which obeys the wave equation

$$\left(\nabla^2 - \frac{1}{c^2} \frac{\partial^2}{\partial t^2} \right) G(\mathbf{r}, t; \mathbf{r}', t') = -4\pi \delta(\mathbf{r} - \mathbf{r}') \delta(t - t') \tag{20.20}$$

In contrast to (20.19), in (20.20) the inhomogeneity on the right-hand side is restricted to a point-like source in space-time. The Green function obeying equation (20.20) depends only on the difference of the coordinates of position and time; thus, $G(\mathbf{r}, t; \mathbf{r}', t') = G(\mathbf{r} - \mathbf{r}'; t - t')$. In the infinite four-dimensional $(\mathbf{r}, t)$-space we can make the ansatz

$$\psi(\mathbf{r}, t) = \int G(\mathbf{r} - \mathbf{r}'; t - t') f(\mathbf{r}', t') \, dV' \, dt'$$

for the solution of (20.19). The Green function $G(\mathbf{r} - \mathbf{r}'; t - t')$ describes the wave originating from the perturbation $f(\mathbf{r}', t')$. This wave function satisfies the differential equation (20.19) only if there are no boundary conditions, which is not the case in general. For the further treatment of the problem it is necessary to turn to mathematics. First, we want to consider the Fourier representation of the δ-function (compare equation (3.27))

$$\delta(\mathbf{r} - \mathbf{r}') \delta(t - t') = \frac{1}{(2\pi)^4} \int_{-\infty}^{+\infty} d^3k \int_{-\infty}^{+\infty} d\omega \, e^{i\mathbf{k} \cdot (\mathbf{r} - \mathbf{r}')} \, e^{-i\omega(t - t')} \tag{20.21}$$

For the time-dependent Green function, we choose the Fourier representation

$$G(\mathbf{r}, t; \mathbf{r}', t') = \int_{-\infty}^{+\infty} d^3k \int_{-\infty}^{+\infty} d\omega \, g(\mathbf{k}, \omega) \, e^{i\mathbf{k}\cdot(\mathbf{r}-\mathbf{r}')} \, e^{-i\omega(t-t')} \qquad \textbf{(20.22)}$$

where the Fourier transform still has to be determined.

Applying the operator $\Box = \nabla^2 - (1/c^2)\partial^2/\partial t^2$ to both sides of equation (20.22), we obtain

$$\Box G(\mathbf{r}, t; \mathbf{r}', t') = \int_{-\infty}^{+\infty} d^3k \int_{-\infty}^{+\infty} d\omega \, g(\mathbf{k}, \omega) \, \Box \, e^{i[\mathbf{k}\cdot(\mathbf{r}-\mathbf{r}')-\omega(t-t')]} \qquad \textbf{(20.23)}$$

Now,

$$\Box \, e^{i[\mathbf{k}\cdot(\mathbf{r}-\mathbf{r}')-\omega(t-t')]} = e^{-i\omega(t-t')} \nabla^2 e^{i\mathbf{k}\cdot(\mathbf{r}-\mathbf{r}')} - \frac{1}{c^2} e^{i\mathbf{k}\cdot(\mathbf{r}-\mathbf{r}')} \frac{\partial^2}{\partial t^2} e^{-i\omega(t-t')}$$

Performing the differentiation we obtain

$$\Box \, e^{i[\mathbf{k}\cdot(\mathbf{r}-\mathbf{r}')-\omega(t-t')]} = \left(\frac{\omega^2}{c^2} - k^2\right) e^{i[\mathbf{k}\cdot(\mathbf{r}-\mathbf{r}')-\omega(t-t')]}$$

and after substituting this result into equation (20.23)

$$\Box G(\mathbf{r}, t; \mathbf{r}', t') = \int_{-\infty}^{+\infty} d^3k \int_{-\infty}^{+\infty} d\omega \, g(\mathbf{k}, \omega) \left(\frac{\omega^2}{c^2} - k^2\right) e^{i[\mathbf{k}\cdot(\mathbf{r}-\mathbf{r}')-\omega(t-t')]} \qquad \textbf{(20.24)}$$

Substituting equations (20.21) and (20.24) for both sides of equation (20.20), the comparison of coefficients yields the Fourier transform:

$$g(\mathbf{k}, \omega) = \frac{1}{4\pi^3} \frac{1}{k^2 - \omega^2/c^2}$$

This is correct insofar as $k^2 \neq \omega^2/c^2$. The treatment of the poles $k = \pm\omega/c$ requires particular attention. For the present, we substitute $g(\mathbf{k}, \omega)$ into (20.22). So, we obtain the following representation of the Green function:

$$G(\mathbf{r}, t; \mathbf{r}', t') = \frac{1}{4\pi^3} \int_{-\infty}^{+\infty} d^3k \int_{-\infty}^{+\infty} d\omega \, \frac{c^2}{c^2 k^2 - \omega^2} e^{i[\mathbf{k}\cdot(\mathbf{r}-\mathbf{r}')-\omega(t-t')]} \qquad \textbf{(20.25)}$$

To avoid the singularities of the integrand at $\omega = \pm ck$ we again recall some tools and definitions of function theory (see Chapter 4):

   (1) The variable z is complex. A function $f(z)$ is called holomorphic if it is differentiable in a partial domain G of the complex plane.

   (2) Cauchy theorem: If $f(z)$ is holomorphic in G and C is a simply closed curve, then

$$\oint_C f(z) \, dz = 0$$

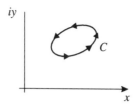

Figure 20.1. Closed integration path taken in the mathematically positive sense.

(3) If $f(z)$ has a pole of the order one at the point z_0 then

$$\text{Res}_{z_0} f(z) = \lim_{z \to z_0} (z - z_0) f(z)$$

is called the *residue* of $f(z)$ in z_0.

(4) The integral of the function $f(z)$ being analytic in the entire, simply connected domain G inside the integration path except at finitely many points $a_1, \ldots, a_n$ is taken in a mathematically positive sense along a closed path (Figure 20.1). This integral equals the sum of the residues at all singular points $a_1, \ldots, a_n$ multiplied by $2\pi i$:

$$\oint_C f(z)\, dz = 2\pi i \sum_{k=1}^{n} \text{Res}_{a_k} f(z) \tag{20.26}$$

With these tools, we can make more detailed statements about the Green function (20.25). In particular, by an appropriate choice of the integration path in (20.25), that is, by an appropriate circumnavigation of the poles in the integration over ω we may guarantee that the wave described by the Green function propagates in a causal manner.

The Green function represents a perturbation caused by a point charge $\mathbf{r}'$ during an infinitesimaly small time interval at time $t = t'$ that propagates as a spherical wave at the velocity c. For a wave propagating in a causal manner we must have

$$G(\mathbf{r}, t; \mathbf{r}', t') = 0 \qquad \text{for } t < t'$$

For $t \geq t'$, G represents a wave propagating into the future. The singularities of $g(\mathbf{k}, \omega)$ are at $\omega = \pm c\, k$. Now, we consider the complex ω-plane: $\omega = \omega_1 + i\omega_2$ in Figure 20.2, and attempt to realize the causal waves by an appropriate choice of the integration path in the complex plane.

For G to vanish for $t < t'$, we have to traverse the poles at $\omega = \pm ck$. This is accomplished by shifting them by an infinitesimal quantity $(-i\epsilon)$, as indicated in the sketch. Mathematically, this means that in equation (20.25) we have to replace ω by $\omega + i\epsilon$:

$$G(\mathbf{r}, t; \mathbf{r}', t') = \frac{1}{4\pi^3} \int_{-\infty}^{+\infty} d^3k \int_{-\infty}^{+\infty} d\omega \, \frac{e^{i\mathbf{k}\cdot\mathbf{R} - i\omega\tau}}{k^2 - (\omega + i\epsilon)^2/c^2} \tag{20.27}$$

where $\mathbf{R} = \mathbf{r} - \mathbf{r}'$ and $\tau = t - t'$.

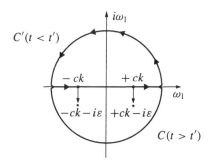

Figure 20.2.   Causal waves are described by the choice of this integration path. These waves are not allowed to vanish. Therefore, the poles of $g(k, \omega)$ must lie in the lower half-plane. In the upper half-plane no poles may occur in order that waves propagating in a noncausal way vanish. The perturbation is present at times t before any waves are emitted.

First, we want to calculate a partial integral of (20.27):

$$I(k) = \int_{-\infty}^{+\infty} d\omega \, \frac{e^{-i\omega\tau}}{k^2 - (\omega + i\epsilon)^2/c^2} \tag{20.27a}$$

The integration paths indicated in the figure satisfy the conditions which have to be set up to apply the Cauchy theorem and the residue theorem. But we have to note that for $\tau = t - t' > 0$ we integrate along the curve C while the integral along C' vanishes necessarily for $t - t' < 0$ (hence, for $\tau < 0$). Always, we choose the path, C or C', which in closing the integral

$$\int_{-\infty}^{\infty} d\omega \ldots \quad \rightarrow \quad \int_{\curvearrowleft} d\omega \ldots \quad \text{or} \quad \int_{\curvearrowright} d\omega \ldots$$

yields the contribution zero, so that

$$\int_{-\infty}^{\infty} d\omega \ldots = \int_{\curvearrowleft} d\omega \ldots$$

or

$$\int_{-\infty}^{\infty} d\omega \ldots = \int_{\curvearrowright} d\omega \ldots$$

For this purpose we look for the absolute value of the integrand of (20.27a) in some detail. With $\omega = \varrho e^{i\varphi} = \varrho(\cos(\varphi) + i \sin(\varphi))$

$$\left| \frac{e^{-i\omega\tau}}{k^2 - \omega^2} \right| = \left| \frac{e^{i\varrho(\cos(\varphi) + i\sin(\varphi))\tau}}{k^2 - \varrho^2(\cos(2\varphi) + i\sin(2\varphi))} \right|$$

$$\underset{\varrho \to \infty}{=} \frac{1}{\varrho^2} e^{\varrho \sin(\varphi)\tau}$$

Therefore, the contribution to the integral over the infinitely distant semicircle is

$$\left| \int_{\cap} d\omega \ldots \right| \underset{\varrho \to \infty}{<} \frac{(\pi \varrho) e^{\varrho \sin(\varphi)\tau}}{\varrho^2} \underset{\varrho \to \infty}{\to} 0 \qquad \text{for } \tau < 0, \text{ because } \sin \varphi > 0$$

or

$$\left| \int_{\cup} d\omega \ldots \right| \underset{\varrho \to \infty}{<} \frac{(\pi \varrho) e^{\varrho \sin(\varphi)\tau}}{\varrho^2} \underset{\varrho \to \infty}{\to} 0 \qquad \text{for } \tau > 0, \text{ because } \sin \varphi < 0$$

So, the integral (20.27a) has to be closed in the upper half-plane for $\tau < 0$ and in the lower half-plane for $\tau > 0$. Only then do the closed integrals equal the integral (20.27a) taken along the ω_1 axis from $\omega_1 = -\infty$ to $\omega_1 = +\infty$.

By displacing the poles of the integrand to $\omega = \pm k - i\epsilon$ into the lower half-plane of the complex ω-plane, the integral closed in the upper half-plane (C'), that is, the integral with $\tau = t - t' < 0$, yields the value zero. This follows directly from Cauchy's theorem

$$\oint f(z) dz = 0$$

because in this case the integrand $f(z) = e^{-iz\tau}/[k^2 - (z + i\varepsilon)^2]$ is holomorphic in the upper half-plane. This is not the case for the integral closed in the lower half-plane (C) because the poles of the integrand now lie in this domain. Here, the residue theorem comes into play.

According to the residue theorem we calculate

$$\oint_C d\omega \frac{c^2 e^{-i\omega\tau}}{c^2 k^2 - (\omega + i\epsilon)^2} = \oint_C d\omega \frac{-c^2 e^{-i\omega\tau}}{\omega^2 - (ck - i\varepsilon)^2}$$

$$= \oint_C d\omega \frac{-c^2 e^{-i\omega\tau}}{(\omega + (ck - i\varepsilon))(\omega - (ck - i\varepsilon))}$$

$$= \oint_C d\omega \, f(\omega, k) \tag{20.27b}$$

Obviously the integrand has two poles of the first order at $\omega_1 = ck - i\varepsilon$ and $\omega_2 = -ck - i\varepsilon$. As we have two poles of the first order, we also obtain two residues, first, for $\omega_1 = ck - i\epsilon$:

$$\text{Res}_{\omega_1} f(\omega, k) = c^2 e^{-i\omega_1 \tau} \lim_{\omega \to \omega_1} \frac{\omega - ck + i\epsilon}{c^2 k^2 - (\omega + i\epsilon)^2}$$

$$= c^2 e^{-i\omega_1 \tau} \lim_{\omega \to \omega_1} \frac{\omega - ck + i\epsilon}{(ck + (\omega + i\epsilon)) \cdot (ck - (\omega + i\epsilon))}$$

$$= c^2 e^{-i\omega_1 \tau} \lim_{\omega \to \omega_1} \frac{(\omega - \omega_1)}{(\omega_1 - \omega)(ck + \omega + i\epsilon)}$$

As $\epsilon \to 0$,

$$\text{Res}_{\omega_1} f(\omega, k) = -\frac{c}{2k} e^{-ick\tau}$$

Analogously, for the other pole $\omega_2 = -ck - i\epsilon$:

$$\text{Res}_{\omega_2} f(\omega, k) = +\frac{c}{2k} e^{ick\tau}$$

Now, due to (20.27),

$$\oint f(\omega, k)\, d\omega = -2\pi i(\text{Res}_{\omega_1} f + \text{Res}_{\omega_2} f)$$

thus,

$$I(k) = -2\pi i \left(-\frac{c}{2k} e^{-ick\tau} + \frac{c}{2k} e^{ick\tau} \right) = +\frac{2\pi c}{k} \sin(ck\tau) \qquad \textbf{(20.27c)}$$

The general minus sign on the right-hand side of the last two equations, compared to the positive sum of the residues in (20.27), arises from the fact that the integration path $C'(\circ)$ is traversed in the mathematically negative sense.

After performing the integration over $d\omega$, we obtain the following intermediate result:

$$G(\mathbf{R}, \tau) = \begin{cases} 0 & \text{for } t < t' \ (\tau < 0) \\ \dfrac{c}{2\pi^2} \displaystyle\int_{-\infty}^{+\infty} d^3k\, e^{i\mathbf{k}\mathbf{R}} \cdot \dfrac{+\sin(ck\tau)}{k} & \text{for } t > t' \ (\tau > 0) \end{cases} \qquad \textbf{(20.28)}$$

So, also the boundary condition $G = 0$ for $t < t'$ is fulfilled.

Now, we still have to perform the integration over d^3k $(\tau > 0)$. For this purpose we introduce spherical coordinates in $\mathbf{k}$-space, where $\mathbf{R}$ gives the direction of the k_z-axis. See Figure 20.3. Because this axis can be chosen arbitrarily, this is always possible. We obtain

$$d^3k = k^2 \sin \Theta_k\, dk\, d\Theta_k\, d\varphi_k \qquad \text{and} \qquad \mathbf{k} \cdot \mathbf{R} = kR \cos \Theta_k$$

This is substituted into the expression (20.28),

$$G(\mathbf{R}, \tau) = +\frac{c}{2\pi^2} \int d\varphi_k\, d\Theta_k\, dk\, k \sin\cdot\Theta_k \sin(ck\tau) e^{ikR\cos\Theta_k}$$

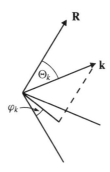

Figure 20.3. $\mathbf{R}$ is chosen to be the polar axis (z-axis) in $\mathbf{k}$-space.

The integration over $d\varphi_k$ yields the factor 2π, and we obtain

$$G(\mathbf{R}, \tau) = -\frac{c}{\pi} \int_{k=0}^{\infty} k \sin(ck\tau) \, dk \int_{\Theta=0}^{\pi} -e^{ikR\cos\Theta_k} \sin\Theta_k \, d\Theta_k \qquad (20.29)$$

First, we calculate the integral $P_\Theta = \int_{\Theta=0}^{\pi} -e^{ikR\cos\Theta_k} \sin\Theta_k \, d\Theta_k$ by means of the substitution $u = ikR\cos\Theta_k$, and we obtain $P_\Theta = -2/(kR)\sin(kR)$.

The substitution into (20.29) yields

$$G(\mathbf{R}, \tau) = \frac{2c}{\pi R} \int_0^{\infty} \sin(kR)\sin(ck\tau) \, dk$$

Now, we substitute $x = ck$ and rewrite the integrand. For this purpose we use the addition theorem and the representation of the cosine by the exponential function, to obtain

$$A = \sin(\tau x) \sin\left(\frac{R}{c}x\right) = \frac{1}{2}\left[\cos\left(\tau x + \frac{R}{c}x\right) - \cos\left(\tau x - \frac{R}{c}x\right)\right]$$

$$= \frac{1}{4}\left[\left(e^{i(\tau - R/c)x} + e^{-i(\tau - R/c)x}\right) - \left(e^{i(\tau + R/c)x} + e^{-i(\tau + R/c)x}\right)\right]$$

With the extension of the domain of integration and the combination of terms, the substitution yields

$$G = \frac{2}{\pi R} \int_0^{\infty} A(x) \, dx = \frac{1}{2\pi R} \int_{-\infty}^{+\infty} \left(e^{i(\tau - R/c)x} - e^{i(\tau + R/c)x}\right) dx$$

But $\int_{-\infty}^{+\infty} e^{iax} \, dx = 2\pi\delta(a)$. Hence, the Green function is

$$G = \frac{1}{R}\left[\delta\left(\tau - \frac{R}{c}\right) - \delta\left(\tau + \frac{R}{c}\right)\right]$$

Since the argument $(\tau + R/c)$ for $\tau > 0$ does not vanish anywhere, the δ-function of this argument cannot contribute to the solution. Thus, there remains

$$G = \frac{1}{R}\delta\left(\tau - \frac{R}{c}\right)$$

or, since $\mathbf{R} = \mathbf{r} - \mathbf{r}'$ and $\tau = t - t'$,

$$G(\mathbf{r}, t; \mathbf{r}', t') = \frac{\delta\left(t - t' - \dfrac{|\mathbf{r} - \mathbf{r}'|}{c}\right)}{|\mathbf{r} - \mathbf{r}'|} \qquad (20.30)$$

This is the required *representation of the time-dependent Green function*. It is also called the *retarded Green function* because it describes the causal behaviour connected with a perturbation of the wave, in the following sense: according to equation (20.22) an excitation (perturbation) happens at time $t = t'$ at the point $\mathbf{r} = \mathbf{r}'$, then this perturbation reaches the point $\mathbf{r}(\neq \mathbf{r}')$ *after time* $t - t' = |\mathbf{r} - \mathbf{r}'|/c$. *The excitation propagates at the speed of light into the future.* Substituting our solution into the initial equation, then

$$\psi(\mathbf{r}, t) = \int dV' dt' \frac{f(\mathbf{r}', t')}{|\mathbf{r} - \mathbf{r}'|} \delta\left(t - t' - \frac{|\mathbf{r} - \mathbf{r}'|}{c}\right) \tag{20.31}$$

Performing the integration over dt', we obtain the *retarded potential*, when according to (20.10a) the charge density is substituted into the function f:

$$\phi(\mathbf{r}, t) = \int dV' \frac{\rho\left(\mathbf{r}', t - |\mathbf{r} - \mathbf{r}'|/c\right)}{|\mathbf{r} - \mathbf{r}'|} \tag{20.32}$$

An analogous calculation for the vector potential (20.10b) yields

$$\mathbf{A}(\mathbf{r}, t) = \frac{1}{c} \int dV' \frac{\mathbf{j}\left(\mathbf{r}', t - |\mathbf{r} - \mathbf{r}'|/c\right)}{|\mathbf{r} - \mathbf{r}'|} \tag{20.33}$$

The potentials with the time dependence $t - |\mathbf{r} - \mathbf{r}'|/c$ are called retarded potentials. Advanced potentials having the time dependence $t + |\mathbf{r} - \mathbf{r}'|/c$ could be obtained if in the integration in the complex plane the poles are displaced into the upper ω-half-plane. Then, only the path $C'(\frown)$ would give a nonvanishing contribution. This would lead us to the advanced Green function

$$G(\mathbf{r}, t; \mathbf{r}', t') = \frac{\delta(-\tau - R/c)}{R} = \frac{\delta\left(t' - t - \dfrac{|\mathbf{r} - \mathbf{r}'|}{c}\right)}{|\mathbf{r} - \mathbf{r}'|}$$

which, obviously, describes a propagation $(t - t') < 0$, that is, a propagation backward in time.

Retarded potentials are used to calculate the fields of moving charges. The time dependence $t - |\mathbf{r} - \mathbf{r}'|/c = t'$ points out that the field observed at the position $\mathbf{r}$ at time t was caused at the *earlier* time $t' < t$ at the point $\mathbf{r}'$. The quotient $|\mathbf{r} - \mathbf{r}'|/c$ is just the time needed by a wave travelling at the speed of light to propagate from $\mathbf{r}'$ to $\mathbf{r}$. So, the principle of causality for perturbations propagating into the future is taken into account by retarded potentials. Furthermore, the principle of superposition becomes evident, because the perturbations originating from several sources superpose linearly.

The advanced potentials play a role in the description of physical processes in the present from their knowledge in the future. Such problems are of some importance in field theory (S-matrix formalism).

We further point out that the physical content of the solutions (20.32) and (20.33) is different from that of the differential equations (20.10a,b). While in the differential equations (20.10a,b) the sign of the time is by no means accentuated (the differential equations (20.10a,b) are invariant under the transformation $t' = -t$), the solutions (20.32) distinguish between the past and the future. The reason is that, in determining the Green function (20.30), we have constructed only retarded perturbations propagating into the future, as required by the principle of causality.

Liénard-Wiechert potentials

To calculate the total charge q in the static case one has to calculate the integral $q = \int \rho(\mathbf{r}') \, dV'$. In the dynamic case this can be done not so easily: to any point of observation there belongs a retarded time depending explicitly on the point of observation. The same holds for the potentials. To obtain a formal independence of the time t' we go back to the representation (20.31) of the retarded potentials in terms of the δ-function. If we consider a charge of minor extension (or a point charge) then let $\mathbf{r}_0(t)$ be the trajectory of this charge and let $\mathbf{v}(t) = d\mathbf{r}_0(t)/dt$ be its velocity. Then, the charge densities and current densities are

$$\rho(\mathbf{r}, t) = e\,\delta(\mathbf{r} - \mathbf{r}_0(t)), \qquad \mathbf{j}(\mathbf{r}, t) = e\mathbf{v}(t)\,\delta(\mathbf{r} - \mathbf{r}_0(t)) \tag{20.34}$$

and from equation (20.32) we obtain, e.g., for the Coulomb potential

$$\phi(\mathbf{r}, t) = e \int dV' \, \frac{1}{|\mathbf{r} - \mathbf{r}'|}\, \delta\left(\mathbf{r}' - \mathbf{r}_0\left(t - \frac{|\mathbf{r} - \mathbf{r}'|}{c}\right)\right)$$

This integral is difficult to calculate. Only if the trajectory $\mathbf{r}_0(t)$ is given explicitly can one easily go further.

Therefore, we go back to the expression (20.31). The Coulomb potential reads

$$\phi(\mathbf{r}, t) = e \int dt' \int dV' \, \frac{1}{|\mathbf{r} - \mathbf{r}'|}\, \delta\left(\mathbf{r}' - \mathbf{r}_0(t')\right)\, \delta\left(t' - t + \frac{|\mathbf{r} - \mathbf{r}'|}{c}\right) \tag{20.35}$$

In this form, the spatial integration can be carried through easily, with the result

$$\phi(\mathbf{r}, t) = e \int dt' \, \frac{1}{|\mathbf{r} - \mathbf{r}_0(t')|}\, \delta\left(t' - t + \frac{|\mathbf{r} - \mathbf{r}_0(t')|}{c}\right) \tag{20.36}$$

For further evaluation we introduce the following substitution:

$$u = t' - t + \frac{|\mathbf{r} - \mathbf{r}_0(t')|}{c} = t' - t + \frac{\sqrt{(x - x_0(t'))^2 + (y - y_0(t'))^2 + (z - z_0(t'))^2}}{c}$$

$$\frac{du}{dt'} = 1 - \frac{\left[(x - x_0(t'))\dot{x}_0(t') + (y - y_0(t'))\dot{y}_0(t') + (z - z_0(t'))\dot{z}_0(t')\right]}{c\sqrt{(x - x_0(t'))^2 + (y - y_0(t'))^2 + (z - z_0(t'))^2}}$$

$$= 1 - \mathbf{n}(t') \cdot \frac{\mathbf{v}(t')}{c}$$

where $\mathbf{n}(t') = (\mathbf{r} - \mathbf{r}_0(t'))/|\mathbf{r} - \mathbf{r}_0(t')|$ points from the charge at the position $\mathbf{r}_0(t')$ to the point of observation (see Figure 20.4). Hence,

$$dt' = \frac{du}{1 - \mathbf{n}(u) \cdot \mathbf{v}(u)/c} \tag{20.37}$$

and (20.36) becomes

$$\phi(\mathbf{r}, t) = e \int du \, \frac{1}{|\mathbf{r} - \mathbf{r}_0(t'(u))|} \cdot \frac{1}{1 - \mathbf{n}(u) \cdot \mathbf{v}(u)/c}\, \delta(u)$$

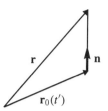

Figure 20.4. The trajectory $\mathbf{r}_0(t')$, the point of observation $\mathbf{r}$, and the unit vector $\mathbf{n}$.

$$= \left. \frac{e}{|\mathbf{r} - \mathbf{r}_0(t')| \cdot (1 - \mathbf{n}(t') \cdot \mathbf{v}(t')/c)} \right|_{u=0 \text{ or } t'=t-|\mathbf{r}-\mathbf{r}_0(t')|/c} \tag{20.38}$$

Similarly, with (20.34) and (20.31) we obtain the vector potential

$$\mathbf{A}(\mathbf{r}, t) = \left. \frac{e\,\mathbf{v}(t')/c}{|\mathbf{r} - \mathbf{r}_0(t')| \cdot (1 - \mathbf{n}(t') \cdot \mathbf{v}(t')/c)} \right|_{u=0 \text{ or } t'=t-|\mathbf{r}-\mathbf{r}_0(t')|/c} \tag{20.39}$$

In this derivation we have taken into account that the part of the integrand in (20.38) which is different from the $\delta(u)$-function depends on u and thus on t'. So, the integral is of the type

$$\int f(u)\,\delta(u)\,du = f(u)|_{u=0} = \left. f(t') \right|_{t'=t-|\mathbf{r}-\mathbf{r}_0(t')|/c}$$

The potentials (20.38) and (20.39) are called Liénard-Wiechert potentials.

They are exact solutions of the equations (20.10a,b) for the potentials of a point particle moving arbitrarily along a trajectory $\mathbf{r}_0(t')$.

Example 20.1:   Exercise: On Kirchhoff's integral representation

Solve the problem of the three-dimensional wave equation with a source term for a given infinite distribution of the wave field and its first time derivative.

Solution   We start the construction of the associated Green function with the identity

$$\varphi \nabla^2 \psi - \psi \nabla^2 \varphi = \nabla \cdot (\varphi \nabla \psi - \psi \nabla \varphi) \tag{20.40}$$

where $\varphi(\mathbf{x})$ and $\psi(\mathbf{x})$ are arbitrary functions of $\mathbf{x}$. Integrating over a certain volume (the integration coordinates are labeled by $\mathbf{x}'$) then

$$\int_V d^3x' \left(\varphi \nabla'^2 \psi - \psi \nabla'^2 \varphi \right) = \int_F dF' \left(\varphi \frac{\partial \psi}{\partial n'} - \psi \frac{\partial \varphi}{\partial n'} \right) \tag{20.41}$$

Let $\psi(\mathbf{x}', t')$ be the desired solution of the three-dimensional wave equation

$$\nabla'^2 \psi(\mathbf{x}', t') - \frac{1}{c^2} \frac{\partial^2}{\partial t'^2} \psi(\mathbf{x}', t') = -4\pi f(\mathbf{x}', t') \tag{20.42}$$

and let $\varphi(\mathbf{x}', t')$ be the *causal Green function* of this equation, which we know to be

$$G(\mathbf{x}, t; \mathbf{x}', t') = \frac{\delta\left(t - t' - \frac{|\mathbf{x} - \mathbf{x}'|}{c}\right)}{|\mathbf{x} - \mathbf{x}'|} \tag{20.43}$$

(compare equation (20.30)). This Green function obeys the equation

$$\left(\nabla^2 - \frac{1}{c^2}\frac{\partial^2}{\partial t^2}\right)G(\mathbf{x}, t; \mathbf{x}', t') = -4\pi\,\delta(\mathbf{x} - \mathbf{x}')\,\delta(t - t') \tag{20.44}$$

and since it is a function of the combinations $\mathbf{x} - \mathbf{x}'$ and $t - t'$ only, it also obeys the equation

$$\left(\nabla'^2 - \frac{1}{c^2}\frac{\partial^2}{\partial t'^2}\right)G(\mathbf{x}, t; \mathbf{x}', t') = -4\pi\,\delta(\mathbf{x} - \mathbf{x}')\,\delta(t - t') \tag{20.45}$$

Substituting these differential equations into the integrals (20.41) and integrating over t' from t_0 to t_1 we obtain

$$\int_{t_0}^{t_1} dt' \int_F dF' \left(G\frac{\partial\psi}{\partial n'} - \psi\frac{\partial G}{\partial n'}\right) \tag{20.46}$$

$$= \int_{t_0}^{t_1} dt' \int_V d^3x' \left\{G(\mathbf{x}, t; \mathbf{x}', t')\left[\frac{1}{c^2}\frac{\partial^2\psi}{\partial t'^2} - 4\pi f(\mathbf{x}', t')\right]\right.$$

$$\left. - \psi(\mathbf{x}, t)\left[\frac{1}{c^2}\frac{\partial^2 G}{\partial t'^2} - 4\pi\delta(\mathbf{x} - \mathbf{x}')\,\delta(t - t')\right]\right\}$$

$$= 4\pi\psi(\mathbf{x}, t) - 4\pi\int_{t_0}^{t_1} dt' \int_V d^3x'\, G(\mathbf{x}, t; \mathbf{x}', t')\, f(\mathbf{x}', t')$$

$$+ \frac{1}{c^2}\int_V d^3x' \int_{t_0}^{t_1} dt'\frac{\partial}{\partial t'}\left(G\frac{\partial\psi}{\partial t'} - \psi\frac{\partial G}{\partial t'}\right)$$

$$= 4\pi\psi(\mathbf{x}, t) - 4\pi\int_V d^3x'\frac{f(\mathbf{x}', t - |\mathbf{x} - \mathbf{x}'|/c)}{|\mathbf{x} - \mathbf{x}'|}$$

$$+ \frac{1}{c^2}\int_V d^3x'\left[G\frac{\partial\psi}{\partial t'} - \psi\frac{\partial G}{\partial t'}\right]_{t'=t_0}^{t'=t_1}$$

Here, we have assumed that the time t lies between t_0 and t_1. Furthermore, from (20.43) one may infer that G vanishes for $t' = t_1$. If the integration volume is chosen to be infinitely large and the function ψ vanishes at infinity, then the surface integral vanishes, and (20.46) reduces to

$$\psi(\mathbf{x}, t) = \int_V d^3x'\frac{f(\mathbf{x}', t - |\mathbf{x} - \mathbf{x}'|/c)}{|\mathbf{x} - \mathbf{x}'|} + \frac{1}{4\pi c^2}\int_V d^3x'\left[G\frac{\partial\psi}{\partial t'} - \psi\frac{\partial G}{\partial t'}\right]_{t'=t_0} \tag{20.47}$$

If the initial function ($t_0 = 0$) is denoted by

$$\psi(\mathbf{x}, 0) = F(\mathbf{x}) \qquad \text{and} \qquad \left.\frac{\partial\psi(\mathbf{x}, t)}{\partial t}\right|_{t=0} = D(\mathbf{x}) \tag{20.48}$$

then (20.47) becomes

$$\psi(\mathbf{x}, t) = \int_V d^3x'\frac{f(\mathbf{x}', t - |\mathbf{x} - \mathbf{x}'|/c)}{|\mathbf{x} - \mathbf{x}'|} + \frac{1}{4\pi c^2}\int_V d^3y\frac{\delta(|\mathbf{y}|/c - t)}{|\mathbf{y}|}D(\mathbf{x} + \mathbf{y}) \tag{20.49}$$

$$-\frac{1}{4\pi c^2} \int_V d^3 y\, F(\mathbf{x}+\mathbf{y}) \frac{\delta'\left(|\mathbf{y}|/c - t\right)}{|\mathbf{y}|}$$

where in the second integral we have substituted the new variable $\mathbf{y} = \mathbf{x}' - \mathbf{x}$. Because

$$\delta\left(\frac{|\mathbf{y}|}{c} - t\right) = c\delta\left(|\mathbf{y}| - ct\right) \tag{20.50}$$

$$\delta'\left(\frac{|\mathbf{y}|}{c} - t\right) = \frac{\partial}{\partial\left(\frac{|\mathbf{y}|}{c}\right)} \delta\left(\frac{|\mathbf{y}|}{c} - t\right) = c^2 \delta'\left(|\mathbf{y}| - ct\right)$$

we further obtain

$$\psi(\mathbf{x}, t) = \int_V d^3 x' \frac{f\left(\mathbf{x}', t - |\mathbf{x} - \mathbf{x}'|/c\right)}{|\mathbf{x} - \mathbf{x}'|} + \frac{1}{4\pi c} \int_V \frac{d^3 y}{|\mathbf{y}|} D(\mathbf{x}+\mathbf{y}) \delta\left(|\mathbf{y}| - ct\right)$$

$$+ \frac{1}{4\pi} \int_V \frac{d^3 y}{|\mathbf{y}|} F(\mathbf{x}+\mathbf{y}) \delta'\left(|\mathbf{y}| - ct\right) \tag{20.51}$$

If one attempts, e.g., to calculate the function $\psi(\mathbf{x}, t)$ at the origin of the coordinate system $\mathbf{x} = 0$, $\psi(0, t)$, then it is suitable to introduce the spherical coordinates $|\mathbf{y}|$ and Ω so that $d^3 y = |\mathbf{y}|^2\, d|\mathbf{y}|\, d\Omega$, and one obtains

$$\psi(0, t) = \int d\Omega \int_0^\infty d|\mathbf{y}|\,|\mathbf{y}| f\left(|\mathbf{y}|, \Omega, t - \frac{|\mathbf{y}|}{c}\right) + \frac{t}{4\pi} \int d\Omega\, D(ct, \Omega)$$

$$+ \frac{1}{4\pi} \int d\Omega \frac{\partial}{\partial t}\left[t\, F(ct, \Omega)\right] \tag{20.52}$$

On the other hand, if one considers a finite integration volume *containing no sources*, and if ψ as well as $\partial\psi/\partial t$ vanish at the initial time t_0 in the entire volume, then from (20.46)

$$\psi(\mathbf{x}, t) = \frac{1}{4\pi} \int_{t_0}^{t_1} dt' \int_F dF'\left(G\frac{\partial\psi}{\partial n'} - \psi\frac{\partial G}{\partial n'}\right)$$

$$= \frac{1}{4\pi} \int_{t_0}^{t_1} dt' \int_F dF'\,\mathbf{n}'\cdot\left(G\nabla'\psi - \psi\nabla'G\right) \tag{20.53}$$

If we introduce for abbreviation $\mathbf{R} = \mathbf{x} - \mathbf{x}'$, then

$$G = \frac{\delta\left(t' + \dfrac{R}{c} - t\right)}{R} \tag{20.54}$$

$$\nabla'G = \frac{\partial G}{\partial R}\nabla'R = \left[-\frac{\delta\left(t' + \dfrac{R}{c} - t\right)}{R^2} + \frac{\delta'\left(t' + \dfrac{R}{c} - t\right)}{cR}\right]\left(-\frac{\mathbf{R}}{R}\right) \tag{20.55}$$

In (20.53) the δ-function may be eliminated if the time interval is extended from $-\infty$ to $+\infty$; then one chooses $t_0 = -\infty$ and $t_1 = +\infty$. Then, from (20.53),

$$\psi(\mathbf{x}, t) = \frac{1}{4\pi} \int_F dF'\,\mathbf{n}'\cdot\left[\frac{\nabla'\psi(\mathbf{x}', t')}{R} - \psi(\mathbf{x}', t')\frac{\mathbf{R}}{R^3} - \frac{\partial\psi(\mathbf{x}', t')}{\partial t'}\frac{\mathbf{R}}{cR^2}\right]_{t'=t-R/c} \tag{20.56}$$

This is *Kirchhoff's integral representation*. It has the following illustrative meaning: Huygen's principle implies that all individual points of a wave front at a certain moment (that is, at a certain instant) may be regarded as a perturbation generating the wave for a later time. Kirchhoff's integral (20.56) expresses this in a mathematically exact manner because $\psi(\mathbf{x}, t)$ is given by the sum of all perturbations (surface integral over quantities depending on the wave field ψ and its derivatives on the surface) at the surface for the retarded time $t' = t - R/c$.

**Example 20.2:   Exercise: Liénard-Wiechert potentials of a point charge
moving with constant velocity**

In principle, it is possible to derive the electromagnetic fields of a point charge moving along an arbitrary trajectory (see Figure 20.5) from the Liénard-Wiechert potentials. Here, the simple case of a charge moving at constant velocity is considered.

(a) Determine the Liénard-Wiechert potentials.

(b) Derive the field intensities from the potentials and discuss the results.

Solution   The Liénard-Wiechert potentials read

$$\Phi(\mathbf{r}, t) = \frac{qc}{|\mathbf{R}|c - \mathbf{R} \cdot \mathbf{v}} \tag{20.57}$$

$$\mathbf{A}(\mathbf{r}, t) = \frac{q\mathbf{v}}{|\mathbf{R}|c - \mathbf{R} \cdot \mathbf{v}}$$

The retarded potentials $\Phi(\mathbf{r}, t)$ and $\mathbf{A}(\mathbf{r}, t)$ do not depend on the state of the charge q at time t, but on the state of the charge at the retarded time t'. $\mathbf{v}$ is the velocity of the particle at the retarded time t'. $\mathbf{R} = \mathbf{r} - \mathbf{r}_0(t')$ denotes the vector of the retarded position to the point of observation P. We choose the coordinate system such that at time $t = 0$ the charge is placed at the origin

$$\mathbf{r}_0(t) = \mathbf{v}t \tag{20.58}$$

Since information originating at the particle cannot propagate faster than the speed of light, the retarded time is defined implicitly by

$$|\mathbf{R}| = |\mathbf{r} - \mathbf{r}_0(t')| = c(t - t') \tag{20.59}$$

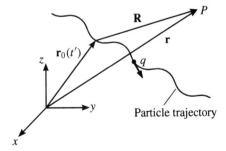

Figure 20.5. A general particle trajectory.

Only for simple particle trajectories may this implicit dependence be transformed into an explicit one. In our simple case, this can be done easily. Substituting $\mathbf{r}_0(t') = \mathbf{v}t'$ into (20.59) leads to

$$r^2 - 2\mathbf{r} \cdot \mathbf{v}t' + v^2 t'^2 = c^2(t^2 - 2tt' + t'^2) \tag{20.60}$$

This quadratic equation has two solutions

$$t'_\pm = \frac{tc^2 - \mathbf{r} \cdot \mathbf{v}}{c^2 - v^2} \pm \sqrt{\left(\frac{tc^2 - \mathbf{r} \cdot \mathbf{v}}{v^2 - c^2}\right)^2 + \frac{r^2 - c^2 t^2}{c^2 - v^2}} \tag{20.61}$$

To fix the sign we set $\mathbf{v} = 0$:

$$t'_\pm = t \pm \frac{r}{c} \tag{20.62}$$

Because $t' < t$ has to be valid strictly, in (20.62) only the negative sign is physically meaningful. We calculate the denominator in (20.57)

$$\begin{aligned}
|\mathbf{R}|c - \mathbf{R} \cdot \mathbf{v} &= c^2(t - t') - (\mathbf{r} - \mathbf{v}t')\mathbf{v} \\
&= (v^2 - c^2)t' + c^2 t - \mathbf{r} \cdot \mathbf{v} \\
&= \sqrt{(tc^2 - \mathbf{r} \cdot \mathbf{v})^2 + (r^2 - c^2 t^2)(c^2 - v^2)}
\end{aligned} \tag{20.63}$$

In the last step we have used (20.62). Hence, for the potentials

$$\Phi(\mathbf{r}, t) = \frac{qc}{\sqrt{(tc^2 - \mathbf{r} \cdot v)^2 + (r^2 - c^2 t^2)(c^2 - v^2)}} \tag{20.64}$$

$$\mathbf{A}(\mathbf{r}, t) = \frac{q\mathbf{v}}{\sqrt{(tc^2 - \mathbf{r} \cdot v)^2 + (r^2 - c^2 t^2)(c^2 - v^2)}}$$

The expression under the root sign may be simplified if the vector $\mathbf{w} = \mathbf{r} - \mathbf{v}t$ is introduced which points from the actual (not retarded) position of the charge to the point of observation P. See Figure 20.6. Thus, for the expression under the root sign we obtain

$$\begin{aligned}
(t^2 c^2 - \mathbf{r} \cdot \mathbf{v})^2 + (r^2 - c^2 t^2)(c^2 - v^2) &= (\mathbf{r} \cdot \mathbf{v})^2 - r^2 v^2 + c^2 \mathbf{w}^2 \\
&= -r^2 v^2 \sin^2 \alpha + c^2 w^2 \\
&= c^2 w^2 \left(1 - \frac{r^2 v^2}{w^2 c^2} \sin^2 \alpha\right) \\
&= c^2 w^2 \left(1 - \frac{v^2}{c^2} \sin^2 \Theta\right)
\end{aligned} \tag{20.65}$$

Here, α is the angle between $\mathbf{r}$ and $\mathbf{v}$, and Θ is the angle between $\mathbf{w}$ and $\mathbf{v}$. In the last step we used the law of sines.

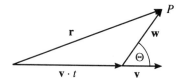

Figure 20.6. Geometry of the problem.

Then, the potentials read

$$\Phi(\mathbf{r}, t) = \frac{q}{w\sqrt{1 - \dfrac{v^2}{c^2}\sin^2\Theta}} \tag{20.66}$$

$$\mathbf{A}(\mathbf{r}, t) = \frac{q\mathbf{v}}{cw\sqrt{1 - \dfrac{v^2}{c^2}\sin^2\Theta}}$$

The field intensities $\mathbf{E}$ and $\mathbf{B}$ are obtained from the potentials according to

$$\mathbf{E} = -\nabla\Phi - \frac{1}{c}\frac{\partial\mathbf{A}}{\partial t} \tag{20.67}$$

$$\mathbf{B} = \nabla \times \mathbf{A}$$

Immediately, we can show that the $\mathbf{B}$-field is related to the $\mathbf{E}$-field in a simple way.

$$\mathbf{B} = \nabla \times \mathbf{A} = \nabla \times \left(\frac{\mathbf{v}}{c}\Phi\right) = \left(\nabla \times \frac{\mathbf{v}}{c}\right) - \frac{\mathbf{v}}{c} \times \nabla\Phi \tag{20.68}$$

$$= -\frac{\mathbf{v}}{c} \times \nabla\Phi = \frac{\mathbf{v}}{c} \times \left(\mathbf{E} + \frac{1}{c}\frac{\partial\mathbf{A}}{\partial t}\right)$$

$$= \frac{\mathbf{v}}{c} \times \mathbf{E}$$

We calculate the gradient of the scalar potential using (20.63):

$$\nabla\Phi = \frac{qc}{\left((tc^2 - \mathbf{r}\cdot v)^2 + (r^2 - c^2t^2)(c^2 - v^2)\right)^{3/2}}\left[(c^2t - \mathbf{r}\cdot\mathbf{v})\mathbf{v} - (c^2 - v^2)\mathbf{r}\right] \tag{20.69}$$

Correspondingly, one finds

$$\frac{\partial\mathbf{A}}{\partial t} = \frac{\mathbf{v}}{c}\frac{\partial\Phi}{\partial t} \tag{20.70}$$

$$= \frac{-qc^2}{\left((tc^2 - \mathbf{r}\cdot v)^2 + (r^2 - c^2t^2)(c^2 - v^2)\right)^{3/2}}\left[(c^2t - \mathbf{r}\cdot\mathbf{v})\mathbf{v} - (c^2 - v^2)\mathbf{r}\right]\cdot\mathbf{v}$$

So, one obtains for $\mathbf{E}$

$$\mathbf{E} = \frac{qc}{\left((tc^2 - \mathbf{r}\cdot v)^2 + (r^2 - c^2t^2)(c^2 - v^2)\right)^{3/2}}(c^2 - v^2)\left[\mathbf{r} - t\mathbf{v}\right] \tag{20.71}$$

In an analogous manner, the vector $\mathbf{w} = \mathbf{r} - \mathbf{v}t$ may be substituted, and one obtains the final expression for the $\mathbf{E}$-field

$$\mathbf{E}(\mathbf{r}, t) = q\frac{1 - v^2/c^2}{(1 - (v^2/c^2)\sin^2\theta)^{3/2}}\frac{\mathbf{w}}{w^3} \tag{20.72}$$

$$\mathbf{B}(\mathbf{r}, t) = \frac{v}{c} \times \mathbf{E}(\mathbf{r}, t)$$

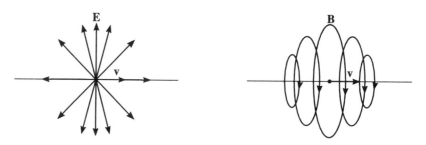

Figure 20.7. Fields of a rapidly moving charge.

For sake of completeness, we have given also the **B**-field once more. For nonrelativistic speeds ($v \ll c$) one obtains known fields for a charge at rest.

It is notable that the vector **E** starts from the actual position of the charge at the nonretarded time t. Actually, this is an extraordinary coincidence, since the retarded position of the charge is the cause of the field. The field of a rapidly moving charge is "compressed." In forward and backward directions the field is reduced by the factor $(1 - (v/c)^2)$. In perpendicular directions it is amplified by a factor $(1 - (v/c)^2)^{1/2}$ (Figure 20.7).

It is highly instructive to derive the result given above with the methods of relativistic electrodynamics. This will be done in Example 23.2.

Biographical notes

Alfred Marie Liénard, b. Apr. 2, 1869, Amiens (Somme)–d. Apr. 29, 1958, Paris. Physicist. 1908–1929, professor; 1919–1929, sous-dir; 1929–1936, dir., Ecole Nationale Supérieure des Mines, Paris. Member of the Académie des sciences, Paris.

Johann Emil Wiechert, b. Dec. 26, 1861, Tilsit–d. March 19, 1928, Göttingen. Geophysicist. 1881–1887, studies; 1889, PhD; 1890, outside lecturer of physics at Königsberg University; 1897, outside lecturer; 1898, associate professor for geophysics and director of the Geophysical Institute; 1905, professor at Göttingen University.

Gustav Robert Kirchhoff, b. March 12, 1824, Königsberg–d. Oct. 17, 1887, Berlin. This physicist was a professor in Breslau (1850), Heidelberg (1854), and Berlin (1875). In 1845 he established the loop rule and the current rule (Kirchhoff's laws) in a much more general version than G.S. Ohm. He also dealt with problems of the theory of oscillations, classical thermodynamics, heat conduction, and diffraction of light. The theoretical and experimental research on emission and absorption of light, made by Kirchhoff and Robert Bunsen in 1859–1860, were of an enormous impact for the fields of astronomy, chemistry, and physics. Kirchhoff's findings led to the establishment of Kirchhoff's law of radiation as well as to the explanation of the Fraunhofer lines in the solar spectrum, and to the invention of spectral analysis. In the treatise "On the theory of light waves" Kirchhoff presented a wave theory of light much better than the one developed by Fresnel.

21 The Hertzian Dipole

Now, we want to consider an application of the Liénard-Wiechert potentials, namely, the oscillating electric dipole with the dipole moment $\mathbf{p} = \mathbf{p}(t)$, first investigated by Heinrich Hertz. First, the expressions for the scalar potential $\phi(r, t)$, equation (20.38), and the vector potential $\mathbf{A}(\mathbf{r}, t)$, equation (20.39), are noted again:

$$\phi(\mathbf{r}, t) = \left. \frac{e}{|\mathbf{r} - \mathbf{r}_0(t')| - \dfrac{\mathbf{r} - \mathbf{r}_0(t')}{|\mathbf{r} - \mathbf{r}_0(t')|} \cdot \dfrac{\mathbf{v}(t')}{c}} \right|_{t' = t - |\mathbf{r} - \mathbf{r}_0(t')|/c} \tag{21.1a}$$

$$\mathbf{A}(\mathbf{r}, t) = \left. \frac{\dfrac{e}{c}\mathbf{v}(t')}{|\mathbf{r} - \mathbf{r}_0(t')| - \dfrac{\mathbf{r} - \mathbf{r}_0(t')}{|\mathbf{r} - \mathbf{r}_0(t')|} \cdot \dfrac{\mathbf{v}(t')}{c}} \right|_{t' = t - |\mathbf{r} - \mathbf{r}_0(t')|/c} \tag{21.1b}$$

Corresponding to Figure 21.1, we imagine the dipole to be a pair of point charges with the charges $\pm e$ located at a distance $\mathbf{r}_0 = \mathbf{r}_0(t')$ from each other. Hence,

$$e\mathbf{v}(t) = -e\frac{\dot{\mathbf{r}}_0(t)}{2} + e\frac{-\dot{\mathbf{r}}_0(t)}{2} = -e\dot{\mathbf{r}}_0(t) = \dot{\mathbf{p}}_0(t) \tag{21.2}$$

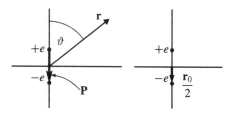

Figure 21.1. The Hertzian dipole: **r** is the vector, constant in time, from the dipole to the point of observation.

429

As in the figure, the vector $\mathbf{r}_0/2$ points from the origin of the coordinate system to the negative charge, so that the dipole moment is $\mathbf{p}(t) = -e\mathbf{r}_0(t)$ (see Chapter 1, in particular, Exercise 1.4).

For a harmonic oscillation of frequency ω the estimate

$$\frac{v}{c} \approx \frac{2r_0^{\max}}{Tc} = \frac{r_0^{\max}\omega}{\pi c} = \frac{2r_0^{\max}}{\lambda}$$

is valid, where $r_0^{\max}$ labels the amplitude of the oscillation. We determine the scalar potential $\phi(\mathbf{r}, t)$ and the vector potential $\mathbf{A}(\mathbf{r}, t)$ from equation (20.38) and (20.39), respectively, introducing

$$\Delta\mathbf{r}_-(t') = \mathbf{r} - \frac{\mathbf{r}_0(t')}{2}, \qquad \Delta\mathbf{r}_+(t') = \mathbf{r} + \frac{\mathbf{r}_0(t')}{2} \tag{21.3}$$

as the vectors of the distance between the point of observation and the negative and positive charges, respectively:

$$\phi(\mathbf{r}, t) = \left. \frac{-e}{|\Delta\mathbf{r}_-(t')| - \dfrac{\Delta\mathbf{r}_-(t')}{|\Delta\mathbf{r}_-(t')|} \cdot \dfrac{\dot{\mathbf{r}}_0(t')}{2c}} \right|_{t'=t-|\Delta\mathbf{r}_-(t')|/c}$$

$$+ \left. \frac{+e}{|\Delta\mathbf{r}_+(t')| - \dfrac{\Delta\mathbf{r}_+(t')}{|\Delta\mathbf{r}_+(t')|} \cdot \dfrac{-\dot{\mathbf{r}}_0(t')}{2c}} \right|_{t'=t-|\Delta\mathbf{r}_+(t')|/c} \tag{21.4}$$

and

$$\mathbf{A}(\mathbf{r}, t) = \left. \frac{\dfrac{-e}{c}\dfrac{\dot{\mathbf{r}}_0(t')}{2}}{|\Delta\mathbf{r}_-(t')| - \dfrac{\Delta\mathbf{r}_-(t')}{|\Delta\mathbf{r}_-(t')|} \cdot \dfrac{\dot{\mathbf{r}}_0(t')}{2c}} \right|_{t'=t-|\Delta\mathbf{r}_-(t')|/c}$$

$$+ \left. \frac{\dfrac{e}{c}\dfrac{(-\dot{\mathbf{r}}_0(t'))}{2}}{|\Delta\mathbf{r}_+(t')| - \dfrac{\Delta\mathbf{r}_+(t')}{|\Delta\mathbf{r}_+(t')|} \cdot \dfrac{-\dot{\mathbf{r}}_0(t')}{2c}} \right|_{t'=t-|\Delta\mathbf{r}_+(t')|/c} \tag{21.5}$$

Now we make the assumption that the distance vectors $\Delta\mathbf{r}_\pm(t') = \mathbf{r}\pm\mathbf{r}_0(t')/2$ may be approximated by $\mathbf{r}$. This is well fulfilled as far as the extension of the dipole $|\mathbf{r}_0|$ is much smaller than the distance between the dipole and the point of observation $\mathbf{r}$, $|\mathbf{r}_0| \ll |\mathbf{r}|$. In this case, t' in the first term is identical with t' in the second term of the equations (21.4) and (21.5). Then, both terms may be combined.

Under the additional assumption that the wavelength λ is large compared to the maximal elongation $r_0^{\max}$ of the charges (limit of long wavelengths) from equation (21.2) one obtains $v/c \ll 1$. Therefore, terms of the order $\mathbf{v}/c$ may be neglected in the denominator of equation (21.5); in the expansion of the denominator in terms of v/c, this corresponds to neglecting the quadratic terms.

From equation (21.5) we obtain

$$\mathbf{A}(\mathbf{r}, t) = \left. \frac{-\dfrac{e}{c}\,\dot{\mathbf{r}}_0(t')}{|\mathbf{r}|} = \frac{\dot{\mathbf{p}}(t')}{c\,r} \right|_{t'=t-r/c} \tag{21.6}$$

According to the Lorentz gauge

$$\operatorname{div}\mathbf{A} + \frac{1}{c}\frac{\partial\phi}{\partial t} = 0$$

Substituting the vector potential then

$$\frac{\partial\phi}{\partial t} = -\operatorname{div}\cdot\frac{\dot{\mathbf{p}}(t-r/c)}{r}$$

and hence,

$$\phi = -\nabla\cdot\int\frac{\dot{\mathbf{p}}(t-r/c)}{r}\,dt = -\nabla\cdot\frac{\mathbf{p}(t-r/c)}{r}$$

because $\mathbf{r}$ is time-independent (compare the figure). Performing the differentiation, the scalar potential is

$$\phi = \frac{\dot{\mathbf{p}}\cdot\mathbf{r}}{cr^2} + \frac{\mathbf{p}\cdot\mathbf{r}}{r^3} \tag{21.7}$$

The first summand comes from retardation. In the near zone, $r \ll \lambda$, retardation can be neglected; then, equation (21.7) becomes the (static) dipole potential $\phi = \mathbf{p}\cdot\mathbf{r}/r^3$.

At large distances the second term, with $1/r^3$, tends to zero more rapidly and the term arising from retardation dominates (Figure 21.2)

$$\phi \approx \frac{\dot{\mathbf{p}}\cdot\mathbf{r}}{cr^2} \sim \frac{1}{r}$$

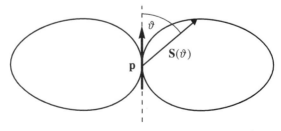

Figure 21.2. Dipole radiation in the far field: The oscillating dipole does not radiate in the direction of the dipole axis, but radiates at maximum in the perpendicular direction.

Substituting the potentials $\mathbf{A}$ and ϕ into the relations

$$\mathbf{H} = \text{curl}\,\mathbf{A} \qquad \text{and} \qquad \mathbf{E} = -\frac{1}{c}\frac{\partial \mathbf{A}}{\partial t} - \nabla\phi$$

we obtain the electric and magnetic field:

$$\mathbf{E} = -\frac{\ddot{\mathbf{p}}}{c^2 r} - \frac{\dot{\mathbf{p}}}{cr^2} + \frac{3(\dot{\mathbf{p}}\cdot\mathbf{r})\mathbf{r}}{cr^4} + \frac{(\ddot{\mathbf{p}}\cdot\mathbf{r})\mathbf{r}}{c^2 r^3} + \frac{3(\mathbf{r}\cdot\mathbf{p})\mathbf{r}}{r^5} - \frac{\mathbf{p}}{r^3}$$

$$\mathbf{H} = \frac{\ddot{\mathbf{p}}\times\mathbf{r}}{c^2 r^2} + \frac{\dot{\mathbf{p}}\times\mathbf{r}}{cr^3} \tag{21.8}$$

Again, restricting ourselves to the two limiting cases then we obtain for the *near zone* ($r \ll \lambda$), because the higher powers of r dominate in the denominator and the retardation term drops out:

$$\mathbf{E} = \frac{3(\mathbf{p}\cdot\mathbf{r})\mathbf{r}}{r^5} - \frac{\mathbf{p}}{r^3} \qquad \text{and} \qquad \mathbf{H} = \frac{\dot{\mathbf{p}}\times\mathbf{r}}{cr^3} \tag{21.9}$$

Here, also as for the potential, it turns out that in the near zone the field intensity corresponds to the field distribution of the static dipole (see Exercise 1.4). In the *far field* ($r \gg \lambda$) all terms with the higher powers of r may be neglected so that there remains only:

$$\mathbf{E} = -\frac{\ddot{\mathbf{p}}}{c^2 r} + \frac{(\ddot{\mathbf{p}}\cdot\mathbf{r})\mathbf{r}}{c^2 r^3} = \frac{1}{c^2 r^3}(\ddot{\mathbf{p}}\times\mathbf{r})\times\mathbf{r} \qquad \text{and} \qquad \mathbf{H} = \frac{\ddot{\mathbf{p}}\times\mathbf{r}}{c^2 r^2} \tag{21.10}$$

It follows directly that in the far zone ($\mathbf{r}, \mathbf{E}, \mathbf{H}$) form an orthogonal system and $\mathbf{E}$ and $\mathbf{H}$ are of equal magnitude, namely,

$$\mathbf{E} = \mathbf{H} \times \frac{\mathbf{r}}{r} \qquad \text{and} \qquad \mathbf{H} = -\mathbf{E} \times \frac{\mathbf{r}}{r} \tag{21.11}$$

So, in the far zone $\mathbf{E}$ and $\mathbf{H}$ are mutually orthogonal outgoing spherical waves, that is, propagating in $\mathbf{r}$-direction (compare the Poynting vector below).

The Poynting vector S of the dipole field

We have $\mathbf{S} = c/(4\pi)\mathbf{E}\times\mathbf{H}$. Substituting the relation (21.11) for the electric field intensity, we obtain for the far zone:

$$\mathbf{S} = \frac{c}{4\pi}\left(\mathbf{H}\times\frac{\mathbf{r}}{r}\right)\times\mathbf{H} = \frac{c}{4\pi}(\mathbf{H}\cdot\mathbf{H})\frac{\mathbf{r}}{r} - \frac{c}{4\pi}\left(\mathbf{H}\cdot\frac{\mathbf{r}}{r}\right)\mathbf{H}$$

Since $\mathbf{H}$ is perpendicular to $\mathbf{r}$, $\mathbf{H}\cdot\mathbf{r} = 0$. So, in the far field we obtain the energy-flux density

$$\mathbf{S} = \frac{c}{4\pi}(\mathbf{H}\cdot\mathbf{H})\frac{\mathbf{r}}{r} = \frac{c}{4\pi}\frac{(\ddot{\mathbf{p}}\times\mathbf{r})^2}{c^4 r^4}\frac{\mathbf{r}}{r}, \qquad S = \frac{|\ddot{\mathbf{p}}|^2}{4\pi c^3 r^2}\sin^2\theta \tag{21.12a}$$

Here, θ is the angle between the axis of oscillation and the position vector $\mathbf{r}$. Hence, in the far zone the energy flux flows in the radial direction; for increasing distance r from the dipole it decreases proportional to $1/r^2$, as can be seen from equation (21.12a). This is important for energy conservation! Furthermore, a $\sin^2 \theta$-dependence holds in the far field. Obviously, for $\theta = 0$, $\sin \theta = 0$, and $S = 0$, that is, *in the far field the dipole does not radiate in the direction of oscillation*. Further, energy is irradiated only if there is an acceleration ($\ddot{\mathbf{p}} \sim \ddot{\mathbf{r}}_0$). The *bremsstrahlung of accelerated charges* is based on this fact. These processes play an important role in heavy-ion physics and also in the acceleration of particles by large accelerators. With equation (21.9), the near-field approximation for the Poynting vector **S** is

$$
\begin{aligned}
\mathbf{S} &= \frac{c}{4\pi}\mathbf{E} \times \mathbf{H} = \frac{c}{4\pi}\left(\frac{3(\mathbf{p}\cdot\mathbf{r})\mathbf{r}}{r^5} - \frac{\mathbf{p}}{r^3}\right) \times \left(\frac{\dot{\mathbf{p}} \times \mathbf{r}}{cr^3}\right) \\
&= \frac{1}{4\pi r^8}\left(3(\mathbf{p}\cdot\mathbf{r})\mathbf{r} - r^2\mathbf{p}\right) \times (\dot{\mathbf{p}} \times \mathbf{r}) \\
&= \frac{1}{4\pi r^8}\left\{3(\mathbf{p}\cdot\mathbf{r})\mathbf{r} \times (\dot{\mathbf{p}} \times \mathbf{r}) - r^2\mathbf{p} \times (\dot{\mathbf{p}} \times \mathbf{r})\right\} \\
&= \frac{1}{4\pi r^8}\left\{3(\mathbf{p}\cdot\mathbf{r})\mathbf{r}\left[r^2\dot{\mathbf{p}} - (\mathbf{r}\cdot\dot{\mathbf{p}})\mathbf{r}\right] - r^2\left[(\mathbf{r}\cdot\mathbf{p})\dot{\mathbf{p}} - (\mathbf{p}\cdot\dot{\mathbf{p}})\mathbf{r}\right]\right\} \\
&= \frac{1}{4\pi r^8}\left\{2(\mathbf{p}\cdot\mathbf{r})r^2\dot{\mathbf{p}} \mp p\dot{p}r^2\left[3\cos\theta + 1\right]\mathbf{r}\right\}
\end{aligned}
\tag{21.12b}
$$

due to $\dot{\mathbf{p}}\cdot\mathbf{r} = (\pm)\dot{p}r\cos\theta$ and $\mathbf{p}\cdot\mathbf{r} = (\pm)pr\cos\theta$.

So, in the near field the Poynting vector possesses a much more complicated behavior than in the far field. For four half steps of the period, Figure 21.3 shows how one may visualize the radiation of energy. The electric field is shown for the instants $\omega t = 0$, $\pi/4$, $\pi/2$, and $3\pi/4$. At $t = 0$, the near field is just that of a static dipole. In the course of time the field lines tie themselves up, tear off, and travel outward. Finally, in the far field one can find only those components obeying equations (21.10), (21.11) and (21.12a). All other components lead only to an oscillation of the corresponding energy contributions in the near-zone region.

To calculate the total amount of energy radiated per time we go back to the magnitude of the Poynting vector in the far field, equation (21.12a).

The average energy of the radiation emitted by the dipoles during one period through a large spherical surface of radius R may be calculated as the integral of the energy flux density over the spherical surface and one period

$$
E = \frac{1}{T}\iint S_R R^2 \, d\Omega \, dt = \frac{1}{T}\int_0^T dt \int_0^\pi \int_0^{2\pi} \frac{\ddot{p}^2}{4c^3\pi}\sin^2\theta \sin\theta \, d\theta \, d\varphi
\tag{21.13}
$$

The integration over the angles yields the factors 2π and $4/3$, so that

$$
E = \frac{2}{3c^3 T}\int_0^T \ddot{p}^2 \, dt
\tag{21.14}
$$

(a)   (b)

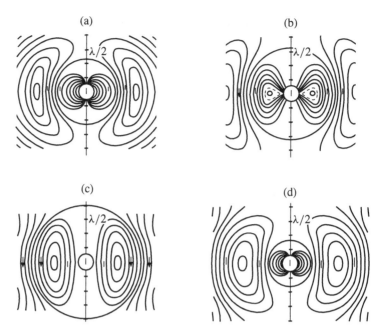

(c)   (d)

Figure 21.3. Representation of the electric field of an oscillating dipole for the instants $\omega t = 0$, $\pi/4$, $\pi/2$, and $3\pi/4$.

where

$$\ddot{p}^2 = \ddot{p}^2 \left(t - \frac{r}{c} \right)$$

To solve the time integral, for the dipole we make the ansatz of a harmonic oscillation

$$p = p_0 \cdot \sin \omega \left(t - \frac{r}{c} \right) \tag{21.15}$$

Hence, $\ddot{p} = -\omega^2 \cdot p_0 \cdot \sin \omega \, (t - r/c)$. Substituting into (21.14) and integrating over time within the limits 0 and T, one obtains

$$E = \frac{\omega^4 p_0^2 \cdot T}{3c^3 T} \tag{21.16a}$$

Since $\omega = 2\pi/T$ and $cT = \lambda$, on the average the energy radiated per second is

$$E = \frac{16}{3} \cdot \frac{\pi^4 c \, p_0^2}{\lambda^4} \tag{21.16b}$$

Applying this formula for atomic dipoles one realizes that short wavelengths are scattered much more than long wavelengths. For example, this explains the blue color of the sky, because the short-wavelength, blue component in the white sunlight is scattered much more (compare Exercise 21.8) and if one takes into account that the dipole also receives (absorbs) like it emits (irradiates).

Radiation of an arbitrarily oscillating charge distribution

Now the field of an arbitrarily oscillating charge distribution will be calculated. We restrict ourselves to the limit of long waves and consider the far field. The vector potential is

$$\mathbf{A}(\mathbf{r}, t) = \frac{1}{c} \int \frac{\mathbf{j}\left(\mathbf{r}', t - |\mathbf{r} - \mathbf{r}'|/c\right)}{|\mathbf{r} - \mathbf{r}'|} dV' \tag{21.17}$$

Because in our case the region of the oscillating charge distribution is small compared to the distance from the point of observation (Figure 21.4), $s \ll r$, we set $|\mathbf{r} - \mathbf{r}'| \approx r$, and from equation (21.17) we obtain

$$\mathbf{A}(\mathbf{r}, t) = \frac{1}{cr} \int \mathbf{j}\left(\mathbf{r}', t - \frac{r}{c}\right) dV'$$

Since $\mathbf{H} = \text{curl } \mathbf{A}$, we obtain in the far zone

$$\mathbf{H} = -\frac{1}{c^2 r} \left(\frac{\mathbf{r}}{r} \times \frac{\partial}{\partial t} \int \mathbf{j}\left(\mathbf{r}', t - \frac{r}{c}\right) dV' \right)$$

Because $\partial \mathbf{E}/\partial t = c \, \text{curl } \mathbf{H}$, by taking the curl (far zone, dominant term) we obtain

$$\frac{\partial \mathbf{E}(\mathbf{r}, t)}{\partial t} = +\frac{1}{c^2 r^3} \left\{ \mathbf{r} \times \left[\mathbf{r} \times \frac{\partial^2}{\partial t^2} \int \mathbf{j}\left(\mathbf{r}', t - \frac{r}{c}\right) dV' \right] \right\}$$

For the change of the charge, we define

$$\dot{\mathbf{q}} = \int \mathbf{j}\left(\mathbf{r}', t - \frac{r}{c}\right) dV' \tag{21.18}$$

so that

$$\mathbf{E} = \frac{(\ddot{\mathbf{q}} \times \mathbf{r}) \times \mathbf{r}}{c^2 r^3}, \qquad \mathbf{H} = \frac{\ddot{\mathbf{q}} \times \mathbf{r}}{c^2 r^2} \tag{21.19}$$

The analogy to the equation (21.11) is obvious. Also, for arbitrary emitters the picture of outgoing spherical waves arises in the far field, as for an oscillating electric dipole. However, the vector $\mathbf{q}$ replacing the dipole moment in (21.10) contains now an additional dependence on the direction due to retardation. In particular, from a comparison of (21.19)

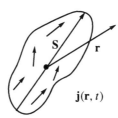

Figure 21.4. The dimension of the oscillating system is s. In the volume V currents are flowing with the density $\mathbf{j}(\mathbf{r}, t)$.

and (21.10) we also see that for an arbitrarily oscillating charge distribution the vectors $(\mathbf{r}, \mathbf{E}, \mathbf{H})$ form an orthogonal system, as in the case of the electric dipole.

For the following consideration we improve our approximation by expanding the distance vector

$$|\mathbf{r} - \mathbf{r}'| = \sqrt{r^2 - 2\mathbf{r} \cdot \mathbf{r}' + r'^2}$$

into a Taylor series

$$|\mathbf{r} - \mathbf{r}'| \approx r - \frac{\mathbf{r} \cdot \mathbf{r}'}{r} + \frac{(\mathbf{r} \times \mathbf{r}')^2}{2r^3} \cdots$$

So, for $\dot{\mathbf{q}}$ we may write

$$\dot{\mathbf{q}} = \int \mathbf{j}\left(\mathbf{r}', t - \frac{r}{c} + \frac{\mathbf{r} \cdot \mathbf{r}'}{rc}\right) dV' \tag{21.20}$$

or

$$\dot{\mathbf{q}} = \int \left(\mathbf{j}\left(\mathbf{r}', t - \frac{r}{c}\right) + \dot{\mathbf{j}}\left(\mathbf{r}', t - \frac{r}{c}\right)\frac{\mathbf{r} \cdot \mathbf{r}'}{cr}\right) dV' \equiv \dot{\mathbf{q}}_1 + \dot{\mathbf{q}}_2$$

The integral $\dot{\mathbf{q}}_1 = \int \mathbf{j}\left(\mathbf{r}', t - r/c\right) dV'$ describes the case (21.18) and (21.19) considered up till now. It is small compared to the second term $\dot{\mathbf{q}}_2$ for weak, rapidly varying currents. For this second term

$$\dot{\mathbf{q}}_2 = \frac{1}{cr} \int (\mathbf{r} \cdot \mathbf{r}') \dot{\mathbf{j}}\left(\mathbf{r}', t - \frac{r}{c}\right) dV'$$

we obtain

$$\dot{\mathbf{q}}_2 = \frac{-1}{2cr} \mathbf{r} \times \int \mathbf{r}' \times \dot{\mathbf{j}}\left(\mathbf{r}', t - \frac{r}{c}\right) dV'$$

Since $\mathbf{j}\, dV' = I\, d\mathbf{l}'$ where I is the current

$$\dot{\mathbf{q}}_2 = \frac{-\mathbf{r}}{2cr} \times \int \mathbf{r}' \times \dot{I}\left(t - \frac{r}{c}\right) d\mathbf{l}'$$

If we further introduce the magnetic moment

$$\mathbf{m}\left(t - \frac{r}{c}\right) = \frac{I\,(t - r/c)}{2c} \int \mathbf{r}' \times d\mathbf{l}' \tag{21.21}$$

we obtain

$$\dot{\mathbf{q}}_2 = -\frac{\mathbf{r}}{r} \times \dot{\mathbf{m}}\left(t - \frac{r}{c}\right)$$

Now we substitute this relation into $\mathbf{S} = c/(4\pi)\mathbf{E} \times \mathbf{H}$, and the Poynting vector becomes

$$\mathbf{S} = \frac{|(\ddot{\mathbf{q}}_2 \times \mathbf{r})|^2}{4c^3 r^4}\frac{\mathbf{r}}{r} = \frac{\ddot{m}^2 \sin^2\theta}{4\pi c^3 r^2}\frac{\mathbf{r}}{r} \tag{21.22}$$

We find the same $\sin^2\theta$-dependence of the Poynting vector as in (21.12a). Here, the angle θ is the angle between the magnetic moment $\mathbf{m}$ and the point of observation $\mathbf{r}$. The realization of an oscillating magnetic dipole may be seen, in principle, from Figure 21.5. The distance

Figure 21.5. On the radiation of a straight antenna.

Figure 21.6. On the radiation of a closed oscillating current.

of the plates is very small, but the closed antenna is large so that r' becomes comparable with the wavelength λ and the next approximation ($\ddot{q}_2$ in (21.20)) becomes important.

For the electric dipole already treated, $\mathbf{j}(\mathbf{r}, t) = e\mathbf{v}(\mathbf{r}')\delta(\mathbf{r} - \mathbf{r}')$, and so

$$\dot{\mathbf{q}}_1 = \int \mathbf{j} \, dV' = e\mathbf{v} = \dot{\mathbf{p}} \tag{21.23}$$

This is in agreement with (21.6).

Considering two metal plates A and B at the distance l connected by a wire (compare Figure 21.6), then $|\mathbf{j} \, dV'| = I \, dl'$ and

$$|\dot{\mathbf{q}}_1| = \left| \int \mathbf{j} \left(\mathbf{r}', t - \frac{r}{c} \right) dV' \right| = \int_A^B I \left(t - \frac{r}{c} \right) dl' = I \left(t - \frac{r}{c} \right) \int_A^B dl'$$

$$|\dot{\mathbf{q}}_1| = I \left(t - \frac{r}{c} \right) l \quad \text{and} \quad |\ddot{\mathbf{q}}_1| = \dot{I} \left(t - \frac{r}{c} \right) l \tag{21.24}$$

We obtain the emitted power N by integrating S from equation (21.22) over the surface of a sphere:

$$N = \frac{2}{3c^3} \left| \ddot{\mathbf{q}}_1 \left(t - \frac{r}{c} \right) \right|^2 = \frac{2\dot{I}^2}{3c^3} l^2 = \frac{2\omega^2 l^2}{3c^3} I^2 \tag{21.25}$$

In the last step, we have supposed the I is a sine-like alternating current ($I = I_0 e^{i\omega t} \rightarrow \dot{I} = i\omega I$) . The time average of the energy radiated during one period is

$$N = \frac{2\omega^2 l^2}{3c^3} \overline{I^2} \equiv R_s \overline{I^2} \tag{21.26}$$

The quantity $R_s = 2\omega^2 l^2/(3c^3)$ is denoted the *radiation impedance*. So, equation (21.26) becomes

$$N = R_s \cdot \overline{I^2} = R_s I_{\text{eff}}^2 \tag{21.27}$$

where the effective current is given by

$$I_{\text{eff}} = \sqrt{\overline{I^2}}$$

Example 21.1: Exercise: Radiation of rotating charges

A charge e moves at a constant angular velocity ω_0 along a circular orbit of radius a (Figure 21.7).

(a) Determine the electromagnetic fields $\mathbf{E}$ and $\mathbf{H}$ in the far zone ($r \gg a$) for nonrelativistic motion of the particle. Investigate the polarization of the radiation. What is the value of the radiant power?

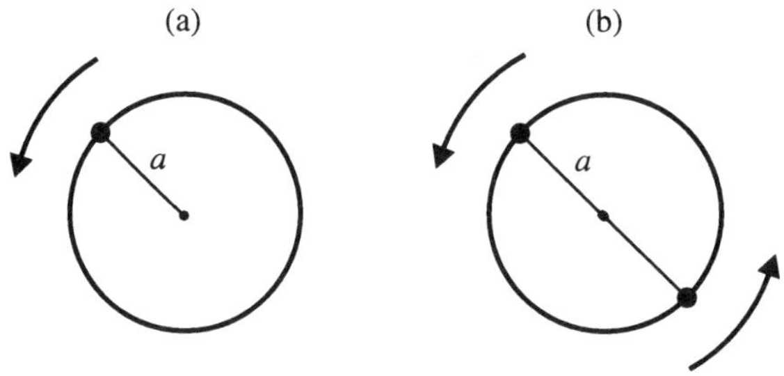

Figure 21.7. Rotating single charge (a) and double charges (b).

(b) Answer the same questions when two identical charges e move along the circle at equal speeds.

Solution   A charge distribution or a current distribution that is periodic in time (period $T = 2\pi/\omega_0$) may be decomposed in terms of a Fourier series:

$$\begin{pmatrix} \rho(\mathbf{r}, t) \\ \mathbf{j}(\mathbf{r}, t) \end{pmatrix} = \sum_{n=-\infty}^{+\infty} \begin{pmatrix} \rho_n(\mathbf{r}) \\ \mathbf{j}_n(\mathbf{r}) \end{pmatrix} e^{-in\omega_0 t} \tag{21.28}$$

$$\begin{pmatrix} \rho_n(\mathbf{r}) \\ \mathbf{j}_n(\mathbf{r}) \end{pmatrix} = \frac{1}{T} \int_0^T \begin{pmatrix} \rho(\mathbf{r}, t) \\ \mathbf{j}(\mathbf{r}, t) \end{pmatrix} e^{in\omega_0 t} \, dt$$

Electrodynamics may be described by the retarded potentials ϕ and $\mathbf{A}$. In the Lorentz gauge $\phi(\mathbf{r}, t)$ and $\mathbf{A}(\mathbf{r}, t)$ are decoupled *linear* functionals of $\rho(\mathbf{r}, t)$ or $\mathbf{j}(\mathbf{r}, t)$, respectively; that is, we may express them as

$$\begin{pmatrix} \phi(\mathbf{r}, t) \\ \mathbf{A}(\mathbf{r}, t) \end{pmatrix} = \int d^3 r' \, dt' \, \frac{1}{|\mathbf{r} - \mathbf{r}'|} \sum_n \begin{pmatrix} \rho_n(\mathbf{r}) \\ \frac{1}{c}\mathbf{j}_n(\mathbf{r}) \end{pmatrix} e^{-in\omega_0 t'} \delta\left(t' + \frac{|\mathbf{r} - \mathbf{r}'|}{c} - t \right) \tag{21.29}$$

$$= \int \frac{d^3 r'}{|\mathbf{r} - \mathbf{r}'|} \sum_n e^{-in\omega_0 t} \begin{pmatrix} \rho_n(\mathbf{r}) \\ \frac{1}{c}\mathbf{j}_n(\mathbf{r}) \end{pmatrix} e^{in\omega_0 |\mathbf{r} - \mathbf{r}'|/c}$$

$$= \sum_n \left\{ \int d^3 r' \begin{pmatrix} \rho_n(\mathbf{r}) \\ \frac{1}{c}\mathbf{j}_n(\mathbf{r}) \end{pmatrix} \frac{e^{in\omega_0 |\mathbf{r} - \mathbf{r}'|/c}}{|\mathbf{r} - \mathbf{r}'|} \right\} e^{-in\omega_0 t}$$

$$= \sum_n \begin{pmatrix} \phi_n(\mathbf{r}) \\ \mathbf{A}_n(\mathbf{r}) \end{pmatrix} e^{-in\omega_0 t} = \sum_n \begin{pmatrix} \phi_n(\mathbf{r}, t) \\ \mathbf{A}_n(\mathbf{r}, t) \end{pmatrix}$$

The ϕ_n and $\mathbf{A}_n$ possessing now a harmonic time-dependence can be treated separately. The magnetic field is $\mathbf{B}_n = \nabla \times \mathbf{A}_n$, and outside the sources we have for the radiation fields ($n \neq 0$):

$$\nabla \times \mathbf{B}_n = \frac{1}{c} \frac{\partial}{\partial t} \mathbf{E}_n(\mathbf{r}, t) = \frac{1}{c} \frac{\partial}{\partial t} \left(-\nabla \phi_n(\mathbf{r}) - \frac{1}{c} \frac{\partial}{\partial t} \mathbf{A}(\mathbf{r}) \right) e^{-in\omega_0 t}$$

$$= \frac{-in\omega_0}{c} \mathbf{E}_n$$

thus,

$$E_n(\mathbf{r}, t) = \frac{ic}{n\omega_0} \nabla \times (\nabla \times \mathbf{A}_n(\mathbf{r}))$$

that is, we need the vector potentials only.

We want to consider the fields far from the source, so $|\mathbf{r}| \gg d$, if d is the extension of the source (see Figure 21.8). Then, approximately $R = |\mathbf{r} - \mathbf{r}'| \approx r - \mathbf{n} \cdot \mathbf{r}'$ with $\mathbf{n} = \mathbf{r}/r$; $\mathbf{k}_n = (\omega_n/c)\mathbf{n}$. Furthermore, we assume that the dimensions of the source are much smaller than the wavelength of the radiation (as, e.g., for an atom), so $\lambda = 2\pi/k_0 \gg d$. But this means that the frequencies occurring are bounded, so $A_n = 0$ for large n. We get $e^{ik_n|\mathbf{r}-\mathbf{r}'|} \approx e^{ik_n r} e^{-i\mathbf{k}_n \mathbf{r}'}$.

In order of magnitude, $e^{-i\mathbf{k}_n \cdot \mathbf{r}'} \approx 1 - id/\lambda$. On the other hand, the frequency of the radiation is bounded below by the eigenfrequency ω_0 of the source, that is, $k_0 < k_n$. Let the observer be at large distance so that $r \gg 1/k_0 > 1/k_n$ (far zone). Then, in order of magnitude

$$|\mathbf{r} - \mathbf{r}'|^{-1} \approx \frac{1}{r}\left(1 + \frac{d}{r}\right), \qquad \left(1 + \frac{d}{r}\right)\left(1 - i\frac{d}{\lambda}\right) \approx 1 + \frac{d}{r} - i\frac{d}{\lambda} \approx 1 - i\frac{d}{\lambda} \tag{21.30}$$

According to the considerations $\mathbf{A}_n(\mathbf{r})$ (see 21.29) can be approximated by

$$\lim_{k_n r \to \infty} \mathbf{A}_n(\mathbf{r}) = \frac{e^{ik_n r}}{rc} \int \mathbf{j}_n(\mathbf{r}') e^{-i\mathbf{k}_n \mathbf{r}'} d^3 r' \tag{21.31}$$

$$= \frac{e^{ik_n r}}{rc} \sum_{m=0}^{\infty} \frac{(-i\omega_n/c)^m}{m!} \int \mathbf{j}_n(\mathbf{r}')(\mathbf{n} \cdot \mathbf{r}')^m d^3 r'$$

$$= \sum_m \mathbf{A}_n^{(m)}(\mathbf{r})$$

Figure 21.8. The problem geometry.

where we have approximated the denominator by $1/r$ due to (21.30). The sum over m converges rapidly due to $k_n d \ll 1$. At first, we consider the individual summands for a fixed value of n:

$$m = 0: \quad \mathbf{A}_n^{(0)}(\mathbf{r}) = \frac{e^{ik_n r}}{rc} \int \mathbf{j}_n(\mathbf{r}') \, d^3 r'$$

Using the continuity equation and the definition of the electric dipole moment, we obtain $\int \mathbf{j}(\mathbf{r}')d^3 r' = -\int \mathbf{r}' (\nabla' \mathbf{j}_n(\mathbf{r}')) \, d^3 r' = -\int \mathbf{r}' (i\omega_n)\rho_n(\mathbf{r}') \, d^3 r' = -i\omega_n \mathbf{p}_n$.

So,

$$\mathbf{A}_n^{(0)}(\mathbf{r}) = -ik_n \frac{e^{ik_n r}}{r} \mathbf{p}_n$$

$$m = 1: \quad \mathbf{A}_n^{(1)}(\mathbf{r}) = (-ik_n) \frac{e^{ik_n r}}{r} \int \mathbf{j}_n(\mathbf{r}')(\mathbf{n} \cdot \mathbf{r}') \, d^3 r'$$

From vector algebra, $\mathbf{j}_n(\mathbf{r}')(\mathbf{n} \cdot \mathbf{r}') = \frac{1}{2}\left[(\mathbf{n} \cdot \mathbf{r}')\mathbf{j}_n + (\mathbf{n} \cdot \mathbf{j}_n)\mathbf{r}'\right] + \frac{1}{2}(\mathbf{r}' \times \mathbf{j}_n) \times \mathbf{n}$.

Similarly, taking into account the definition of the magnetic dipole moment of a charge distribution $\mathbf{m}_n = 1/(2c) \int d^3 r' \, (\mathbf{r}' \times \mathbf{j}_n(\mathbf{r}'))$ using (21.30) we can show

$$\frac{1}{2} \int \left[(\mathbf{n} \cdot \mathbf{r}')\mathbf{j}_n + (\mathbf{n} \cdot \mathbf{j}_n)\mathbf{r}'\right] d^3 r' = -\frac{ik_n c}{2} \int \mathbf{r}'(\mathbf{n} \cdot \mathbf{r}')\rho_n(\mathbf{r}') \, d^3 r'$$

and

$$\mathbf{A}_n^{(1)}(\mathbf{r}) = -\frac{k_n^2}{2} \frac{e^{ik_n r}}{r} \int \mathbf{r}'(\mathbf{n} \cdot \mathbf{r}')\rho_n(\mathbf{r}') \, d^3 r' + ik_n \frac{e^{ik_n r}}{r}(\mathbf{n} \times \mathbf{m}_n)$$

Later on we shall see that the first term is connected with the electric quadrupole moment, and we shall not need the summands for $m > 1$.

Now, we sum $A_n^{(m)}$ over n:

$$\mathbf{A}_n^{(0)}(\mathbf{r}, t) = \sum_n \mathbf{A}_n^{(0)}(\mathbf{r})e^{-i\omega_0 nt}$$

$$= \frac{1}{r} \sum_n \left(-i\frac{\omega_n}{c}\right) e^{-i\omega_0 n(t-r/c)} \mathbf{p}_n(\mathbf{r})$$

$$= +\frac{1}{c}\frac{\partial}{\partial t}\frac{1}{r} \sum_n \mathbf{p}_n(\mathbf{r})e^{-i\omega_0 n(t-r/c)} = +\frac{1}{rc}\frac{\partial}{\partial t}\mathbf{p}\left(\mathbf{r}, t - \frac{r}{c}\right)$$

$$= +\frac{1}{rc}\dot{\mathbf{p}}_{ret}$$

$$\mathbf{A}_n^{(1)}(\mathbf{r}, t) = \frac{1}{r} \sum_n i\frac{n\omega_0}{c}(\mathbf{n} \times \mathbf{m}_n)e^{-in\omega_0(t-r/c)} - \frac{1}{r} \sum_n \frac{(n\omega_0)^2}{2c^2} \int \mathbf{r}'(\mathbf{n} \cdot \mathbf{r}')\rho_n(\mathbf{r}') \, d^3 r' \, e^{-in\omega_0(t-r/c)}$$

$$= -\frac{1}{rc}\mathbf{n} \times \dot{\mathbf{m}}_n + \frac{1}{2rc^2}\frac{\partial^2}{\partial t^2} \int \mathbf{r}'(\mathbf{n} \cdot \mathbf{r}')\rho\left(\mathbf{r}', t - \frac{r}{c}\right) d^3 r'$$

With $\nabla r = \mathbf{r}/r$ and $\nabla 1/r = -\mathbf{n}/r^2$ and due to (21.31), in the limit $r \to \infty$:

$$\mathbf{H}_n = \nabla \times \mathbf{A}_n = \frac{1}{r}\mathbf{A}_n \times \mathbf{n} + ik_n \times \mathbf{A}_n$$

$$\Rightarrow \quad ik_n \times \mathbf{A}_n$$

$$E_n = \frac{i}{k_n} \nabla \times H_n = -\nabla \times (n \times A_n) = -n(\nabla A_n) + (n\nabla)A_n$$

$$\Rightarrow \quad -n(ik_n A_n) + ik_n A_n$$

$$= -ik_n \times (n \times A_n) = -ik_n n \times (n \times A_n) = H_n \times n$$

and therefore also

$$H = \nabla \times A$$

$$\Rightarrow \sum_n (ik_n \times A_n)e^{-in\omega_0 t} = -\frac{1}{c}\frac{\partial}{\partial t}\left(n \times \sum_n A_n e^{-in\omega_0 t}\right) = \frac{1}{c}\dot{A} \times n$$

and analogously, $E = H \times n$.

Here, omitting the index *ret* we give only H:

$$H = \frac{1}{rc^2}\left[(n \times \ddot{p}) + (\ddot{m} \times n) \times n + \frac{1}{6c}\dddot{Q}(n) \times n + \ldots\right] \tag{21.32}$$

Here,

$$n \times \int r'(n \cdot r')\rho_n(r')\, d^3r' = \frac{1}{3}n \times Q_n(n)$$

with the quadrupole tensor

$$Q_{ik}^n = \int \left(3x_i x_k - r^2\delta_{ik}\right)\rho_n(r)\, d^3r \qquad \text{and} \qquad Q_n(n) = \hat{Q}_n \cdot n$$

(a) Let the electron move along a circular orbit of radius a in the (x, y)-plane (Figure 21.9):
$r(t) = a(e_x \cos \omega_0 t + e_y \sin \omega_0 t)$, so $d = er(t)$, $m \equiv 0$, Q is neglected. Then, $\ddot{d} = -\omega_0^2 d$.
Due to the choice of the polar coordinates $n = e_r$ and

$$e_x = \sin\theta \, \cos\varphi \, e_r + \cos\theta \, \cos\varphi \, e_\theta - \sin\theta \, e_\varphi$$

$$e_y = \sin\theta \, \sin\varphi \, e_r + \cos\theta \, \sin\varphi \, e_\theta + \cos\theta \, e_\varphi$$

we have (21.32)

$$H = -\frac{\omega_0^2}{rc^2}(d \times n) = -\frac{ea\omega_0^2}{rc^2}\left(\cos\omega_0 t_0 e_x + \sin\omega_0 t_0 e_y\right) \times e_r$$

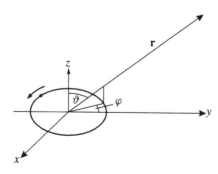

Figure 21.9. The problem geometry.

$$\mathbf{H} = -\frac{ea\omega_0^2}{rc^2} \left[\cos\theta \cos(\omega_0 t_0 - \varphi)\mathbf{e}_\varphi - \sin(\omega_0 t_0 - \varphi)\mathbf{e}_\theta \right], \qquad t_0 = t - \frac{r}{c} \qquad (21.33)$$

and

$$\mathbf{E} = \mathbf{H} \times \mathbf{n} = -\frac{ea\omega_0^2}{rc^2} \left[-\sin(\omega_0 t_0 - \varphi)\mathbf{e}_\varphi + \cos\theta \cos(\omega_0 t_0 - \varphi)\mathbf{e}_\theta \right]$$

For $\theta = 0, \pi$ the wave is circularly polarized (the observer sees the circular orbit from above), for $\theta = \pi/2$ the wave is linearly polarized (the observer sees the electron orbit from the side).

In all other cases, one has elliptical polarization. The average radiant power per solid angle in direction $\mathbf{n}$ is

$$\frac{dI_n}{d\Omega} = r^2 \langle |\mathbf{S}\,\mathbf{n}| \rangle$$

where

$$\mathbf{S} = \frac{c}{4\pi}\mathbf{E} \times \mathbf{H} = \frac{c}{4\pi}(\mathbf{H} \times \mathbf{n}) \times \mathbf{H} = \frac{c}{4\pi}H^2\mathbf{n}$$

With (21.33) one obtains

$$\frac{dI_n}{d\Omega} = \frac{e^2 a^2 \omega_0^4}{8\pi c^3}(\cos^2\theta + 1)$$

and the total radiant power

$$I = \int \frac{dI_n}{d\Omega}\,d\Omega = 2\pi \frac{e^2 a^2 \omega_0^4}{8\pi c^3} \int_0^\pi d\theta\,\sin\theta(\cos^2\theta + 1) = \frac{2}{3}\frac{e^2 a^2 \omega_0^4}{c^3}$$

So, the radiation losses of a particle on a circular orbit increase as the fourth power of the angular velocity. In fact, the accessible final energy of synchrotron accelerators is limited by this effect. However, the radiation formula has then to be corrected relativistically.

(b) Let the coordinates of the two particles be $\mathbf{r}_1(t) = -\mathbf{r}_2(t) = \mathbf{r}(t)$; then, the lowest contribution is quadrupole-like.

Now

$$2e(\mathbf{n} \cdot \mathbf{r})\mathbf{r} = 2ea^2 \left[(\mathbf{n} \cdot \mathbf{e}_x \cos\omega_0 t + \mathbf{n} \cdot \mathbf{e}_y \sin\omega_0 t)(\mathbf{e}_x \cos\omega_0 t + \mathbf{e}_y \sin\omega_0 t) \right]$$
$$= 2ea^2 \sin\theta \cos(\omega_0 t - \varphi)$$
$$\times \left[\sin\theta \cos(\omega_0 t - \varphi)\mathbf{e}_r + \cos\theta \cos(\omega_0 t - \varphi)\mathbf{e}_\theta + \sin(\omega_0 t - \varphi)\mathbf{e}_\varphi \right]$$

and

$$\frac{1}{3}\mathbf{Q}(\mathbf{n}) \times \mathbf{n} = 2e(\mathbf{n} \cdot \mathbf{r}(t_0)) \times \mathbf{e}_r$$
$$= ea^2 \sin\theta[-\cos\theta(1 + \cos 2(\omega_0 t_0 - \varphi))]\mathbf{e}_\varphi + \sin 2(\omega_0 t_0 - \varphi)\mathbf{e}_\theta$$

and the associated field is

$$\mathbf{H} = \frac{1}{2c^3 r}\,\dddot{\mathbf{Q}}(\mathbf{n}) \times \mathbf{n}$$
$$= -\frac{4ea^2\omega^3}{c^3 r}\sin\theta \left[\cos\theta \sin(2\omega_0 t_0 - 2\varphi)\mathbf{e}_\varphi + \cos(2\omega_0 t_0 - 2\varphi)\mathbf{e}_\theta \right]$$

The emitted wave has twice the frequency of the circular motion. The radiant power can be calculated analogously to (a):

$$\frac{dI'_n}{d\Omega} = \frac{2e^2a^4\omega^6}{c^5\pi}\sin^2\theta(\cos^2\theta + 1), \qquad I' = \frac{32}{5}\frac{e^2a^4\omega^6}{c^5}$$

Due to negative interference, the radiant power is reduced by the factor

$$\frac{I'}{I} = \frac{48}{5c^2}a^2\omega^2 \ll 1$$

Example 21.2: Exercise: The Rutherford atom model

The Rutherford atom model assumes that electrons are moving around the nucleus in circular orbits. Estimate the lifetime T of the hydrogen atom within this classical model.

Solution The balance condition for the motion of an electron (H-atom) on a circle of radius R reads $m\omega^2 R = e^2/R^2$, that is, $\omega^2 R^2 = e^2/(mR)$. The energy of the electron on this circular orbit is

$$E = \frac{1}{2}mv^2 - \frac{e^2}{R} = -\frac{1}{2}\frac{e^2}{R}$$

The energy loss is $-I$, so $dE/dt = -I$. Using the result of the Exercise 21.1 (a) one obtains

$$\frac{e^2}{2R^2}\frac{dR}{dt} = -\frac{2}{3}\frac{e^2}{c^3}\omega^4R^2 = -\frac{2}{3}\frac{e^2}{c^3}\frac{e^4}{m^2R^4} \quad \Rightarrow \quad \frac{dR}{dt} = -\frac{4}{3}\frac{e^4}{c^3m^2}\frac{1}{R^2}$$

With $R(t_0) = a$ (Bohr radius), $R(t) = R$ this yields the integral

$$t - t_0 = -\frac{1}{4}\frac{m^2c^3}{e^4}(R^3 - a^3)$$

With $t_0 = 0$ and the requirement $R(T) = 0$, the decay time is $T = m^2c^3a^3/(4e^4) = 3.2 \times 10^{-13}$s. Therefore, in the framework of the classical physics the stability of the atoms cannot be understood. This and other problems led finally to the development of quantum mechanics.

Example 21.3: Exercise: Radiation of a uniformly charged symmetric top

A uniformly charged symmetric top (radius R, charge Q) rotates at constant angular velocity ω about its longitudinal axis. How much energy is radiated per unit time?

Solution No electromagnetic waves are emitted, because the charge density and the current density are constant in time due to the rotational symmetry of the charge distribution.

Example 21.4: Exercise: Motion of a point charge in the field of a stationary point charge

A point charge moves in the field of a stationary point charge.

(a) What are the equations of motion (relativistic, plane polar coordinates)?

(b) Solve the equations of motion.

Solution   We suppose that one of the charges is stationary because the generalization that both charges are moving about the common center of gravity is extremely complicated. Then, the finite propagation velocity of electromagnetic signals has to be taken into account. This problem has never been solved completely (analytically). But, if one of the particles is so heavy that the center of gravity and the heavy particle coincide, then this particle may be viewed to be stationary (e.g., electron and proton), and the problem becomes accessible for an analytic solution.

(a) Neglecting the radiation loss the (relativistic) equations of motion are

$$\frac{d\mathbf{p}}{dt} = q\mathbf{E}, \qquad \frac{dE_{\text{kin}}}{dt} = q\mathbf{v} \cdot \mathbf{E}$$

Here, $\mathbf{p} = \gamma m_0 \mathbf{v}$ is the momentum ($\gamma = (1 - v^2/c^2)^{-1/2}$) and $E_{\text{kin}} = \gamma m_0 c^2$ is the kinetic energy of the light particle of mass m_0, charge q, and velocity $\mathbf{v}$. t denotes the laboratory time. $\mathbf{E} = q'\mathbf{e}_r/r^2$ is the electric field of the heavy particle. We use cylindrical coordinates because the motion is planar. Then

$$\mathbf{r} = r\mathbf{e}_r, \qquad \mathbf{v} = \dot{\mathbf{r}} = \dot{r}\mathbf{e}_r + r\dot{\varphi}\mathbf{e}_\varphi$$

and so

$$q\mathbf{E} = qq'\frac{\mathbf{e}_r}{r^2} = \frac{d}{dt}\left(\gamma m_0 \dot{r}\mathbf{e}_r + \gamma m_0 r\dot{\varphi}\mathbf{e}_\varphi\right)$$

$$= \frac{d}{dt}(\gamma m_0 \dot{r})\,\mathbf{e}_r + \gamma m_0 \dot{r}\dot{\varphi}\mathbf{e}_\varphi + \frac{d}{dt}(\gamma m_0 r\dot{\varphi})\,\mathbf{e}_\varphi - \gamma m_0 r\dot{\varphi}^2\mathbf{e}_r$$

We compare the components in φ- and r-direction,

$$\frac{d}{dt}(\gamma m_0 r\dot{\varphi}) + \gamma m_0 \dot{r}\dot{\varphi} = 0$$

$$\frac{d}{dt}(\gamma m_0 \dot{r}) - \gamma m_0 r\dot{\varphi}^2 = qq'\frac{1}{r^2} \tag{21.34}$$

Multiplying the first equation by r, we find that $(d/dt)(\gamma m_0 r^2 \dot{\varphi}) = 0$ or

$$\gamma m_0 r^2 \dot{\varphi} = l \tag{21.35}$$

is constant in time. Of course, the second equation of motion $\dot{E}_{\text{kin}} = q\mathbf{v} \cdot \mathbf{E}$ yields the conservation of energy

$$\gamma m_0 c^2 + \frac{qq'}{r} = E \tag{21.36}$$

(b) Now we turn again to the radial equation which is multiplied by γ and rewritten by means of (21.36)

$$m_0 \left\{\gamma \frac{d}{dt}(\gamma \dot{r}) - r\gamma^2 \dot{\varphi}^2\right\} = \frac{qq'}{m_0 c^2 r^2}\left[E - \frac{qq'}{r}\right]$$

On the left-hand side we introduce the derivative with respect to φ by $d/dt = \dot{\varphi}d/d\varphi$; then for the orbit $r(\varphi)$ we have

$$m_0 \left\{\gamma\dot{\varphi}\frac{d}{d\varphi}\left(\gamma\dot{\varphi}\frac{dr}{d\varphi}\right) - r(\gamma\dot{\varphi})^2\right\} = \frac{qq'}{m_0 c^2 r^2}\left[E - \frac{qq'}{r}\right]$$

Due to (21.35) the product $\gamma\dot{\varphi}$ is just $l/(m_0 r^2)$,

$$l^2 \left\{ \frac{d}{d\varphi} \frac{1}{r^2} \frac{dr}{d\varphi} - \frac{1}{r} \right\} = \frac{qq'}{c^2} \left[E - \frac{qq'}{r} \right]$$

Introducing finally $x = 1/r$, then the differential equation is simplified appreciably:

$$\frac{d^2}{d\varphi^2} x + \alpha^2 x = -\frac{qq'}{l^2 c^2} E \tag{21.37}$$

with

$$\alpha^2 = 1 - \left(\frac{qq'}{lc} \right)^2 \tag{21.38}$$

This equation may be solved at once:

$$x(\varphi) = \frac{1}{r(\varphi)} = -\frac{qq'}{l^2 c^2 \alpha^2} E + A \cos(\alpha\varphi + \varphi_0) \tag{21.39}$$

So, the orbit is not closed ($\alpha \neq 1$), but it is a rosette. Against all appearance, the integration constant A is fixed already by the angular momentum l and the total energy E of the particle; namely, we have not yet used that

$$\gamma^2 \left(1 - \frac{v^2}{c^2} \right) = 1 \tag{21.40}$$

Noting

$$\gamma^2 \frac{v^2}{c^2} = \frac{\gamma^2}{c^2} \left[\dot{r}^2 + r^2 \dot{\varphi}^2 \right] = \frac{1}{c^2} \gamma^2 \dot{\varphi}^2 \left[\left(\frac{dr}{d\varphi} \right)^2 + r^2 \right]$$

$$= \left(\frac{l}{m_0 c} \right)^2 \left(\left(\frac{dx}{d\varphi} \right)^2 + x^2 \right) \tag{21.41}$$

using (21.36) we obtain finally the condition

$$\left(\frac{dx}{d\varphi} \right)^2 + x^2 - \frac{1}{l^2 c^2} [E - qq'x]^2 = -\left(\frac{m_0 c}{l} \right)^2 \tag{21.42}$$

Substituting the solution (21.39) into (21.42), after a long computation we obtain

$$A = \pm \frac{1}{lc\alpha^2} \sqrt{E^2 - \alpha^2 m_0^2 c^4} \tag{21.43}$$

where the distinct signs correspond to the two orientations of the conic sections. Of course, φ_0 corresponds to the orientation of the orbit.

For two mutually attracting particles ($qq' < 0$), the orbit is given by

$$\frac{lc\alpha^2}{r} = \sqrt{1 - \alpha^2} E \pm \sqrt{E^2 - \alpha^2 m_0^2 c^4} \cos(\alpha\varphi + \varphi_0) \tag{21.44}$$

A necessary condition for the existence of periodic solutions is $\alpha^2 > 0$ or $l > |qq'/c|$; so, the angular momentum cannot become arbitrarily small. For quantum mechanical reasons the electron has the orbital angular momentum $l = n\hbar (n \in |N_0)$. So, if there is an electron in the field of a point charge Ze, then we must have $\hbar > Ze^2/c$ or $Z < \hbar c/e^2 \approx 137$, that is, for point-like atomic nuclei with $Z \gtrsim 137$ there are no solutions.

Furthermore, one realizes that for $E > mc^2$ the electron is free, that is, it removes arbitrarily far from the nucleus.

For $\alpha m_0 c^2 \leq E < m_0 c^2$ the electron is bound. The case $E = \alpha m_0 c^2$ corresponds to a circular orbit.

Example 21.5:   Exercise: Electron in an electromagnetic field

An electromagnetic wave described by a vector potential

$$\mathbf{A} = (0, A(x - ct), 0)$$

strikes an electron at rest. Here, $A(x - ct)$ is an arbitrary function vanishing as $t \rightarrow -\infty$. The components of the velocity of the electron are required. The problem has to be solved relativistically.

Solution   The electron obeys the equation of motion

$$\dot{\mathbf{p}} = e \left(\mathbf{E} + \frac{1}{c} \mathbf{v} \times \mathbf{B} \right) \tag{21.45}$$

The electric and magnetic field are expressed by the vector potential

$$\mathbf{E} = -\frac{1}{c} \frac{\partial \mathbf{A}}{\partial t}, \qquad \mathbf{B} = \operatorname{curl} \mathbf{A} \tag{21.46}$$

To calculate the derivatives of the vector potential, we substitute $u = x - ct$. Then we set

$$A' = \frac{dA}{du}$$

So, the partial derivatives are

$$\frac{\partial}{\partial x} A(u) = A' \qquad \text{and} \qquad \frac{\partial}{\partial t} A(u) = -cA'$$

For the total derivative with respect to time, we obtain

$$\frac{d}{dt} A(u) = \dot{A} = (\dot{x} - c)A' \qquad \text{and} \qquad \frac{dA^2}{dt} = 2AA'(\dot{x} - c)$$

Expressing the fields in equation (21.46) by these expressions, then

$$\mathbf{E} = (0, A', 0), \qquad \mathbf{B} = (0, 0, A')$$

For the Lorentz force term we obtain

$$\mathbf{v} \times \mathbf{B} = (\dot{y} A', -\dot{x} A', 0)$$

Substituting this into the equation of motion (21.46), then

$$\dot{\mathbf{p}} = \frac{e}{c} (\dot{y} A', (c - \dot{x}) A', 0) = \frac{e}{c} (\dot{y} A', -\dot{A}, 0)$$

so there is no component of the force in z-direction.

For the y-component, we get

$$\ddot{y} = -\frac{e}{mc} \dot{A}$$

or integrating ($\dot{y}(0) = 0$); $\dot{y} = -e/(mc)A$.

Then, for the x-component,

$$\ddot{x} = \frac{e}{mc}\dot{y}A' = -\left(\frac{e}{mc}\right)^2 AA'$$

$$\ddot{x} = -\left(\frac{e}{mc}\right)^2 \frac{1}{2(\dot{x}-c)}\frac{d}{dt}A^2$$

From this,

$$2\ddot{x}(\dot{x}-c) = -\left(\frac{e}{mc}\right)^2 \frac{d}{dt}A^2$$

and after integration (with $\dot{x}(0) = 0$):

$$\dot{x}^2 - 2c\dot{x} + \left(\frac{e}{mc}\right)^2 A^2 = 0$$

Solving for $\dot{x}$,

$$\dot{x} = c - \sqrt{c^2 - \left(\frac{e}{mc}\right)^2 A^2}$$

where the positive sign of the root has to be excluded since we must have $\dot{x} < c$.

Then $(e/(mc))^2 \ll c^2$, so that the root may be expanded:

$$c\sqrt{1 - \left(\frac{e}{mc^2}A\right)^2} = c\left(1 - \frac{1}{2}\left(\frac{e}{mc^2}A\right)^2 \pm \cdots\right)$$

Hence, we obtain

$$\dot{x} \approx \frac{1}{2c}\left(\frac{eA}{mc}\right)^2$$

The z-component is zero. So, the y-component of the velocity depends linearly on the vector potential, the x-component depends quadratically on the vector potential. If the vector potential is of oscillatory character, then the electron oscillates in y-direction, and the velocity component in x-direction has always a positive sign (pressure of the wave in x-direction).

Example 21.6: Radiation loss of a harmonically oscillating charge

A positive charge is attached to a spring (Figure 21.10) in such a way that a radiating harmonic oscillator results. We want to show that small radiation losses with a minor reaction on the motion of the oscillator may be decribed by introducing a frictional force proportional to the third derivative of the elongation.

Figure 21.10. A charge is attached to a spring.

The charge moves along the x-axis. If, for the moment, the radiation is not taken into account then the equation of motion for the mass point attached to the spring of eigenfrequency ω_0 is

$$m\frac{d^2x(t)}{dt^2} + m\omega_0^2 x(t) = 0 \tag{21.47}$$

The general solution is

$$x(t) = \text{Re}(x_0 e^{-i\omega_0 t}) \tag{21.48}$$

where x_0 may be complex. Then, the complex amplitude for the dipole moment of this motion is

$$p = ex_0$$

and the total irradiated energy per unit of time follows according to equation (21.16a)

$$E_{\text{radiated}} = \frac{c}{3}k^4|p|^2 = \frac{ck^4}{3}e^2|x_0|^2 = \frac{\omega_0^4}{3c^3}e^2|x_0|^2 \tag{21.49}$$

The time rate of change of the energy of the oscillating harmonic oscillator is obtained from (21.47) by multiplication by dx/dt:

$$m\frac{dx}{dt}\frac{d^2x}{dt^2} + m\omega_0^2 x\frac{dx}{dt} = 0$$

leading to

$$\frac{d}{dt}\left[\frac{m}{2}\left(\frac{dx}{dt}\right)^2 + \frac{m\omega_0^2}{2}x^2\right] = 0 \tag{21.50}$$

To see how the radiation loss (21.49) may be taken into account in the equation of motion (21.47), we recognize first that the time average of $(dx/dt)(d^3x/dt^3)$ is given by

$$\overline{\frac{dx}{dt}\frac{d^3x}{dt^3}} = \overline{[-|x_0|\omega_0 \sin(\omega_0 t + \varphi)][|x_0|\omega_0^3 \sin(\omega_0 t + \varphi)]} = -\frac{1}{2}|x_0|^2\omega_0^4 \tag{21.51}$$

Therefore, the radiation loss (21.49) may be written also in the form

$$E_{\text{radiation loss}} = -\frac{2}{3}\frac{e^2}{c^3}\overline{\frac{dx}{dt}\frac{d^3x}{dt^3}} \tag{21.52}$$

The decrease of the oscillator energy per unit of time has to be equal to the radiation loss per unit of time,

$$-\frac{d}{dt}\left[\frac{m}{2}\left(\frac{dx}{dt}\right)^2 + \frac{m\omega_0^2}{2}x^2\right] = E_{\text{radiation loss}} \tag{21.53}$$

or

$$\frac{d}{dt}\left[\frac{m}{2}\dot{x}^2 + \frac{m\omega_0^2}{2}x^2\right] = \frac{2}{3}\frac{e^2}{c^3}\dot{x}\,\dddot{x} \tag{21.54}$$

The assumption that the radiation damping has only a minor influence on the motion enters essentially in the form of the right-hand side of equation (21.53) because the irradiated energy $E_{\text{radiation loss}}$ has been calculated with the imperturbed solution $x(t)$ of the oscillating mass m. We also have used the approximation that in (21.52) the time average has been replaced by the actual value. After differentiating the left-hand side of (21.53) we obtain

$$m\dot{x}\ddot{x} + m\omega_0^2 x\dot{x} = \frac{2}{3}\frac{e^2}{c^3}\dot{x}\,\dddot{x} \tag{21.55}$$

and furthermore,

$$m\frac{d^2x}{dt^2} = -m\omega_0^2 x + \frac{2}{3}\frac{e^2}{c^3}\frac{d^3x}{dt^3} \tag{21.56}$$

This differential equation for the motion of the harmonically oscillating, charged mass point takes into account the radiation loss to a good approximation as far as the radiation loss is small. The damping of the motion caused in this way is small if the parameter

$$\frac{\text{energy loss per period}}{\text{stored energy}} = \frac{\dfrac{\omega_0^4 e^2 |x_0|^2}{3c^3}}{\dfrac{m}{2}\omega_0^2|x_0|^2} \Bigg/ \frac{\omega_0}{2\pi} = \frac{4\pi}{3}\frac{\omega_0 e^2}{mc^3}$$

is small. Then, the motion of oscillating mass is damped only over many cycles of the oscillation.

Example 21.7:   Exercise: Natural line width due to radiation loss

We consider an isolated system which emits dipole radiation mainly with the frequency ω_0. Due to the radiation, the energy of the system is diminished permanently. This implies that also frequencies $\omega = \omega_0 + \Delta\omega$ adjacent to ω_0 are emitted by the system. $\Delta\omega$ is called the *natural width* of the emission line. Show that for a radiating harmonic oscillator of mass m and charge e, in case of weak damping, the natural line width is given by

$$\Delta\omega = \frac{2}{3}\frac{e^2\omega_0^2}{mc^3}$$

Solution   According to the Example 21.6, the effect of the radiation loss upon the motion of the oscillating charge due to its dipole radiation can be taken into account by the differential equation

$$m\frac{d^2x}{dt^2} = -m\omega_0^2 x + \frac{2}{3}\frac{e^2}{c^3}\frac{d^3x}{dt^3} \tag{21.57}$$

To find the solution, we try the ansatz $x = x_0 e^{-i\omega t}$, leading us to the characteristic equation

$$-m\omega^2 = -m\omega_0^2 + i\frac{2}{3}\frac{e^2}{c^3}\omega^3 \tag{21.58}$$

This equation of third degree in ω is solved by perturbation theory. Since the radiation loss (second term on the right-hand side) has been assumed to be small, the solution in zero order is

$$\omega = \omega_0 \tag{21.59}$$

In first order we get

$$-m\omega^2 = -m\omega_0^2 + i\frac{2}{3}\frac{e^2}{c^3}\omega_0^3$$

yielding

$$\omega^2 = \omega_0^2 \left[1 - i \frac{2}{3} \frac{e^2}{mc^3} \omega_0 \right] \qquad \text{or} \qquad \omega = \omega_0 \left[1 - i \frac{1}{3} \frac{e^2}{mc^3} \omega_0 \right] \tag{21.60}$$

So, the corresponding complex solution for $x(t)$ is

$$x(t) = x_0 e^{-i\omega_0 t} e^{-(1/3)(e^2/(mc^3))\omega_0^2 t} \equiv x_0 e^{-i\omega_0 t} e^{-\gamma t}, \qquad \gamma = \frac{1}{3} \frac{e^2}{mc^3} \omega_0^2 \tag{21.61}$$

The real part gives the elongation at time t. Now, we assume that the actual elongation vanishes for $t < 0$, and for $t \geq 0$ is given by the real part of (21.61):

$$\overline{x}(t) \equiv \text{Re}\, x(t) = 0, \qquad x < t$$
$$\overline{x}(t) = \text{Re}\, x(t) = |x_0| e^{-\gamma t} \cos \omega_0 t, \qquad x \geq t \tag{21.62}$$

To find the frequency spectrum of this motion, we have to perform a Fourier decomposition of $x(t)$:

$$\overline{x}(t) = \frac{1}{\sqrt{2\pi}} \int_{-\infty}^{\infty} d\omega\, f(\omega) e^{-i\omega t} \tag{21.63}$$

Since $\overline{x}(t)$ is real, for $f(\omega)$ there is a crossing relation (compare Example 19.2 (equation (19.25))):

$$f^*(\omega) = f(-\omega) \tag{21.64}$$

and therefore,

$$\begin{aligned}
\overline{x}(t) &= \frac{1}{\sqrt{2\pi}} \int_{-\infty}^{0} d\omega\, f(\omega) e^{-i\omega t} + \frac{1}{\sqrt{2\pi}} \int_{0}^{\infty} d\omega\, f(\omega) e^{-i\omega t} \\
&= \frac{1}{\sqrt{2\pi}} \int_{0}^{\infty} d\omega\, f(-\omega) e^{+i\omega t} + \frac{1}{\sqrt{2\pi}} \int_{0}^{\infty} d\omega\, f(\omega) e^{-i\omega t} \\
&= \frac{1}{\sqrt{2\pi}} \int_{0}^{\infty} d\omega\, [f(\omega) e^{-i\omega t} + f^*(\omega) e^{+i\omega t}] = \frac{2}{\sqrt{2\pi}} \text{Re} \int_{0}^{\infty} d\omega\, f(\omega) e^{i\omega t}
\end{aligned} \tag{21.65}$$

Disregarding the phase, the contribution of the frequency ω to the motion (21.62) is given by $|f(\omega)|$. Because the energy of an oscillator is proportional to x^2, one is mostly interested in $|f(\omega)|^2$. The Fourier amplitude for the oscillator with the initial conditions included in (21.62) is

$$\begin{aligned}
f(\omega) &= \frac{1}{\sqrt{2\pi}} \int_{-\infty}^{\infty} dt\, \overline{x}(t) e^{i\omega t} \\
&= \frac{1}{\sqrt{2\pi}} \int_{0}^{\infty} \frac{|x_0|}{2} [e^{(-\gamma+i\omega_0+i\omega)t} + e^{(-\gamma-i\omega_0+i\omega)t}]\, dt \\
&= \frac{1}{\sqrt{2\pi}} \frac{|x_0|}{2} \left[\frac{e^{(-\gamma+i\omega_0+i\omega)t}}{-\gamma + i\omega_0 + i\omega} + \frac{e^{(-\gamma-i\omega_0+i\omega)t}}{-\gamma - i\omega_0 + i\omega} \right]\Big|_{0}^{\infty} \\
&= \frac{1}{\sqrt{2\pi}} |x_0| \frac{\gamma - i\omega}{\omega_0^2 - \omega^2 + \gamma^2 - i2\gamma\omega}
\end{aligned} \tag{21.66}$$

Hence, the square of the absolute value of the Fourier amplitude is

$$|f(\omega)|^2 = \frac{|x_0|^2}{2\pi} \frac{\gamma^2 + \omega^2}{(\omega_0^2 - \omega^2 + \gamma^2)^2 + 4\gamma^2\omega^2} \tag{21.67}$$

For small damping ($\gamma \ll \omega_0$) there is a resonance $\omega = \sqrt{\omega_0^2 + \gamma^2} \approx \omega_0$. Therefore, we consider the deviation $\omega' = \omega - \omega_0$ from the resonance frequency and expand (21.67) under the assumption that $\gamma/\omega_0 \ll 1$ and $\omega'/\omega_0 \ll 1$, that is, adjacent to the resonance. The various terms in (21.67) yield

$$\gamma^2 + \omega^2 = \gamma^2 + (\omega_0^2 + \omega'^2) = \omega_0^2 \left(1 + O\left(\frac{\gamma}{\omega_0}\right)\right)$$

$$(\omega_0^2 - \omega^2 + \gamma^2)^2 + 4\gamma^2\omega^2 = [\omega_0^2 - (\omega_0 + \omega')^2 + \gamma^2]^2 + 4\gamma^2(\omega_0 + \omega')^2$$

$$= (-2\omega_0\omega' - \omega'^2 + \gamma^2)^2 + 4\gamma^2(\omega_0^2 + 2\omega_0\omega' + \omega'^2)$$

$$= (4\omega_0^2\omega'^2 + 4\gamma^2\omega_0^2)\left[1 + O\left(\frac{\gamma}{\omega_0}\right)\right] \qquad (21.68)$$

Therefore, in a small neighbourhood of $\omega = \omega_0$ (or $\omega' = 0$), we obtain

$$|f(\omega)|^2 \approx \frac{|x_0|^2}{2\pi} \frac{1}{4(\omega'^2 + \gamma^2)} \qquad (21.69)$$

The maximum value is at $\omega' = 0$ and has the magnitude $|x_0|^2/(8\pi\gamma^2)$; half the maximum value is at $\omega' = \pm\gamma$. Therefore, the full width at half maximum (abbreviation: the width of the resonance) is

$$\Delta\omega = 2\gamma = \frac{2e^2\omega_0^2}{3mc^3} \qquad (21.70)$$

This characterizes the frequency distribution in the motion (21.62). Now, we may summarize our results in the following way: in general, a function of time vanishing for $t < 0$ and of the form $e^{-\gamma t}\cos(\omega_0 t + \varphi)$ for $t > 0$, where weak damping ($\gamma \ll \omega_0$) has been assumed, contains a lot of Fourier frequencies (frequency band). The distribution of the frequencies has a resonance at $\omega = \omega_0$ and a width $\Delta\omega = 2\gamma$. In quantum mechanics we will see that this result may be interpreted as the uncertainty relation of energy and time $\Delta E \Delta t \approx \hbar/2$.

Example 21.8: Exercise: Scattering of light from a polarizable molecule

Calculate the total cross section for the scattering of light from a polarizable molecule. How does the scattering cross section depend on the wavelength λ and the polarizability α?

Solution   The dipole moment induced in the molecule by the electric field $\mathbf{E}(t) = \mathbf{E}_0 e^{-i\omega t}$ of the light is

$$\mathbf{p}(t) = \alpha\mathbf{E}(t) = \alpha\mathbf{E}_0 e^{-i\omega t} = \mathbf{p}_0 e^{-i\omega t} \qquad (21.71)$$

Here, $\mathbf{p}_0 = \alpha\mathbf{E}_0$ is the complex amplitude of the dipole moment. The energy flux of the light incident on the molecule is $cE_0^2/(8\pi)$, and the scattered energy per unit of time (compare to equation (21.16a)) is $\frac{1}{3}ck^4p_0^2$. Hence, the scattering cross section σ (scattered energy per unit of time/incident energy flux) is given by

$$\sigma = \frac{\frac{c}{3}k^4 p_0^2}{\frac{c}{8\pi}E_0^2} = \frac{8\pi}{3}k^4\alpha^2 = \frac{4(2\pi)^5}{3}\frac{\alpha^2}{\lambda^4} \qquad (21.72)$$

This scattering cross section was derived by Lord Rayleigh. He used it to explain the blueness of the sky and the redness of the sunrise and the sunset. To understand this somewhat better we notice first that the polarizability of a nitrogen molecule (the main constituent of the air) can be regarded to be approximately equal to that of a conducting sphere of radius $1.2\,\text{Å} = 1.2 \cdot 10^{-8}\,\text{cm}$. This was calculated in Example 6.2, and we found in the limit $\epsilon \to \infty$

$$\alpha = a^3 = 1.7 \cdot 10^{-24}\,\text{cm}^3 \tag{21.73}$$

Red light has a wavelength of about

$$\lambda_{\text{red}} \approx 6500\,\text{Å} = 6.5 \cdot 10^{-5}\,\text{cm} \tag{21.74}$$

The scattering cross section (21.72) is

$$\sigma_{\text{red}} = \frac{4(2\pi)^5}{3} \frac{(1.7 \cdot 10^{-24})^2}{(6.5 \cdot 10^{-5})^4}\,\text{cm}^2 = 2.1 \cdot 10^{-27}\,\text{cm}^2 \tag{21.75}$$

Under normal atmospheric conditions at sea level, the number of molecules per cm^3 is

$$n = \frac{6 \cdot 10^{23}}{22.4 \cdot 10^3\,\text{cm}^3} = 2.7 \cdot 10^{19}\,\text{cm}^{-3} \tag{21.76}$$

Therefore, the *mean free path* of the red light (that is, the mean path it travels without being scattered) we find to be

$$L_{\text{red}} = \frac{1}{n\sigma} = \frac{\text{cm}^3}{2.7 \cdot 10^{19}} \frac{1}{2.1 \cdot 10^{-27}\,\text{cm}^2} = 1.8 \cdot 10^7\,\text{cm} = 180\,\text{km} \tag{21.77}$$

Blue light lies in the region of wavelengths of about

$$\lambda_{\text{blue}} \approx 4700\,\text{Å} = 4.7 \cdot 10^{-5}\,\text{cm} \tag{21.78}$$

and in this case the mean free path is

$$L_{\text{blue}} = 180\,\text{km} \left(\frac{4700}{6500}\right)^4 = 49\,\text{km} \tag{21.79}$$

The sky is blue because the light comming directly from the sun is scattered as soon as it enters the atmosphere. The small mean free path of the blue light component compared to the red one indicates that the scattering process for the blue light is much more effective than for the red one.

The redness of the sunset or the sunrise can be explained similarly: At sunrise and sunset, the light has to cover a longer path through the atmosphere (in particular, through the dense zones). As demonstrated by the mean free paths the blue light is scattered much more strongly than the red one. One can say the red light remains.

These estimations give the principles of the process of scattering of light in the atmosphere. But it should be mentioned that statistical fluctuations and air pollution (vapor, dust) play a role, too.

Example 21.9:   Exercise: Radiation pattern of a simple antenna

As an example of the radiation of an oscillating charge distribution we want to investigate a simple model of an antenna. The geometry of the system is shown schematically in Figure 21.11. The antenna consists of two thin wires of length $d/2$ each. The signal to be transmitted is fed into the space in between both rods, for example, with the help of a coaxial cable. We assume that the signal possesses a simple harmonic time dependence. Then, the general case may be constructed by superposition of various Fourier components. Analogously, we assume a current distribution exhibiting a harmonic

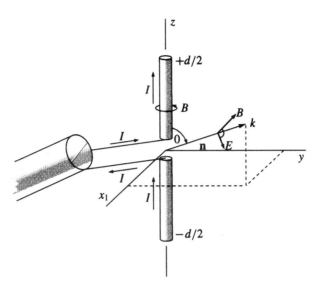

Figure 21.11. Geometry of the considered system.

space and time distribution along the wire of the antenna. The current has to vanish at the endpoints of the antenna. Thus, we may write

$$\mathbf{j}(\mathbf{r}', t') = \delta(x)\delta(y)I_0 \exp(-i\omega t') \sin\left(k\left(\frac{d}{2} - |z|\right)\right)\mathbf{e}_3 \tag{21.80}$$

so that the input signal at the point $|z| = 0$ becomes

$$I(t') = I_0 \exp(-i\omega t') \sin\left(k\frac{d}{2}\right) \tag{21.81}$$

To investigate the radiation field of the antenna, we start from the general form of the vector potential

$$\mathbf{A}(\mathbf{r}, t) = \frac{1}{c} \int \frac{\mathbf{j}(\mathbf{r}', t - |\mathbf{r} - \mathbf{r}'|/c)}{|\mathbf{r} - \mathbf{r}'|} dV' \tag{21.82}$$

We restrict ourselves to the investigation of the far zone of the radiation field and approximate $|\mathbf{r} - \mathbf{r}'| \approx |\mathbf{r}| = r$; that is,

$$\mathbf{A}(\mathbf{r}, t) = \frac{1}{cr} \int \mathbf{j}\left(\mathbf{r}', t - \frac{r}{c}\right) dV' \tag{21.83}$$

The magnetic induction is calculated from

$$\mathbf{B} = \nabla \times \mathbf{A} = \frac{1}{c}\nabla \times \left\{\frac{1}{r}\int \mathbf{j}\left(\mathbf{r}', t - \frac{r}{c}\right) dV'\right\}$$

$$= \frac{1}{c}\left\{\frac{1}{r}\nabla \times \int \mathbf{j}\left(\mathbf{r}', t - \frac{r}{c}\right) dV' - \left(\int \mathbf{j}\left(\mathbf{r}', t - \frac{r}{c}\right) dV'\right)\nabla\frac{1}{r}\right\} \tag{21.84}$$

We are interested only in the radiation field. The gradient term is proportional to $1/r^2$ and vanishes in the far zone compared to the first term. Hence,

$$\mathbf{B}_{rad} = \frac{1}{c}\frac{1}{r}\nabla \times \int \mathbf{j}\left(\mathbf{r}', t - \frac{r}{c}\right) dV' = \frac{1}{cr}\nabla \times \mathbf{g}(t') \tag{21.85}$$

For the curl we may write

$$\left[\nabla \times \mathbf{g}(t')\right]_i = \varepsilon_{ijk}\frac{\partial g_k}{\partial t'} = \varepsilon_{ijk}\frac{\partial g_k}{\partial t'}\frac{\partial t'}{\partial x_j} = -\frac{1}{cr}\varepsilon_{ijk}x_j\frac{\partial g_k}{\partial t'} \tag{21.86}$$

With $\partial t'/\partial x_j = -1/(cr)x_j = -n_j/c$ and using (21.83), we obtain

$$\mathbf{B}_{rad} = -\frac{1}{c^2 r}\mathbf{n} \times \frac{\partial}{\partial t'}\int \mathbf{j}(\mathbf{r}', t - \frac{r}{c}) dV'$$

$$= -\frac{1}{c}\mathbf{n} \times \frac{\partial \mathbf{A}(\mathbf{r}, t)}{\partial t'} \tag{21.87}$$

The substitution $\partial/\partial t' \to \partial/\partial t$ is allowed because in our approximation $t' = t - r/c$ with $r = $ const. Corresponding to the time dependence of the current $\mathbf{j}(\mathbf{r}, t')$ also $\mathbf{A}(\mathbf{r}, t)$ depends harmonically on t:

$$\mathbf{A}(\mathbf{r}, t) = \mathbf{A}(\mathbf{r})\exp(-i\omega t) \tag{21.88}$$

so after differentiating

$$\mathbf{B}_{rad} = i\frac{\omega}{c}\mathbf{n} \times \mathbf{A} = ik\mathbf{n} \times \mathbf{A} \tag{21.89}$$

This relation holds generally always for the magnetic field intensity in the far field. The chosen harmonic time-dependence does not restrict the generality since any time-dependence may be represented by superposition of many Fourier components. With this time-dependence let us return to equation (21.82)

$$\mathbf{A}(\mathbf{r}, t) = \mathbf{A}(\mathbf{r})\exp(-i\omega t) = \frac{1}{c}\int \frac{\mathbf{j}(\mathbf{r}')e^{-i\omega t}e^{ik|\mathbf{r}-\mathbf{r}'|}}{|\mathbf{r}-\mathbf{r}|} dV' \tag{21.90}$$

The distance $|\mathbf{r} - \mathbf{r}'|$ appearing now in the phase factor is written as

$$|\mathbf{r} - \mathbf{r}'| \simeq r\left(1 - 2\frac{\mathbf{r}\cdot\mathbf{r}'}{r^2}\right)^{1/2} \simeq r\left(1 - \frac{\mathbf{r}\cdot\mathbf{r}'}{r^2}\right) = r\left(1 - \frac{\mathbf{n}\cdot\mathbf{r}'}{r}\right) \tag{21.91}$$

Thus, (21.89) becomes

$$\mathbf{A}(\mathbf{r}, t) = \frac{e^{ikr}}{cr}\int \frac{\mathbf{j}(\mathbf{r}')e^{-ik(n\cdot\mathbf{r}')}}{(1 - (n\cdot\mathbf{r})/r)}$$

$$\simeq \frac{e^{ikr}}{cr}\int \mathbf{j}(\mathbf{r}')e^{-ikn\cdot\mathbf{r}'\cos\theta} \tag{21.92}$$

Here, in the denominator we have neglected $(\mathbf{n}\cdot\mathbf{r}'/r)$ as small compared to one, always. This cannot be done in the phase. We write $\mathbf{n}\cdot\mathbf{r}' = nr'\cos\theta$. See Figure 21.12.

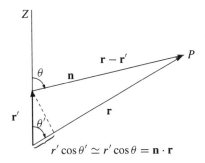

Figure 21.12. Definition of the vectors.

Up till now all considerations have been of general character. The only assumption, $|\mathbf{r}| \gg |\mathbf{r}'|$, restricts the validity of the relations derived to the Lorenz field. Now, we come back to the antenna, and substitute the current (21.80) into (21.92):

$$\mathbf{A}(\mathbf{r}) = \frac{I_0 e^{ikr}}{cr} \int_{-d/2}^{+d/2} \sin\left(\frac{kd}{2} - k|z|\right) e^{-ikz\cos\theta} \, dz \, \mathbf{e}_3 \tag{21.93}$$

Elementary integration yields

$$\mathbf{A}(\mathbf{r}) = \frac{2I_0 e^{ikr}}{ckr} \left[\frac{\cos\left(\frac{kd}{2}\cos\theta\right) - \cos\frac{kd}{2}}{\sin^2\theta} \right] \mathbf{e}_3 \tag{21.94}$$

Since $\mathbf{A}$ has only components in z-direction, we may write immediately for $\mathbf{B}_{rad}$

$$|B_{rad}| = |ik\mathbf{n} \times \mathbf{A}| = k|A_3| \sin\theta \tag{21.95}$$

On average in time we obtain for the square of the absolute value

$$\langle |B_{rad}|\rangle^2 = \frac{1}{2} k^2 A_3^2 \sin^2\theta$$

$$= \frac{2I_0^2}{c^2 r^2} \left(\frac{\cos\left(\frac{kd}{2}\cos\theta\right) - \cos\left(\frac{kd}{2}\right)}{\sin\theta} \right)^2 \tag{21.96}$$

According to (21.91) the Poynting vector is given by

$$\langle \mathbf{S}\rangle_{rad} = \frac{c}{4\pi} \langle \mathbf{B}^2\rangle_{rad} \cdot \mathbf{n} \tag{21.97}$$

The mean energy emitted into the solid angle $d\omega$ is

$$\left\langle \frac{dE}{d\omega} \right\rangle = r^2 \langle \mathbf{S}\rangle_{rad} \cdot \mathbf{n}$$

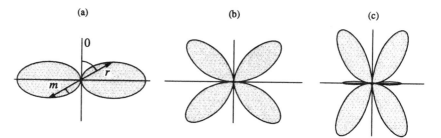

Figure 21.13. Radiation pattern for different m: (a) $m = 1$. (b) $m = 2$. (c) $m = 3$.

$$= \frac{I_0^2}{2\pi c} \left(\frac{\cos\left(\frac{kd}{2}\cos\theta\right) - \cos\frac{kd}{2}}{\sin\theta} \right)^2 \tag{21.98}$$

The angular distribution of the radiation depends on $kd/2$. Now, we consider the cases for which a simple relation exists between the wavelength of the input oscillation $\lambda = 2\pi/k$ and the size of the antenna. Let d be an integral multiple of half the wavelength:

$$d = m\frac{\lambda}{2} = m\frac{\pi}{k} \tag{21.99}$$

Then,

$$\left\langle \frac{dk}{d\omega} \right\rangle = \frac{I_0^2}{2\pi c} \left(\frac{\cos\left(\frac{m\pi}{2}\cos\theta\right) - \cos\left(\frac{m\pi}{2}\right)}{\sin\theta} \right)^2 \tag{21.100}$$

For odd m we obtain a relatively simple angular distribution:

$$\left\langle \frac{dE}{d\omega} \right\rangle = \frac{I_0^2}{2\pi c} \left(\frac{\cos\left(\frac{m\pi}{2}\cos\theta\right)}{\sin\theta} \right)^2, \qquad m = 1, 3, 5, 7, \ldots$$

$$\left\langle \frac{dE}{d\omega} \right\rangle = \frac{2I_0^2}{\pi c} \left(\frac{\cos^4\left(\frac{\pi}{2}\cos\theta\right)}{\sin^2\theta} \right)^2, \qquad m = 2 \tag{21.101}$$

In Figure 21.13 the three lowest radiation patterns $m = 1, 2, 3$ are given. The most intensive cone is always the one lying closest to the antenna. For $m \to \infty$ it coincides with the antenna. So, for an infinitly long conductor electromagnetic energy may be propagated along the conductor, but no radiation is emitted into the empty space.

One can see that $m = 1$ is the part corresponding to the dipole radiation, $m = 2$ corresponds to the quadrupole radiation, and $m = 3$ resembles octupole radiation. Nevertheless, the calculation performed is exact in the sense that no multipole expansion of the field has been done.

Biographical notes

Heinrich Rudolph Hertz, b. Feb. 22, 1857, Hamburg–d. Jan, 1, 1894, Bonn. His research on the propagation of electric waves in 1887–1888 corroborated the forecasts of Maxwell's electromagnetic theory of light. The "Hertzian waves" discovered by Hertz became one of the physical foundations of modern radio technology. In 1887 Hertz furnished proof of the impact of ultrared light on electric discharge. This proof, in turn, led W. Hallwachs to the discovery of the photoelectric effect. In 1892 Hertz observed that cathode rays came through thin layers of metal, and Ph. Lenard was thus able to find out their nature. The exact definition of hardness was given by Hertz, too. [BR]

John William Strutt Rayleigh, Lord, b. Nov. 12, 1842, Langford (n. Maldon, Essex)–d. June 30, 1919, Witham (Essex). After having studied in Cambridge, he received the M.A. in 1868. From 1876 to 1884 he was professor for experimental physics in Cambridge. In 1887 he became professor for mathematical physics at the Royal Institute in London. Together with William Ramsey, Rayleigh in 1895 discovered the element argon. In 1904 he was awarded the Nobel Prize for physics. In 1873 he had taken over his father's peerage. Most of Rayleigh's numerous treatises on problems of acoustics, optics, and electrophysics had been published in the "Philosophical Transactions" of the Royal Society. His main works were the "Theory of Sound" (two vols, London 1877–1878), and the "Scientific Papers" (London, 1900–1903).

22 Covariant Formulation of Electrodynamics

In this chapter we want to give a view of the treatment of electrodynamics in the theory of relativity. For this purpose we will recapitulate the relativistic formalism. We use the convention that indices appearing twice are summed: Latin indices $i, j, k, \ldots$ can take values from 1 to 3; Greek indices $\lambda, \mu, \nu, \ldots$ run from 1 to 4.

The Lorentz transformation

In the lectures on Mechanics I, the four-dimensional Minkowski space of the points $(x_\mu) = (x_1, x_2, x_3, x_4) = (x_1, x_2, x_3, ict) = (\mathbf{x}, ict)$ was introduced. In the transition between two coordinate systems $K(x_\mu)$ and $K'(x'_\mu)$ moving uniformly with respect to each other, the components of the position vector in the Minkowski space transform according to a Lorentz transformation. This transformation is linear and homogeneous, and may be described by a transformation matrix $\hat{C} = (c_{\mu\nu})$,[1]

$$x'_\mu = \sum_{\nu=1}^{4} c_{\mu\nu} x_\nu = c_{\mu\nu} x_\nu \tag{22.1}$$

We consider the special case of two coordinate systems whose axes are parallel. The origin of K' should move at the velocity v along the x_3-axis.

---

[1] Here, we have used Einstein's sum convention: ν appears twice and consequently must be summed.

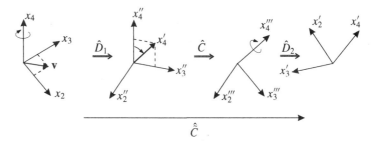

Figure 22.1. The general Lorentz transformation $\hat{\tilde{C}}$ can be built up out of the special Lorentz transformation $\hat{C}$ and the spatial rotations $\hat{D}_1$ and $\hat{D}_2$.

Then, with the abbreviations $\beta = v/c$ and $\gamma = 1/\sqrt{1-\beta^2}$ the transformation matrix is

$$\hat{C} = (c_{\mu\nu}) = \begin{pmatrix} 1 & 0 & 0 & 0 \\ 0 & 1 & 0 & 0 \\ 0 & 0 & \gamma & i\gamma\beta \\ 0 & 0 & -i\gamma\beta & \gamma \end{pmatrix} \tag{22.2}$$

Geometrically, one may see that all transformations of the (homogeneous) Lorentz group can be obtained from a special case by means of two spatial rotations: first spatial rotation $\hat{D}_1$ rotates the new x_3''-axis into the direction of the translational velocity v so that now the Lorentz transformation may be applied (rotation in the $x_3''x_4''$-plane). This ensures that the x_4'''-axis will coincide with the x_4'-axis. So, the transition from the K''' system to the desired K' system requires only a further spatial rotation $\hat{D}_2$.

Thus, any Lorentz transformation $\hat{\tilde{C}}$ may be represented as the product of the special Lorentz transformation $\hat{C}$ and two pure spatial rotations (Figure 22.1):

$$\hat{\tilde{C}} = \hat{D}_2 \hat{C} \hat{D}_1 \tag{22.3}$$

The $\hat{\tilde{C}}$ obtained in this way must be an element of the Lorentz group because $\hat{C}$ as well as the spatial rotations $\hat{D}_1$, $\hat{D}_2(v=0)$ belong to this group. Therefore, it is sufficient to consider the behavior of a quantity under the transformation (22.3).

Properties of the transformation matrix $(c_{\mu\nu})$

The Lorentz transformation describes rotations in the four-dimensional Minkowski space; thus, the matrix is orthogonal, and the following relations are valid. The magnitude of the position vector remains conserved:

$$x_\nu' x_\nu' = x_\nu x_\nu \tag{22.4}$$

Hence, $x'_\nu x'_\nu = c_{\nu\sigma} x_\sigma \cdot c_{\nu\tau} x_\tau = c_{\nu\sigma} c_{\nu\tau} x_\sigma x_\tau = x_\nu x_\nu$, so we have orthogonality of the columns

$$c_{\nu\sigma} c_{\nu\tau} = \delta_{\sigma\tau} \tag{22.5}$$

The inverse transformation follows from $x'_\sigma = c_{\sigma\nu} x_\nu$ by multiplication by $c_{\sigma\mu}$ and summation:

$$c_{\sigma\nu} c_{\sigma\mu} x_\nu = \delta_{\nu\mu} x_\nu = x_\mu = c_{\sigma\mu} x'_\sigma \tag{22.6}$$

So, $(c_{\mu\sigma})^{-1} = (c_{\sigma\mu}) = (c_{\mu\sigma})^t$, the inverse matrix is equal to the transposed matrix. From (22.6) one obtains the orthogonality of rows:

$$x_\nu x_\nu = c_{\sigma\nu} x'_\sigma c_{\tau\nu} x'_\tau = x'_\tau x'_\tau \tag{22.7}$$

$$c_{\sigma\nu} c_{\tau\nu} = \delta_{\sigma\tau} \tag{22.8}$$

By means of the multiplication theorem and of the relation

$$\det A^T = \det A$$

we may derive the determinant

$$1 = \det(\delta_{\sigma\tau}) = \det(c_{\nu\sigma} c_{\nu\tau}) = \det(c_{\sigma\nu})^t \det(c_{\nu\tau}) = (\det(c_{\mu\nu}))^2$$

thus,

$$\det(c_{\mu\nu}) = \pm 1 \tag{22.9}$$

With the determinant $+1$ we restrict ourselves to the proper Lorentz transformation since $\det(c_{\mu\nu}) = -1$ contains a reflection (inversion).

Four-vectors and four-tensors

The 4^k quantities $T_{\alpha_1 \ldots \alpha_k}$ ($\alpha_i = 1, 2, 3, 4$) form a tensor of rank k if they transform under an orthogonal transformation according to the equation

$$T'_{\alpha_1 \ldots \alpha_k} = c_{\alpha_1 \beta_1} \cdots c_{\alpha_k \beta_k} T_{\beta_1 \ldots \beta_k} \tag{22.10}$$

Any single index of a tensor transforms like the components of a vector (see equation (22.1)).

Example 22.1:   Examples

$k = 0$: A tensor of rank 0 is a scalar, which is invariant under Lorentz transformations, for example, the length of the position vector x_μ.

$k = 1$: A 4-tuple with the transformation properties

$$T'_\mu = c_{\mu\nu} T_\nu \tag{22.11}$$

is a *four-vector*. An example is the vector (x, y, z, ict) of the coordinates of a space-time point.

Product

As can be seen immediately from equation (22.10) according to

$$U_{\alpha_1 \ldots \alpha_m \beta_1 \ldots \beta_n} = S_{\alpha_1 \ldots \alpha_m} T_{\beta_1 \ldots \beta_n} \tag{22.12}$$

it is possible to form a tensor of rank $(m + n)$ by the componentwise multiplication of two tensors of rank m and n.

Contraction

If one sums over two indices of a tensor due to the orthogonality relation (22.5), one obtains

$$T'_{\alpha\alpha\gamma} = c_{\alpha\mu} c_{\alpha\nu} c_{\gamma\lambda} T_{\mu\nu\lambda} = \delta_{\mu\nu} c_{\gamma\lambda} T_{\mu\nu\lambda} = c_{\gamma\lambda} T_{\mu\mu\lambda} \tag{22.13}$$

In general, one can read off from equation (22.13): equating (and summing) two indices of a tensor of rank k one obtains a tensor of rank $(k-2)$. This procedure is denoted *contraction* of the tensor.

An example is the scalar product of two four-vectors $A_\mu B_\mu$ which can be regarded as the contraction of the tensor $T_{\mu\nu} = A_\mu B_\nu$, so it forms a scalar invariant.

Tensor analysis

The three-dimensional ∇ operator may be extended to a four-quantity. Due to the chain rule and equation (22.6),

$$\frac{\partial}{\partial x'_\mu} = \frac{\partial x_\sigma}{\partial x'_\mu} \frac{\partial}{\partial x_\sigma} = c_{\mu\sigma} \frac{\partial}{\partial x_\sigma} \tag{22.14}$$

which means the operator $(\partial/\partial x_\mu)$ behaves like a four-vector.

The application of this operator to a four-scalar ϕ yields the *four-gradient*: $(\partial\phi/\partial x_\mu)$. Because

$$\frac{\partial A'_\mu}{\partial x'_\mu} = c_{\mu\sigma} \frac{\partial}{\partial x_\sigma} (c_{\mu\tau} A_\tau) = \frac{\partial A_\sigma}{\partial x_\sigma} \tag{22.15}$$

the sum $\partial A_\mu / \partial x_\mu$, the *four-divergence*, is a *Lorentz scalar*.

If we substitute the corresponding derivation $\partial/\partial x_\mu$ into the four-divergence, instead of A_μ, then we obtain the scalar operator (four-dimensional Laplacian operator)

$$\Box = \frac{\partial}{\partial x_\mu} \frac{\partial}{\partial x_\mu} = \triangle - \frac{1}{c^2} \frac{\partial^2}{\partial t^2} \tag{22.16}$$

Volume element

The volume element $d^4x = dx_1 dx_2 dx_3 dx_0$ with $dx_0 = dx_4/i = c\,dt$, chosen to be real, remains conserved in a Lorentz transformation. It is a Lorentz scalar because

$$d^4x' = \frac{\partial(x_0', x_1', x_2', x_3')}{\partial(x_0, x_1, x_2, x_3)} d^4x \equiv \det(c_{\mu\nu}) d^4x = d^4x \tag{22.17}$$

since the Jacobian $\partial(x_0', x_1', x_2', x_3')/\partial(x_0, x_1, x_2, x_3) \equiv \det(c_{\mu\nu})$ is $+1$ in the case of the proper Lorentz transformation (see equation (22.6)).

The field equations and the field tensor

Einstein set up two postulates as the foundation of the special theory of relativity:

(1) The same laws of nature hold in all systems uniformly moving with respect to each other.

(2) The velocity of light has the same value in all systems moving uniformly with respect to each other, independent of the velocity of the source relative to the observer.

From requirement (2) one deduces that the transition between two systems is described by a Lorentz transformation. Then, the first postulate reads: For all laws of nature there is a *covariant formulation*. Covariance of an equation means that its form does not change under a Lorentz transformation. According to equation (22.10) this implies that the equations must be relations between tensors of equal rank.

In principle, it is possible to derive the transformation property of the electric and magnetic field quantities directly and to show that, e.g., that Maxwell's equations are covariant (contrary to the laws of classical mechanics, where relativistic changes of the fundamental Newton equations have been neccessary). In the following, we will restrict ourselves, starting from the first postulate, to bring the equations of electrodynamics into covariant form. Therefore, we will consider the fields in vacuum.

The continuity equation

$$\nabla \cdot \mathbf{j} + \frac{\partial \rho}{\partial t} = 0$$

is written in the form

$$\frac{\partial j_1}{\partial x_1} + \frac{\partial j_2}{\partial x_2} + \frac{\partial j_3}{\partial x_3} + \frac{\partial(ic\rho)}{\partial(ict)} = 0 \tag{22.18}$$

Then, introducing the *current four-vector*

$$(j_\mu) = (\mathbf{j}, ic\rho) \tag{22.19}$$

the covariant notation of the continuity equation as four-divergence of the current density becomes obvious:

$$\frac{\partial j_\mu}{\partial x_\mu} = 0 \tag{22.20}$$

The fact that j_μ is really a four-vector can be derived directly from the experimental finding of charge conservation, since if $dq = \rho\, dx_1\, dx_2\, dx_3$ as well as $i\, d^4x = dx_1\, dx_2\, dx_3\, dx_4$ are scalar invariants, ρ must transform like the forth component of a four-vector. Similarly, one concludes $\mathbf{j} = \rho\mathbf{v}$. Recognizing the four-vector character of j_μ, at once, we have under control the transformation property of the current density $\mathbf{j}$ and the charge density ρ in a transition from one inertial frame to another one. Then,

$$j'_\nu = c_{\nu\mu} j_\mu \tag{22.21}$$

Hence, the current density and the charge density are closely connected. For example, according to (22.21) the z-component j'_3 is

$$j'_3 = \gamma(j_3 - v\rho)$$

a plausible result easy to interpret: a moving system K' is composed of the charges (j_3) moving in the original system K, and the static charges (ρ) in the system K that appearing to have been moved due to the motion of the system K'.

The potential equations

$$\Delta\mathbf{A} - \frac{1}{c^2}\frac{\partial^2\mathbf{A}}{\partial t^2} = -\frac{4\pi}{c}\mathbf{j} \tag{22.22}$$

$$\Delta\phi - \frac{1}{c^2}\frac{\partial^2\phi}{\partial t^2} = -4\pi\rho \tag{22.23}$$

with the *Lorentz gauge*

$$\nabla\cdot\mathbf{A} + \frac{1}{c}\frac{\partial\phi}{\partial t} = 0 \tag{22.24}$$

lead us to introduce the *four-potential*:

$$(A_\mu) = (\mathbf{A}, i\phi) \tag{22.25}$$

Then, the two equations (22.22) and (22.23) can be summarized into one:

$$\Box A_\mu = -\frac{4\pi}{c} j_\mu \tag{22.26}$$

with the gauge

$$\frac{\partial A_\mu}{\partial x_\mu} = 0 \tag{22.27}$$

These examples already demonstrate that the equations of electrodynamics obtain a simple and clear form in terms of four-vector notation.

Now, we pass from the potentials to the field intensities **E** and **B** defined by

$$\mathbf{E} = -\nabla\phi - \frac{1}{c}\frac{\partial \mathbf{A}}{\partial t} \tag{22.28}$$

$$\mathbf{B} = \nabla \times \mathbf{A} \tag{22.29}$$

written componentwise, these equations read

$$E_1 = i\frac{\partial A_4}{\partial x_1} - i\frac{\partial A_1}{\partial x_4}, \qquad B_1 = \frac{\partial A_3}{\partial x_2} - \frac{\partial A_2}{\partial x_3}$$

$$E_2 = i\frac{\partial A_4}{\partial x_2} - i\frac{\partial A_2}{\partial x_4}, \qquad B_2 = \frac{\partial A_1}{\partial x_3} - \frac{\partial A_3}{\partial x_1}$$

$$E_3 = i\frac{\partial A_4}{\partial x_3} - i\frac{\partial A_3}{\partial x_4}, \qquad B_3 = \frac{\partial A_2}{\partial x_1} - \frac{\partial A_1}{\partial x_2}$$

The unified form of these equations leads to the definition of the *field tensor*

$$F_{\mu\nu} = \frac{\partial A_\nu}{\partial x_\mu} - \frac{\partial A_\mu}{\partial x_\nu} \tag{22.30}$$

The field tensor $(F_{\mu\nu})$ is an antisymmetric tensor of second rank and, hence, has 6 independent components, E_i and B_i.

Equation (22.30) represents the four-dimensional generalization of the rotation. In three dimensions, as an antisymmetric tensor with three components, it possesses "by chance" the properties of a three-vector, apart from the behavior under inversion.

Written explicitly, the field tensor has the form

$$F_{\mu\nu} = \begin{pmatrix} 0 & B_3 & -B_2 & -iE_1 \\ -B_3 & 0 & B_1 & -iE_2 \\ B_2 & -B_1 & 0 & -iE_3 \\ iE_1 & iE_2 & iE_3 & 0 \end{pmatrix} \tag{22.31}$$

So, we obtain the result that in a notation with four-quantities the electromagnetic field is no longer described by two seperate vectors but by a single tensor. This implies that in a transition between systems moving with respect to each other the components **E** and **B** transform in a mixed way. If they transformed as three-components of two independent four-vectors then their components would mix only among themselves. Thus, a pure electric field in the unprimed system will appear as a mixture of an electric and a magnetic field in the primed system.

Now, we choose again the special Lorentz transformation (22.2), that is, parallel axes and motion of K' in z-direction. Then, from

$$F'_{\mu\nu} = c_{\mu\sigma}c_{\nu\tau}F_{\sigma\tau}$$

with the $c_{\mu\nu}$ from equation (22.2) we obtain the transformation equations

$$E_1' = \gamma(E_1 - \beta B_2), \qquad B_1' = \gamma(B_1 + \beta E_2)$$
$$E_2' = \gamma(E_2\beta B_1), \qquad B_2' = \gamma(B_2 - \beta E_1) \qquad \textbf{(22.32)}$$
$$E_3' = E_3, \qquad B_3' = B_3$$

This result may also be written in vector form:

$$\mathbf{E}_\perp' = \gamma\left(\mathbf{E}_\perp + \frac{\mathbf{v}}{c} \times \mathbf{B}\right), \qquad \mathbf{B}_\perp' = \gamma\left(\mathbf{B}_\perp - \frac{\mathbf{v}}{c} \times \mathbf{E}\right)$$

$$\mathbf{E}_\parallel' = \mathbf{E}_\parallel, \qquad \mathbf{B}_\parallel' = \mathbf{B}_\parallel \qquad \textbf{(22.33)}$$

Here, $\parallel$ and $\perp$ mean parallel and perpendicular to $\mathbf{v}$, respectively. One verifies directly that each of the first two equations (22.32) are just the 1,2-components of each of the first equations (22.33). For small relative velocities, if $\gamma \approx 1$, in the rest frame of an electron moving at the velocity $\mathbf{v}$ one obtains the Lorentz force

$$\mathbf{F} = e\mathbf{E}' \approx e\left(\mathbf{E} + \frac{\mathbf{v}}{c} \times \mathbf{B}\right)$$

The second set of equations (22.33) for the magnetic induction yields the Biot-Savart law, as will be illustrated by the Example 22.2.

Example 22.2:   Exercise for repetition: Lorentz transformations

From the lectures on Mechanics it is known that a Lorentz transformation (LT) can be written in the form $x_\mu' = a_{\mu\nu}x_\nu$ or $x' = Ax$ with Einstein's convention $a_{\mu\nu}x_\nu \equiv \sum_{\nu=1}^{4} a_{\mu\nu}x_\nu$ and the definition $\underline{x} = x_\nu = (x_1, x_2, x_3, x_4) = (\mathbf{x}, ict)$, $A = (a_{\mu\nu})$.

(a) Let $\underline{x}$ be given in the frame of reference Σ. Determine A for the transformation $\underline{x} \to \underline{x}' \in \Sigma'$, where the system Σ' should move in z-direction at a velocity v with respect to Σ. Such transformations are denoted as boosts: here, we have a boost in the z-direction.

(b) Let A be an arbitrary LT $\underline{x}' = A\underline{x}$. Derive the relation $A^T = A^{-1}$ from $\underline{x}'^2 = x_\mu'x_\mu' = x_\nu x_\nu = \underline{x}^2$; that is, A is orthogonal, and hence $\det A = \pm 1$. Why is $|A_{44}| \geq 1$?

(c) Show that the LT form a group (notation $\mathcal{L}$).

(d) Show that any matrix $L \in \mathcal{L}$ may be expressed by $L = DL_\bigcirc$ with $D = 1, P, T, PT$. Here,

$$P = \begin{pmatrix} 1 & 0 & 0 & 0 \\ 0 & 1 & 0 & 0 \\ 0 & 0 & 1 & 0 \\ 0 & 0 & 0 & -1 \end{pmatrix}, \qquad T = \begin{pmatrix} -1 & 0 & 0 & 0 \\ 0 & -1 & 0 & 0 \\ 0 & 0 & -1 & 0 \\ 0 & 0 & 0 & 1 \end{pmatrix}$$

$$PT = -1, \qquad \det L_\bigcirc = 1$$

as well as $(L_\bigcirc)_{44} \geq 1$. The set of all $L_\bigcirc$ is denoted by $\mathcal{L}_\uparrow^+$. Show that $\mathcal{L}_\uparrow^+$ forms a group (the so-called restricted Lorentz group). What is the physical meaning of P, T, and PT?

(e) What kinds of coordinate transformations are contained in $L_\uparrow^+$? Is $\mathcal{L}$ an Abelian group?

(f) How are four-vectors and four-tensors defined? What is a scalar?

Solution   **(a)** With $\beta = v/c$, $\gamma = (1 - \beta^2)^{-1/2}$

$$x_1' = x_1, \qquad x_2' = x_2, \qquad x_3' = (x_3 + ix_4\beta)\gamma$$

$$x_4' = ict' = ic\frac{t - vx_3/c}{\sqrt{1 - v^2/c^2}} = (x_4 - ix_3\beta)\gamma$$

or in matrix notation

$$\underline{x}' = \begin{pmatrix} x_1' \\ x_2' \\ x_3' \\ x_4' \end{pmatrix} = \begin{pmatrix} x_1 \\ x_2 \\ (x_3 + ix_4\beta)\gamma \\ (x_4 - ix_3\beta)\gamma \end{pmatrix} = \begin{pmatrix} 1 & 0 & 0 & 0 \\ 0 & 1 & 0 & 0 \\ 0 & 0 & \gamma & i\beta\gamma \\ 0 & 0 & -i\beta\gamma & \gamma \end{pmatrix}\begin{pmatrix} x_1 \\ x_2 \\ x_3 \\ x_4 \end{pmatrix} = A\underline{x} \qquad (22.34)$$

(b) $\underline{x}'^2 = x_\mu' x_\mu' = A_{\mu\nu}x_\nu A_{\mu\sigma}x_\sigma = x_\nu A_{\nu\mu}^T A_{\mu\sigma}x_\sigma = x_\nu x_\nu = \underline{x}^2 \Rightarrow A_{\nu\mu}^T A_{\mu\sigma} = \delta_{\nu\sigma}$, or $A^T = A^{-1}$. Therefore, further $1 = \det 1 = \det(AA^T) = \det A \det A^T = (\det A)^2$; hence, $\det A = \pm 1$. Since **x**, t are real quantities, the components $A_{i4}(i = 1, 2, 3)$ must be complex and $A_{ij}(ij = 1, 2, 3)$, A_{44} must be real. Due to $1 = (A^T A)_{44} = A_{44}^2 - |A_{14}|^2 - |A_{24}|^2 - |A_{34}|^2$ we obtain $|A_{44}| \geq 1$.

(c) $L_1 \in \mathcal{L}, L_2 \in \mathcal{L} \Leftrightarrow L_1^T = L_1^{-1}, L_2^T = L_2^{-1} \Rightarrow (L_1 L_2)^T = L_2^T L_1^T = L_2^{-1} L_1^{-1} = (L_1 L_2)^{-1} \Rightarrow (L_1 L_2) \in \mathcal{L}$; associativity, unit element, and the inverse are given.

(d) It is $P^2 = T^2 = (PT)^2 = 1$. Let $A \in \mathcal{L}$.

Case 1: $\det A = 1$, $A_{44} \geq 1 \Rightarrow A_0 = A \in \mathcal{L}_\uparrow^+$, $D = 1$, $A \in \mathcal{L}_\uparrow^+$

Case 2: $\det A = 1$, $A_{44} \leq -1 \Rightarrow A_0 = PTA \in \mathcal{L}_\uparrow^+$, $D = PT$, $A \in \mathcal{L}_\downarrow^+$

Case 3: $\det A = -1$, $A_{44} \geq 1 \Rightarrow A_0 = PA \in \mathcal{L}_\uparrow^+$, $D = P$, $A \in \mathcal{L}_\uparrow^-$

Case 4: $\det A = -1$, $A_{44} \leq -1 \Rightarrow A_0 = TA \in \mathcal{L}_\uparrow^+$, $D = T$, $A \in \mathcal{L}_\downarrow^-$

P means spatial inversion, T time reflection, and PT inversion in space and time. We note that 1, P, T, and PT form a discrete subgroup of $\mathcal{L}$. Because in $\mathcal{L}_\uparrow^+$ associativity is given trivially, $\underline{1} \in \mathcal{L}_\uparrow^+$ and $(A^{-1})_{44} = (A^T)_{44} > A_{44} \geq 1$, only closure has to be shown; that is, $A \in \mathcal{L}_\uparrow^+$, $B \in \mathcal{L}_\uparrow^+ \Rightarrow AB \in \mathcal{L}_\uparrow^+$, so $(AB)_{44} \geq 1$. But now $(AB)_{44} = A_{41} B_{14} + A_{42} B_{24} + A_{43} B_{34} + A_{44} B_{44} \geq (A_{44}B_{44}) - |A_{41}B_{14}| - |A_{42}B_{24}| - |A_{43}B_{34}| = [(1 + |A_{14}|^2 + |A_{24}|^2 + |A_{34}|^2)(1+|B_{41}|^2+|B_{42}|^2+|B_{43}|^2)]^{1/2}-|A_{41}B_{14}|-|A_{42}B_{24}|-|A_{43}B_{34}| \geq 1$ as can be seen easily. The total Lorentz group can be reconstructed from $\mathcal{L}_\uparrow^+$ if for any $L_0 \in \mathcal{L}_\uparrow^+$ the elements PL_0, TL_0, and PTL_0 are formed. One also writes $\mathcal{L} = \mathcal{L}_\uparrow^+ \otimes \{1, P, T, PT\}$ or $\mathcal{L}_\uparrow^+ = \mathcal{L}/\{1, P, T, PT\}$.

(e) Besides boosts, $\mathcal{L}_\uparrow^+$ also contains rotations (so that $\mathcal{L}$ cannot be Abelian) and arbitrary products of both.

(f) A point $x_\mu = (x_1, x_2, x_3, x_4)$ in Minkowski space is denoted by the expression $X = \underline{x}_\mu \underline{e}_\mu$, referred to the basis $\underline{e}_1, \underline{e}_2, \underline{e}_3, \underline{e}_4$. Here, $\{\underline{e}_i, i = 1, 2, 3\}$ is an orthonormal system in position space, $\underline{e}_4$ points in the t-direction. Of course, the point X is a fixed quantity, only its description can change if the frame of reference is changed. In a Lorentz transformation the components x_μ of X are transformed according to $x_\mu \to x'_\mu = L_{\mu\nu}x_\nu$. Since the point X itself remains unchanged this corresponds to the basis transformation $\underline{e}'_\mu = L_{\mu\nu}\underline{e}_\nu$. Then,

$$x'_\mu \underline{e}'_\mu = L_{\mu\nu}x_\nu x_\nu L_{\mu\sigma}\underline{e}_\sigma = x_\nu L^T_{\nu\mu}L_{\mu\sigma}\underline{e}_\sigma = x_\nu \underline{e}_\nu = X.$$

An arbitrary four-vector $\underline{q}$ is defined by $\underline{q} = q_\mu \underline{e}_\mu$. A LT $\underline{e}_\mu \to \underline{e}'_\mu$ corresponds to a transformation $q_\mu \to q'_\mu = L_{\mu\nu}q_\nu$ of the components of $\underline{q}$, that is, in a LT a four-vector transforms like the position vector $\underline{x}$. Because, in general, in utilizing the covariant notation one does not want to make use of the unit vectors $\underline{e}_\mu$, a four-vector is defined by the transformation properties of its components. Analogously, one proceeds in defining tensors: a pattern of numbers $T_{\alpha_1\alpha_2\cdots\alpha_n}$, $\alpha_i = 1, 2, 3, 4$ $(i = 1, \ldots, n)$ is called a tensor of rank n if in a Lorentz transformation L it transforms as

$$T'_{\beta_1\ldots\beta_n} = L_{\beta_1\alpha_1}L_{\beta_2\alpha_2}\cdots L_{\beta_n\alpha_n}T_{\alpha_1\ldots\alpha_n}$$

So, a four-vector is a tensor of first rank. Scalar quantities do not transform in a LT; they do not possess indices. So, in a LT for a scalar one has $s = s'$. From a vector q_μ, e.g., one obtains a scalar by forming the product (scalar product)

$$(q')^2 = (q'_\mu q'_\mu) = L_{\mu\nu}q_\nu L_{\mu\sigma}q_\sigma = q_\gamma q_\gamma = q^2$$

Example 22.3: Electric and magnetic fields of a relativistically moving point charge

An electric point charge q moves along a straight line at the velocity $\mathbf{v} = v\mathbf{e}_3$. The charge is at rest in the moving system K'. What charge is seen by an observer at rest in K? Figure 22.2 shows the geometric situation and the impact parameter b. Let the origin of both coordinate systems coincide at time $t = t' = 0$. Therefore, in K' the point P has the coordinates $x'_1 = b, x'_2 = 0, x'_3 = -vt'$ and lies at the distance

$$r' = \sqrt{b^2 + (vt')^2}$$

It is neccessary to express r' in the system K. The only coordinate which has to be transformed is the time:

$$t' = \gamma \left(t - \frac{v}{c^2}x_3\right) = \gamma t \tag{22.35}$$

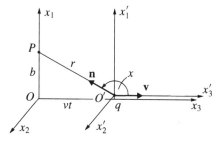

Figure 22.2. A point charge q, at rest in the moving system K', is observed from the point P at rest in the system K. b is the impact parameter.

because for the observer P, $x_3 = 0$ and $b = b'$. In the system K' there are only electric field intensities E'_ν ($\nu = 1, 2, 3$). At the point P in K' these E'_i are

$$\mathbf{E}' = \{E'_1, E'_2, E'_3\} = \frac{q}{r'^2}\frac{\mathbf{r}'}{r'}$$

or explicitly

$$E'_1 = +\frac{qb}{r'^3}, \qquad E'_2 = 0, \qquad E'_3 = -\frac{qvt'}{r'^3}$$

$$B'_1 = 0, \qquad\qquad B'_2 = 0, \qquad\quad B'_3 = 0 \tag{22.36}$$

From the system K', the field intensities are transformed according to the inverse equations (22.32). At the same time, we express the primed coordinates by the unprimed ones and obtain in K:

$$E_1 = \gamma(E'_1 + \beta B'_2), \qquad E_2 = \gamma(E'_2 - \beta B'_1), \qquad E_3 = E'_3$$

$$B_1 = \gamma(B'_1 - \beta E'_2), \qquad B_2 = \gamma(B'_2 + \beta E'_1), \qquad B_3 = B'_3 \tag{22.37}$$

and explicitly with (22.36),

$$E_1 = \gamma E'_1 = +\frac{qb\gamma}{(b^2 + \gamma^2 v^2 t^2)^{3/2}}, \qquad E_2 = \gamma E'_2 = 0$$

$$E_3 = E'_3 = -\frac{q\gamma vt}{(b^2 + \gamma^2 v^2 t^2)^{3/2}} \tag{22.38}$$

$$B_1 = 0, \qquad B_2 = \gamma\beta E'_1 = \beta E_1, \qquad B_3 = B'_3 = 0$$

The relativistic effects on the field intensities are visible most clearly when $v \to c$. Then, due to the moving charge at the point P a magnetic induction appears along the y-axis (B_2) the intensity of which, for $\beta \to 1$, equals the electric field intensity E_1. Really, according to the Biot-Savart law we expect

$$\mathbf{B} \simeq \frac{q}{c}\frac{\mathbf{v} \times \mathbf{r}}{r^3} \tag{22.39}$$

and realize that the B_2-component of equation (22.38) for small velocities equals the corresponding expression in (22.39). So, *the transformation laws of relativity theory yield the Biot-Savart law, including realtivistic corrections.*

Furthermore, we see that the maximum E_1-component of the electric field intensity reached at $t = 0$ (the instant of passing the observer P) is given by

$$(E_1)_{t=0} = +\frac{q}{b^2}\gamma \tag{22.40}$$

For higher velocities, $\gamma = 1/\sqrt{1 - \beta^2} \gg 1$, and, therefore, E_1 is very much larger than the static value the point charge being at rest at 0 would generate at the position P (Figure 22.3).

But one must bear in mind the duration of the action of this field intensity is much diminished, namely,

$$\Delta t = \frac{1}{\gamma}\Delta t' = \frac{1}{\gamma}\frac{b}{v} \tag{22.41}$$

As $v \to c$ the order of magnitude of the impulse of force $E_1\Delta t = +(q/b^2)\gamma(1/\gamma)(b/v) = +q/(b \cdot v)$ remains conserved. The figure illustrates the time dependence of the components of the electric field intensity. For $\beta \to 1$ the observer, in fact, sees only a transverse electric field (E_1) and

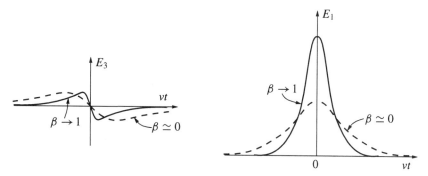

Figure 22.3. The components of the field intensity during the course of time in passing a point charge.

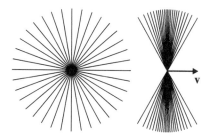

Figure 22.4. Transversality of the electric field of a moving charge.

an orthogonal magnetic induction (B_2) of equal strength. The longitudinal component E_3 is positive during the approach of the charge ($t < 0$) and negative during its departure. So, its time average is zero. Taking into account the inertia of the device of the observer P then this component is of no importance: in fact, the device is not affected by $E_3(t)$.

Furthermore, the spatial distribution of the fields relative to an instantaneous observer in the laboratory system at the same (instantaneous) position of the charge is of interest. From (22.38) follows

$$\frac{E_3}{E_1} = -\frac{vt}{b} = \cot \psi$$

and hence, corresponding to Figure 22.2, $\mathbf{E} \sim \mathbf{n}$. According to that figure, $\mathbf{n}$ is a unit vector from the instantaneous position of the charge to the observer P. So, the electric field is a central field as always. With $b = r \sin \psi$ and $vt = -r \cos \psi$ we may combine E_1, E_2 and E_3 from (22.38) to

$$\mathbf{E} = \frac{q\mathbf{r}}{r^3 \gamma^2 (1 - \beta^2 \sin^2 \psi)^{3/2}} \tag{22.42}$$

There, $r^2 = b^2 + v^2 t^2$. The $\mathbf{E}$-field is directed radially, but the force lines are no longer isotropic (due to the additional term $(1 - \beta^2 \sin^2 \psi)^{-3/2}$). They are illustrated in Figure 22.4. Due to the motion of the electric charge its electric field becomes more transverse, in the limit $v \to c$ even strictly transverse. Hence, a charge passing at near velocity of light is seen by an observer like a short transverse $\mathbf{E}$-field flash, that is, like a transverse impulse of force.

Invariants of the field

With the help of the tensor notation two quantities may be found by contraction which do not change in the transition to other inertial systems. First,

$$F_{\mu\nu}F_{\mu\nu} = 2(B^2 - E^2), \qquad \text{so} \quad \mathbf{B}^2 - \mathbf{E}^2 = \text{invariant}$$

A further invariant is obtained by contraction of the field tensor with the "completely antisymmetric unit tensor of fourth rank" defined by

$$\epsilon_{\kappa\lambda\mu\nu} = \begin{cases} 0 & \text{if two indices are equal} \\ +1 & \text{if } (\kappa\lambda\mu\nu) \text{ is an even permutation of (1234)} \\ -1 & \text{if } (\kappa\lambda\mu\nu) \text{ is an odd permutation of (1234)} \end{cases} \qquad \textbf{(22.43)}$$

One may be convinced easily that $\epsilon_{\kappa\lambda\mu\nu}$ is a tensor of rank 4 because

$$\epsilon'_{\kappa'\lambda'\mu'\nu'} = c_{\kappa'\kappa}c_{\lambda'\lambda}c_{\mu'\mu}c_{\nu'\nu}\epsilon_{\kappa\lambda\mu\nu}$$

Due to the orthonormality relations for the $c_{\mu\nu}$ and the relations (definitions) in equation (22.43) one obtains at once that for $\epsilon'_{\kappa'\lambda'\mu'\nu'}$ the relations analogous to (22.43) are valid.

Now,

$$\epsilon_{\kappa\lambda\mu\nu}F_{\kappa\lambda}F_{\mu\nu} = -8i\mathbf{E}\cdot\mathbf{B}$$

So, the scalar product $\mathbf{E}\cdot\mathbf{B}$ is Lorentz invariant.

Of course, the invariance of $B^2 - E^2$ and $\mathbf{E}\cdot\mathbf{B}$ can be proved also directly from the transformation equations (22.32). This is left for the reader. It can be shown that the field has no further invariants.

Maxwell's equations

With the help of the field tensor $F_{\mu\nu}$ the four Maxwell equations can be brought to a covariant form. It turns out that pair of them combines to one four-vector equation. The equations read

$$\frac{\partial F_{\mu\nu}}{\partial x_\nu} = \frac{4\pi}{c}j_\mu \qquad \textbf{(22.44a)}$$

$$\frac{\partial F_{\mu\nu}}{\partial x_\lambda} + \frac{\partial F_{\nu\lambda}}{\partial x_\mu} + \frac{\partial F_{\lambda\mu}}{\partial x_\nu} = 0 \qquad \textbf{(22.44b)}$$

The last equation holds automatically for the antisymmetric tensor $F_{\mu\nu} = \partial A_\nu/\partial x_\mu - \partial A_\mu/\partial x_\nu$. It represents an identity, called the *Jacobi identity*. Really,

$$\frac{\partial}{\partial x_\lambda}\left(\frac{\partial A_\nu}{\partial x_\mu} - \frac{\partial A_\mu}{\partial x_\nu}\right) + \frac{\partial}{\partial x_\mu}\left(\frac{\partial A_\lambda}{\partial x_\nu} - \frac{\partial A_\nu}{\partial x_\lambda}\right) + \frac{\partial}{\partial x_\nu}\left(\frac{\partial A_\mu}{\partial x_\lambda} - \frac{\partial A_\lambda}{\partial x_\mu}\right) = 0$$

because the terms cancel pairwise. As can be verified by the substitution of the equations (22.20) and (22.32), equation (22.44a) contains the two Maxwell equations which connect fields and charges:

$$\mu = 1, 2, 3: \quad \nabla \times \mathbf{B} - \frac{1}{c}\frac{\partial \mathbf{E}}{\partial t} = \frac{4\pi}{c}\mathbf{j}; \qquad \mu = 4: \quad \nabla \cdot \mathbf{E} = 4\pi\rho$$

Equation (22.44b) yields the two homogeneous Maxwell equations:

$$\nabla \times \mathbf{E} + \frac{1}{c}\frac{\partial \mathbf{B}}{\partial t} = 0; \qquad \nabla \cdot \mathbf{B} = 0$$

One obtains only four distinct nontrivial equations because the left-hand side of equation (22.44b) vanishes if two indices are equal, due to the antisymmetry of $F_{\mu\nu}$. Because nothing further is changed by a permutation of $(\lambda\mu\nu)$, it is sufficient to consider the combinations (234), (341), (412), and (123).

Each term of equation (22.44b) represents a tensor of third rank. By using the antisymmetric unit tensor $\epsilon_{\kappa\lambda\mu\nu}$ a four-vector equation can be found which also contains the homogeneous Maxwell equations:

$$\epsilon_{\kappa\lambda\mu\nu}\frac{\partial F_{\lambda\mu}}{\partial x_{\nu}} = 0 \tag{22.45}$$

The plane light-wave

In space free of currents and charges, $j_{\mu}(x) = 0$. Then, according to (22.44a), $\partial F_{\mu\nu}/\partial x_{\nu} = 0$, and from (22.44b)

$$\frac{\partial}{\partial x_{\lambda}}\frac{\partial}{\partial x_{\lambda}}F_{\mu\nu} = -\frac{\partial}{\partial x_{\mu}}\frac{\partial}{\partial x_{\lambda}}F_{\nu\lambda} - \frac{\partial}{\partial x_{\nu}}\frac{\partial}{\partial x_{\lambda}}F_{\lambda\mu} = 0$$

so

$$\frac{\partial}{\partial x_{\lambda}}\frac{\partial}{\partial x_{\lambda}}F_{\mu\nu} = 0 \tag{22.46}$$

or

$$\Box\, F_{\mu\nu} = 0$$

This equation allows for plane waves of the type

$$F_{\mu\nu}(x) = f_{\mu\nu}\, e^{ik_{\sigma}x_{\sigma}} \tag{22.47}$$

Substituted into (22.46),

$$k_{\lambda}k_{\lambda}\, F_{\mu\nu}(x) = 0$$

Because $F_{\mu\nu}(x) \neq 0$,

$$k_{\lambda}k_{\lambda} = 0 \tag{22.48}$$

This has to be valid in all Lorentz systems. Consequently,

$$\{k_\lambda\} = \{k_1, k_2, k_3, k_4\}$$

must be a four-vector transforming according to

$$k'_\mu = c_{\mu\nu} k_\nu \tag{22.49}$$

Now, we want to identify more precisely the four-wave vector k_μ. Usually, a plane light wave is described by

$$F_{\mu\nu}(\mathbf{x}, t) = f_{\mu\nu} \cdot e^{i(\mathbf{k}\cdot\mathbf{x} - \omega t)} = f_{\mu\nu} e^{i k_\sigma x_\sigma} \tag{22.50}$$

where the elements $\pm(B_0)_k$, $\pm i(E_0)_k$ ($k = 1, 2, 3$) of $f_{\mu\nu}$ are the constant amplitudes. Obviously, we must have $\{k_\mu\} = \{\mathbf{k}, i\omega/c\}$. Hence, the relation (22.49) is identical to the dispersion relation

$$\mathbf{k}^2 = \frac{\omega^2}{c^2} \tag{22.51}$$

Due to the covariance of the wave equation, in a moving system K',

$$F'_{\mu\nu}(\mathbf{x}', t') = f'_{\mu\nu} \, e^{i(\mathbf{k}'\cdot\mathbf{x}' - \omega' t')} = f'_{\mu\nu} e^{i k'_\sigma x'_\sigma} \tag{22.52}$$

We want to think about how the two waves (22.50) and (22.52) describing the same wave are connected at the space-time point $\mathbf{x}, t$. Waves (22.50) and (22.52) are identical, only they are described in different Lorentz systems. At the same space-time point they are connected via the transformation law

$$F'_{\mu\nu}(\mathbf{x}', t') = c_{\mu\sigma} \, c_{\nu\tau} \, F_{\sigma\tau}(\mathbf{x}, t) \tag{22.53}$$

So, the tensor transformation is valid at the same space-time point $\mathbf{x}, t$ where on the left-hand side the point $\mathbf{x}, t$ is expressed by the primed coordinates $\mathbf{x}', t'$. The relation alone can be fulfilled identically only if the phase is the same on both sides

$$\mathbf{k}' \cdot \mathbf{x}' - \omega' t' = \mathbf{k} \cdot \mathbf{x} - \omega t$$

or with $\mathbf{k} = (\omega/c)\mathbf{n}$, where $\mathbf{n}$ is the vector of the wave normal.

$$\omega'(\mathbf{n}' \cdot \mathbf{x}' - ct') = \omega(\mathbf{n} \cdot \mathbf{x} - ct)$$

As we have seen above, both sides of this equation may be written as the scalar product $k_\mu x_\mu$ of the vector $\{x_\mu\} = \{\mathbf{x}, ict\}$ with the four-wave vector $\{k_\mu\} = \{\mathbf{k}, i\omega/c\} = (\omega/c)\{\mathbf{n}, i\} = \{\mathbf{k}, ik_0\}$. Here, $k = |\mathbf{k}| = \omega/c = k_0$, because $k_\mu k_\mu = \mathbf{k} - \omega^2/c^2 = 0$ is a light vector. This follows from the fact that $F_{\mu\nu}(\mathbf{x}, t)$ obeys the wave equation, $\Box F_{\mu\nu}$ or $(\partial/\partial x_\lambda)(\partial/\partial x_\lambda)F_{\mu\nu}(x_\rho) = 0 = k_\lambda k_\lambda F_{\mu\nu}(x_\rho)$, leading directly to the dispersion relation (22.49), so $k_\lambda k_\lambda = 0$. So, the invariance of the phase of the plane wave appears again as a scalar product of two four-vectors.

From the invariance of the phase $k_\mu x_\mu$ the effects of *light abberation, Doppler shift*, and *reflection* from moving mirrors may be derived.

Since $\{k_\mu\}$ must be a four-vector, according to the general Lorentz transformation from an inertial system to an inertial system moving with $\boldsymbol{\beta} = \mathbf{v}/c$, one obtains immediatly the transformation equations (22.50), that is, $k'_\mu = c_{\mu\nu} k_\nu$, explicitly

$$ik_0^\| = -i\gamma\beta k_\| + \gamma i k_0 = i\gamma \left(k_0 - \frac{\boldsymbol{\beta} \cdot \mathbf{k}}{\beta} \beta \right)$$

$$k'_\| = \gamma(k_\| + i\beta(ik_0))$$

$$\mathbf{k}'_\perp = \mathbf{k}_\perp$$

so

$$k'_0 = \gamma(k_0 - \boldsymbol{\beta} \cdot \mathbf{k}) \tag{22.54}$$
$$k'_\| = \gamma(k_\| - \beta k_0)$$

Here,

$$\mathbf{k}'_\perp = \mathbf{k}_\perp, \qquad \gamma = \frac{1}{\sqrt{1 - \beta^2}}, \qquad k_\| = \mathbf{k} \cdot \frac{\mathbf{v}}{v}, \qquad \mathbf{k}_\perp = \mathbf{k} - \left(\mathbf{k} \cdot \frac{\mathbf{v}}{v} \right) \frac{\mathbf{v}}{v}$$

Summarized as four-vectors (compare Exercise 22.3)

$$\{k_\mu\} = \{\mathbf{k}_\perp, k_\|, k_0\} = \left\{ \mathbf{k}_\perp, \frac{\mathbf{k} \cdot \boldsymbol{\beta}}{\beta}, k_0 \right\} = \left\{ k_{\perp 1}, k_{\perp 2}, \frac{\mathbf{k} \cdot \boldsymbol{\beta}}{\beta}, k_0 \right\}$$

The Lorentz transformation of $\{\mathbf{k}, ik_0\}$ has exactly the same form as that of the space-time vector $\{\mathbf{x}, ix_0\}$ as it must be according to the invariance of the scalar product $k_\mu x_\mu$. Further, for light waves

$$|\mathbf{k}| = k_0 = \frac{\omega}{c}, \qquad |\mathbf{k}'| = k'_0 = \frac{\omega'}{c}$$

Hence, from the first equation in (22.54) one obtaines at once the *Doppler shift formula*

$$\omega' = \gamma \, \omega (1 - \beta \, \cos \Theta) \tag{22.55}$$

and from all equations in (22.54) the aberration equation (see Figure 22.5)

$$\tan \Theta' = \frac{k'_\perp}{k'_\|} = \frac{k_\perp}{\gamma(k_\| - \beta k_0)} = \frac{\sqrt{k^2 - k^2 \cos^2 \Theta}}{\gamma(k \cos \Theta - \beta k)} = \tag{22.56}$$
$$= \frac{k\sqrt{1 - \cos^2 \Theta}}{\gamma k(\cos \Theta - \beta)} = \frac{\sin \Theta}{\gamma(\cos \Theta - \beta)}$$

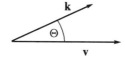

Figure 22.5. The definition of the angle Θ in the aberration equation.

which expresses the change of the direction of a light wave in inertial systems moving with respect to each other. Equation (22.55) is the usual Doppler relation modified by the factor $\gamma = 1/\sqrt{1 - \beta^2}$. This factor shows that, relativistically, there is also a transverse Doppler shift even if $\Theta = \pi/2$, this was experimentally detected by *Ives* and *Stilwell* by observing the light of moving atoms (atomic radiation).

Example 22.4:   Relativistic Doppler effect and aberration

Derive explicitly the relativistic Doppler shift and the formula for aberration.
Note: Consider the behavior of a plane light wave under a Lorentz transformation.

Solution   The plane light wave is described by the ansatz

$$F_{\mu\nu}(\mathbf{x}, t) = f_{\mu\nu} e^{i(\mathbf{k} \cdot \mathbf{x} - \omega t)} \tag{22.57}$$

where $f_{\mu\nu}$ stands for the constant amplitudes $\pm B_{0k}, \pm i E_{0k}$. Since the wave equation is covariant, in the moving system

$$F'_{\mu\nu}(\mathbf{x}', t') = f'_{\mu\nu} e^{i(\mathbf{k}' \cdot \mathbf{x}' - \omega' t')} \tag{22.58}$$

Equation (22.58) transforms under Lorentz transformation like a tensor of second rank

$$F'_{\mu\nu}(\mathbf{x}', t') = c_{\mu\sigma} c_{\nu\tau} F_{\sigma\tau}(\mathbf{x}, t) \tag{22.59}$$

Using equation (22.58) one observes that the phases must be identical, so

$$\mathbf{k}' \cdot \mathbf{x}' - \omega' t' = \mathbf{k} \cdot \mathbf{x} - \omega t \tag{22.60}$$

With $\mathbf{k}' = (\omega'/c)\mathbf{n}'$ this is equivalent to

$$\omega'(\mathbf{n}' \cdot \mathbf{x}' - ct') = \omega(\mathbf{n} \cdot \mathbf{x} - ct) \tag{22.61}$$

Introducing the four vectors $\{k_\mu\} = \{\mathbf{k}, i\omega/c\} = \{\mathbf{k}, ik_0\}$ and $\{x_\mu\} = \{\mathbf{x}, ict\}$ equation (22.61) can be written as a scalar product. The four vector $\hat{k}$ defined in this way can be transformed into an inertial system moving at $\boldsymbol{\beta} = \mathbf{v}/c$ with respect to the unprimed system.

$$\mathbf{k}'_\perp = \mathbf{k}_\perp, \qquad k'_\parallel = \gamma(k_\parallel - \beta k_0), \qquad k'_0 = \gamma(k_0 - \boldsymbol{\beta} \cdot \mathbf{k}) \tag{22.62}$$

Here, $\gamma = 1/\sqrt{1 - \beta^2}$. Furthermore, from the figure one can see

$$k_\parallel = \mathbf{k} \cdot \frac{\mathbf{v}}{v}, \qquad \mathbf{k}_\perp = \mathbf{k} - \left(\mathbf{k} \cdot \frac{\mathbf{v}}{v}\right) \frac{\mathbf{v}}{v} \tag{22.63}$$

With $k'_0 = \gamma(k_0 - \boldsymbol{\beta} \cdot \mathbf{k})$, we obtain from $k'_0 = \omega'/c$:

$$\frac{\omega'}{c} = \gamma\left(\frac{\omega}{c} - \frac{\omega}{c}\beta \cos \Theta\right) \quad \Longleftrightarrow \quad \omega' = \gamma\omega(1 - \beta \cos \Theta) \tag{22.64}$$

Equation (22.64) describes the relativistic Doppler shift. From Figure 22.5 one can deduce the connection

$$\tan \Theta' = \frac{k'_\perp}{k'_\parallel} \tag{22.65}$$

With the help of equations (22.62) and (22.63), this can be expressed by the angle Θ:

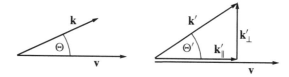

Figure 22.6. The angles Θ and Θ' in the aberration equation, where Θ' is the angle after a Lorentz transformation.

$$k'_\parallel = \gamma \left(\mathbf{k} \cdot \frac{\mathbf{v}}{v} - \beta k_0 \right) = \gamma \frac{\omega}{c}(\cos \Theta - \beta) \tag{22.66}$$

$$\mathbf{k}_\perp \cdot \mathbf{k}_\perp = k^2 - \left(\mathbf{k} \cdot \frac{\mathbf{v}}{v} \right)^2 = \left(\frac{\omega}{c} \right)^2 - \left(\frac{\omega}{c} \right)^2 \cos^2 \Theta = \left(\frac{\omega}{c} \sin \Theta \right)^2$$

$$|\mathbf{k}_\perp| = \frac{\omega}{c} \sin \Theta \tag{22.67}$$

Substituting equations (22.66) and (22.67) into (22.65) yields the aberration formula

$$\tan \Theta' = \frac{\sin \Theta}{\gamma (\cos \Theta - \beta)} \tag{22.68}$$

which according to the figure gives the change of direction of the wave number vector, that is, the direction of propagation, in the moving system.

The energy-momentum tensor

For the density of the Lorentz force (force per unit of volume) acting on a charge distribution one can write

$$\mathbf{f} = \rho \mathbf{E} + \frac{\mathbf{j}}{c} \times \mathbf{B} \tag{22.69}$$

Expressed by the field tensor, the first component reads

$$f_1 = \rho E_1 + \frac{1}{c}(j_2 B_3 - j_3 B_2) = \frac{1}{c}(F_{14} j_4 + F_{12} j_2 + F_{13} j_3) = \frac{1}{c} F_{1\nu} j_\nu$$

where the term $F_{11} j_1$ has been supplemented because $F_{11} = 0$. More generally,

$$f_k = \frac{1}{c} F_{k\nu} j_\nu, \qquad k = 1, 2, 3 \tag{22.70}$$

Since a four-vector appears on the right-hand side, the left-hand side also may be completed as a four-vector. We may write

$$f_\mu = \frac{1}{c} F_{\mu\nu} j_\nu \tag{22.71}$$

Computing f_4, one finds for the *vector of the force density*

$$\{f_\mu\} = \left\{ \mathbf{f}, \frac{i}{c} \mathbf{E} \cdot \mathbf{j} \right\}$$

The expression $\mathbf{E} \cdot \mathbf{j} = \rho \mathbf{v} \cdot \mathbf{E} = \mathbf{v} \cdot \mathbf{f} = (d\mathbf{r}/dt) \cdot \mathbf{f} = (d\mathbf{r}/dt) \cdot (\mathbf{F}/V)$ in the fourth component denotes the power density, that is, the mechanical work done by the electric field on the charges per unit time and unit volume. The vector $\{f_\mu\}$ describes the change of the density of mechanical momentum and of the mechanical energy. Therefore, we will try to derive the conservation laws for these quantities.

Using the Maxwell equation (22.44a), we find

$$f_\mu = \frac{1}{4\pi} F_{\mu\nu} \frac{\partial F_{\nu\lambda}}{\partial x_\lambda} \tag{22.72}$$

the right-hand side can be split into two terms,

$$4\pi f_\mu = \frac{\partial}{\partial x_\lambda}(F_{\mu\nu} F_{\nu\lambda}) - F_{\nu\lambda} \frac{\partial F_{\mu\nu}}{\partial x_\lambda} \tag{22.73}$$

The second summand can be further rewritten by using the antisymmetry of the field tensor and in the third step substituting the Maxwell equation (22.44b):

$$F_{\nu\lambda} \frac{\partial F_{\mu\nu}}{\partial x_\lambda} = \frac{1}{2} F_{\nu\lambda} \frac{\partial F_{\mu\nu}}{\partial x_\lambda} + \frac{1}{2} F_{\lambda\nu} \frac{\partial F_{\mu\lambda}}{\partial x_\nu} = \frac{1}{2} F_{\nu\lambda} \left(\frac{\partial F_{\mu\nu}}{\partial x_\lambda} + \frac{\partial F_{\lambda\mu}}{\partial x_\nu} \right)$$

$$= -\frac{1}{2} F_{\nu\lambda} \frac{\partial F_{\nu\lambda}}{\partial x_\mu} = -\frac{1}{4} \frac{\partial}{\partial x_\mu}(F_{\nu\lambda} F_{\nu\lambda}) = -\frac{1}{4} \delta_{\mu\lambda} \frac{\partial}{\partial x_\lambda}(F_{\sigma\tau} F_{\sigma\tau})$$

Hence, we obtain for the force density

$$f_\mu = \frac{1}{4\pi} \frac{\partial}{\partial x_\lambda} \left(F_{\mu\nu} F_{\nu\lambda} + \frac{1}{4} \delta_{\mu\lambda} F_{\sigma\tau} F_{\sigma\tau} \right) \quad \text{or} \quad f_\mu = \frac{\partial T_{\mu\lambda}}{\partial x_\lambda} \tag{22.74}$$

with the symmetric *energy-momentum tensor*

$$T_{\mu\lambda} = \frac{1}{4\pi} \left(F_{\mu\nu} F_{\nu\lambda} + \frac{1}{4} \delta_{\mu\lambda} F_{\sigma\tau} F_{\sigma\tau} \right) \tag{22.75}$$

Explicitly, the elements of this tensor are

$$(T_{\mu\nu}) = \begin{pmatrix} T_{11} & T_{12} & T_{13} & -icg_1 \\ T_{21} & T_{22} & T_{23} & -icg_2 \\ T_{31} & T_{32} & T_{33} & -icg_3 \\ -icg_1 & -icg_2 & -icg_3 & u \end{pmatrix} \tag{22.76}$$

Here,

$$\mathbf{g} = \frac{1}{4\pi c} \mathbf{E} \times \mathbf{B} = \frac{1}{c^2} \mathbf{S} \tag{22.77}$$

is the electromagnetic momentum density connected with the Poynting vector via the factor c^2 and

$$u = \frac{E^2 + B^2}{8\pi} \tag{22.78}$$

is the energy of the field.

The elements

$$T_{ij} = \frac{1}{4\pi} \left(E_i E_j + B_i B_j - \frac{1}{2} \delta_{ij} (E^2 + B^2) \right) \tag{22.79}$$

form the three-dimensional Maxwellian stress tensor (compare with the previous results from Chapter 13, equation (13.31)). From equation (22.72) we can see that the trace of the energy-momentum tensor vanishes since

$$\mathrm{Tr}(\hat{T}) = T_{\mu\mu} = \frac{1}{4\pi} \left(F_{\mu\nu} F_{\nu\mu} + 4 \cdot \frac{1}{4} F_{\sigma\tau} F_{\sigma\tau} \right) = 0 \tag{22.80}$$

Conservation laws

By volume integration over the spatial components of the four-force density one obtains the rate of mechanical momentum transfer. With equation (22.71),

$$\frac{\partial P_k}{\partial t} = \int_V f_k \, dV = \int \frac{\partial T_{ki}}{\partial x_i} \, dV - \int \frac{\partial}{\partial t} g_k \, dV$$

or

$$\frac{d}{dt}(\mathbf{P} + \mathbf{G}) = \int_V \nabla \cdot \hat{T} \, dV = \oint_{\text{surface}} \mathbf{n} \cdot \hat{T} \, da \tag{22.81}$$

This is the *conservation law of momentum*, where we have identified $\mathbf{G} = \int_V \mathbf{g} \, dV$ with the electromagnetic field momentum: The rate of the total momentum (mechanical field momentum) equals the momentum flowing through the surface (zero for closed systems).

The remaining fourth component is

$$f_4 = \frac{i}{c} \mathbf{E} \cdot \mathbf{j} = -\frac{i}{c} \nabla \cdot \mathbf{S} + \frac{1}{ic} \frac{\partial u}{\partial t} \tag{22.82}$$

integrated, it yields the *conservation law of energy*

$$\int \left(\mathbf{E} \cdot \mathbf{j} + \frac{\partial u}{\partial t} \right) dV = \frac{\partial}{\partial t}(W + U) = -\int_V \nabla \cdot \mathbf{S} \, dV = -\oint \mathbf{n} \cdot \mathbf{S} \, da$$

Here, W denotes the mechanical energy and U the field energy in the volume V. Again, the right-hand side vanishes for closed systems.

In deriving the laws of conservation from equation (22.71) it is noticable that the elements (F_{14}, F_{24}, F_{34}) of the energy-momentum tensor are interpreted as $-ic\mathbf{g}$ while (F_{41}, F_{42}, F_{43}) have the meaning of $-(i/c)\mathbf{S}$. so, due to the symmetry of the tensor the relation $\mathbf{S} = c^2 \mathbf{g}$ follows from general considerations. In fact, we are dealing with the mass-energy relation $E = mc^2$ referred to the energy flux and the momentum density (mass-current density) of the electromagnetic field.

23 Relativistic-Covariant Lagrangian Formalism

In this chapter we want to discuss the relativistic-covariant formulation of the Lagrange equation of Mechanics. Therefore, we will remind the reader briefly of the essential aspects of the Lagrangian formulation of point mechanics. This theory is based on *Hamilton's principle*. It tells us that the time integral over the Lagrange function $L(q_1, q_2, \ldots; \dot{q}_1, \dot{q}_2, \ldots; t)$ should be an extreme value, that is,

$$\delta \int_{t_1}^{t_2} dt\, L(q_1, q_2, \ldots; \dot{q}_1, \dot{q}_2, \ldots; t) = 0 \tag{23.1}$$

Here, q_ν are the generalized coordinates, and $\dot{q}_\nu$ are the associated generalized velocities. *To be an extreme value* means that variations of the kind

$$q_i(t) \rightarrow q_i(t) + \delta q_i(t) \tag{23.2}$$

where

$$\delta q_i(t_1) = \delta q_i(t_2) = 0$$

do not change the time integral (23.1) (Figure 23.1). Necessarily, this leads to the *Euler-Lagrange equations*:

$$\frac{\partial L}{\partial q_i} - \frac{d}{dt} \frac{\partial L}{\partial \dot{q}_i} = 0, \qquad i = 1, 2, \ldots \tag{23.3}$$

These are the equations of motion of a system having $i = 1, 2, \ldots$ degrees of freedom. The quantities

$$\pi_i = \frac{\partial L}{\partial \dot{q}_i} \tag{23.4}$$

are called *generalized momenta*.

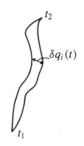

Figure 23.1. The integration paths.

The Hamiltonian function is defined by

$$H = \sum_i \pi_i \dot{q}_i - L(q_v, \dot{q}_v, t) \tag{23.5}$$

The Hamiltonian equations

$$\dot{\pi}_i = -\frac{\partial H}{\partial q_i}, \qquad \dot{q}_i = \frac{\partial H}{\partial \pi_i} \tag{23.6}$$

are equivalent to the Lagrange equations (23.3). This is illustrated in the following example:

Example 23.1: A charged particle in the Coulomb field

The Lagrangian function for a charged particle in the Coulomb field is

$$L = T - V = \frac{1}{2}m\mathbf{v}^2 - \frac{k}{r} = \frac{1}{2}m(\dot{r}^2 + r^2\dot{\vartheta}^2 + r^2\sin^2\vartheta\,\dot{\varphi}^2) - \frac{k}{r} \tag{23.7}$$

Here, $k = \pm Ze^2$, if the central charge is $\pm Ze$. The conjugate momenta are

$$\pi_r = \frac{\partial L}{\partial \dot{r}} = m\dot{r}, \qquad \pi_\vartheta = \frac{\partial L}{\partial \dot{\vartheta}} = mr^2\dot{\vartheta}, \qquad \pi_\varphi = \frac{\partial L}{\partial \dot{\varphi}} = mr^2\sin^2\vartheta\,\dot{\varphi} \tag{23.8}$$

and the Lagrange equations are

$$\frac{d}{dt}(m\dot{r}) + \frac{d}{dr}\left(\frac{k}{r}\right) - mr\dot{\vartheta}^2 - mr\sin^2\vartheta\,\dot{\varphi}^2 = 0$$

$$\frac{d}{dt}(mr^2\dot{\vartheta}) - \frac{\partial}{\partial\vartheta}\left(\frac{1}{2}mr^2\sin^2\vartheta\,\dot{\varphi}^2\right) = 0 \tag{23.9}$$

$$\frac{d}{dt}(mr^2\sin^2\vartheta\,\dot{\varphi}) = 0$$

With (23.8) the generalized velocities $\dot{r}$, $\dot{\vartheta}$, and $\dot{\varphi}$ may be expressed by the corresponding momenta

$$\dot{r} = \frac{\pi_r}{m}, \qquad \dot{\vartheta} = \frac{\pi_\vartheta}{mr^2}, \qquad \dot{\varphi} = \frac{\pi_\varphi}{mr^2\sin^2\vartheta} \tag{23.10}$$

So, we obtain the Hamiltonian function

$$H = \sum_\alpha \pi_\alpha \dot{q}_\alpha - L = \pi_r \dot{r} + \pi_\vartheta \dot{\vartheta} + \pi_\varphi \dot{\varphi} - L$$

$$= \frac{1}{2m}\left(\pi_r^2 + \frac{\pi_\vartheta^2}{r^2} + \frac{\pi_\varphi^2}{r^2 \sin^2 \vartheta}\right) + \frac{k}{r} \tag{23.11}$$

Hence, the Hamiltonian equations are

$$\dot{r} = \frac{\pi_r}{m}, \qquad\qquad \dot{\pi}_r = \frac{\pi_\vartheta^2 + \dfrac{1}{\sin^2 \vartheta}\pi_\vartheta^2}{mr^3} + \frac{k}{r^2}$$

$$\dot{\vartheta} = \frac{\pi_\vartheta}{mr^2}, \qquad\qquad \dot{\pi}_\vartheta = \frac{\cot \vartheta\, \pi_\varphi^2}{mr^2 \sin^2 \vartheta} \tag{23.12}$$

$$\dot{\varphi} = \frac{\pi_\varphi}{mr^2 \sin^2 \vartheta}, \qquad \dot{\pi}_\varphi = 0$$

and, moreover $dH/dt = 0$. If the various pairs of equations (23.12) are combined, one obtains the Lagrange equations for the coordinates r, ϑ, and φ. For example, the first equation in (23.12) for $\dot{\pi}_r$ reads explicitly

$$\frac{d}{dt}(m\dot{r}) = \frac{m^2 r^4 \dot{\vartheta}^2 + m^2 r^4 \sin^2 \vartheta\, \dot{\varphi}^2}{mr^3} - \frac{d}{dr}\left(\frac{k}{r}\right)$$

$$= mr\dot{\vartheta}^2 + mr\sin^2 \vartheta\, \dot{\varphi}^2 - \frac{d}{dr}\left(\frac{k}{r}\right) \tag{23.13}$$

which agrees exactly with the first Lagrange equation. The last of the equations in (23.12) means that $\pi_\varphi = $ const. (that is, φ is a cyclic coordinate). Furthermore, due to $dH/dt = 0$ also $H = $ const.

Example 23.2:   A charged particle in an external magnetic field

Let e be the charge of a particle. Let $\phi(\mathbf{r}, t)$ be the position- and time-dependent Coulomb potential and $\mathbf{A}(\mathbf{r}, t)$ the corresponding vector potential. The kinetic energy of the particle is $T = (1/2)m\mathbf{v}^2 = (1/2)m(v_x^2 + v_y^2 + v_z^2)$. The interaction energy is

$$V = e\phi(\mathbf{r}, t) - \frac{e}{c}\mathbf{A}\cdot\mathbf{v} = e\phi - \frac{1}{c}\mathbf{j}\cdot\mathbf{A} \tag{23.14}$$

which leads to the Lagrangian function

$$L = \frac{1}{2}m\mathbf{v}^2 - e\phi + \frac{e}{c}\mathbf{v}\cdot\mathbf{A} \tag{23.15}$$

this is confirmed by checking whether this Lagrangian function leads to the correct equation of motion. It is sufficient to check the x-degree of freedom:

$$\frac{d}{dt}\frac{\partial L}{\partial v_x} - \frac{\partial L}{\partial x} = 0 \tag{23.16}$$

Obviously, with (23.15)

$$\frac{\partial L}{\partial x} = -e\frac{\partial \phi}{\partial x} + \frac{e}{c}\frac{\partial \mathbf{A}}{\partial x}\cdot\mathbf{v}$$

$$\frac{\partial L}{\partial v_x} = mv_x + \frac{e}{c}A_x \tag{23.17}$$

and therefore, (23.16) can be written as

$$\frac{dmv_x}{dt} = -e\frac{\partial \phi}{\partial x} + \frac{e}{c}\frac{\partial \mathbf{A}}{\partial x} \cdot \mathbf{v} - \frac{e}{c}\frac{dA_x}{dt} \tag{23.18}$$

Furthermore, we calculate

$$\frac{dA_x}{dt} = \frac{\partial A_x}{\partial t} + \frac{\partial A_x}{\partial x}\frac{dx}{dt} + \frac{\partial A_x}{\partial y}\frac{dy}{dt} + \frac{\partial A_x}{\partial z}\frac{dz}{dt}$$

$$= \frac{\partial A_x}{\partial t} + \frac{\partial A_x}{\partial x}v_x + \frac{\partial A_x}{\partial y}v_y + \frac{\partial A_x}{\partial z}v_z \tag{23.19}$$

and

$$\frac{\partial \mathbf{A}}{\partial x} \cdot \mathbf{v} = \frac{\partial A_x}{\partial x}v_x + \frac{\partial A_y}{\partial y}v_y + \frac{\partial a_z}{\partial z}v_z \tag{23.20}$$

so that equation (23.18) becomes

$$\frac{dmv_x}{dt} = e\left(-\frac{\partial \phi}{\partial x} - \frac{1}{c}\frac{\partial A_x}{\partial t}\right) + \frac{e}{c}\left(\frac{\partial A_y}{\partial x} - \frac{\partial A_x}{\partial y}\right) \cdot v_y - \frac{e}{c}\left(\frac{\partial A_x}{\partial z} - \frac{\partial A_z}{\partial x}\right) \cdot v_z$$

$$= eE_x + \frac{e}{c}\left(B_z v_y - B_y v_z\right) = e\left(\mathbf{E} + \frac{1}{c}\mathbf{v} \times \mathbf{B}\right)_x \tag{23.21}$$

Similar equations hold for the other components. Therefore, the Lagrange equations can be summarized in vector form

$$\frac{d}{dt}(m\mathbf{v}) = e\left(\mathbf{E} + \frac{1}{c}\mathbf{v} \times \mathbf{B}\right) \tag{23.22}$$

These are exactly the Newton equations for a charged mass point in an electromagnetic field. Hence, (23.15) is the corresponding (correct) Langrangian function. Further, we set up the *Hamiltonian function*: The conjugate momenta $\pi_\nu = \partial L/\partial v_\nu$ ($\nu = 1, 2, 3$) may be calculated from (23.15) and combined in vector form

$$\boldsymbol{\pi} = \{\pi_x, \pi_y, \pi_z\} = m\mathbf{v} + \frac{e}{c}\mathbf{A}(\mathbf{r}, t) \tag{23.23}$$

so that

$$\mathbf{v} = \frac{1}{m}\left(\boldsymbol{\pi} - \frac{e}{c}\mathbf{A}(\mathbf{r}, t)\right) \tag{23.24}$$

Obviously, the relation between the velocity vector $\mathbf{v}$ and the momentum vector $\boldsymbol{\pi}$ for a particle in an electromagnetic field is distinct from that for a free particle. The additional term $(e/c)\mathbf{A}(\mathbf{r}, t)$ arises alone from the external magnetic field.

Now, the Hamiltonian function is

$$H = \sum_\alpha \pi_\alpha \dot{q}_\alpha - L, \qquad H = \boldsymbol{\pi} \cdot \mathbf{v} - L = \frac{1}{2m}\left(\boldsymbol{\pi} - \frac{e}{c}\mathbf{A}\right)^2 + e\phi \tag{23.25}$$

The first term is the kinetic energy $mv^2/2$, and the second term is the potential energy arising from the electric field $\mathbf{E}$. Obviously, there is no potential energy which could come from the magnetic field, because the Lorentz force $(e/c)\mathbf{v} \times \mathbf{B}$ is always perpendicular to the direction of motion ($\mathbf{B} \perp \mathbf{v}$) and therefore does not do work on the mass point along its motion.

The Hamiltonian equations are

$$\mathbf{v} = \frac{\partial H}{\partial \boldsymbol{\pi}} = \frac{1}{m}\left(\boldsymbol{\pi} - \frac{e}{c}\mathbf{A}\right)$$

$$\dot{\boldsymbol{\pi}} = -\frac{\partial H}{\partial \mathbf{r}} = -\left\{\frac{\partial H}{\partial x}, \frac{\partial H}{\partial y}, \frac{\partial H}{\partial z}\right\} = \frac{1}{m}\left(\boldsymbol{\pi} - \frac{e}{c}\mathbf{A}\right) \cdot \frac{e}{c}\frac{\partial \mathbf{A}}{\partial \mathbf{r}} - e\frac{\partial \phi}{\partial \mathbf{r}}$$

$$= \frac{e}{mc}\left\{\left(\pi_x - \frac{e}{c}A_x\right)\frac{\partial A_x}{\partial x} + \left(\pi_y - \frac{e}{c}A_y\right)\frac{\partial A_y}{\partial x} + \left(\pi_z - \frac{e}{c}A_z\right)\frac{\partial A_z}{\partial x} - c\frac{\partial \phi}{\partial x}\right.,$$

$$\left(\pi_x - \frac{e}{c}A_x\right)\frac{\partial A_x}{\partial y} + \left(\pi_y - \frac{e}{c}A_y\right)\frac{\partial A_y}{\partial y} + \left(\pi_z - \frac{e}{c}A_z\right)\frac{\partial A_z}{\partial y} - c\frac{\partial \phi}{\partial y},$$

$$\left.\left(\pi_x - \frac{e}{c}A_x\right)\frac{\partial A_x}{\partial z} + \left(\pi_y - \frac{e}{c}A_y\right)\frac{\partial A_y}{\partial z} + \left(\pi_z - \frac{e}{c}A_z\right)\frac{\partial A_z}{\partial z} - c\frac{\partial \phi}{\partial z}\right\} \qquad (23.26)$$

and furthermore,

$$\frac{dH}{dt} = -\frac{1}{m}\left(\boldsymbol{\pi} - \frac{e}{c}\mathbf{A}\right) \cdot \frac{e}{c}\frac{\partial \mathbf{A}}{\partial t} + e\frac{\partial \phi}{\partial t} \qquad (23.27)$$

The two equations (23.26) may be combined to the Lagrange equation (23.18) and (23.22), respectively. Equation (23.27) expresses the fact that the total energy is a constant of motion only if the vector potential $\mathbf{A}$ as well as the Coulomb potential ϕ are time-independent.

Covariant Lagrangian formalism for relativistic point charges

In the case that the particle velocities are comparable to the velocity of light we must use the relativistically covariant formulation of Hamilton's principle. *Covariant* (under a Lorentz transformation) means that all laws of physics take the same form in any (Lorentz) system. It follows that the Lagrangian function and the action integral have to be scalars. Therefore, we write

$$W = \int_{s_1}^{s_2} L(x_\mu, u_\mu, s)\, ds \qquad \textbf{(23.28)}$$

Here, L is a Lorentz scalar. $\hat{x} \equiv \{x_\mu\}$ is the space-time vector, and $\hat{u} \equiv \{u_\mu\} = \{dx_\mu/ds\}$ is the four-velocity. Also, the integration parameter s must be a Lorentz scalar. It is suitable to choose s as

$$ds = \sqrt{-d\hat{r} \cdot d\hat{r}} = \sqrt{c^2\, dt^2 - dx^2 - dy^2 - dz^2} = c\, dt\sqrt{1 - \frac{\mathbf{v}^2}{c^2}} \qquad \textbf{(23.29)}$$

the *arc length* in the Minkowski space. Because $v/c \ll 1$ (23.28) has to transform into the action integral of the nonrelativistic limit,

$$L(x_\mu, u_\mu, s)\, ds \quad \overset{\text{nonrel limit}}{\Longrightarrow} \quad L_{\text{nonrel}}(\mathbf{r}, \mathbf{v}, t)\, dt \qquad \textbf{(23.30a)}$$

or

$$L(x_\mu, u_\mu, s)c\sqrt{1 - \frac{\mathbf{v}^2}{c^2}} \quad \overset{\text{nonrel limit}}{\Longrightarrow} \quad L_{\text{nonrel}}(\mathbf{r}, \mathbf{v}, t) \tag{23.30b}$$

We will give some examples of this.

Example 23.3: The free particle

The nonrelativistic Lagrangian function of a free particle is simply the kinetic energy minus the rest energy (which can be regarded as potential energy, but as a constant may be dropped out nearly always).

$$L_{\text{nonrel}} = \frac{1}{2}m_0\mathbf{v}^2 - m_0c^2 \tag{23.31}$$

The nonrelativistic Lagrangian function, which must be a Lorentz scalar and has to fulfill the condition (23.31), is obviously

$$L(x_\mu, u_\mu, s) = -m_0c \tag{23.32}$$

since

$$-m_0c \cdot c\sqrt{1 - \frac{\mathbf{v}^2}{c^2}} \approx \frac{m_0}{2}\mathbf{v}^2 - m_0c^2$$

This means that the relativistically correct Lagrangian function for free point particles which has to be used in the classical, canonical formalism is (not written in a covariant form),

$$L(r, \vartheta, \varphi) = -m_0c^2\sqrt{1 - \frac{\mathbf{v}^2}{c^2}} \tag{23.33}$$

Example 23.4: A charged particle in an external electromagnetic field

In this case the nonrelativistic Lagrangian function is

$$L_{\text{nonrel}} = T - V = \frac{1}{2}m_0\mathbf{v}^2 - m_0c^2 - e\left(\phi(\mathbf{r}, t) - \frac{1}{c}\mathbf{A} \cdot \mathbf{v}\right) \tag{23.34}$$

where ϕ and $\mathbf{A}$ are the Coulomb potential and the vector potential, respectively, forming together the four-vector (four-potential)

$$\hat{A} = \{A_\mu\} = \{\mathbf{A}, i\phi\} \tag{23.35}$$

see Chapter 22, equation (22.25). The electric and magnetic field intensities are calculated in the known manner:

$$\mathbf{E} = -\nabla\phi - \frac{1}{c}\frac{\partial \mathbf{A}}{\partial t}, \qquad \mathbf{B} = \nabla \times \mathbf{A} \quad \text{or} \quad F_{\mu\nu} = \frac{\partial A_\nu}{\partial x_\mu} - \frac{\partial A_\mu}{\partial x_\nu} \tag{23.36}$$

The correct relativistic Lagrangian function, which is seen at once to be a Lorentz scalar, is obviously

$$L(x_\mu, u_\mu, s) = -m_0c + \frac{e}{c}\sum_{\mu=1}^{4} A_\mu u_\mu \tag{23.37}$$

In fact, we rewrite (23.37) as

$$L(x_\mu, u_\mu, s) = -m_0 c - \frac{e\phi}{c\sqrt{1 - \dfrac{\mathbf{v}^2}{c^2}}} + \frac{e}{c}\frac{\mathbf{A}\cdot\mathbf{v}}{c\sqrt{1 - \dfrac{\mathbf{v}^2}{c^2}}} \tag{23.38}$$

which shows clearly the nonrelativistic limit (23.34) using (23.36) from the preceeding example. In a special Lorentz system in which the three-dimensional vector $\mathbf{A}$ vanishes, that is, there is no magnetic field (compare (23.36)) (23.37) and correspondingly (23.34) reduce to the Lagrangian function for a charged particle in the Coulomb potential.

Now, we want to derive the relativistic Lagrange equations. For this purpose we have to vary (23.28), that is, to grasp the consequences of

$$\delta \int_{s_1}^{s_2} L(x_\mu, u_\mu, s)\, ds = 0 \tag{23.39}$$

in a computational manner. In the nonrelativistic theory, the variation δx_i and $\delta \dot{x}_i = \delta v_i$ (except for the limits of integration) have been arbitrary. In the relativistic theory this is no longer the case: δx_μ and δu_μ are not arbitrary because u_μ has always to satisfy the equation

$$-\sum_{\mu=1}^{4} u_\mu u_\mu = -\frac{1}{ds^2}\sum_{\mu=1}^{4} dx_\mu dx_\mu = \frac{ds^2}{ds^2} = 1 \tag{23.40}$$

To avoid this difficulty, we replace the invariant parameter s by the (invariant) affine parameter τ, which may be given by any differentiable function of s:

$$\tau = \tau(s) \tag{23.41}$$

and define the new, not normalized *four-velocity* by

$$w_\mu = \frac{dx_\mu}{d\tau} = \frac{dx_\mu}{ds}\frac{ds}{d\tau} = u_\mu \frac{ds}{d\tau} \tag{23.42}$$

the norm of which is

$$-\sum_{\mu=1}^{4} w_\mu w_\mu \equiv -w^2 = -u^2\left(\frac{ds}{d\tau}\right)^2 = \left(\frac{ds}{d\tau}\right)^2 \tag{23.43}$$

The freedom of the norm allows for free variation with respect to w_μ. It follows that

$$\frac{ds}{d\tau} = \sqrt{-\hat{w}^2} \tag{23.44}$$

and therefore with (23.42)

$$w_\mu = u_\mu \sqrt{-\hat{w}^2} \tag{23.45}$$

One recognizes that though there is a constraining condition (23.42) for u_μ there is none for w_μ. Therefore, the variational principle (23.39) is rewritten as

$$\delta \int_{\tau_1}^{\tau_2} L(x_\mu, w_\mu, \tau)\, d\tau = 0 \tag{23.46}$$

where

$$L(x_\mu, w_\mu, \tau) = L(x_\mu, w_\mu, s)\sqrt{-\hat{w}^2} \tag{23.47}$$

The variations δx_μ and δw_μ are now arbitrary apart of the usual constraint

$$\delta x_\mu(\tau_1) = \delta x_\mu(\tau_2) = \delta w_\mu(\tau_1) = \delta w_\mu(\tau_2) = 0 \tag{23.48}$$

Now, the variation (23.46) can be carried out, and we obtain

$$\delta \int_{\tau_1}^{\tau_2} L \, d\tau = \int_{\tau_1}^{\tau_2} \delta L \, d\tau = \int_{\tau_1}^{\tau_2} \left(\frac{\partial L}{\partial x_\mu} \delta x_\mu + \frac{\partial L}{\partial w_\mu} \delta w_\mu \right) d\tau = 0 \tag{23.49}$$

Here, we have used Einstein's summation convention. With

$$\frac{\partial L}{\partial w_\mu} \delta w_\mu = \frac{d}{d\tau} \left(\frac{\partial L}{\partial w_\mu} \delta x_\mu \right) - \frac{d}{d\tau} \left(\frac{\partial L}{\partial w_\mu} \right) \delta x_\mu$$

(23.46) becomes

$$\delta \int_{\tau_1}^{\tau_2} L \, d\tau = \int_{\tau_1}^{\tau_2} \left(\frac{\partial L}{\partial x_\mu} - \frac{d}{d\tau} \frac{\partial L}{\partial w_\mu} \right) \delta x_\mu \, d\tau + \left. \frac{\partial L}{\partial w_\mu} \delta x_\mu \right|_{\tau_1}^{\tau_2} = 0 \tag{23.50}$$

The last term vanishes because of (23.48), so that due to the arbitrary variation δx_μ we obtain the *relativistic Lagrange equations*

$$\frac{\partial L}{\partial x_\mu} - \frac{d}{d\tau} \frac{\partial L}{\partial w_\mu} = 0, \qquad \mu = 1, 2, 3, 4 \tag{23.51}$$

Since L is a Lorentz scalar, both terms in (23.51) are Lorentz vectors. So, the Lagrange equations are of covariant form.

Example 23.5: A free particle

According to Example 23.3 and (23.47), the Lagrangian function is

$$L(x_\mu, w_\mu, \tau) = -m_0 c \sqrt{-\hat{w}^2} \tag{23.52}$$

leading at once to

$$\frac{\partial L}{\partial x_\mu} = 0, \qquad \frac{\partial L}{\partial w_\mu} = m_0 c \frac{w_\mu}{\sqrt{-\hat{w}^2}} \tag{23.53}$$

and to the Lagrange equations corresponding to (23.51)

$$\frac{d}{d\tau} \left(m_0 c \frac{w_\mu}{\sqrt{-\hat{w}^2}} \right) = 0, \qquad \mu = 1, 2, 3, 4 \tag{23.54}$$

Evidently, (23.54) means $(d/ds)(m_0 c u_\mu) = 0 \equiv (d/ds) p_\mu$, that is, the four-momenta are constant.

Example 23.6: Charged particles in an external electromagnetic field

According to Example 23.4 and (23.47), the Lagrangian function is

$$L(x_\mu, w_\mu, \tau) = -m_0 c \sqrt{-\hat{w}^2} + \frac{e}{c} A_\mu w_\mu \tag{23.55}$$

so that

$$\frac{\partial L}{\partial x_\mu} = \frac{e}{c}\frac{\partial A_\nu}{\partial x_\mu}w_\nu, \qquad \frac{\partial L}{\partial w_\mu} = m_0 c\frac{w_\mu}{\sqrt{-\hat{w}^2}} + \frac{e}{c}A_\mu \tag{23.56}$$

and the Lagrangian equations (23.51) read

$$\frac{e}{c}\frac{\partial A_\nu}{\partial x_\mu}w_\nu - \frac{d}{d\tau}\left(m_0 c\frac{w_\mu}{\sqrt{-\hat{w}^2}} + \frac{e}{c}A_\mu\right) = 0 \qquad (\mu = 1, 2, 3, 4) \tag{23.57}$$

With

$$\frac{dA_\mu}{d\tau} = \frac{dA_\mu}{ds}\frac{ds}{d\tau} = \frac{\partial A_\mu}{\partial x_\nu}\frac{dx_\nu}{ds}\sqrt{-\hat{w}^2} = \frac{\partial A_\mu}{\partial x_\nu}u_\nu\sqrt{-\hat{w}^2} = \frac{\partial A_\mu}{\partial x_\nu}w_\nu$$

we obtain

$$\frac{d}{ds}\left(m_0 cu_\mu\right) = \frac{e}{c}F_{\mu\nu}u_\nu, \qquad \mu = 1, 2, 3, 4 \tag{23.58}$$

where

$$F_{\mu\nu} = \frac{\partial A_\nu}{\partial x_\mu} - \frac{\partial A_\mu}{\partial x_\nu}$$

are the components of the antisymmetric field tensor

$$F_{\mu\nu} = \begin{pmatrix} 0 & B_3 & -B_2 & -iE_1 \\ -B_3 & 0 & B_1 & -iE_2 \\ B_2 & -B_1 & 0 & -iE_3 \\ iE_1 & iE_2 & iE_3 & 0 \end{pmatrix} \tag{23.59}$$

(compare equations (22.31), (22.32)). The right-hand side of (23.58) is the Lorentz force, as may be seen from equation (20.49)

$$f_\mu = \frac{e}{c}F_{\mu\nu}u_\nu \tag{23.60}$$

Due to the antisymmetry of $F_{\mu\nu}$,

$$f_\mu u_\mu = \frac{e}{c}F_{\mu\nu}u_\nu u_\mu = 0 \tag{23.61}$$

This implies that the Lorentz force is perpendicular to the four-velocity.

It is instructive to consider the space and time components of equation (23.58) separately. We start with the first space component ($\mu = 1$, that is, the x-component):

$$\frac{d}{ds}\left(m_0 c\frac{v_1}{c\sqrt{1 - \mathbf{v}^2/c^2}}\right) = \frac{e}{c}F_{1\nu}u_\nu = \frac{e}{c}\frac{(B_3 v_2 - B_2 v_3 + E_1 c)}{\sqrt{1 - \mathbf{v}^2/c^2}}$$

Evidently, this may be also written in the following way:

$$\frac{d}{dt}(mv_1) = e\left(\mathbf{E} + \frac{\mathbf{v}}{c}\times\mathbf{B}\right)_1 \tag{23.62}$$

Here, $m = m_0/\sqrt{1 - \mathbf{v}^2/c^2}$ is the relativistic mass. Similar relations hold also for the other components so that (23.58) is seen as

$$\frac{d(m\mathbf{v})}{dt} = e\left(\mathbf{E} + \frac{1}{c}\mathbf{v} \times \mathbf{B}\right) \tag{23.63}$$

that is, as the Newtonian (relativistic) equation of motion for Lorentz forces. The fourth component (time component) of (23.58) yields

$$\frac{d}{ds}\left(m_0 c \frac{1}{\sqrt{1 - \mathbf{v}^2/c^2}}\right) = \frac{e}{c}F_{4\nu}u_\nu$$

One obtains

$$\frac{d(mc^2)}{dt} = e\mathbf{E} \cdot \mathbf{v} \tag{23.64}$$

The rate of the kinetic energy equals the power (work per unit of time) which has be done against the electric field $\mathbf{E}$. The magnetic field intensity $\mathbf{B}$ does not appear because $(e/c)\mathbf{v} \times \mathbf{B}$ is perpendicular to $\mathbf{v}$ and therefore no work is done against this part of the force.

Example 23.7: Motion of a charged particle in a steady uniform magnetic field

In this case (23.63) reduces to

$$\frac{d}{dt}(m\mathbf{v}) = \frac{e}{c}\mathbf{v} \times \mathbf{B} \tag{23.65}$$

Let the charge of the test particle be e and let the direction of the uniform magnetic field be the z-axis. With $E = mc^2$ (23.65) is written as

$$\frac{E}{c^2}\frac{d\mathbf{v}}{dt} = \frac{e}{c}\mathbf{v} \times \mathbf{B} \tag{23.66}$$

since according to (23.64) the energy of the test particle is constant. Written explicitely (23.66) reads

$$\dot{v}_x = \omega v_y, \qquad \dot{v}_y = -\omega v_x, \qquad \dot{v}_z = 0 \tag{23.67}$$

where we have set $\omega \equiv ecB_z/E$. Then,

$$\frac{d}{dt}(v_x + iv_y) = -i(v_x + iv_y)\omega$$

and hence,

$$v_x + iv_y = ae^{-i\omega t} \tag{23.68}$$

with a constant (complex) amplitude a. Then, (23.68) may be written as

$$v_x = v_0 \cos(\omega t + \alpha), \qquad v_y = -v_0 \sin(\omega t + \alpha)$$

where α is the initial phase (at $t = 0$) and v_0 the velocity of the particle in the (x, y)-plane (remaining constant):

$$v_0^2 = v_x^2 + v_y^2 \tag{23.69}$$

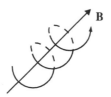

Figure 23.2. The trajectory of a charged particle in a uniform magnetic field is a helical line.

So, the complete integration of (23.67) reads

$$x = x_0 + R \sin(\omega t + \alpha)$$
$$y = y_0 + R \cos(\omega t + \alpha) \tag{23.70}$$
$$z = z_0 + v_0 t$$

with

$$R \equiv \frac{v_0}{\omega} \tag{23.71}$$

one can see from (23.70) that the test charge moves along a helical line whose axis is determined by the magnetic field (here the z-axis) and whose radius R is given by (23.71); see Figure 23.2. ω is the circular frequency of the motion, which in the nonrelativistic limit ($E = m_0 c^2$) is given by

$$\omega_{\text{nonrel}} = \frac{e B_z}{m_0 c} \tag{23.72}$$

Example 23.8:   Exercise: Motion of a charged particle in uniform, steady E- and B-fields

Discuss the motion of a charged particle in uniform, steady electric (**E**) and magnetic (**B**) fields.

Solution   Now, we will consider a generalization of the last example. Let a point charge e be in a uniform, steady electric and magnetic field. However, this will be treated in a nonrelativistic approximation. The uniform, constant **B**-field points in z-direction, the (y, z)-plane is spanned by **B** and **E** (Figure 23.3). In nonrelativistic approximation, (23.63) reads

$$m_0 \dot{\mathbf{v}} = e \left(\mathbf{E} + \frac{1}{c} \mathbf{v} \times \mathbf{B} \right) \tag{23.73}$$

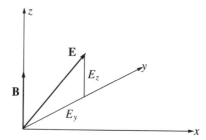

Figure 23.3. The vectors **E** and **B** lie in the (y, z)-plane.

In the given geometry

$$m_0\ddot{x} = \frac{e}{c}\dot{y}B$$
$$m_0\ddot{y} = eE_y - \frac{e}{c}\dot{x}B \qquad\qquad (23.74)$$
$$m_o\ddot{z} = eE_z$$

The third of the equations (23.74) may be integrated immediately

$$z(t) = \frac{eE_z}{2m_0}t^2 + v_{0z}t \qquad\qquad (23.75)$$

Analogous to the approach in the preceding example, we obtain

$$\frac{d}{dt}(\dot{x}+i\dot{y}) + i\omega(\dot{x}+i\dot{y}) = i\frac{e}{m_0}E_y \qquad\qquad (23.76)$$

with $\omega = eB/(m_0c)$. The circular frequency ω is the same as in the last example (see equation (23.72)). The solution of (23.76) is

$$\dot{x}+i\,\dot{y} = ae^{-i\omega t} + \frac{cE_y}{B} \qquad\qquad (23.77)$$

Here, a is arbitrary. Now, we choose it to be real. Then, equation (23.77) may be decomposed according to

$$\dot{x} = a\cos\omega t + c\frac{E_y}{B}, \qquad \dot{y} = -a\sin\omega t \qquad\qquad (23.78)$$

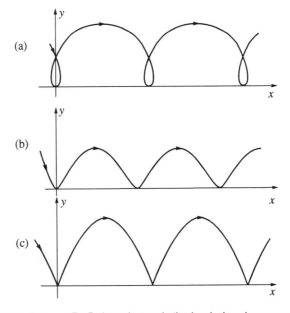

Figure 23.4. If $a = -cE_y/B$, the trajectory in the (x, y)-plane becomes a cycloid.

One can state that the components of the velocity in x- and y-directions are periodic functions in time. The mean velocities in these two directions are

$$\bar{\dot{x}} = \frac{cE_y}{B}, \qquad \bar{\dot{y}} = 0 \tag{23.79}$$

This mean velocity of the test particle is denoted the *electric drift velocity* (Figure 23.4). It is perpendicular to the crossed electric and magnetic fields and does not depend on the sign of the charge. In general, the drift velocity is given by

$$\mathbf{v}_{\text{drift}} = \frac{c[\mathbf{E} \times \mathbf{B}]}{B^2} \tag{23.80}$$

With the initial conditions $x(0) = y(0) = 0$ the equation (23.78) may be integrated

$$x(t) = \frac{a}{\omega} \sin \omega t + c\frac{E_y}{R}t, \qquad y(t) = \frac{a}{\omega}(\cos \omega t - 1) \tag{23.81}$$

This is a parametric representation of the so-called *trochoid* which contains the illustrated trajectory.

Example 23.9:   Exercise: Relativistic motion of a charge in a uniform, steady E-field

Determine the relativistic motion of a charge e in a uniform, steady electric field $\mathbf{E}$ pointing in direction of the x-axis.

Solution

We choose the coordinate system such that the motion proceeds in the (x, y)-plane. Then, the equation of motion (23.63) takes the form

$$\dot{p}_x = eE, \qquad \dot{p}_y = 0 \tag{23.82}$$

This leads to

$$p_x = eEt, \qquad p_y = p_0 \tag{23.83}$$

if p_0 is initial momentum of the particle. From the relativistic energy-momentum theorem we obtain

$$\begin{aligned} T_{\text{kin}} &= c(m_0^2c^2 + p^2)^{1/2} = (m_0^2c^4 + c^2p_0^2 + (ceEt)^2)^{1/2} \\ &= (T_0^2 + (ceEt)^2)^{1/2} \end{aligned} \tag{23.84}$$

Here, T_0 is the kinetic energy at the beginning of the motion ($t = 0$). From $\mathbf{p} = m\mathbf{v}$ and $T_{\text{kin}} = mc^2$, we obtain the relation $\mathbf{v} = \mathbf{p}c^2/T_{\text{kin}}$. For the x-component this means

$$\frac{dx}{dt} = \frac{p_xc^2}{T_{\text{kin}}} = \frac{c^2eEt}{\sqrt{T_0^2 + (ceEt)^2}} \tag{23.85}$$

The integration yields (if the integration constant is chosen to be zero)

$$x(t) = \frac{1}{eE}\sqrt{T_0^2 + (ceEt)^2} \tag{23.86}$$

Correspondingly, for the y-component

$$\frac{dy}{dt} = \frac{p_yc^2}{T_{\text{kin}}} = \frac{p_0c^2}{\sqrt{T_0^2 + (ceEt)^2}} \tag{23.87}$$

and

$$y(t) = \frac{p_0 c}{eE} \operatorname{arcsinh} \frac{ceEt}{T_0} \tag{23.88}$$

The trajectory in the (x, y)-plane is obtained by eliminating t:

$$x = \frac{E_0}{cE} \cosh \frac{eE_y}{p_0 c} \tag{23.89}$$

We see that a point charge in a uniform electric field moves along a catenary.

In the nonrelativistic approximation ($\mathbf{p}_0 = m_0 \mathbf{v}_0$; $E_0 = m_0 c^2$), from an expression of (23.89) we obtain

$$x = \frac{cE}{2mv_0^2} y^2 + \text{const} \tag{23.90}$$

This implies that for $|\mathbf{v}| \ll c$ the charge moves along a parabola.

Example 23.10: Exercise: Relativistic motion of a charge in parallel, uniform E- and B-fields

Determine the relativistic motion of a charge in parallel, uniform electric and magnetic fields.

Solution   The direction of the fields points in direction of the z-axis (Figure 23.5). Since the component of the force resulting from the magnetic field is perpendicular to the latter (see equation (23.63)), the magnetic field has no influence on the motion in z-direction. So, from (23.86) we obtain

$$z(t) = \frac{T_{\text{kin}}}{eE}, \qquad T_{\text{kin}} = \sqrt{T_0^2 + (ceEt)^2} \tag{23.91}$$

For the motion in the (x, y)-plane, the equation of motion yields

$$\dot{p}_x = \frac{e}{c} B v_y, \qquad \dot{p}_y = -\frac{e}{c} B v_x \tag{23.92}$$

leading to

$$\frac{d}{dt}(p_x + ip_y) = -i\frac{eB}{c}(v_x + iv_y) = \frac{-icBc}{T_{\text{kin}}}(p_x + ip_y) \tag{23.93}$$

Here, we have used again the relation $\mathbf{v} = \mathbf{p}c^2 / T_{\text{kin}}$. From (23.91),

$$p_x + ip_y = p_t e^{-i\varphi} \tag{23.94}$$

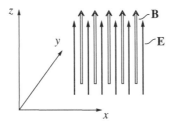

Figure 23.5. Parallel, uniform electric and magnetic fields pointing in z-direction.

p_t is the (constant) component of the momentum in the (x, y)-plane, and ϕ is determined by

$$d\varphi = eBc\frac{dt}{T_{\text{kin}}}, \qquad ct = \frac{T_0}{eE} \sinh\frac{E}{B}\varphi \tag{23.95}$$

Furthermore, from (23.93), (23.94) and (23.95)

$$p_x + ip_y = \frac{T_{\text{kin}}}{c^2}(\dot{x} + i\dot{y}) = \frac{eB}{c}\frac{d(x+iy)}{d\varphi} = p_t e^{-i\varphi} \tag{23.96}$$

and hence,

$$x = \frac{cp_t}{eB}\sin\varphi, \qquad y = \frac{cp_t}{eB}\cos\varphi \tag{23.97}$$

Equation (23.91) may also be rewritten

$$z = \frac{T_0}{eE}\cosh\left(\frac{E}{M}\varphi\right)$$

So, the motion is determined completely. The charge moves along a helical line of radius $R = cp_t/(eB)$. The angular velocity of the motion decreases, $\dot{\varphi} = eBc/E_{\text{kin}}$, whereas the velocity component in z-direction approaches the velocity of light.

Conjugate momenta

Because in the nonrelativistic case the generalized momenta π_i are defined as the particle derivatives of the Lagrangian function with respect to the generalized velocities $\dot{q}_i$, it is obvious in the relativistic theory that we should define the conjugate momenta by

$$\pi_\mu = \frac{\partial L}{\partial w_\mu} \tag{23.98}$$

So, with this definition and with equation (23.63) (see Example 23.5) one can find the conjugate momenta for a free particle

$$\pi_\mu = m_0 c u_\mu = m_0 v_\mu \tag{23.99}$$

or for a particle in an external electromagnetic field

$$\pi_\mu = m_0 c u_\mu + \frac{e}{c}A_\mu = m_0 v_\mu + \frac{e}{c}A_\mu \tag{23.100}$$

One sees that the conjugate momentum (23.99) is identical to the four-momentum of a free particle. On the other hand, according to (23.102), for a particle in the electromagnetic field there is an additional term proportional to the four-potential, that is, $(e/c)A_\mu$. Further, it is noticeable that the space components of the conjugate momentum-four vector are identical to the conjugate momenta in the nonrelativistic limit. The time component π_4 equals the total energy divided by c. The latter is called the Hamiltonian function; so,

$$H = \pi_4 c \tag{23.101}$$

With the definition (23.98) the Lagrange equations (23.99) may be written as

$$\frac{d\pi_\mu}{d\tau} = -\frac{\partial L}{\partial x_\mu}$$

(23.102)

If L does not depend on x (cyclic coordinate), from (23.102) we obtain at once

$$\frac{d\pi_\mu}{d\tau} = 0 \quad \Rightarrow \quad \pi_\mu = \text{const}$$

(23.103)

For example, this is the case for a free particle (compare (23.63)). Therefore, the momenta π_1, π_2, π_3, and the energy π_4 are constants of motion for a free particle.

For a particle in an electromagnetic field (Example 23.1) only π_4, that is, the energy, is a constant of motion, but only if ϕ as well as also **A** are, that is, if all components of the four-potential A_μ, do not depend explicitly on time t.

The relativistic Hamilton equations

After having formulated Hamilton's principle in a relativistic covariant form, we briefly give the Hamiltonian formulation of the basic equations: the Lagrangian function L is a Lorentz scalar. Therefore, the Lagrange equations (23.62) resulted in relativistic covariant form. *The Hamiltonian function H is the total energy, and therefore not a Lorentz scalar.* Under Lorentz transformations H transforms like the fourth component of a four-vector. Therefore the equations of motion in the Hamiltonian formulation are not Lorentz covariant. They are valid always only for a special, given Lorentz system.

24 Systems of Units in Electrodynamics

SUPPLEMENT

Traditionally, various distinct systems of units have been used in electrodynamics. Because thereby not only the magnitude of the units used are different but also the form of the equations, in using an expression one carefully has to note in which sytem of units it has been given. Throughout this book the *Gaussian system* has been used, which is particularly useful and common in physics. On the other hand, in the technical and experimental domain nowadays the *MKSA system* is applied predominantly. It has been made obligatory by the International Standardization Organization (ISO) under the name "Système International d'Unités" (SI). For official and business use the employment of SI units is provided by law. In this section we will present the relation between Gaussian units and SI units, and practical "rules of translation" will be supplied.

A basic difference lies in the fact that in the Gaussian system all physical quantities are reduced to the three dimensional mechanical base quantities length, mass, and time (measured in centimeters, grams, and seconds, the cgs systems). The SI system uses the mechanical units meters, kilograms, and seconds. More importantly, it introduces an additional electric base quantity, the Ampere (A) as the unit of current. As a consequence, in many equations dimensional conversion factors occur.

The first difference in the systems of units was explained already at the beginning of Chapter 1: for the constant of proportionality in Coulomb's force law

$$\mathbf{F} = k_1 \frac{q_1 q_2}{r^3} \mathbf{r} \tag{24.1}$$

one chooses $k_1 = 1$ in the Gaussian system and $k_1^* = (4\pi\epsilon_0)^{-1}$ in the SI system. (*Note:* In the following, all SI quantities will be marked by an asterisk.) ϵ_0 is the permittivity constant in free space:

$$\epsilon_0 \simeq \frac{1}{4\pi \times 9 \times 10^9} \frac{\text{As}}{\text{Vm}} \tag{24.2}$$

A further constant of proportionality appears in the force law for the magnetic interaction between currents. Ampère's law for the force between elements of a conductor reads

$$d\mathbf{F}_{12} = k_2 I_1 I_2 \frac{d\mathbf{s}_1 \times (d\mathbf{s}_2 \times \mathbf{r}_{12})}{r_{12}^3} \ . \tag{24.3}$$

Because [current] = [charge/time], the dimensions of the constants k_1 and k_2 are related to each other: The dimension of the ratio k_1/k_2 is the square of a speed. This universal constant occurring here turns out to be the speed of light. The Gaussian system takes into account this by

$$k_2 = \frac{1}{c^2} \tag{24.4}$$

In the SI system the magnetic permeability constant of free space μ_0 is introduced by

$$k_2^* = \frac{\mu_0}{4\pi} \tag{24.5}$$

The numerical value of μ_0 is fixed by the following *exact* relationship:

$$\mu_0 = 4\pi \times 10^{-7} \frac{\text{Vs}}{\text{Am}} \tag{24.6}$$

Because $k_1/k_2 = k_1^*/k_2^* = c^2$,

$$\frac{1}{\sqrt{\mu_0\epsilon_0}} = c = 2.99792458 \times 10^8 \frac{\text{m}}{\text{s}} \tag{24.7}$$

One should note also that this relation holds exactly because the meter is defined via the given value for the speed of light. Equation (24.2) is slightly incorrect: the factor $9 = 3^2$ has to be replaced, in fact, by the square of $2.99....$ In the following, this will hold always when the numbers 3 or 9 appear in the conversion factor.

A further constant of proportionality k_3 occurs in the definition of the magnetic field intensity. For the Lorentz force on a charge moving in a magnetic field one writes

$$\mathbf{F} = k_3 \frac{q}{c} \mathbf{v} \times \mathbf{B} \tag{24.8}$$

with

$$k_3 = 1 \quad \text{(Gaussian system)} \qquad \text{or} \qquad k_3^* = c \quad \text{(SI system)} \tag{24.9}$$

In Gaussian system the unit of **B** is again a combination of the mechanical base units, and it is denoted by *Gauss*.

$$1\,\text{G} = 1\,\frac{\text{dyn}}{\text{stat Coul}} = 1\,\frac{\text{dyn}}{\sqrt{\text{dyn}}\,\text{cm}} = 1\,\frac{\sqrt{\text{g cm/s}^2}}{\text{cm}} = 1\,\sqrt{\frac{\text{g}}{\text{cm s}^2}} \qquad (24.10)$$

On the other hand, in the SI system the unit of **B** is the *Tesla*, which with $1\,\text{V} = 1\,\text{Nm/C}$ may be expressed as follows:

$$1\,\text{T} = 1\,\frac{\text{N}}{(\text{m/s})\,\text{C}} = 1\,\frac{\text{Nm}}{\text{C}}\,\frac{\text{s}}{\text{m}^2} = 1\,\frac{\text{Vs}}{\text{m}^2} \qquad (24.11)$$

The conversion factor between Tesla and Gauss follows from (24.8) by equating the actions of force

$$B = c\frac{q^*}{q}B^* \qquad (24.12)$$

so that with (24.10) and (24.11)

$$1\,\text{Tesla} \rightarrow 3 \times 10^8\,\frac{\text{m}}{\text{s}}\,\frac{\text{C}}{3 \times 10^9 \text{stat Coul}}\,1\,\text{T} = 10^{-1}\,\frac{\text{m}}{\text{s}}\,\frac{\text{G}}{\text{dyn}}\,\text{C}\,\frac{\text{Ns}}{\text{Cm}}$$

$$= 10^{-1}\,\frac{\text{N}}{\text{dyn}}\text{G} = 10^4\,\text{G} \qquad (24.13)$$

A further difference between the systems of units consists in the fact that the SI system clearly distinguishes between the magnetic field intensity **H** and the magnetic induction **B**, and also between the electric field intensity **E** and the electric displacement **D**. One sets

$$\mathbf{D}^* = \epsilon\epsilon_0\mathbf{E}^*, \qquad \mathbf{B}^* = \mu\mu_0\mathbf{H}^* \qquad (24.14)$$

ϵ and μ are the permittivity constant and permeability constant, respectively, of the medium (independent of the system of units). Contrary to the Gaussian system, here, $\mathbf{D}^*$ and $\mathbf{E}^*$, or $\mathbf{B}^*$ and $\mathbf{H}^*$ have distinct values in free space ($\epsilon = \mu = 1$) and also distinct dimensions, namely,

$$[E^*] = \frac{\text{V}}{\text{m}}, \qquad [D^*] = \frac{\text{As}}{\text{m}^2}, \qquad [B^*] = \frac{\text{Vs}}{\text{m}^2}, \qquad [H^*] = \frac{\text{A}}{\text{m}} \qquad (24.15)$$

The described differences between the systems of units lead to a distinct form for nearly all equations of electrodynamics. Table 24.1 lists Maxwell's equations and some other important relations in both systems. The transition between the Gaussian system and the SI system is facilitated by Table 24.2.

The rules of translation may be used to rewrite equations from one system of units to the other one. This may be checked easily for the equations from Table 24.1.

Finally, Table 24.3 lists how various units may be converted into each other. One should recall again that the factors 3 and 9 are valid only approximately.

Table 24.1. Important equations of electrodynamics, expressed in different systems of units.

	Gaussian	SI
Maxwell's equations	$\nabla \times \mathbf{H} = \dfrac{1}{c}\dfrac{\partial}{\partial t}\mathbf{D} + \dfrac{4\pi}{c}\mathbf{j}$	$\nabla \times \mathbf{H}^* = \dfrac{\partial}{\partial t}\mathbf{D}^* + \mathbf{j}^*$
	$\nabla \times \mathbf{E} = -\dfrac{1}{c}\dfrac{\partial}{\partial t}\mathbf{B}$	$\nabla \times \mathbf{E}^* = -\dfrac{\partial}{\partial t}\mathbf{B}^*$
	$\nabla \cdot \mathbf{D} = 4\pi\varrho$	$\nabla \cdot \mathbf{D}^* = \varrho^*$
	$\nabla \cdot \mathbf{B} = 0$	$\nabla \cdot \mathbf{B}^* = 0$
Energy density	$w = \dfrac{1}{8\pi}(\mathbf{E}\cdot\mathbf{D} + \mathbf{B}\cdot\mathbf{H})$	$w = \dfrac{1}{2}(\mathbf{E}^*\cdot\mathbf{D}^* + \mathbf{B}^*\cdot\mathbf{H}^*)$
Poynting vector	$\mathbf{S} = \dfrac{c}{4\pi}\mathbf{E}\times\mathbf{H}$	$\mathbf{S} = \mathbf{E}^*\times\mathbf{H}^*$
Force law	$\mathbf{f} = \varrho\mathbf{E} + \dfrac{1}{c}\mathbf{j}\times\mathbf{B}$	$\mathbf{f} = \varrho^*\mathbf{E}^* + \mathbf{j}^*\times\mathbf{B}^*$

Table 24.2. Correspondence between various measured quantities in the Gaussian and SI systems.

Quantity	Gaussian	SI	Quantity	Gaussian	SI
Charge density	ϱ	$\dfrac{1}{\sqrt{4\pi\epsilon_0}}\varrho^*$	Current density	$\mathbf{j}$	$\dfrac{1}{\sqrt{4\pi\epsilon_0}}\mathbf{j}^*$
Electric field intensity	$\mathbf{E}$	$\sqrt{4\pi\epsilon_0}\mathbf{E}^*$	Magnetic field density	$\mathbf{H}$	$\sqrt{4\pi\mu_0}\mathbf{H}^*$
Electric displacement	$\mathbf{D}$	$\sqrt{\dfrac{4\pi}{\epsilon_0}}\mathbf{D}^*$	Magnetic induction	$\mathbf{B}$	$\sqrt{\dfrac{4\pi}{\mu_0}}\mathbf{B}^*$
Electric polarization	$\mathbf{P}$	$\dfrac{1}{\sqrt{4\pi\epsilon_0}}\mathbf{P}^*$	Magnetization	$\mathbf{M}$	$\sqrt{\dfrac{\mu_0}{4\pi}}\mathbf{M}^*$
			Magnetic flux	ϕ	$\sqrt{\dfrac{4\pi}{\mu_0}}\phi^*$
Resistance	R	$4\pi\epsilon_0 R^*$			
Capacity	C	$\dfrac{1}{4\pi\epsilon_0}C^*$	Inductance	L	$4\pi\epsilon_0 L^*$

Table 24.3. Conversion factors between Gaussian and SI units.

Quantity	Symbol	SI unit	Gaussian unit
Charge	q	1 Coulomb (C)	$3 \cdot 10^9$ stat Coul
Current	I	1 Ampère (A)	$3 \cdot 10^9$ stat Amp
Voltage	U	1 Volt (V)	$\frac{1}{3} \cdot 10^{-2}$ stat Volt
Electric field intensity	**E**	$1\ \dfrac{V}{m}$	$\frac{1}{3} \cdot 10^{-4}\ \dfrac{\text{stat Volt}}{\text{cm}}$
Electric displacement	**D**	$1\ \dfrac{As}{m^2}$	$4\pi \cdot 3 \cdot 10^5\ \dfrac{\text{stat Volt}}{\text{cm}}$
Magnetic field density	**H**	$1\ \dfrac{A}{m}$	$4\pi \cdot 10^{-3}$ Oersted (Oe)
Magnetic induction	**B**	$1\ \dfrac{Vs}{m^2} = 1$ Tesla (T)	10^4 Gauss (G)
Magnetic flux	ϕ	1 Weber (Wb)	10^8 Maxwell (Mx)
Resistance	R	1 Ohm (Ω)	$\frac{1}{9} \cdot 10^{-11}\ \dfrac{s}{\text{cm}}$
Capacitance	C	1 Farad (F)	$9 \cdot 10^{11}$ cm
Inductance	L	1 Henry (H)	$\frac{1}{9} \cdot 10^{-11}\ \dfrac{s^2}{\text{cm}}$

25 About the History of Electrodynamics

History of electrodynamics

The development of electromagnetic theory differs remarkably from that of mechanics. The reason is probably that mechanics has changed little in time. After creating a self-contained logical conceptual system, a language to describe forces and movement, and after the finding of the basic laws, that permitted one to calculate the different mechanical experiences in a precise way, it was essentially complete.

Electricity was different. The many different phenomena (magnetic stones, strange forces by rubbing amber on cat fur, electric fishes, sheet lightning and thunderstorms, and luminous phenomena) did not seem related to each other in the beginning. The great merit of electromagnetic theory is, on the one hand, the knowledge that all these phenomena are related, and on the other hand, the creation of new (technical) applications, that do not appear as such in nature and that have changed the world in an unforseen way.

The third, but historically most important benefit of the electromagnetic theory was the creation of a stringent conceptual system, which arose from Faraday's thoughts and culminated in Maxwell's equations. It was a new, logically self-contained theory of physical phenomena that was quite different from mechanics. This new theory meant a departure from the mechanical world picture and contributed essentially to today's development of the concepts of force and field.

The history of electrodynamics can be roughly divided into a prescientific period of time (approximately until the 16th century), an era of the origin of magnetostatics and electrostatics, a period when it was a success to create permanent flowing electric currents,

---

*All footnotes are given at the end of this chapter.

and the epoch of combining all electric and magnetic phenomena into electrodynamics. The last epoch brought the conceptual clarification.

Although certain electric and magnetic phenomena, like forces that emanated from amber or from magnetic stones, were already known (since the 4th century BC), there were no systematic surveys about electromagnetic phenomena until 1600. It is true that electric fishes were used as a remedy, and it was known since the Romans, that the magnetic force could even penetrate metal. And since the Chinese used the terrestrial magnetic field to determine directions (approximately from 300 AD) magnetism had even its first technical application (the compass). But in the western world a kind of superstition prevailed about the magnetism, which had its expression in tales from the magnetic mountain and from free floating pictures of gods and saints. By the way, the name electricity derives from $\eta\lambda\epsilon\kappa\tau\rho o\nu$, which, derived from $\eta\lambda\epsilon\kappa\tau o\rho$ (shine), means Ag-Au-alloys (coin metal) as well as amber. Soon it was noticed, that besides amber there were a few other materials that became electric by friction. Nowadays we regard these frictional electricity phenomena as complicated. Indeed, they are easily perturbed.

The year 1600 was a milestone in the history of physics, in the form of a book about magnets by *William Gilbert*[2] (1544–1603). The book contained the results of careful experiments, it showed that positive and negative magnetic poles cannot be separated; Gilbert measured magnetic fields with magnetic needles (so to speak a precursor of the Gaussian method) and even showed, that the terrestrial magnetic field could be interpreted as that of a magnetic ball. Only the law of the quantitative force $\sim \left(1/r^2\right)$ was missing in this important work. Likewise, his knowledge about the electric phenomena was incomplete.

Electricity scarcely progressed; there was little interest in it. It was the bloom, even in essential features the completion, of mechanics as the first stringent natural sciences. About 1660 *Otto von Guericke*[3] (1602–86) constructed an influence machine; with it he was able to generate electric sparks and study electric repulsion. But he could not give rise to a genuine clarification of the electric process.

There was no new advance in electrostatics until 1730–60, when in 1733 *Dufay*[4] discovered the existence of two types of electricity (positive and negative), and realized, that like electric charges repel and that unlike charges attract each other. He managed this by transferring electricity to isolated metal objects. Then he observed how electrified fields attract non-electrified ones, how they electrify them and then repel them. Metals, electrified with glass, are repelled by electrified glass, but attracted by electrified resin. Metals, electrified by resin, are repelled by resin, but attracted by glass. He concluded, that there were two types of electricity. About 1745 *Kleist*[5] experimented in the chimney (Pomerania) with the influence machine and discovered the so called tube of amplification, which *Musschenbrock*[6] examined further in Leyden. *Benjamin Franklin's*[7] (1732–90) systematic studies led him to conclusions about the tube and about electrostatic induction (1747). He did not yet know about the two types of electricity; he rather explained the effect of positive and negative charges with a surplus or a lack of electricity, respectively. He also explained thunderstorms correctly in this model. Positive and negative charges were finally noticed clearly by *Carl Wilcke* (1732–96) and *Symmer* (1759). The discovery of the polarization of dielectrics by Wilcke in 1758 belongs to the type of premature realization that was soon lost again. Scientists at the time were not yet able to judge its importance. Franklin,

Wilcke, and Symmer showed a clear, qualitative comprehension of the essential process in electrostatics, except for the polarizing behavior of the intermediate medium, which was not cleared up until Faraday.

We do notice: it was a long, quite elaborate way until these phenomena, only indirectly accessible to the human mind, were organized and somewhat cleared up. The basic law of electrostatics, the Coulomb law of forces, was not yet discovered. Only *Priestley*[8] concluded, by the lack of an electrostatic power effect inside conductors, that the force decreases as $1/r^2$. That was in 1767.

Coulomb[9] used the torsion balance (1785) and proved that the electric and magnetic forces are proportional to $1/r^2$. At this time, *Euler*[49] and *Lagrange* had formally completed mechanics. Thus it is no wonder, that the results of electrostatics (still without permanent flowing currents) were somewhat in the shadow of the great events of mechanics (compare volume 2 of the series, last chapter).

The production of electric currents: Magnetostatics

It was known, that the phenomena appearing in amber, thunderstorms, and electric fishes belong together, yet the connection with magnetism was still unknown. For that a new discovery was necessary, namely the production of permanent flowing electric currents: This led to a series of exciting successive discoveries, so that the first half of the 19th century must be considered as the epoch of the breakthrough of understanding electric phenomena, as the 17th century had been the bloom of mechanics.

L. Galvani[10] occupied himself with electric phenomena in animals and noticed in 1780, that under certain conditions frog's legs twitch. Although without direct external electric influences, such phenomena appeared in the animal. He (still) held animal electricity responsible for that. Only *A. Volta*[11] noticed that it did not depend on the animal organism. As a master of the electric method of measurement (he developed the electroscope provided with a condenser), he pursued galvanic experiments and introduced the idea of the metallic electricity, which appears, when there is a liquid between two metals (1794). Later he was able to prove the existence of electricity between two metals without any liquid. With the *voltaic pile*, consisting of a series of galvanic elements, he succeeded in an experimental breakthrough: now it was possible to produce permanent currents with a tension, that was considerable at that time. Soon it was proved, that galvanism and electricity were one and the same thing.

Now that there were permanent flowing currents, a series of important discoveries followed inevitably, which led to the idea of the electric structure of matter and to the combination of electricity and magnetism. The instant decomposition of water in oxygen and hydrogen was observed (but only under clean conditions). Other materials were also electrically analyzed in a similar way. About 1807 *Davy*[12] described the elements K, Na, Ba, Sr, Ca electrolytically. The idea arose that chemical affinity can be proscribed to an electric force between the particles. *Berzelius*[13] extended this idea, the gist of which contains our view of heteropolar chemical binding.

The observations of the magnetic effects of current and electromagnetic induction brought a great progress to deeper understanding of electric phenomena. Till then electric and magnetic appearances had been seen as separate, but analogous to each other. Now it was realized that they are indissolubly connected.

In July 1820 *H.C. Oersted*[14] observed the deflection of a magnetic needle at a wire, which had been made glowing by electric current. This was the beginning of a series of discoveries that were made quickly one after the other, and that made the decade 1820–30 probably the most important period in the development of our knowledge about electricity. Oersted noticed, that the deflection of the needle depended on the direction of the current, and he showed that nonmagnetic materials brought between the current and the magnetic needle, did not influence the phenomena. He also soon noticed that the wire did not have to be incandescent, and he demonstrated the magnetic effect from wire to wire. In the same year *Arago*[15] in Paris and *Seebeck*[16] in Berlin realized the magnetization of certain materials by current. *Ampère*[17] made precise the concepts of electric tension and current strength, and formulated the rule of the force effect of current on magnets and the forces between two currents. He also introduced the idea of current elements in the smallest material particles as the origin of the magnetic behavior of certain materials. Almost simultaneously, *Biot*[18] and *Savart*[19] measured the interval dependence of the magnetic force of a current-carrying wire and found its proportion to be $1/r$. In 1826 *Georg Simon Ohm*[20] (at that time a teacher at a college in Cologne) examined the dependence of the current strength on the length and cross section of the wire and noticed the proportion of current to the cross section and the voltage gradient.

A second, greater breakthrough in the combination of electricity and magnetism came with *M. Faraday's*[21] discovery of electromagnetic induction (1831–32). He realized, that a moved magnet induces a current impulse in a nearby wire; the same result is obtained using a moving, current-carrying wire or a wire with current of variable strength. Encouraged and stimulated by this, he carried on with his ideas about the nature of the electric and magnetic field. He viewed the essential part of the electric phenomena as a state of stress which is effective from place to place in the medium between charges and flows. He introduced the idea of the line of force. The concept of field initiated here. In 1833 *C.F. Gauss*[22] began in Göttingen the systematic measuring of magnetic quantities; *W. Weber*[23] joined him, and together they founded a logical system of measurements for electric and magnetic quantities. Finally, in 1841 *Joule*[24] discovered the heat action of electric current and its laws.

One should notice, that this was a swift series of important and instructive discoveries which began in 1820 with the production of steady-state currents. This rich decade was a blooming period of physics, which can be related as well to the external circumstances: the French Revolution was over, little by little there was peace again, and new classes worked their way up and helped to form the intellectual life of the nations.

Precise theoretical-mathematical studies of electric and magnetic processes were made in close cooperation between mathematicians and physicists in a similar manner. Physics suggested the creation of new mathematic disciplines, for example potential theory. The first basic theorems of potential theory were tackled through the theory of gravitation. *Laplace*[25] (1782), for instance, had gravitation in mind when he proved that, outside the mass density $\rho(\mathbf{r}')$,

$$\Delta V = 0$$

if

$$V(\mathbf{r}) = \int \frac{\rho(\mathbf{r}')dV'}{|\mathbf{r} - \mathbf{r}'|}$$

was the so defined potential of gravitation; *Poisson*[26] acted similarly, when he generalized to

$$\Delta V(\mathbf{r}) = -4\pi\rho(\mathbf{r})$$

In contrast, *Green*[27] thought of examples of electromechanics when deriving the theorems named after him (1828). Gauss' general theorems and laws facilitated a reasonable description of electric and magnetic phenomena in 1839.

At the same time (around 1842) the general energy conservation law was discovered. Logically this led to the energy of the potential (potential energy), and finally the energy conditions during induction were clarified. *F. Neumann*,[28] *W. Kirchhoff*,[29] and *H. Helmholtz*[30] participated authoritatively.

The culmination of the theoretic formulation, and thus the clear conceptual ordering of the electromagnetic phenomena, came with the theory of James Clark *Maxwell*[31] (1861–64). He based the theorems that now are named after him on Faraday's idea of the lines of force. From that also he calculated the possibility of electromagnetic waves.

Maxwell's theory not only represents the résumé of the phenomena known by then; at the same time it is the theory of light, of which the propagation velocity was determined by constants in the Maxwell equations by *Weber* and *Kohlrausch*.[32] Thus electrophysics and optics were connected, and moreover, optics became a special subfield of electrophysics. This was an extraordinary result of theoretical physics.

Maxwell's theory was not immediately well received on the European continent. The reason was probably, that these thoughts—as already noticed—rely on Faraday, who was unfamiliar to mechanically trained thoughts. Even such an important scientist as *L. Boltzmann*[33] tried to explain the correlations, explained in Maxwell's equations, with complicated mechanic images and models. Yet the field theory was victorious over all others. In 1888 *Heinrich Hertz*[34] finally confirmed it experimentally when he discovered electrically produced waves, thus proving, that these waves have the same characteristics as light waves.

The impetuous development of physics of the electromagnetism began with Galvani's discovery and was complete after five generations of physics. Again we present them in a tabular survey:

(1) Born 1727–1749: Galvani, Volta, Laplace.

(2) Born 1770–1779: Seebeck, Oersted, Biot, Ampere, Gauss, Davy, Berzelius.

(3) Born 1786–1804: Arago, Ohm, Savart, Faraday, Green, F. Neumann, Weber.

(4) Born 1818–1831: Joule, Helmholtz, Kirchhoff, Maxwell.

(5) Born 1853–1862: H.A. Lorentz, J.J. Thomson, H. Hertz, M. Planck, Ph. Lenard.

The generation mentioned under point 5 was already essentially part of the basic clean up of atomic structure. This would not have been possible without the former cleaning up of electromagnetism.

The electromagnetic phenomena had unforseen technical applications, which changed our world lastingly. We recall only two technical branches here:

(1) Communications:

1809: *Sömmering's*[35] model of an electrochemical telegraph.

1820: Ampère's model of a magnetoelectric telegraph.

1834: Gauss–Weber practically useful telegraph.

1860: *Philip Reis*[36] telephone transmits sounds.

1877: *Bell's*[37] telephone transmits speech.

1895: *Marconi's*[38] wireless telegraphy.

(2) Production of electric energy:

1832: The first magnetoelectric machine is built.

1866: The first machine with an electromagnet.

1867: The efficient machine of Siemens.

The history of optics

We now want to give a short survey of optics and thus clarify, first of all, when and in what scientific-historical context the essential appearances of light were seen correctly.

So to speak the sine law expresses the property of radiation most clearly. It goes back to *W. Snell (Snellius)*[39]. *Isaac Newton*[40] discovered that white light was composed of colored light. The knowledge of the wave nature and with it the coherent periodicity of the light propagation process advanced somewhat clumsily. The wave theory of light goes back to *Ch. Huygens;*[41] the periodicity and the first measurement of the progress of waves goes back to *Newton*. It was not until 150 years later that *A. Fresnel*[42] was able to clarify the traversing of a wave through isotropic (and antisotropic) materials theoretically. Soon after the discovery of the Malus it became clear, that polarized light has to mean it is a transverse wave. Already Faraday supposed that light is an electromagnetic wave; Maxwell inferred this connection from his theory. It was Heinrich Hertz, who furnished the experimental proof, by producing waves by electromagnetic means, and who proved that they have the properties of light waves. The cause of the fast development of optics is on one hand the connection of this discipline with the big, general question of the nature of light, and on the other hand, and quite considerably, the recurring task of the perfection of optical instruments. A lot of ideas resulted from the latter task.

We now want to show the historical aspects of the development of light theory in more detail. The refraction law was first, from which the light ray followed as a useful notion. Already Cl. Ptolomäus (round 100 AD) indicated the connection of the angle of incidence and the angle of emergence in tabulations. Around 1611 *Johannes Kepler*[43] used their proportionality for minor angles. In 1637 *René Descartes*[44] published the sine law, which was already known by W. Snell, and he interpreted it as follows: the light particles should get an additional impulse during their entry in a dense medium, so that they are faster in the denser medium. *P. Fermat*[45] (1665) derived the refraction law from an integral principle (Fermat's principle), according to which the optical path (or the time that light needs to travel this path) should be a minimum. Accordingly Fermat's light velocity in a denser medium was smaller. Soon there were important technical applications: the microscope and the telescope (with convergent and divergent lense) were discovered in Holland round 1600. *Galilei*[46] copied the telescope and discovered with it the Jovian moons and sunspots. Thus he had an additional, essential support for the world system of *Kopernikus* and against the permanency of the celestial bodies. In 1611 *Kepler* constructed a telescope and two convergent lenses. Soon reflecting telescopes were constructed as well, so by Newton. *W. Herschel*[47] used them and thus founded fixed star research around 1790.

In its beginnings, the *science of color* goes back to *Grimaldi*[48] (1650). *Newton* carried on a series of experiments, from which he deduced, that the spectral colors are the inde-composable components of the light's incandescent body. Around 1670 he gave lectures about it; in the following years he communicated his knowledge gradually with the Royal Society; his book about optics appeared only in 1704. Newton believed that the chromatic aberration of lenses was unavoidable, because he presumed the proportionality of dispersion and refraction. *L. Euler*[49] had a different opinion and supposed that there was the possibility of achromatism. Around 1760 also *Dolland*[50] and *Euler* produced achromatic lenses, which were improved essentially 60 years later (1820) by *J. Fraunhofer*.[51] Fraunhofer studied systematically the dispersion of different spectacle glasses, and thus he found dark lines in the solar spectrum, which he used to exactly characterize spectral colors. The result of Fraunhofer's works is an essential improvement of telescope lenses, and thus, the possibility of measuring stellar parallaxes.

It is remarkable, that the entire clarification of the wave nature and the *periodicity of light* took almost 150 years, from 1675 to approximately 1820. It took that long to realize, that the wave nature and periodicity belong together.

Newton discovered that monochromatic light shows a periodic process. He concluded this from careful experiments to study the colors of thin plates. First he wanted to explain the phenomena as ether oscillations; but in his already mentioned volume about optics (1704) he speaks about light particles, which get periodic impulses of light reflection and periodic impulses of light passing. After all, the interval of the impulses is what we call nowadays, a half wavelength. He even observed, that the length of intervals λ_1 and λ_2 behave in different media like the sine of the refraction angle, thus, $\sin\alpha_1 / \sin\alpha_2 = \lambda_1/\lambda_2$.

They were almost twice as big for red light than for violet light. Obviously Newton measured the wavelength of light; he even explained the constant natural colors of bodies with the assumption of a determined size for the least components of these bodies.

The wave nature of light was still little established by Grimaldi. Only *Ch. Huygens*[41] made a remarkable progress with his *traitè de la lumière* in 1678. He explained systematically the rectilinear propagation, the reflection and the refraction out of the wave theory, and he solved quite difficult refraction problems with the help of wave surfaces that run perpendicularly to the ray direction. Huygens wanted to give a mechanical explanation of light and its different appearances. The reducing of all phenomena to mechanics seemed to him the only correct way in physics. He thus ignored diffraction and interference; neither did he make use of the periodicity of the wave process.

Huygens thus didn't use the phenomena of diffraction and interference at all, which we consider nowadays as characteristic for the wave nature. Newton knew the periodicity, but he also did not use it to found the general wave character of light. Probably because of the extraordinary authority of Newton, the wave nature and the periodicity stood side by side for over a century without being combined. Newton simply declined the wave theory.

The enemies of the wave structure of light considered it as a very unreasonable demand: Neither for Newton the light went around the corner and consequently the wave character was excluded. Also, Huygens' interpretation of the double refraction at the *Iceland spar*, where he presumed two types of the oscillatory medium, seemed too artificial to Newton. Probably Newton's aversion to the hypothetical ideas of the action between bodies was involved, for with his law of gravitation he gave an example of the mathematical version of the law of force in a masterly manner, without having to make vague presumptions of yet unobserved processes. Huygens' ether theory, with the spreading waves, seemed too hypothetical and complicated to Newton. On the other hand, Huygens declined the particle aspect of light, because it seemed to contradict the indifference of light rays traveling in different directions, the permeability to light of many bodies, and the high light velocity, which just had been observed by *Olaf Römer*.[52]

The wave theory of light wasn't carried on until past 1800. So around 1801 *Th. Young*[53] explained qualitatively the interference phenomenon with the help of periodical waves. *A. Fresnel* improved Huygens' principle and applied it to the diffraction phenomenon and to the interference phenomenon. Moreover, he regarded scalar waves. He as well as Fraunhofer examined diffraction. Thus, Fraunhofer succeeded in measuring the length of light waves, exactly, up to four digits.

Now we come to the *polarization phenomena*: such already existed in the double refraction at the Iceland spar. Newton was on the right track for such an interpretation, when he noticed, that the double refraction indicated a lateral direction of the light particles, the effect of which on the crystal he saw by analogy with the effect of magnetic poles. *Malus*[54] discovered in 1808 polarization of reflected rays and recognized the connection of polarization and double refraction. In 1812 *Young* interpreted this by transverse waves, and *A. Fresnel* completed this theory in 1817 by the observation of transverse waves in elastic ether. Thus, he explained reflection, refraction, and double refraction quantitatively.

His elastic light theory leads—not quite inevitably—essentially to the same equations, which we derive from Maxwell's equations nowadays. The direction of vibration corresponds to the dielectric displacement vector **D**; Fresnel's vibration plane stood perpendicularly on the polarization plane. Other light theories presumed directions of vibration, which corresponded to the electric field vector **E**, or respectively, to the magnetic induction **B**. The theory of *F. Neumann* belongs to the latter group: according to it the vibration plane and the polarization plane coincide.

Next was the rotation of light as an electromagnetic wave. *Faraday* supposed a connection, which he confirmed by measuring the rotation of the polarization plane by passing through a magnetic body (1845). In 1861–62 *J.S. Maxwell* gave the complete mathematical

Table 25.1. History of the developments in optics.

Ray	Color	Velocity	Wave	Practical optics
1611 Kepler's optics				Telescope, microscope
1626 Snell died	≈1650 Grimaldi		≈1650 Grimaldi	Jove moons, sun spots
1665 Fermat principle	≈1670 Newton's lectures	1676 Römer	1678 Huygens (wave)	Saturn moons and rings
	1704 Newton's optics		1704 Newton (periodic impulses)	Reflecting telescope, Plant cells, Protozoen
	≈1750 Euler, Dollond	1728 Bradley		Achromatic lenses, Fixed star research, Uranus
				Planet Ceres
			1801 Young (wave)	
			1808 Malus	
			1817 Young (transverse waves)	
			1819 Fresnel (diffraction)	
	≈1820 Fraunhofer		≈1820 Fraunhofer (lattice)	λ-measurement
1833 Hamilton			1822 Fresnel (elastic theory)	Improvement of the telescope
≈1840 Gauss			1845 Faraday (rotation)	Stellar parallax
≈1870 Abbe		1849 Fizeau	1856 Weber–Kohlrausch	Cell division
		1850 Foucault	1861–62 Maxwell	Perfection of the microscope
			1888 Hertz	Spectroscopy

formulation of this idea. His theory predicted the existence of transverse electromagnetic waves, of which the velocity was calculated by measuring the electromagnetic quantity. In 1856 this measurement was carried out by *W. Weber* and *F. Kohlrausch*, and yielded the velocity of light. *H. Hertz* generated electromagnetic waves by electric vibrations and showed that they have the same properties as light waves.

Thus, the light was discerned as an electromagnetic wave, but it took some time until this knowledge was accepted. Little by little it was noticed, that the similarity of the electromagnetic field with the tensions and deformations of elastic bodies was just an analogy, and that the electromagnetic wave had to be seen as completely nonmechanical. Thus, tests seeking to model light waves as mechanical ether waves came to an end.

We summarize: The development of the comprehension of the light phenomena concentrated on two short periods. There was the decade between Newton's experiments on dispersion and the publication of his book about optics. In this period were also Huygens' contributions and considerable progresses in practical optics. (Huygens and his brother constructed telescopes, discovered Saturn's halo, and a Saturn moon; Newton constructed a reflecting telescope.) On the other hand there were the impetuous two centuries between the wave theory by Young and its completion by Fresnel. It is remarkable, that the development of optics preceded directly the certainly quite impetuous development of electromagnetism, which began about 1820 with Oersted's discovery.

Table 25.1 resumes the most important stages in a tabular form. The general beginnings of the theory of the optical image by *Hamilton*[55], *Gauss, Abbe*[56] are added. Abbe contributed essentially to the technical progress of optics by completely calculating the detection of the path of rays in the microscope. The column *practical optics* contains the construction of important instruments and discoveries in astronomy and biology, which were made by it. Thus we express modestly, that physics (here the optics) provides other sciences with important foundations: the invention of the telescope revolutionized astronomy, the invention of the microscope initiated a new epoch in biology and medicine. Optics provided even decisive impulses for the further progress of physics; remembering just the importance of the spectroscopy for the atomic structure.

Notes

[1] In this chapter, we will follow to a large extent the appendices concerning the history of science in Friedrich Hund's books on electrodynamics and optics (Bibliographisches Institut VEB, Leipzig, 1951). Bibliographical data of individual scientists are partly taken from the *Lexikon der Schulphysik*, ed. by O. Höfling, Aulis Verlag Deubner & Co (1922), the *Brockhaus Enzyklopädie*, F. A. Brockhaus Wiesbaden, and the *Fachlexikon ABC Mathematik*, Verlag Harri Deutsch.

[2] *William Gilbert*, b. May 24, 1544, Colchester–d. Nov. 30, 1603, London, British natural scientist and physician. Having been a GP in London from 1573 onward, he became personal physician to Elizabeth I in 1601 and after her death to King James I of England. In his fundamental work "De magnete, magneticisque corporibus et de magno magnete

Tellure physiologia nova" (London 1600, facsimile edition Berlin 1892, engl. transl. and annotations by S. P. Thompson in the collectors series in science, 1958) Gilbert united the discoveries of earlier authors into an impressive theory about magnetism and geomagnetism, and added several new observations and findings. In the works second book a special chapter on "corpora electrica" is included, on matter like amber (electrum) which are able to attract light bodies after having been rubbed. Some of Gilbert's contemporaries, e. g. Kepler and Galilei, were impressed by the work. His essay "De monde nostro sublunari philosophia nova" was published posthumously (Amsterdam, 1561).

[3] *Otto von Guericke*, orig. Gericke, b. Nov. 20, 1602, Magdeburg–d. May 11, 1686, Hamburg, statesman, politician, and physicist. He became a councilman in 1626. Later on he worked as an engineer in Swedish and Saxonian service, and in 1646 he became one of the four majors of the city of Magdeburg.

During the Reichstag at Regensburg 1654 Guericke did several experiments with an air pump which he had invented around 1650. The electoral prince from Mainz bought the gadgets and had them brought to Wiesbaden, where a professor of mathematics named K. Schott repeated the tests and described them in an appendix to his "Mechanika hydraulio-pneumatica" (1657). It was a couple of years later that Guericke invented the Magdeburg hemispheres and experimented with them. On these Schott gave a report in "Technica Curiosa" (1664). Independently of the Italian physicists, Guericke realized the nature of air pressure and used barometrical observations to forecast storms. At his instigation, comparative barometrical observations took place in Berlin, Hamburg, and Magdeburg. His friends pressed for him to write a physical-astronomical textbook which was more or less finished in 1663 but not published until 1672 in Amsterdam. It was titled "Experimenta nova (ut vocantur) Magdeburgia de vacuo spatio" (repr. 1962). In the fourth book of this work Guericke described experiments with a sulphurous sphere using it as a model of "Weltkörper" to show their inherent forces (*virtutes mundanae*). Thus, it was not really an electrostatic generator, although Guericke observed the phenomena of electrical repulsion, conduction, electrostatic influence, and the discharging effect of a flame in this context— without realizing the importance of these facts. Guericke also wrote a "Geschichte der Belagerung, Eroberung und Zerstörung Magdeburgs" ("history of the siege, conquest, and destruction of Magdeburg" ed. by F. W. Hoffmann, 1860).

[4] *Charles-Francois de Cisternay Dufay* (Du Fay), b. Sept. 14, 1698, Paris–d. July 16, 1739, Paris, French physicist. At the instigation of R. A. F. de Réaumur Dufay was admitted to the Académie des sciences in 1723 after a successful military career. Dufay worked in various fields (archeology, botany, chemistry, mechanics, optics, and planar geometry), invented among other things a fire pump, discovered several colorants, and analyzed the properties of phosphorous. His most important works dealt with electricity. He discovered, among other things, the conductivity of flames. Dufay was the first to realize the nature of electrostatical repulsion and to differentiate the two types of electrical charge (1733). The tests done by him became the basis for further research and systematization in the field of electricity.

[5]*Ewald Jürgen von Kleist*, b. March 7, 1715, Zeblin (Pomerania)–d. Aug. 24, 1759, Frankfurt/O. Member of the family of Heinrich von Kleist, a famous German poet. Kleist was dean of cathedral, later chairman of the court in Pomerania. He electrified a nail which stuck in a medicine tube. The tube, however, must have been slightly damp, and Kleist got an electric shock when holding the tube with one hand and the nail with the other. With the tube containing alcohol or mercury, the effect was even stronger. These observations, made in 1754, were announced to several scientists and stirred a sensation. One year later the same test was made in Leyden (Netherlands) and thus the tube which was now much used in experiments was called the *Leyden jar*, although the historically correct name would have to be *Pomeranian jar* or *Kleist's jar*. It was soon covered with different metals and the respective electric shocks and effects of the sparks were studied.

[6]*Pieter van Musschenbroek*, b. March 14, 1692, Leyden–d. Sept. 19, 1761, Leyden. In 17th century Netherlands experimental and observing natural sciences flowered. The physicist Christian Huygens, the microscopist Antoni von Leeuwenkoek, the anatomist Hermann Boerhave, and the experimenter Willem Jacob's Gravesande were outstanding representatives. Musschenbroek's father, his uncle Samuel, and his brother Jan, being scholarly mechanics, built telescopes and microscopes, and improved air pumps and equipment used for lecturing. Coming from such a background, Musschenbroek became one of the best experimenters in his time. He was a student of medicine, natural sciences, and mathematics, received a doctorate in medicine in 1715 and another one in philosophy in 1719. After that he got a professorship in mathematics and physics in Duisburg, and in 1723 one in Utrecht. In 1739 he got a chair in his home town Leyden where from 1717 Gravesande taught, too. Both tried to spread empirical sciences and were pioneers of Newton's theories.

In 1731 Musschenbroek's latin translation of the *Saggi* by the Accademia del Cimento was published. In this work the disciples of Galilei had set down their experiments on thermometer and barometer. Musschenbroek provided the book with annotations. Like Gravesande and Desaguliers, he wrote important textbooks, above all *Elementa physicae* (published 1729), *Beginsels der Naturkunde* (1739), and an *Introductio ad philosophiam naturalem*. He quite early invented a pyrometer which worked by expansion of a metal rod and was thus able to measure higher temperatures than a thermometer. This gadget was later improved by Johann Heinrich Lambert. Musschenbroek determined specific weights, he tested the resistance of wooden and of metallic rods, and he made experiments on friction, capillary force, mirrors, magnetism, and electricity. He scrutinized Willebrord Snellius' measurement of the angle of the meridian and attended to meteorological observations. In the course of this, he collected various gadgets.

Once Musschenbroek tried to electrify water in an isolating glassbottle by putting a metal wire into the bottle. The wire conducted electricity from the conductor of an electrostatic generator into the water. His assistant Cunõus who held the bottle and happened to touch the wire got a heavy electric shock. Musschenbroek repeated the experiment and at the beginning of 1746 told Réaumur about it.

[7]*Benjamin Franklin*, b. Jan. 17, 1706, Boston–d. Apr. 17, 1790, Philadelphia, American statesman and writer. A former printer, he published a Philadelphia newspaper from 1730

and an almanac from 1732 to 1757 (*Poor Richard's Almanac*). In accordance with his lenient Puritanism he taught an orderly and economical life leading to success (autobiography until 1757). Franklin became famous for his works on physics (theory of electricity, lightning conductor, capacitor). Being a member of the Pennsylvania Assembly from 1751 to 1763, he represented Pennsylvania against the Penn family from 1757 to 1762 and, between 1764 and 1775, several colonies against the London government. Originally negotiating between colonies and the mother country, he was at work for independence of the North American colonies from 1775. He joined the Continental Congress and signed the Declaration of Independence in 1776. From 1776 to 1785 Franklin lived as a legate in France and achieved big success with the French American alliance and the peace of 1783. He contributed a lot to the Constitution of 1787, which he signed, too.

Franklin was renowned as America's most important philosopher of the Enlightenment in Europe. He was admitted to the Académie Française. As a writer, he preferred short forms like maxim, essay, satire, and a clear, humorous, and ironic style. The *American Philosophical Society* was founded by Franklin.

[8]*Joseph Priestley*, b. March 13, 1733, Fieldham (near Leeds)–d. Feb. 6, 1804, Northumberland (Penn.). According to Priestley his first aquaintance with natural sciences took place between 1750 and 1752. During these years, he was taught mathematics twice a week by Mr. Haggerstone, a disciple of Colin MacLaurins. At the same time, he read William Jacob's Gravesande's *Mathematische Elemente der Physik*. As a young preacher, he simultaneously gave lectures on the usage of globes. In 1758 he founded a school in Nantwich, supplied it with an air pump and an electrostatic generator, and encouraged the pupils to experiment with these instruments. In 1761 Priestley was been offered a position as a language teacher in Warrington which he accepted. On a visit to London he met B. Franklin. Although Priestley had to teach every day, he wrote a brilliant *History of Electricity* in a very short time. It included some of his own observations. In 1767 he became preacher in Leeds. There he wrote a history of optics. At a nearby brewery he observed the gases developing during fermentation. Priestley began to heat matter in a tube and caught the escaping gases in a pneumatic basin in which mercury served as confining liquid. As a librarian to Lord Shelbourne from 1773 to 1780 he had some spare time. In 1774 Priestley discovered the properties of oxygen after having developed the gas from mercury oxide. During a stay in Paris, he told Antoine-Laurent Lavoisier about it. Other gases discovered by Priestley are protoxide of nitrogen, hydrochloric, ammonia, and sulphur dioxide. While experimenting with carbon dioxide, Priestley understood the process of assimilation in plants. Humphry Davy said he didn't know a better work to lead students on their way to own discoveries than Priestley's six books on the different kinds of air.

However, Priestley, working as a preacher in Birmingham from 1781, wrote important books on other topics as well—e.g., theology, ecclesiastical history, politics, educational theory and philosophy. He was involved in ideological quarrels, too. When celebrating the anniversary of the French Revolution in 1791, an incited mob destroyed his church and his house. Priestley fled to Hackney (near London). From there he went to America in 1794. He settled a five days jouney upstream Philadelphia at the Susquehanna. There he discovered fluorine and carbon monoxide.[HE].

[9]*Charles Augustin Coulomb*, b. June 14, 1736, Angoulème–d. Aug, 23, 1806, Paris, French physicist. Corresponding member of the Académie des Sciences since 1774, he became a full member of its successor organization (Institut National) in 1795. Until 1776 Coulomb held the position of Lieutenant-Colonel du génie in Martinique, and after that he was Inspecteur général de l'Université in Paris.

Besides works engineering sciences Coulomb published tests on torsional elasticity, the results of which he used to construct a torsion balance in 1784. This gadget is often named after H. Cavendish although the concept was developed by John Michel (b. 1724, d. 1793) in 1750. The improved version of Coulomb was the first reliable gauge for quantitative electrostatic and magnetostatic measurements. It enabled him to deduce electrostatic and magnetostatic basic laws which were then named after him (Coulomb's laws). These laws imply that two quantities of electricity (two magnetic pole strengths thought of as points) attract and repel each other with a force that is directly proportional to their product and inversely proportional to the square of their difference to each other. It was only during 19th century that the factors of proportionality of the surrounding medium (permittivity and permeability constant) which had some influence on the laws were fully understood. Because of a technical similarity between Coulomb's laws and Newton's law of gravitation the electromagnetic and electrodynamic effects, too, have still been explained and mathematized as instantenously acting long-range forces in the second half of 19th century. Even in nuclear physics of the 20th century, Coulomb forces were thought to be the only active ones within the sphere of the atom. This changed, however, when the importance of the interaction forces independent of charge were realized. The practical unit of the quantity of electricity is named after Coulomb, too.

[10]*Luigi Galvani*, b. Sept. 9, 1737, Bologna–d. Dec. 4, 1798, Bologna. In the year 1780 the anatomist experimented with an electrostatic generator and a freshly preserved frog with a knife set to its legs nerves. Galvani noticed a twitch of the frog's legs during the spark-over. Even at that time there was nothing new about this discovery. Both the electric sensitivity of frogs legs and the phenomenon of *electric recoil* (today known as electromagnetic induction) had been known. To Galvani, however, the observation came as a surprise, and he began extensive experiments. In the course of these he made a really new and important discovery: even if the frog's legs were just linked to different metals and then connected, they twitched. In 1791 Galvani finally published his findings in the journal at the Institute for Anatomy on Bologna university. His essay *De viribus electricitatis in motu musculari commentarius* caused a sensation because the *essence of vitality* was thought to have been almost found. The era of galvanism began.

Galvani himself tried to understand the phenomenon of *animal electricity* or *frog electricity* by analogy to the Leyden jar. Alessandro Volta became the leading scientist in the new field.

[11]*Alessandro Volta*, (Count from 1810) b. Feb. 18, 1745, Como–d. March 5, 1827, Como, italian physicist. In 1775 he invented the electrophonis which had in a way already been used by J. C. Wilcke in 1762. From that one Volta developed a very sensitive instrument to detect low charges (1782), and the condenser. In 1781, he constructed a straw-electroscope. Already in 1792 (about the same time as J. C. Reil, b. 1758, d. 1813) Volta supposed that

all that would be necessary to achieve the animal electricity discovered by Galvani, was a contact between two first class conductors (metallic conductors) and one second-class conductor (electrolytic conductor) or one between two second-class conductors and one of the first class. During the following years he extended this theory of contiguity, and in 1800 he published his most momentous discoveries, the *Seulenapparat* and the *Tassenkrone*. Both gadgets were the earliest kinds of galvanic batteries producing electricity of higher voltage by other means than that of using electrostatic generators. In addition, Volta carried out experiments with thermal expansion of gases and vapors and invented several electrical gadgets no longer in use today. To honor him, the unit of electromotive force is called *Volt*. The Volta museum has been existing in Como since 1927.

[12] *Humphry Davy*, b. Dec. 17, 1778, Penzance (Cornwall)–d. May 29, 1829, Geneva. D. became lecturer in 1801. Only a year later he got a chair at the Royal Institution which— though newly founded by Count Rumford (Benjamin Thomson)—quickly became famous for Davy's brilliant lectures. He was ennobled in 1812 and married into a rich family the same year. Financially independent, he travelled a lot. Thus, Michael Faraday, who had begun to work for the Royal Institution as a laboratory assistant (Davy's "biggest discovery") slowly but surely became successor to Davy, who was a member since 1803, secretary from 1807 to 1812, and president of the Royal Society from 1820 to 1827.

Davy became famous for his discoveries in the field of chemistry and electrochemistry as well as for inventing the safty lamp for mining (1815). However, besides having been a successful experimenter, Davy was an aesthete, a brilliant orator and socializer, a poet and a man of great intellect. His essays are masterpieces of scientific prose. Electrolysis of both potash and soda and of the respective leaches had been unsuccessful in water solution. Davy used an alcohol flame and a fan to melt waterfree potash in a platinum spoon. He connected the spoon with a positive pole and a platinum wire with the negative pole of a battery. When dipping the wire into the electrolytic melt, Davy saw a bright light on it, and when reversing the polarity, little bubbles rising from the melt: liquid potassium. Davy discovered the new element on October 6, 1807. Some days later he discovered sodium. The following year saw the first description of barium, strontium, calcium, and magnesium.

[13] *Jöns Jacob Freiherr von Berzelius*, b. Aug. 20, 1779, Väversunda (Östergötland)–d. Aug. 7, 1848, Stockholm, Swedish chemist. He was ennobled in 1818. The title *baron* was bestowed on him in 1835. From 1807 to 1832 he had a chair in Stockholm. Chemistry owes to Berzelius important advances in the field of methods of analysis both organic and anorganic. He discovered the elements cerium (1803), selenium (1817), lithium (1817, together with L. A. Arfvedson), thorium (1828), and was the first scientist to show silicon, zirkon, and tantalum in free form. His important theoretical works on chemical proportions are based on precise determinations of atomic weight. According to Berzelius's dualistic electrochemical theory which is important for valency rules, chemical bonds come about by groups of atoms with positive and negative electrical charge. The chemical sign language still in use today goes back to Berzelius. Important German chemists (L. Gmelin, H. G. Magnus, E. Mitscherlich, H. Rose, F. Wöhler) were his disciples. Both his *Lehrbuch der Chemie* (textbook of chemistry) (3 vols, 1808— 1818) and his *Lehrbuch der organischen Chemie* (textbook of organic chemistry) (3 vols, 1827— 1830) have been published in numerous

editions in all world languages. These textbooks together with Berzelius's critically judging 'Jahres-Berichte über die Fortschritte der physischen Wissenschaften' (annual reports about the progresses in the physical sciences) (once a year from 1822 to 1848, 22 vols) had a big influence on the development of chemistry in his time.[BR].

[14] *Hans Christian Oersted*, b. Aug. 14, 1777, Rudkjöbing (Langeland)–d. March 9, 1851, Copenhagen. Very early in life, the pharmacist's son was interested in his father's books and his laboratory. In 1794 he became a student of medicine, physics, and astronomy at Copenhagen university where his essay "On the limits of poetic and prose language" won a prize. In 1799 Oersted did a doctorate in the faculty for medicine on a topic of Kant's philosophy.

As he was interested in literature and philosophy he aquainted himself with romanticism and natural philosophy while traveling through Germany in 1801–02. According to his diary he was together "with Ritter dayin, dayout" (Jena 1802). Experimenting together with Johann Wilhelm Ritter he tried to find a link between electricity and magnetism. They often used the voltaic pile to detect magnetic properties. As they forgot to close the battery wire and analyze the electric current, however, their attempts led to nowhere.

After having worked together, they kept in touch. Correspondence lasted up to Ritter's death in 1810. In *Ansichten der chemischen Naturgesetze* (view of the chemical laws of nature) (1812) he firmly supported dynamism. He took it to be "confinement of the horizons" not to see, as Ritter did, galvanism, electricity, magnetism, light, and warmth as "different activities of the general force of nature." He was therefore sure of an interaction between electricity and magnetism. Early in 1820 after having read on "electricity, galvanism, and magnetism" he was attracted to the idea again and decided to do further experiments. Most of his hopes Oersted placed in the discharge of a voltaic battery by means of a metal wire of high resistance. He was sure the conflict between positive and negative electricity (for which he used the image of two hostile troops getting muddled up in a fight) would warm the wire. But perhaps, Oersted suggested, only a part of this activity led to warmth and the rest to something different. So Oersted was not surprised by the magnetic action of current itself but by its quantity.

Oersted published the discovery of electro-magnetism on July 21, 1820, in Latin. In a four-page essay he told the scientific world: "The first experiments on the subject I try to elucidate have been done during lectures on electricity, galvanism, and magnetism, which I read last winter. These tests seemed to illuminate that it was possible to change the direction of a magnetic needle by a galvanic apparatus with a closed galvanic circle and not—as some famous physicist had tried several years ago—with an open one. The two opposite ends of a galvanic apparatus have to be thought of as connected by a metal wire. For the sake of brevity I am going to call this wire "connecting conductor" or "connecting wire" whereas the effect it has will be called "electric conflict." A straight part of the connecting wire is to be put horizontally above a simple and freely moving magnetic needle until the two run parallel. When all this is done, the magnetic needle will begin to move like this: underneath that part of the connecting wire which comes from the negative pole of the galvanic apparatus it is going to deviate into western direction. If the distance between wire and needle is less than 5/4 inches, deviation is about 45 degrees. The actual deviation, however, depends on the strength of the apparatus. . . ."

The discovery, told in these words, stirred a sensation. In Geneva and Paris, in Heidelberg, Kiel, Göttingen, Munich, and Leipzig, in London and St. Petersburg, an academies, universities, at courts and in private mansions, Oersted's experiments were talked about and repeated. Georg Wilhelm Muncke told that "the news had hardly been presented to the public when not only all physicists, but a lot of natural scientists, physicians, and amateurs as well as even those to whom scientific research is unknown, showed a burning enthusiasm for the new discovery. This reaction can justly be compared to the enthusiasm evoked by the first aerostatic machines (i.e., airships called "montgolfiers") which achieved something hitherto thought of as impossible to do."

Ludwig Wilhelm Gilbert who published the Annals of physics admitted in 1820 of having mistrusted Oersted initially, the latter having been a disciple of Ritter. Wilhelm Ostwald, too, took the discovery made by Oersted to have been pure chance or even absurd. "This physicist, having become famous for discovering deviation of magnetic needles by electricity, may have been an even worse philosopher of nature than Ritter. The major discovery he managed to make shows the strange ways nature goes to reveal its secrets. At the same time it shows that even these people may succeed in making a rare find."

Oersted himself, Hendrik Steffens, Friedrich Schelling, and others emphasized that the world of thought of romantic philosophy of nature itself had inspired their experiments with electricity and magnetic needle. In 1832 the Bayerische Akademie der Wissenschaft (Bavarian Academy of Sciences) celebrated Faraday's discovery of electromagnetic induction with a speech made by its president Friedrich Schelling. Its main topic was the description of three discoveries in the field of physics and the attempt to prove their interdependence. These discoveries were the beginning of electrochemistry by Johann Wilhelm Ritter in 1798, the discovery of electromagnetism by Oersted in 1820, and the discovery of electromagnetic induction by Faraday in 1831. In an appendix to the published version, Schelling wrote: "The intention was to show the necessary correlation between the discoveries mentioned and to demonstrate how they were necessarily developed out of each other and, like that, could be more or less predicted by natural scientists using their brains."

After having founded the field of electromagnetism, Oersted carried out experiments with compressibility of water and invented the piezometer, a gadget gauging this property in fluids. Furthermore, he was the first to produce aluminium chloride by breaking aluminium oxide down into its constituents and thereby led Friedrich Wöhler to the discovery of metallic aluminium in 1827. Oersted, too, had already got small quantities of this metal by distillation of aluminium amalgam.

However, it was his most urgent desire to teach. He gave a lot of lectures—up to four hours a day. As a teacher Oersted was very popular because of his modest behaviour and his lectures which were easily understandable—although he often was pricelessly absentminded.

He expounded his philosophy of life (which was mainly formed by romantic philosophy of nature) in several much-read books. Oersted had an influence on L. A. Colding's theories on conservation of energy and played an important role in the Danish world of thoughts.

[15]*Dominique François Jean Arago*, b. Feb. 26, 1786, Estagel (n. Perpignan)–d. Oct. 2, 1853, Paris, French astronomer, physicist, and politician. Together with J.B. Biot Arago carried out measurings of latitudes in Spain and Scotland. From 1805 he and A. von Humboldt

were close friends. In 1809 Arago became member of the Institut National. Two years later he discovered the rotation of polarization plane of light by pieces of rock crystal. In 1824 he explained the sparkling of stars as an interference perturbation brought about by disturbances of air, and discovered rotation magnetism which Faraday later proved to be effected by induction. As a moderate Republican, he held the War and Navy Dept. in the 1848 provisional government.[BR].

[16]*Thomas Johann Seebeck*, b. Apr. 9, 1770, Reval–d. Dec. 10, 1831, Berlin. The physicist was a friend of Hegel and Goethe whom he gave advice concerning optical studies. Since 1818 he was a member of the Berliner Akademie der Künste (Academy of Arts). In 1821 he discovered the so-called Seebeck-effect, in 1818 (among other things) the rotation of the polarization plane of light in sugar solution and the (almost perfect) reproduction of illumination color with the help of damp silver chloride.

His son Ludwig Friedrich Wilhelm August, b. Dec. 27, 1805, Jena–d. March 19, 1849, Dresden, became principal of the *technische Bildungsanstalt* (now *Technische Universität*) Dresden in 1843. He became well-known for his tests on the importance of overtones for tone color, and for experiments on color-blindness.[BR].

[17]*André Marie Ampère*, b. Jan. 20, 1775, Poleymieux (n. Lyon)–d. June 10, 1836, Marseille. Ampère first was a teacher of physics in Bourg and Lyon, then he had a chair at the École Polytechnique and the Collège de France. Later on he became inspector general of the French universities. Three years after Amedeo Avogadro he came to the same conclusions about ideal gases (i.e., equal volumes contain the same number of molecules under identical conditions). In 1820 he repeated Oersted's experiments concerning magnetic effects of galvanic current. Shortly after that he put forward a theory about electrodynamic interdependence of conductors carrying a current. The Amperian float rule was developed by him as well as the explanation of the pheomenon of magnetism by so-called amperian molecular current. In 1822, he suggested (together with Jacques Babinet) an electromagnetic telegraph. In the field of mathematics, too, Ampère achieved a great deal: he wrote essays on a theory of partial differential equations and on probability theory. Being interested in literature and philosophy besides natural sciences, he published a book in two thick volumes titled *Essai sur la philosophie des sciences ou exposition analytique d'une classification naturelle de toutes les connaissances humaines* (1834). Ampère was a man of great sensitivity. Although a successful scientist, he was susceptible to melancholy due to the early death of his father and his first wife. At the same time he was known as a very kind and polite man. These characteristics can be traced in his voluminous correspondence and extensive diaries.

[18]*Jean-Baptiste Biot*, b. Apr. 21, 1774, Paris–d. Feb. 3, 1862, Paris, mathematician, astronomer, geodesist, physicist, chemist, and historicist of natural sciences. First he was an office apprentice, then a gunner for the northern army, and after that he was among the first to study at the École Polytechnique. Here he became a student and protege of Gaspard Mouge and Pierre-Simon Laplace. From 1800 to 1862 he worked as professor of mathematical physics at the College de France. In addition to this, he had a chair for astronomy at Paris University from 1809 to 1848. His textbooks in several volumes show his ability to give good descriptions. Already in 1803 the historian wrote a history of natural sciences

during the Revolution. Later he worked on Newton and on the astronomy of Egypt and China.

Together with Gay-Lussac, Biot made a flight in a montgolfier in 1804 to measure the decrease of geomagnetic effects in greater hights. The following year he and Arago did some tests on refraction in gases and after that B. did geodetic measurings in Spain. He developed the theory of chromatic polarization, discovered several new aspects of rotation of polarization plane of light in liquids and steams, and invented the polarimeter. With Félix Savart he found the law of the electromagnetic effect of elements of currents.

[19] *Félix Savart*, b. June 30, 1791, Mezières–d. March 16, 1841, Paris. The French physician and physicist was professor of physics at the Collège de France and did a lot of experiments in the field of acoustics. In 1820 he classified the oscillation frequency of tones with a pinion siren and (together with Biot) deduced from experiments the basic law of electromagnetism, which is named for them.

[20] *Georg Simon Ohm*, b. March 16, 1789, Erlangen–d. July 6, 1854, Munich. The discovery of an amazingly simple law which sorted out a bewildering variety of electrical phenomena has justly been praised as a relieving feast. It helped immensely to open the young science of galvanism to exact mathematical treating and thus established a basis for the astonishing development of the science of electricity in the 19th century.

Although young Ohm, son of a master fitter incredibly intersted in mathematics and philosophy, stopped studying mathematics, physics, and philosophy after eighteen months in order to work as a teacher in Switzerland (1806), he managed to finish his studies in Erlangen in 1811 with a doctorate and a postdoc thesis in mathematics. After having taught as an outside lecturer in Erlangen for several years, he went to Bamberg where he wrote his first book titled *Grundlagen zu einer zweckmäßigen Behandlung der Geometrie* (Basics of a purposeful treatment of geometry). Due to this publication he was appointed senior primary school teacher for mathematics and physics at the so-called Jesuiten-Kollegium Köln in 1817. Hoping for an academic career, he left Cologne after nine years of successful teaching and research and went to Berlin. There he held lectures at the Artillerie- und Ingenieurschule. In 1833 several petitions to the king of Bavaria led to his getting a chair at the newly-founded polytechnical school in Nürnberg. From 1839 to 1849 he was headmaster of the school which by now is named for him. It was only in 1852—he was 63 by then—when he was appointed full professor of physics at Munich university. His hopes had finally come true.

The experiments which in the end led to his law had begun in Cologne already where a big collection of physical gadgets had been at his disposal. Due to both inconstancy of the voltaic columns used and a confusion of terms the search for internal correlation between intensity of current, voltage, and resistance in an electric cell was a failure at first. Only when Johann Christian Poggendorf advised him to use a thermo couple (invented by Thomas Johann Seebeck in 1821) as a source of energy of stabile voltage, Ohm had the big success which immortalized his name. In the famous essay "Bestimmung des Gesetzes, nach welchem Metall die Contactelektrizität leiten. . ." (Determination of the law according to which metals conduct the contact electricity) (1826) he summarized a table of measured intensities of voltage as follows: "The numbers given above may as well be represented by the equation $X = a/(b + x)$, with X being the intensity of the magnetic effect on a conductor with the length x, while a and b stand for constant quantities dependent on exciting force and pipe resistance of the other parts of the chain." (Today it is common practice to write $I = U/(R_i + R_a)$ instead.)

In the even better-known essay from 1827 "Die galvanische Kette, mathematisch bearbeitet" (Mathematical treatment of the galvanic chain) Ohm's law got the shape in which it became known. Based on his empirical findings and analogous to the laws of heat transport (Joseph Fourier, Siméon-Denis Poisson) he with mathematical methods developed the equation $S = A/L$ "which is universally valid and already by its shape reveals constancy of the quantity of current at every point of the chain." From that he discarded all theories (e.g. Oersted's theory of conflict) suggesting a change of intensity of current in an unbranched chain. Ohm expressed his extremely simple law in words as well: "The quantity of current [S] in an electric cell is directly proportional to the sum of all voltages [A], and in inverse proportion to the whole reduced length of the chain [L]." By *reduced length* he understood "the sum of all quotients . . . built by the real lengths belonging to homogeneous parts and the product of the respective conductivity and cross-sections." Moreover, both the regularity concerning current branching and the difference between internal and external resistance of a circuit were known to him.

At first Ohm's pioneering discoveries were discarded rather than agreed to. One of the few to realize the impact of Ohm's law quite early was Gustav Theodor Fechner. In 1831 the physicist from Leipzig "(could) not help but attach to Ohm the credit for having established a new era of galvanism with those few letters of a simple formula." The awarding of the Copley medal by the London Royal Society in 1841 finally helped Ohm on the road to worldwide success. By then, Ohm had turned to other fields in physics. Acoustics owes to this research—among other things—the introduction of the Fourier analysis to characterize tone colors, and Ohm's law of acoustics (later falsely attributed to Helmholtz). Moreover, during his time in Munich Ohm worked on problems of crystal optics and molecular physics.

[21]*Michael Faraday*, b. Sept. 22, 1791, Newington Butts–d. Aug. 25, 1867, London. After having attended the evening lectures of Humphry Davy at the Royal Institution the dearest wish of the self-educated man (i.e., to do research) came true: in February 1813 Faraday became lab assistant at the Royal Institution.

Soon enough he got a better position. Davy was absent quite often, so that Faraday had to see to talks of guest lecturers and to prepare experiments. In 1816 he published the first scientific works, a chemical analysis of quicklime. In 1823 he presented chlorine in liquid form, a year later he came across benzole and butylene while distilling fat oils. To this he added the field of technical physics. From 1820 to 1822 he worked on the development of stainless steel, whereas from 1825 to 1829 he dealt with lenses with specific optical properties. After Oersted's discovery (electromagnetism) Davy and Faraday repeated the experiments. At first they had wrongly believed the interaction between electric current and magnetic needle to be the known conventional forces along the tie line. In August 1821, however, Faraday changed his mind. The following month he constructed a gadget showing quite clearly that the forces stood vertically towards the tie line. In the experiment an electric conductor rotated around a fixed magnet and, vice versa, a mobile magnet rotated around a fixed electric conductor as well. Although it was a very simple one, this apparatus may justly be seen as the first electric motor. Now Faraday tried to prove the opposite effect as well, i.e., the electric effect of a magnet. A remarkable entry is to be found in his notebook from 1822 already: "Convert magnetism into electricity."

To actually achieve this conversion was an important objective of his research work during the following years. According to the notes in his lab journal, he had built the right experimental setup as early as 1825 and 1828. However, the sensitivity of the gauge was too little. With a setup today known as a transformer he finally managed to discover the effect of electromagnetic induction. This was on August 29, 1831. In the following months he analysed the effect in detail. Having no knowledge of mathematics, Faraday used the lines of magnetic force to illustrate his findings. According to Faraday the direction of induced current depended on how the lines of force running from North Pole to South Pole were cut by the wire. Although he used them as a theoretical model, as an illustration only, Faraday after while became convinced of the physical reality of these lines. So, slowly but surely and against the inhibitions of most of the other scientists, Faraday coined the terms of magnetic and electric field of force.

Success in science changed his social status and position, but not his character. Faraday stayed the same modest and kind man he had been before. In 1824 he was appointed member of the Royal Society, a year later director of the Royal Institution, and little by little he became honorary member of 92 scientific societies and academies.

Other than most of his colleagues, Faraday believed in the theory of dynamism. To him, forces (i.e., energies) were the primary conditions of nature, and he firmly believed in mutual convertibility of magnetism, electricity, light, heat, galvanism, etc. This view had led directly to the discovery of electromagnetic induction, and it both influenced his interpretation of electrolytic experiments (Faraday's law) and led to finding the Faraday effect in 1845 as well as to research on magnetism.

Faraday explicitly talked about different manifestations or shapes of conditions of force and about conversion of one force of nature into another. This conviction runs through all his life's work. Already in 1839, years before Julius Robert Mayer and others established the energy principle, Faraday formulated an energetic reason against an apparatus (invented by Volta) which was meant to explain electrolytic voltage as coming into existence solely by a contact of metals without any chemical changes. "This would indeed be a creation of force .... However, there never is a creation of force without a respective exhaustion of something else, something feeding the force."

Early in 1862, after the discovery of spectrographic analysis by Kirchhoff and Bunsen, Faraday, still being convinced of an internal correlation between all natural forces, tried to find some kind of influence of magnetic fields on spectral lines. The attempt was unsuccessful, as is known today, due to the inadequate resolving power of the Steinheil spectrometer he used. From about 1849 on Faraday tried to achieve conversion of gravitation into other forces. These experiments were as unsuccessful as the theoretical attempts of Einstein to combine gravitation and electrodynamics to a single theory. Faraday wrote with resignation: "This is the end of my experiments for now. The results were negative. But I nevertheless can not help to feel a strong relation between gravity and electricity, although the tests up to now did not prove it."

[22]*Carl Friedrich Gauss*, b. Apr. 30, 1777, Braunschweig–d. Feb. 23, 1855, Göttingen. Gauss is said to be one of the most important mathematicians in world history, equal to Archimedes and Newton. Like them he did not only influence his own field of science but neighboring disciplines like physics, astronomy, and geodesy as well.

Coming from a very poor background, he had had the good fortune that his talent was realized and supported quite early. As a sixth-former of Braunschweig Gymnasium (highschool) he was introduced to the Duke of Braunschweig, Karl Wilhelm Ferdinand, and became his protégé. The duke helped him to attend the Collegium Carolinum in Braunschweig from 1792 to 1795 and Göttingen University (from 1795 to 1798) and afterwards to get a long–time research job without any duties. In the beginning Gauss was not sure whether to study classical languages or mathematics instead. But on March 30, 1796, he decided in favor of mathematics. It was the day he succeeded in proving the possibility of constructing the regular *17-Eck* with a pair of compasses and a ruler.

However, it was not geometry but arithmetics that became his "queen of mathematics." Already in 1801 he put himself on a level with the most important mathematicians of all times. It was the year he published *Disquisitiones arithmeticae* in which he created the modern theory of numbers. Outside his own discipline he became known for his sensational achievements in the field of astronomy. In 1801 Giuseppe Piazzi from Naples had discovered the planetoid Ceres Ferdinandea and had managed to measure three points of its trajectory before it disappeared again. After an unsuccessful attempt of some astronomers, Gauss— with the help of a method of approximation developed by him—succeeded in calculating the elliptic trajectory so exact that Ceres could be found again. With the same astonishing precision did Gauss afterwards determine the trajectories of Pallas, Juno, and Vesta which had all been discovered by his friend Wilhelm Olber between 1802 and 1807.

Gauss presented his methods of calculating in his principal work on astronomy called "Theoria motus corporum coelestium in sectionibus conicis solem ambientium" which was published in 1809. This work applied new standards to calculating astronomy. In it Gauss also described his method of smallest squares which made it possible "to infer the most probable values of unknown quantities from a bigger number of observation data depending on them."

Although the russian government tried to win him over to Petersburg academy with generous offers, Gauss stayed in Braunschweig until the death of his patron in 1806. A year later he has been offered a chair for astronomy at Göttingen university. On request of the Hannoverian government he from 1818 onward took part in land-surveying for several years. The readings became more exact than ever before because Gauss' invention of the heliotrope made possible measurings of big geodesic triangles. The survey and its analysis was followed by experiments about the similarity of mapped faces (1822) and about bent faces (1828). With both these and the "Untersuchungen über Gegenstände der höheren Geodesie" (Research about arcticles of higher geodesy) Gauss established the basis for this field of science.

In his second essay on "Theorie der biquadratischen Reste" (1831) Gauss extended the idea of complex numbers. Before him, Leonard Euler (1777) and the Danish Caspar Wessel (1797) had calculated with the term $\sqrt{-1} = i$ and had shown complex numbers in a plane of numbers (later named after Gauss). But the important point was that Gauss saw these numbers as equal to real numbers and demonstrated their impact on arithmetics and analysis. So to the three proofs of existence for the fundamental theorem of algebra he had offered before he added a fourth one with the help of complex numbers in 1849. Gauss had already given first stringent proof of this law (known as early as 17th century) according to which an algebraic equation of nth degree has n solutions (roots) in his doctoral dissertation in 1799.

When in 1831 the young physicist Wilhelm Weber got a chair in Göttingen, the two scholars quickly became friends and successfully worked together in the field of physics. More often than Weber, Gauss had the basic ideas whereas Weber did the actual set-ups and experiments. Their work was published in *Resultate aus den Beobachtungen des magnetischen Vereins* which had been founded especially for this reason. In 1832 the important essay "Intensitas vis magneticae terrestris ad mensuram absolutam revocata" was published. In it Gauss developed an absolute magnetic system of measurement based on the fundamental units of length, time, and mass known today as Gaussian system of units. The essay "Allgemeine Theorie des Erdmagnetismus" (1838/39) was extended in 1840 by "Allgemeine Lehrsätze in Beziehung auf die im verkehrten Verhältnis des Quadrats der Entfernung wirkenden Anziehungs- und Abstoßungskräfte." While the first essay develops the potential of geomagnetism, the second (as a result of the first) gives a mathematical explanation of potential theory highly significant for the field of physics.

In connection with these works Gauss and Weber made a pathbreaking invention in telecommunication technics in 1833, namely the electromagnetic telegraph. This equipment connected the Göttingen observatory in which Gauss worked and lived with the physical cabinet of Weber, and it enabled the two scientists to exchange short messages. However, experiments on a larger scale were impossible due to lack of money. Nevertheless, Gauss invented—after having become aware of the phenomena of induction—an earth indicator. This gadget in a version improved by Weber was used to measure geomagnetic inclination.

After Weber left Göttingen in 1838, Gauss returned to mathematics again. It is a pity that he gave up electrodynamics after such promising beginnings as for example the essay "Ableitung der Zusatzkräfte, die zu der gegenseitigen Wirkung ruhender Elektrizitätsteilchen noch hinzukommen, wenn sie in gegenseitiger Bewegung sind, aus der nicht instanten (zeitlos schnellen), sondern auf ähnliche Weise wie beim Licht sich fortpflanzender Wirkung."

[23] *Wilhelm Eduard Weber*, b. Oct. 24, 1804, Wittenberg–d. June 23, 1891, Göttingen. Together with his brother the physicist published the fundamental "Wellenlehre auf Experimente gegründet" in 1825. In 1831 he was appointed to a chair in Göttingen. Together with C. F. Gauss he developed both the first electromagnetic telegraph and a number of measuring instruments, e.g., inductorium (1837), electrodynamometer (1840 and 1846) and reflecting galvanometer (1846 and 1852). In 1835 Weber discovered the elastic aftereffect, and one year later he and his younger brother, the anatomist Eduard Friedrich (b. March 10, 1806, Wittenberg–d. May 10, 1871, Leipzig), published a work titled *Mechanik der menschlichen Gehwerkzeuge*. Being one of the *Göttinger Sieben* he was dismissed in 1837. In 1843 he went to Leipzig as a professor and six years later he was again called to Göttingen. After 1846 Weber published electrodynamical mensurations which became the basis for international units of measurement.

[24]*James Prescott Joule*, b. Dec. 24, 1818, Salford (Lancashire)–d. Oct. 10, 1889, Sale (Cheshire). Joule owned a big brewery. Without any kind of scientific education he did physical experiments at a young age already and he set great score by quantitative measurings. His application for a chair in St. Andrews, Scotland, was unsuccessful. So he carried on with his studies in private. Some time later he became secretary for, and finally president of, the Literary and Philosophical Society in Manchester.

In his work Joule put the main emphasis on heat. In 1840 he found the law of current heat. In 1847 he reported his findings at the British Association in Oxford where William Thomson succeeded in interesting the public in Joule's ideas. Until 1878 several essays on the mechanical equivalent of heat were published. Measurings with several different experimental setups substantiated the energy conservation law.

Joule experimented with changes of longitude of metals in the magnetic field, as well. He did quantitative tests on change of temperature when gas is compressed or untensioned, respectively. Together with William Thomson he published the essay "On the air engine" in 1852. The effect today named after Joule and Thomson is made use of for liquidation of gases.[HE].

[25]*Pierre-Simon Laplace*, b. March 28, 1749, Beaumont-en-Auge (Normandy)–d. March 5, 1827, Arcueil (n. Paris). Coming from a rural background, Laplace attended the Benedictine high school in Beaumont. There he developed a lasting interest in the poet Jean-Baptiste Racine. At Caen university he prepared himself for a priest's office. In addition to this, he was a student of mathematics and published a work on integral calculus in the *Turin Memoirs* from 1769. At the age of 22 he went to Paris where d'Alembert helped him to get a job as teacher of mathematics at the Ecole Militaire. From then on he devoted himself to his life's work as mathematician, astronomer, and physicist. In 1722 he wrote two essays on differential equations in service of astronomy, and in 1724 another two on probability calculus. These essays marked the fields of science he worked in all his life with imperturbable persistence. Laplace devoted all his powers to the application of Newton's law of gravitation to all trajectory paths in the solar system. Next to Lagrange he became the greatest scientist in France. Together with the chemist Claude-Louis Berthollet he founded the Societé d'Arcueil, a hotbed for young researchers looking for stimuli and instructions. Jean-Baptiste Biot has written about the concern Laplace felt for the young people and their careers. He led a very simple life. Thus he had the money to exquisitely equip the observatories at Paris and Marseille. He was a member of the Académie as well as (like Lagrange) a member of the commission for measures and length. At the same time he taught at the Ecole Normale. In Napoleon's service Laplace was Secretary of the Interior for six weeks in 1799. He got through the changing times almost unmolested. It is said that he said before dying "Ce que nous connaissons est peu de chose, ce que nous ignorons est immense" (There is hardly anything we do know, but there is an immense lot we do not know). He died the same day as Alessandro Volta, almost exactly a hundred years after Newton.

Laplace summarized his research in three great works. They deal with the representation of the system of the world, with the mechanics of the sky, and with probability calculus. *Exposition du système du monde* (1796) describes the system of the world without using formula, and shows Laplace as a brillant writer. Especially one part of the work became famous, namely the theory about the genesis of the solar system: it just could not have happened by pure chance that all planets are moving around the sun equidirectionally on almost the same plane in almost perfectly circular orbits, that moons rotate around their planets in the same direction and on about the same plane, and that all known rotary motions occur in the same dirction and on almost the same plane, too. Like the central cores of nebula which can be seen through telescopes, our solar system had been a rotating nebula, too, expanded even farther than the trajectory of the farthest planet. This nebula had become cooler and thickened towards the center; thus, it rotated more quickly, and from time to time concentric gaseous rings had come off the equatorial belt. From them planets came into being while from the remaining center the sun developed. Buffon, on the other hand, had supposed that a comet might have fallen onto the sun at an angle and like that might have teared off some matter. According to Laplace this theory could not explain the circular orbits. He saw comets as foreign bodies. Because Immanuel Kant had given a talk on the hypothesis of the nebula as well, these ideas are today called Kant-Laplace-hypothesis.

Between 1799 and 1825 the five volumes of *Mécanique céleste* were published. This principal work included everything that had been achieved since Newton. Three examples may be given as an illustration: the moon does not move around the earth in an exact ellipse because not only earth, but sun, too, has an impact (either accelerating or slowing down) on it. Determination of the movement of a star attracted by two celestial bodies was a problem solved by Alexis-Claude Clairaut: Earth's axis does not stay parallel to itself but moves in slow circles. D'Alembert proved the effect sun and moon had on the not perfectly spherical earth to be the reason for this movement. The moon always turns the same side towards earth. According to a calculation made by Lagrange, this phenomenon can be explained by an axis of moon in direction of earth lengthened by earth's attraction. In addition to results of other scientists, a lot of his own findings were put down in these books, e.g., parts from the essay on permanency of the axes of planetary orbits (1773), from the theory of the motion of planets (1785), from the studies on tides (1790).

The first two volumes (published in 1799) deal with a general theory of motion and the shape of stars and the tides. Volumes three and four (1802, 1805, resp.) comment on individual planets, moons, and comets, while the last volume (1825) deals with refraction of radiation in atmosphere as well as other problems and gives a brief outline of the history of celestial mechanics. The rich content is presented concisely. Since Newton's discovery of universal gravitation all mathematicians tended to reduce all known phenomena in the system of the world to this law, Laplace wrote in his preface. "It is my aim to present these theories strewn over a big number of works under one common point of view.... Looked at in the most general way, astronomy is nothing else but a question of mechanics." With virtuoso mastery of infinitesimal calculus the motion of stars is analyzed in detail, including all mutual attractions. The most simple way to see irregularities is to suppose a planet moving in an elliptical orbit with the real planet oscillating around the ficticious one.

Whereas Newton only mastered the elliptical motion, Laplace had in mind the influence of several third bodies as well. He proved that this constellation, too, led to an elliptical orbit the elements of which changed slowly but surely (although the large trajectory axis was preserved), ran through the whole sky, and after thousands of years got back to its former position. In the middle of a labyrinth of increasing and decreasing speed, of changing trajectories, distances, and inclinations, Laplace discovered the invariable. The solar system underwent periodical variations from a middle condition. For Laplace, this represented order and duration.

Again and again he managed to explain phenomena as governed by a law which were until then said to be inexplicable. To give two examples: Edmond Halley had inferred that the moon must have rotated more slowly in the past from examining old and new eclipses. This was due to the effect of the sun in combination with secular aberration of the terrestrial orbit, Laplace proved. Jupiter was known to accelerate whereas Saturn slowed down. It was already feared that Jupiter might fall into the sun whereas Saturn might leave the solar system. Laplace realized that the double speed of Jupiter was about the same as the fivefold speed of Saturn, and that this fact led to mutual irregularities with long intervals exactly consistent with the observations and yet were proof of the world's constancy.

Laplace calculated the distance between sun and earth in a new way, namely from the value of irritations effected on moon by the sun which are dependent on their respective distance. The moon was like a treasure trove for him, anyway, and he managed to explain two irregularities by flattening of the earth. Laplace developed better tables of the moon and thus served seafaring and its longimetry well. By comparing the movements of Jupiter's moons with each other, he discovered relations now called Laplace's law. He calculated the rotational speed of the ring of Saturn as well as the flattening of Saturn. From observations of the tide done in Brest at his instigation, he ascertained the mass of the moon. One chapter deals with the impact the moon's force of attraction had on earth's atmosphere. All over his work the mathematical performance and achievements are outstanding. When dealing with the force of attraction of a spheroid on an outer point, Laplace introduced the spherical harmonic and established the partial differential equation (now named after him), the solutions of which are potential functions.

Laplace's creativity was of benefit to mathematical physics, too and Antoine-Laurent Lavoisier even managed to interest him in the field of experimental physics. Together they determined the specific heat of several matter with the help of an ice calorimeter and measured the amount of electricity absorbed by bodies during evaporation. Moreover, they wrote reports together in 1780 and 1781. Caloric theory and the theory saying that heat consists of movement of a bodies molecules are compared (without one of them being favored).

In 1816 Laplace improved Newton's formula for the speed of sound. The latter led to results which were too low, and Laplace found the reason for this. Sound waves were caused by compression of air which produces heat and a higher elasticity. So a corresponding factor had to be added to the formula. However, the most beautiful contribution to theoretical physics Laplace made with his work on capillary action. When calculating refraction in atmosphere and double refraction in crystals he had imagined light as fine particles. Now he considered phenomena to be seen when liquids rise in narrow tubes, or when drops formate, as an expression of molecular forces, too. In the process he distinguished between cohesion and adhesion. Johann Diderik van der Waals continued this work. Laplace developed a formula for longitudinal oscillation in rods and for water waves, too. An academic epitaph read "Nothing in research into nature was unknown to him." He has rendered development of probability calculus very great service: based on preliminary works since Pierre Fermat and Blaise Pascal, Laplace summarized this field in *Theorie analytique des probabilités*

(1812). He had developed this tool while dealing with problems which allowed approximate solutions only, like the theory of observation errors. Two examples may show his activities in the field of statistics: the birth rate of boys is higher than that of girls. Why is that so? Country folk sent more girls to the urban orphanages where the relation of boys to girls was 39 : 38. This reduction explains the number of baptisms. To ascertain the number of French population, Laplace tried to determine the relation between number of inhabitants and birth rate. From 30 departments a number of villages was chosen which had exact details as to the number of inhabitants and births from 1800 to 1802. 28 inhabitants were necessary for one child. If therefore the French birth rate was one million, the country had a population of 28 millions (which indeed corresponded to estimations made later on). Laplace revised his books' new editions to the very end. According to him, it was nothing but "le bon sens réduit au calcul." In an "Essai philosophique sur les probabilités" (published in 1814) which was based on a lecture held at the Ecole Normale in 1795 he explained these topics intelligible to all. The introduction dealt with a fiction which has become famous as "Laplace's world spirit:" "An intelligence which would know all forces of animate nature at one given moment ... could include in one formula both the notion of the largest bodies of the universe and the lightest atoms; nothing would be uncertain, future as well as past would be present in the eyes of this intelligence .... All efforts made in the search for truths strive for the human spirit of this intelligence. They will, however, always be endlessly far away from this aim. Probability calculus refers partly to this ignorance and partly to our knowledge." In the 19th century quite often only the first sentence was quoted and Laplace was thus seen as holding a firm belief in absolute determinacy of natural processes. This point of view, however, did not keep in mind that Laplace had just established the discipline of statistics and had been a very unassuming scientist.

[26] *Siméon-Denis Poisson*, b. June 21, 1781, Pithiviers/Loiret–d. Apr. 25, 1840, Sceaux. Almost three hundred essays on mathematical, astronomical, and physical problems show the creativity and scientific understanding of this small and frail scholar. Due to bad health there was a delay in his school education. His family would have liked him to become a surgeon whereas he himself disliked this profession and more and more preferred the exact sciences. When admitted to the Ecole Polytechnique in 1798, both his extraordinary theoretical talents and his manual clumsiness were immediately noticed. The far-sighted principal freed him from compulsory exercise in drawing. Encouraged by the most eminent scholars (Lagrange, Laplace, Mouge, Berthollet), Poisson soon moved up from student to teacher at this unique school for mathematical sciences of nature and engineering. In the beginning of 1800 he got a chair for efficient mechanics at the newly founded faculty for sciences. He worked as an astronomer for the Bureau of Longitudes for a long time and in 1827 succeeded Laplace as mathematician. Together with his teachers and future colleagues Lagrange and Laplace Poisson made several contributions to the impressive theory of celestial mechanics. Moreover, he was an architect in less far-reaching fields of mathematical physics like the science of electricity. Based on a dualistic theory of electricity according to which heterogeneous electric charges attract each other in accordance with the reciprocal quadratic law of distance and homogeneous ones are repellent, he was able to give solutions for numerous problems. Like this he found an ingenious proof for the

formula of surface forces of charged conductors and calculated local charge densities on the surfaces of two spherical conductors dependent on their distance from each other. His results corresponded perfectly with Coulomb's experimental findings. Moreover, Poisson worked on problems of magnetism, theory of capillary, theory of heat (Poisson's equation), and theory of elasticity (Poisson's ratio). With Louis Navier and Jean Victor Poncelet Poisson was one of the first who taught to apply the principle "of living forces to calculate machines in motion." Mathematics owes important contributions to variational calculus, differential geometry, and probability calculus (Poisson's distribution) to him.

[27]*George Green*, b. July 14, 1773, Nottingham–d. March 31, 1841, Sneinton. The English mathematician and physicist rendered outstanding services to mathematical substantiation of the theories of magnetism and electricity. His main work is titled "Essay on the application of mathematical analysis of the theories of electricity and magnetism" (Nottingham, 1828; also in Erelle's *Journal für die reine und angewandte Mathematik* vols. 44 and 47, Berlin, 1852–53, and in *Ostwalds Klassiker der exakten Wissenschaften*, vol. 61, Leipzig, 1895). He established a mathematical theorem which is important for dealing with potential functions. It is now known as Green's symmetric theorem. This work was followed by essays on analogies of the laws of equilibrium concerning liquids and electricity, on reflection and refraction of sound as well as light, and on wave motion in channels. Green's *Mathematical papers* have been published by Ferres (London, 1871).

[28]*Franz Ernst Neumann*, b. Sept. 2, 1798, Joachimsthal (Uckermark)–d. May 23, 1895, Königsberg (Prussia). The mineralogist and physicist had a chair for mathematical physics in Königsberg. Having brought the department into being, Neumann is the founder of this discipline in Germany. G. Kirchhoff was one of his disciples. Neumann published fundamental works on wave science of light and on pre-maxwellian electrodynamics. In crystallography he introduced zone representations by linear projection.

[29]*Gustav Robert Kirchhoff*, b. March 12, 1824, Königsberg–d. Oct. 17, 1887, Berlin. "Nothing unusual in Kirchhoff's life corresponds to his remarkable genius, his career was rather the usual one for a German professor. Big events took place inside his head only" (Ludwig Boltzmann). In Königsberg Kirchhoff was educated by Franz Neumann, the founder of theoretical physics in Germany. He was still a student and aged 21 when his first essay was published in the Annals of physics (Kirchhoff's law). After having qualified as a lecturer in Berlin in 1848, Kirchhoff in 1850 became professor in Breslau. There he met the chemist Robert Bunsen, who had got a chair, too. Their friendship later turned out to be quite important for natural sciences. When in 1852 Bunsen was appointed to a chair in Heidelberg, he persuaded Kirchhoff to change places, too, which the latter did about two years later. After having turned down several other appointments, Kirchhoff went to Berlin university in 1875. There he worked together with Helmholtz.

At the end of the 1850s, Bunsen had tried to use the colorings of flames brought about by different salts in the Bunsen burner for analytical purposes. The bare eye, however, is unable to differentiate the shades. Bunsen therefore used colored glass to filter the light. Kirchhoff advised him to take a dispersing system instead.

These experimental beginnings led to two discoveries of eminent importance. Together Kirchhoff and Bunsen developed the spectrochemical analysis (1859–60) which is able to detect even the smallest traces of elements. Just a couple of days later, Bunsen found cesium in Dürkheim sparkling water, and rubidium in the Saxonian village of Leipidolitz. The discovery of several other new elements followed.

Even before the development of spectrographic analysis, Kirchhoff had come across a curious phenomenon. "I designed a solar spectrum and had the sunbeams shine through a strong flame of sodium chloride before falling on the gap. When the light was subdued enough, two light lines appeared instead of two dark lines D; when, however, the intensity of light exceeded a certain limit, the two dark lines D could be seen much more clearly...."

From this observation, namely, that each body absorbs those spectral lines it emitted, he deduced Kirchhoff's law: emissivity and absorptance are in constant proportion for each wave length and temperature. This proportion J is a (mathematical) function of wave length and temperature. Physically spoken J stands for the emissivity of the black body: "The quantity named J is a function of wave length and temperature. It is a task of great importance to find this function. Several difficulties stand in the way of its experimental determination. Nevertheless, there is reason to hope that it might be found by experiment because this function doubtless is of a simple form, like all functions are which are not dependent on properties of individual bodies, and which are known by now."

So in 1860 already Kirchhoff had referred emphatically to the universal importance of the black body's emission characteristic. It was not until 1900 that Max Planck succeeded in finding the function's exact shape (Planck's radiation formula). The universal radiation curve here includes the universal constant h.

[30] *Hermann von Helmholtz*, b. Aug. 31, 1821, Potsdam–d. Sept. 8, 1894, Berlin-Charlottenburg. Although devoted to physics quite early in life, he studied medicine to earn a living as medical officer. Having been influenced by Johannes Müller, he studied physiology and made several important discoveries in this field, like the origin of nerve fibres (1842), and the rate of speed of nerve stimulation (1850). Helmholtz formulated the energy principle in his essay "Über die Erhaltung der Kraft" which he presented for the first time on July 23, 1847, in a speech at the Berliner Physikalische Gesellschaft. His thoughts were based on kinetic heat theory and the assumption of newtonic central forces between atoms.

During the 1880s the energy conservation law was regarded as the most important law of nature, high spot and keystone of physics. Being one of its founders, Helmholtz gained a fame not unspoilt by arguments about its priority. In the middle of the century its importance was familiar only to a few colleagues. However, Helmholtz's invention of the ophthalmoscope in 1850 caused a sensation. He got appointments to chairs in Königsberg (1849), Bonn (1855), and Heidelberg (1858). At these universities H. taught physiology, a science from which he and his friend Emil Du Bois-Reymond "drove the forces of life," i.e., he shaped physiology into an exact science.

Not until 1871 did Helmholtz get a chair for physics at Berlin university. There the largest faculty for physics in the whole new German empire was established especially for him (1878). "It is not easy for the younger generations of physicists living here and now to form an idea of the paramount position Helmholtz held in science during the last decades of his life. There has most probably been no physical problem he had not thought out and formed an opinion about how to deal with it" (Willy Wien, 1921). In particular these were the theory of hydromechanics (Helmholtz's equations for vorticity, 1859), physiological optics, the mathematical approach to problems of acoustics, experimentally and theoretically large-scale works to test Maxwell's electrodynamics (from 1873 onward), thermo-chemistry (1882–83), tests on minimum principles (1884–94), etc.

The conservation property of Helmholtz' vortices ("vortices can't come into being, and they can't die") led William Thomson to construct a corresponding model of the atom (*vortex atom*). In general Helmholtz's attitude towards atomicity was rather negative, in spite of the well-known quotation from 1881: "If we suppose atoms to be the basis of chemical elements, we can't help but deciding that electricity, both positive and negative, is divided into certain elementary quanta."

With an exemplary sense of duty did the Chancellor of German physics look after his tasks which included everything from looking after experimental works to socializing at Court. Like Du Bois-Reymont Helmholtz tried to include natural sciences in the contemporary world of thought. In 1888 he became principal of the newly-founded Physikalisch-Technische Reichsanstalt Berlin-Charlottenburg. On the occasion of his 70th birthday, Kaiser Wilhelm II sent a congratulatory message reading "Dedicating your whole life to the benefit of mankind, you have made a great number of wonderful discoveries. Your mind, which has always been striving for the purest and highest ideals, in its flight outstripped all the hussle of politics and parties connected with it."

[31] *James Clerk Maxwell*, b. June 13, 1831, Edinburgh–d. Nov. 5, 1879, Cambridge. His father whom he loved very much was lord of the manor and an eccentric. After the early death of his mother, whose family name was Maxwell, his father sent Maxwell to the best schools. Maxwell was a student of mathematics and physics in Edinburgh for three years. In 1854 he graduated in Cambridge. Here he published his first essay just one year later. It already dealt with what was later on called Maxwell's equations. In 1856 he got a chair in Aberdeen, and from 1860 onward he worked at King's College, London. Like Hermann Helmholtz, Maxwell dealt with physiology of color vision, and he extended Young's three-color theory. Maxwell's work on electrodynamics were epoch-making. In them he shaped the rather intuitive thoughts of Michael Faraday into exact mathematical structures and established the physics of fields. The complete Maxwell equations were published in the *Philosophical Magazine* (1862), entitled "On Physical Lines of Force." When introducing displacement current M. went beyond Faraday: according to Maxwell a changing electric field in a condenser has to show magnetic effects—like electric current does. It was this assumption that led to the possibility of transversal electromagnetic waves. In 1864 Maxwell wrote about mathematically calculated spread velocity: "This velocity is so nearly that of light, that it seems we have strong reason to conclude that light itself (including radiant beat, and other radiation if any) is an electromagnetic disturbance in the form of wave propagated through the electromagnetic field according to electromagnetic laws."

The "Treatise," published in two volumes in 1873, summarized all of Maxwell's earlier works. Maxwell's equations were here presented in a much more complex form, and it was not until Heinrich Hertz and Oliver Heaviside that the original version was used again. It took several decades for the Maxwell equations to become fully understood and acknowledged. But from then on Maxwell's electrodynamics and mechanics (still based on Newton) became the heart of classical physics. Ludwig Boltzmann who had done quite a lot to introduce Maxwell's equations, admired their beauty and symmetry in part 2 of his "Vorlesungen über Maxwells Theorie." As an epigraph for this work he quoted Goethe: "War es ein Gott, der diese Zeilen schrieb?" (Has a God been writing these lines?)

In the field of kinetic theory of gases, Maxwell pioneered new developments, too. He took up the approaches of August Karl König and Rudolf Clausius again. But whereas they had only looked at the mean velocity of molecules, Maxwell was interested in the individual velocity of every single material particle. He discovered what today is called Maxwell's distribution of velocity, and thus established statistical physics. To Ludwig Boltzmann these essays gave the impression of being revelations. In the time that followed, the two of them built up the field, criticizing and encouraging each other. Being a pioneer of kinetic theory of gases, Maxwell was a convinced supporter of atomistics. In a programmatic speech given in 1871 at the British Association for the Advancement of Science, he expressed his conviction that atoms are absolutely constant realities, and that therefore atomic standards should be taken as basic units for mass, length, and time.

Maxwell stopped teaching at King's College in 1865 for health reasons. Due to his estate he was financially independent. Free of academic duties he continued research as a scholar and wrote the thick manuscripts of his works which were published in the early 1870s. When he was offered professorship at St. Andrews, Scotland's oldest university, he turned the offer down. But when Cambridge university founded a chair for experimental physics and—for the first time ever in England—equipped it with a big teaching lab, Maxwell took on the job which was so important for science in Britain. (Until then the physical lab of William Thomson at Glasgow university had been the only one in Britain.) Construction and equipment of the so-called Cavendish Laboratory (after the main financial backer) took up a lot of time. In the long run, however, Maxwell founded modern education and the famous tradition of experimental physics in Cambridge.

[32] *Friedrich Wilhelm Georg Kohlrausch*, b. Oct. 14, 1840, Rinteln–d. Jan. 17, 1910, Marburg. The physicist was the son of Rudolf Kohlrausch (b. 1809–d. 1858) who had carried out the "Zurückführung der Stromintensitätsmessungen auf mechanische Maße" (putting measurings of intensity of current down to mechanical measurements) together with Wilhelm Weber in 1856, and so had done basic work for the development of the science of electricity. Kohlrausch had chairs in Göttingen (1866), Zürich (1870), and Straßburg (1888). After Helmholtz' death in 1895 Kohlrausch became principal of the Physikalisch-Technische Reichsanstalt in Berlin-Charlottenburg. He did important experiments with conductivity of electrolytes, found the law of the independent migration of ions (which is named after him), and precisely determined electromagnetic basic quantities. In 1870, Kohlrausch wrote a *Leitfaden der praktischen Physik* (Introduction to practical physics). Entitled *Lehrbuch der praktischen Physik* (3 vols, 1968) this work served as a model for all similar textbooks.

[33] *Ludwig Boltzmann*, b. 1844, Vienna–d. 1906, Duino (n. Triest). He was a student of physics at Vienna university, received a doctorate in 1866, qualified as a university lecturer in 1867, and got a chair for mathematical physics at Graz university in 1869. After that he had chairs in Vienna, Munich, and Leipzig alternately. Svante Arrhenius, Walther Nernst, Fritz Hasenöhrl, and Lise Meitner were only some of his students. Boltzmann was quite a versatile physicist. As a young man he had successfully worked in the field of experimental

physics (15 years earlier than Hertz did he bear out Maxwell's theory of light by proving the correlation which Maxwell had asked for between refractive index and permittivity in sulphur). Ernst Mach, being an outstanding experimental physicist himself, called him "an experimenter hardly to be surpassed." Although towards the end of his life he tended to deal with rather philosophical problems (materialism vs. idealism), his main interest had always been theoretical physics: "For its praise no sacrifice is too big, it is the be-all and end-all of my existence."

The central problem of his theoretical life work was to trace back thermodynamics to mechanics. For this it was necessary to eliminate the contradiction between reversability of mechanical processes and one-sidedness of thermodynamic processes. He achieved this aim by connecting entropy S with probability of state W: an increase in entropy in a closed system corresponds to a transition from a less probable to a more probable state. The formula (to which Max Planck gave the form used today) which as embodiment of Boltzmann's life work decorates his tombstone on Vienna Central Cemetry reads $S = k \ln W$ (k being the Boltzmann atomic constant). Fritz Hasenöhrl commented on the formula: "The sentence that entropy is in proportion to the logarithm of probability is one of the deepest and most beautiful sentences of theoretical physics, in fact of all natural sciences." Development of modern physics would indeed very likely not have happened without this formula. It was the starting-point of quantum theory both in the version formulated by Max Planck in 1900 and in the extended version formulated by Albert Einstein (1905). Some other important achievements of Boltzmann are the Maxwell-Boltzmann law of energy distribution (derived from statistical considerations), and the theoretical foundation of the law of total radiation of black bodies, the latter having been discovered empirically by Josef Stefan, one of Boltzmann's teachers (Stefan-Boltzmann law, 1884).

Boltzmann was a resolute advocate of atomic theory. Meeting with just little response or even disapproval of renowned colleagues such as Wilhelm Ostwald, Ernst Mach, and Max Planck, was a heavy disappointment he could not forget. Although it certainly was a very productive working hypothesis in chemistry, his opponents refused to see atoms as something real. Boltzmann, however, insisted on their having a real existence and modified Galilei's famous words for this purpose: "I am sure that I can say about molecules: and they are moving nevertheless." Although he was the one to predict the appearance of disequilibria as a necessary result of the statistical theory (and who knew about Brownian movement), he did not live to see the final victory of atomic theory which was initiated by Einstein's theory of fluctuations.

Boltzmann had various interests in the field of arts as well. He had taken lessons in music with Anton Bruckner, he was an ardent admirer of classical literature (Shakespeare, Goethe, and esp. Schiller), and last but not least he has written one of the most humorous German travel stories, "Reise eines deutschen Professors ins Eldorado" (the journey of a German professor to Eldorado). It was perhaps due to a beginning decrease in mental abilities that Boltzmann, then aged 62, committed suicide during a summer vacation in Duino in 1906. Anyway, lack of appreciation can hardly have been the reason because by then he was quite famous.

[34]*Heinrich Hertz*, b. Feb. 22, 1857, Hamburg–d. Jan. 1, 1894, Bonn. Hertz started studying engineering in 1876 and was more than happy when his parents allowed him to study

physics instead. In 1878 he went to Berlin to Helmholtz and immediately won a gold medal at a university competition. Two years later he did a doctorate "On induction in rotating spheres," became assistant to Helmholtz, and worked on evaporation and condensation of liquids, on elasticity (where he among other things gave a clear definition of "hardness"), and on glow discharge. He looked into the physical nature of cathode rays, too, which had been quite a mystery to his contemporaries. But he could find neither the electrostatic deflection nor the reciprocal effect of the deflection already known, the effect cathode rays had on magnetic needles. Hertz' results led Helmholtz and several other scientists especially. in Germany to wrong attempts at an interpretation (like longitudinal ether waves). Hertz' last experimental setup in 1891 is once again devoted to this field. His observation that thin layers of metal are permeable for cathode beams helped his disciple Philipp Lenard (Lenard window) to finally find the solution.

In Kiel Hertz qualified as a university lecturer with "Experiments with glow discharge" which he had already done in Berlin. The year was 1883. He now intensively worked on problems of electrodynamics again. He proved the system of equations based on long-range forces to "surely be incomplete in its present state" and argued that the system—once consequently completed—had to result in Maxwell's theory. In 1886 Hertz succeeded Ferdinand Braun as professor at Karlsruhe polytechnic. Here he began the observations which finally led to the discovery of the Hertzian Waves. What he had had in mind with these experiments "was to check the fundamental hypotheses of the Faraday-Maxwellian theory." The older type of electrodynamics knew an electric field induced by a temporal change of a magnetic field only. Maxwell had added a symmetrical equation, namely a magnetic field induced by the temporal change of an electric field, the latter having a considerable impact only at very high frequencies of the changing electric field. Until then electric oscillation had been unknown. This is why pre-Maxwellian theories had managed to cope with the observations made. After the generation of very fast electric oscillations, Hertz discovered that they detached themselves from the oscillating circuit. On Nov. 13, 1886, he found the transmission of waves from a primary electric circuit to a secondary one, being 1.5 m apart. He had thus constructed sender and receiver of electric waves. On Dec. 2, 1886, he managed to tune the resonances of both circuits. As proof of electromagnetic oscillations Hertz in most cases used the optical observation of spark gaps. This led to the discovery of the photoelectric effect (1887) which was later on scrutinized in detail by Wilhelm Hallwachs, Lenard, and others. In the following time Hertz experimented with the physical nature of Hertzian waves and demonstrated their reflection (on metallic surfaces), their refraction (with prisms of pitch), their transversality, and their polarization. Thus it was a proven fact that electromagnetic waves and light waves corresponded with each other and differed in wave length only.

According to Hertz his discovery was important because it proved Maxwell's theory: "The totality of experiments described here for the first time ever proved the temporal propagation of a supposed force at a distance. This fact is the philosophical and in some sense the most important outcome of the experiments." However, he did not realize the possible technical uses, i.e., the beginning radiotechnology.

In 1899, Hertz succeeded Rudolf Clausius as professor of physics in Bonn. Here he worked on theoretical rather than experimental problems. In 1890 he gave a clear and concise description of electrodynamics in the essay "Über die Grundgleichungen der Elektrodynamik für ruhende Körper" (On the basic equations of electrodynamics of bodies at rest). In Germany at least this essay led to the final breakthrough of Maxwell's theories. From then on, Maxwell's theory was the system of equations given by Hertz. The problem of "Elektrodynamik für bewegte Körper" (electrodynamics of moving bodies) he was not yet able to solve. Although Hertzian mechanics—designed without the notion of force—was fascinating as logical system, it has never been used to any efficient purpose whatsoever. The "extraordinary, beautiful and beautifully written" (Sommerfeld) introduction to his *Prinzipien der Mechanik* (principles of mechanics) deals with epistemological foundations of physics and is quoted quite often this very day.

[35] *Samuel Thomas von Sömmering*, b. 1755–d. 1830, lived in Frankfurt as a physician and was a member of the Münchener Akademie. While doing anatomic studies, Sömmering realized that the blind spot in the eye was an insensitive part of the retina. Moreover, Sömmering was involved in the development of a telegraph. He proposed to use the electrolysis of water instead of a spark-over for telegraphing. ("*Über einen elektrischen Telegraphen,*" memorandum of the Munich Academy II, 1809–10). The isolation of wires by wrapping silk around them is an invention of Sömmering, too.

[36] *Johann Philipp Reis*, b. Jan. 7, 1834, Gelnhausen (n. Frankfurt/M)–d. Jan. 14, 1874, Friedrichsdorf (n. Homburg v.d.H.). Being a baker's son, the pioneer-to-be of telecommunication did not have the means for studying natural sciences. Thus Reis became apprentice in a painter's shop and educated himself in his favourite subjects mathematics, physics, and languages. After having intensified his studies with Rudolph Böttger and Adolf Poppe in Frankfurt, he in 1858 joined Garnier's public school in Friedrichsdorf as a teacher for natural sciences.

Step by step he managed to put into effect an idea he had already had almost ten years earlier, namely "to directly communicate tone language itself into the distance with the help of a galvanic current." Wilhelm Wertheim and others had researched the effect of tones on iron rod made in a current-carrying reel of wire. While Reis could use this effect, he did not have a useful pickup able to transform acoustic impulses into electric ones. To solve this problem, Reis said, he thoroughly studied the mechanics of the human ear: "Led to it by my physics lessons in 1860, I took up a work on hearing tools which I had begun much earlier. Soon afterwards I had the pleasure to see that my efforts had been successful. I had managed to invent a gadget making it possible both to demonstrate clearly and vividly the functioning of hearing tools and to reproduce sounds of all kinds through a galvanic current into any distance. This instrument I called 'telephone.'"

The centerpiece of an improved telephone—Reis had presented an earlier version at the "Physikalischer Verein zu Frankfurt" (Physical Society) where he had held a notable lecture on Oct. 26, 1861—consisted of a wooden cube with a conical drill-hole the small exit of which was covered by a membrane made of a pig's intestine. In the middle of this membrane Reis had stuck on a little platinum tile which was connected to the receiver by a galvanic element, whereas a platinum wire which lay on the platinum tile was directly connected to the receiver. With this apparatus Reis managed to reproduce the sounds of several instruments and "to a certain degree even the human voice" as contemporaries believably stated. The sensation stirred by it—in 1863 even Kaiser Franz Joseph and König Max (of Bavaria) had Reis demonstrate the telephone for them—was, however, nothing more than general curiosity for all technical oddities. Most of the scholars were sceptical; Poggendorff denied publication of one of Reis's essays in the *Annalen der Physik* because he took the invention for mere fooling around. It was a successful demonstration at the gathering of natural scientists in Gießen (1864) that finally led to some scientific acceptance.

To enable physicists to repeat the "interesting experiments with reproduction of sounds in distant stations", Reis and Wilhelm Albert, a mechanic, built and sold the gadgets in Frankfurt. However, they sold but a few—especially because the apparatus was technically inadequate and could only be handled with a lot of skill. In his further works, Reis did not succeed in improving the invention: he did not realize that the current may fluctuate according to the rhythm of the oscillation of the membrane but must be disrupted by it. His personal life ended in a tragic illness: due to tuberculosis, the man who had founded electric telephony, the means to transmit language, was unable to speak for the last months of his life.

Exactly two years later Graham Bell, professor for physiology of vowels at Boston university, applied for a patent on a transmitting apparatus which was the real beginning of telephoning.[HE].

[37] *Alexander Graham Bell*, b. March 3, 1847, Edinburgh–d. Aug. 1, 1922, Baddeck (Nova Scotia), teacher for deaf-mutes and professor for physiology of voices. In 1876 he invented a simple telephone with good electric fitting of the converter which in principle is still in use today. This explains why Bell is regarded as father of telephony instead of Reis, who had transmitted languages by electric means before Bell.

Bell's patent for telephones has been marketed by the Bell Telephone Company which was taken over by the American Telephone and Telegraph Company later on.

[38] *Guglielmo Marconi*, b. Apr. 25, 1874, Bologna–d. July 20, 1937, Rome. An unusual experimental skill, intelligence, and imagination made the second son of a Scottish mother a pioneer of wireless telegraphy. While his mother supported him from the very beginning, his father was quite reluctant when Marconi, after having dealt with physics and chemistry in school already, began to experiment with Hertzian waves for the transmission of information in 1894. By disconnecting the primary circuit of a coil (which fed a Righi-oscillator) with a key button, Marconi was able to generate and send out electromagnetic waves electrically. By 1896 Marconi had improved the sender to an extent that it managed to transfer signals over a distance of several miles. A very long wire stretched vertically and with a spark gap switched on served as an antenna. The weak spot of the gadget was the receiver, a battery-fed tube with a filling of powdered metal. As soon as electromagnetic waves hit the powder it became conductant and had to be brought back to its initial state by knocking before it could register the next signal. This apparatus (named coherer by Oliver Lodge) had been developed in Paris in 1890 by professor Edouard Branly. After an improvement was made by Marconi it served as the only useable receiver in wireless telegraphy for more than a decade.

In 1896 Marconi managed to give some quite impressive demonstrations in London and at the Bristol Channel and one year later in the harbour of La Spezia, too. The public became aware of the young inventor. In England the Wireless Telegraph Company (later named after Marconi) was founded. The Italian government, too, assured him of its support. So now he had more funds to improve transmission technique.

Already in March 1899 Marconi was able to bridge the Channel between England and France. The same year the managed a radio contact between two war-ships 74 miles apart. On Dec. 12, 1901; he for the first time ever succeeded in transmitting signals across the Atlantic Ocean with a new system partly based on inventions made by Nikola Tesla. This success produced clear evidence that the long waves he had used followed the curvature of the earth and thus made possible world wide radio communication. Wireless telegraphy began its rapid triumphal march. The new technology, closely linked to physics, owed its first bigger advancements not only to Marconi and Tesla, but to Karl Ferdinand Braun as well. Marconi and Braun received the Nobel prize for physics in 1909.

[39] *Willebrord Snellius (Snell von Rojen)*, b. 1580, Leyden–d. Oct. 30, 1626, Leyden, Dutch mathematician and physicist. He was the first to carry out a measurement of angles to determine the shape of the earth by ascertaining the length of the graduated arc with the help of triangulation. In the course of this, he found a solution of the resection as well. In the publication of his findings (*De terrae ambitus vera quantitate*, 1617) he introduced himself as "Eratosthenes Batavus." Independent of Descartes, he found the law of refringency (about 1620).

[40] *Isaac Newton*, b. Jan. 4, 1643, Woolsthorpe–d. March 31, 1727, Kensington. Newton was born after his father's death and grew up with his mother and grandmother. He attended the local village school, and afterwards the grammar school in Grantham, a small town nearby. His uncle, a priest, did not force him to take over the paternal farm, but encouraged him to follow his strong leanings towards mathematical studies, experiments, and manual construction. So at the age of 18 Newton attended Cambridge university and was lucky enough to meet Isaac Barrow, a maths teacher who fostered his talent. When in 1665 England was haunted by the Plague, Newton spent two years at home, in Woolsthorpe. Newton himself admitted to have been in the prime of his creativity at that time. His achievements both in infinitesimal calculus and in mechanics and optics had their roots in these years.

Prompted by Barrow, Newton dealt with optics. He tried to construct a telescope of sharper focus but failed to grind a lens without diffusion of the colors. To check the generation of these colors he tested the light with a prism. He had a sunray fall into the darkened room through a hole, converged it onto the prism with a lens, and was surprised that the band of colors appearing behind the prism was five times longer than it was wide. When holding a second prism reversely behind the first, the band of colors was converged again to a white pool of light. Another experiment revealed the reason for the lengthening of the colored picture. Newton put a board with a hole in it vertically into the band of colors in such a way that just a single colored ray came through the hole, and then he put a second prism in the way of the single-wave length ray. The ray was deviated once more but did not lose its color. Repetition of this experiment with each color showed that red rays were deviated the least and purple ones the most. These results correlated with the diffusion behind the first prism. Newton concluded that white light consists of rays of different refrangibility, and the color is inextricably linked with the refrangibility. Thus it seemed like there were no lenses without dispersion. So Newton decided to construct a reflection telescope instead. In order to get a grindable reflecting matter he began experimenting with smelting. After lengthy grinding he completed the first two reflection telescopes in 1668 and 1671. The second one connected him with the Royal Society.

In 1669 Barrow decided to devote himself to theology and arranged for Newton to become his successor as professor for mathematics. The position granted a modest outcome and time for research. So Newton edited the best geographical textbook, *Geographia generalis* by Bernhard Varenius, and continued experiments in the fields of mathematics, optics, and chemistry. Starting at Descartes and John Wallis, he founded infinitesimal calculus and the theory of series, and determined curvature and point of inflection of several curves. Variable quantities were named Fluenten by him, the speed of their change he called Fluxionen. He determined the first from the latter and vice versa.

In the field of optics, he dealt with the rainbow-colored rings which appear when two lenses lay on top of each other. He made detailed measurings and related each color with the depth of the layer of air between the two lenses (Newton's rings). The color fringes of the shadows of narrow bodies discovered by Grimaldi were scrutinized by Newton. He structured optics as a system of theories and did not publish it until 1704. The book ends with 31 detailed questions about unsolved problems. His aim was to penetrate the phenomena of nature with measures and ciphers.

Newton had his biggest success in astronomy. Galilei had maintained that a body, once put in motion, moves on accelerationless. A change in state of motion would only occur if a new force had an impact. The attraction of the sun seemed to lead the planets on their orbit while a counteracting centrifugal force seemed to prevent them from falling towards the sun. From Kepler's third law which links the orbital periods of planets to their respective distance to the sun Newton concluded that the attraction of the sun declined with the distance squared. "Drawing further conclusions from this, I compared the force necessary to keep the moon in its orbit to the gravitation of the earth-surface and realized that they are almost perfectly equal." So Newton assumed the attraction which makes bodies fall to extend to the moon and to deflect it from its straight motion. Since the moon is located in a distance of 60 radia of the earth, gravitational pull there was said to be 3,600 times weaker than on the earth-surface. It was possible to calculate the moon's "falling distance" per second from the size of its orbit and its orbital period. The result multiplied by 3,600 would have to equal the distance covered during the first second of a body's fall on Earth's surface. The number his calculation resulted in, however, was too small; apparently Descartes' whirls had some impact on the deceleration of the fall of the moon.

In 1682 Newton learned that Earth was bigger than hitherto supposed. A calibration done in France by Jean Picard led to the new result. So the moon was farther away than had been assumed, its orbit and the falling distance per second were longer. Now the equation to the potential fall region of bodies on Earth was correct. Descartes' whirls had no impact, and the motion of planets and moons could be registered mathematically. In unique creativity Newton completed his main work, the *Philosophiae naturalis principia mathematica*. While thinking about it permanently, he in a sudden inspiration (purportedly when watching an apple falling down vertically) realized that to calculate the gravitational pull of the mass of Earth the planet had to be thought of as unified in its center. Gravitation acts on a second sphere along the tie line of the two centers. The forces have to be in relation to the masses. Newton presented the manuscript at the Royal Society meeting on April 28, 1686, and the astronomer Edmond Halley was willing to carry out and finance the publication. He had known Newton for some time and knew the work to give solutions to problems he himself and other scientists had searched in vain. The work consisted of three parts. The first and second are based on Newton's axioms and deal with the motion of bodies mathematically, while the third part comments on the world system in a manner more readily comprehensible. The principle of universal gravitation was demonstrated, according to which two bodies attract each other in relation to their masses and in inverse proportion to their squared distance. Not only sun and planet, or planet and moon influenced each other, but sun and moon, or planet and planet, as well. Deviations of the lunar orbit and perturbations of the planets' orbits came from the changing positions towards each other. Newton differentiated the terms mass and weight. At the surface of Earth the centrifugal force of the rotating planet counteracts attractive forces. One and the same stone has to be heavier at the North Pole than it is at the Equator. While weight is reduced by centrifugal forces, mass stays the same. Newton imagined two pipes starting in the center of Earth and ending at the Pole and the Equator, respectively. The water in the pipe near the Pole would be heavier because its weight was not reduced by centrifugal force. For the water pipes to remain in equilibrium, the equatorial one had to be lengthened, the necessary relation between the arms being 230 : 229. Earth would have to be flattened at the Poles.

The sea-level rises and falls. It is high tide both on that side of the earth which rotates around the moon and on the one turned away from it, as well. To explain this phenomenon, Newton imagined the earth and its water on both sides as three bodies. The attractive force of the moon becoming less with increasing distance, the water on the side nearer to the moon is attracted stronger than the earth. The earth again is more attracted than the water on the side that is turned away. If sun and moon are in one line with earth (at new moon and at full moon) heavy spring tides occur while if they form a right angle with earth (at half moon) the compensating influences lead to neap tides.

Being convinced that comets respond to general attraction, too, Newton assumed that they rotated around the sun in flat, long ellipses. Halley, who made a list of 24 cometary orbits, noticed those of 1531, 1607, and 1682 to tally. It had been the same comet in all three cases and its return could be predicted for 1759.

In the years following these laborious tasks, Newton gave undivided attention to chemistry and theology. At that time english scholars acknowledged two fountains of knowledge: nature and the divine revelations of the Bible. Newton persistently worked at a biblical chronology, and at an interpretation of Daniel's prophecies and the apocalypse, and he studied the Church Fathers. John Locke attested that concerning the field of theology and knowledge of the Holy Scriptures Newton was second to none. It was his greatest pleasure to think that his achievement in natural sciences might help to strengthen the belief in a divine creature. He saw physics as a "contemplation of God on the basis of phenomena taking place."

Newton came back to the *Principia*. Its theory of the moon did not satisfy him. The moon rotates around the earth in 30 days—sometimes faster, sometimes more slowly. It is accelerated by the sun while moving towards it, and slowed down while moving away from it. However, due to influences of the different planets the motion of the moon is much more complicated. Nobody had observed the positions of the moon in greater detail than John Flamsteed from Greenwich. In 1694 he gave a list of them to Newton.

Soon after this an event occurred that snatched Newton away from research for some time. England was swamped by bad coins, and it seemed the best idea to mint new ones. Who was to take on the responsible office of supervisor? Having studied in Cambridge, the secretary of the Treasury Charles Montague remembered the specialist for metal alloying there and called the 53-year-old researcher to London. Newton gave the financial reform his undivided attention. It took three years to complete the task which was a success technically and financially. Newton got a promotion and was ennobled by Queen Anne.

The Mint had multiplied its output. Now Newton had some spare time again. In 1700 he invented the mirror sextant. The following year he published *Tables of quantities of heat and degrees of temperature*. In 1703 he constructed an enormously strong burning glass with seven lenses. It was the same year he became president of the Royal Society. Since the new king was prepared to pay the printing of a work, Newton remembered Flamsteed's catalogue of stars and had it published.

Earlier on Newton had had tiresome discussions with Robert Hooke. When finally publishing the *Fluxionsrechnung* in 1704 it came to some annoying arguments with Leibniz on the invention of differential calculus. However, there was light as well as shadow. In 1713 and 1726 Newton published extended new editions of *Principia*. Both times he was helped by supporters. The first time it was professor Roger Cotes from Cambridge, the second time Henry Pemberton, a London physicist. Their correspondence with Newton documents their untiring joint work.

The formative influence Newton had on eighteenth-century physics became visible more and more. The theory of gravitation proved its productiveness. On the continent it was fostered by Daniel Bernoulli and Leonhard Euler, by the French Alexis-Claude Clairaut and Jean le Rond d'Alembert, by Joseph-Louis Lagrange, and especially by Pierre-Simon Laplace, and in Germany by Carl Friedrich Gauss and Friedrich Wilhelm Bessel.

Although his achievements led to everlasting fame, hardly anything is known about the human being behind the physicist. David Brewer was the first to give a closer look to this part of Newton. In the twentieth century an intensified research has begun.[HE].

[41] *Christian Huygens*, b. Apr. 14, 1629, Den Haag–d. July 8, 1695, Den Haag, Dutch physicist and mathematician. Having studied jurisprudence initially, Huygens turned to mathematical research instead and published a treatise on probability calculus in 1657. At the same time he invented a pendulum clock. In March 1655 he discovered the first moon of Saturn, and a year later the Orion fog, and the shape of the rings of Saturn. At that time he was familiar with collision law and the law of central motion, too, but did not publish them until 1669—and even then not being able to give proof for them. In 1663 he was elected member of the Royal Society. Two years later, being a member of the newly founded French academy of sciences, he moved to Paris. From there he went back to the Netherlands in 1681. Having published two essays already—"Horologium oscillatorium" in 1657 and "Systema Saturnium, sive de causis mirandorum Saturni phaenomenon" in 1659—his main work came out 1673: "Horologium oscillatorium" (The pendulum clock)

contained both the description of an improved clock construction and a theory of physical pendulae.

Moreover it contains treatises on cycloids like isochrones and important theorems about central motion and central force. In 1675 he invented a clock with balance spring. The essay "Tractatus de lumine" was written in 1690 and contains the first kind of wave theory (law of collision) of light, with the help of which Huygens developed a theory of double-refraction of icelandic calcareous spar. It explains the spherically shaped spreading of action around the source of light by means of Huygen's principle.

[42]*Augustin Jean Fresnel*, b. May 10, 1788, Broglie (n. Bernay/Eure)–d. July 14, 1827, Ville d'Avray (n. Paris). In 1815 the French physicist substantiated the wave theory of light in detail for the first time. His experimental and theoretical works deal with diffraction, interference, polarisation, double-refraction, and aberration of light. To demonstrate interference Fresnel among other things invented a double mirror and a biprism. On a suggestion of Th. Young, Fresnel in 1821 developed a theory of transverse light waves. Furthermore, he constructed annular lenses still in use today in lighthouses.

[43]*Johannes Kepler*, b. Dec. 27, 1571, Weil der Stadt–d. Nov. 15, 1630, Regensburg. Having become acquainted with the heliocentric conception of the world while studying theology in Tübingen, Kepler became a dedicated supporter of this theory. In 1594 he went to Graz where he wrote an almanac with the usual astronomical, astrological, and meteorological forecasts. Although his first prophesies came out a success, he never stopped doubting astrology as a "foolish daughter of astronomy." However, success spurred him on, and his first scientific work, "Mysterium Cosmographicum" was published in 1596: "It had been mainly three phenomena the causes of which I tried to find out untiringly," Kepler said, "namely, number, size, and motion of planets." "I dared to do this" Kepler explained in the preface "because of the beautiful correspondence between motionless things, sun, fixed stars, space, and God Father, Son, and Holy Spirit." As far as motionless things are concerned, the sky is an image of the Holy Trinity. "Since motionless things acted like this, I had no doubt that moving ones would have this kind of harmony, too," Kepler reasoned, having in mind the six planets known at that time. After a long time of research and reasoning he was sure that the five regular or platonic bodies inevitably led to the existence of six planets. If the five regular bodies, i.e., cube, tetrahedron, dodecahedron, icosahedron, and octahedron, are built into one another in such a way that a sphere inscribed in one body equals the circumsphere of the next, the result are four spherical shells between the bodies. Adding the circumsphere of the biggest and the inscribed sphere of the smallest body leads to a total of six spherical shells or spheres. For 25-year-old Kepler this explanation more or less solved the secret of the universe. Moreover, it answered the question why there were six planets and not "twenty, or a hundred," and it solved the problem of the space between planets.

Kepler sent his treatise to Tycho Brahe who was impressed by Kepler's creative mathematical talent. When moving to Prague as imperial mathematician in 1559, Brahe convinced Kepler to follow him in 1600. Even before the telescope was invented, Brahe had gathered observational data with hitherto unknown exactness and had encouraged Kepler to analyze the motion of Mars: "I am sure it was a work of Providence .... Mars alone enables us to solve the secrets of astronomy which otherwise would be hidden forever" (Kepler).

Kepler's intention was to solve the problem of Mars's orbit in eight days. He even placed a bet on it. The eight days turned out to last several years, but at last Kepler managed to find a correspondence from observation and theory which was totally new: Kepler's laws—in their impact comparable to Galilei's law of falling bodies—established the beginning of modern natural sciences. Striving to understand the construction of the universe in impressive images and similes, Kepler realized the necessity to "link the phenomena perceived externally to the internal ideas and to assess their correspondence with each other."

So Kepler again and again checked if his ideas were in accord with the observations. At last he found a plausible formulation which led to a maximum allowance of eight sexagesimal minutes: "After the divine gift has given us Tycho Brahe as an observer so careful that his observations show the mistake in the calculations to amount to eight minutes, it befits us to gratefully acknowledge and make use of this divine favour, i.e., we have to take a lot of trouble to finally find the true form of celestial motion." Saying this, Kepler discarded his earlier ideas and looked for a new model. As a christian neoplatonist and neopythagorean he was convinced that God had created universe according to a geometrical building plan into which the observations fit perfectly. The essential turn towards Nuova Scienza consisted most of all in this demand made by Kepler. He himself rightly stated: "These eight minutes showed the way to renewal of astronomy." After having found the second law, he found what is now called Kepler's first law soon after Easter 1605. As had already been the case with Galilei's laws of falling bodies, it was much more important "to link the phenomena perceived externally to the internal ideas" than to arrive at the exact result.

As early as 1604, Kepler had published treatises on optics. When Galilei announced the invention of a telescope, and the first observations made with it, in his "Sidereus Nuncius" (1610), Kepler instead of being jealous wrote an enthusiastic "Dissertatio cum Nuncio Sidereo." The year 1611 saw the publication of Kepler's dioptrics which communicates a refraction law valid for small angles of incidence, the theory of Kepler's telescope (with two convex lenses). "The great figures stand on the stage of history with reversed roles: Galilei is the hero of the day because he holds the telescope which in reality belongs to Kepler instead, the latter being the optician, not the former" (Franz Hammer).

Kepler, who had succeeded Tycho Brahe as imperial mathematician in 1601, had the best years of his life in Paris. In 1611 he had to cope with heavy strokes of fate (like the death of his first wife). In 1613 he moved to Linz where he stayed till 1626. Here he published an extensive textbook (1618–20) on the Copernican system, the contents being structured in questions and answers. In 1619 "Harmonices mundi" came out. No other work of his reveals more clearly the way Kepler thought, felt, and foresaw. For him, emotional forces, too, were means to understand the secret of Creation. Everywhere in the planetary system Kepler saw harmonies "fitted to each other according to the highest decision and kind of carrying each other, all being parts of the same edifice." Kepler heard a celestial music of the spheres, "a constant polyphonious music (perceived by intellect rather than ear)."

It took quite a lot of work to complete the new tables of planets based on the new conception of the world. As these "Tabulae Rudolphinae" proved to be more exact than all other calculations, they got wider acceptance as well. However, their correspondence to experimental data was not attributed to the improved theory, but to improved mathematics instead.

Kepler had the "Tabulae" printed in Ulm. Afterwards he made several journeys and spent some time in Sagan (Silesia) where Wallenstein resided. In 1630 he went to the Regensburg Kurfürstentag in order to demand his salary of several years which the Kaiser hadn't paid till then. Kepler died there. He himself has written his epitaph which is difficult to translate: "Mensus eram coelos, nunc terrae metior umbras. Mens coelestis erat, corporis umbra iacet." (My mind strode across the skies, now I am measuring the depths of Earth; the mind was celestial, while here the earthly body rests.)

[44] *René Descartes*, b. March 31, 1596, La Haye–d. Feb. 11, 1650, Stockholm. Descartes was the son of a senior official at Brittany Parliament. He was educated at a Jesuit school and became a student of law. From 1618 onward he took part in several campaigns. Afterwards he travelled Europe for several years before settling in the Netherlands in 1628. In 1649 he moved to Sweden as a teacher of philosophy. Descartes's main achievement in the field of mathematics was the establishment of analytical geometry as put down in his *Géometrie* (1637). This work had an impact on further development of infinitesimal calculus.

[45] *Pierre de Fermat*, b. Aug. 8/17(?)/1601, Beaumont der Lomagne (n. Montauban)–d. Jan. 12, 1665, Castres (n. Toulouse). Since 1631 the French mathematician had been senior official at Toulouse Parliament. His contemporaries got to know about his mathematical discoveries by copies and hints in letters only. He worked out strict infinitesimal methods (determination of extreme values and tangentials, quadratures, etc.), dealt with problems of probability, and deduced the law of refraction of optics from the minimum principle (Fermat's principle): a ray of light reacts to reflection and refraction in such a way that it needs the shortest time possible for the complete distance.

While studying the mathematicians in the ancient world, Fermat came across axial geometry (beginnings of analytical geometry). His contemporaries lacked the understanding necessary to follow Fermat's insights concerning the theory of numbers. Tests on properties of divisibility led to Fermat's small law: if a is an integral number which cannot be divided by a prime number p, then $a^{p-1} - 1$ can be divided by p. Fermat's big law (or assumption, rather) says that the equation $a^m + b^m = c^m$ in integral numbers a, b, c cannot be solved for integral exponents m (m bigger than 2, as e.g. $3^2 + 4^2 = 5^2$). Before the first World War the mathematician P. Wolfskehl had offered 100,000 Mark for a complete proof. The price money, however, was devalued by inflation. Even until today the assumption could not be proved completely, i.e., for all numbers m, but the attempts supported the theory of numbers, anyway. Fermat's assumption that all numbers $F^q = 2^n + 1$ with $n = 2^q$ were prime numbers was proved wrong by L. Euler in 1732: $q = 5$ leads to $F^5 = 2^{32} + 1 = 6416700417$, which is no prime number.

[46] *Galileo Galilei*, b. Feb. 15, 1564, Pisa–d. Jan. 8, 1642, Arceti (n. Florence). After having got a classical and linguistic education at Vallombrosa Benedictine convent, Galilei attended Pisa university in 1581 where his father wanted him to study medicine. That Galilei realized the laws of pendulum motion in 1583 already while watching a swinging chandelier in Pisa cathedral is most probably nothing but a legend. One thing is certain though, and

that is that after having been interested in aristotelian philosophy and ancient physicians initially, Galilei soon turned towards mathematics, Euklid, and especially Archimedes instead. Following the latter, Galilei wrote his first independent treatises in 1585–86 on the center of gravity of bodies and on the construction of a hydrostatic scale.

On Guidobaldo del Monte's (an influential mathematician and physicist) recommendation the 25-year-old scholar got a chair of mathematics at Pisa university. There his duty mainly was to give lectures on Euklid's geometry and on the *sphaera*, i.e., on elements of the ptolemaic (geocentric) planetary system. Moreover, he dealt with the mathematization of the Aristotelian qualitatively oriented theory of motion in the same way Archimedes had mathematized statistics.

Galilei had always been especially interested in the problem of the falling body. He started criticizing the Aristotelian theory of motion according to which a heavy body is falling faster than a lighter one. Galilei was not the first to utter such a criticism. In *De motu* (On motion), a treatise written in 1590 but not published in his lifetime, Galilei came to the conclusion that a body would fall the faster the higher its specific weight was. He said he had proven several times that lead fell faster than wood from the top of a high tower. So it is but a legend, too, that Galilei had proved "all bodies to fall at the same speed" with experiments on Pisa tower already in 1590. In reality it took him two decades to find the correct phrase, and he did not find the law of falling bodies by experiment, but by thinking and using mathematics instead.

In 1592 Galilei went to the venetian university of Padua. The eighteen years he spent there were to become the happiest time of his life. In this epoch he succeeded in establishing his new theory of motion as a real new science, a nuova scienza, and in making important astronomical discoveries. Besides, he worked on technical problems.

The year 1609 marks the beginning of Galilei's astronomical discoveries. In the summer of 1609 he heard about the invention of a telescope enabling man to clearly recognize things from a great distance. He later said that "After I had heard about this, I went to Padua ... and thought about this problem. I solved it in the first night after my return, and the very next day I built such an instrument. Immediately afterwards I started building an even better one. Six days later I took the improved gadget to Venice where all the distinguished nobles of the Republic were most amazed by it."

He presented the instrument to the Venetian government and got a generous reward for it. So Galilei had not invented the telescope as such, but his version surely was better than any of the earlier ones. Anyway, he was one of the first to watch the skies with it, and thus, due to his excellent talent for observation, made surprising discoveries which were published in the latin essay "Sidereus Nuncius" (star message, 1610). He saw, among other things, that the moon has mountains and valleys just like earth, that the Milky Way consists of an enormous number of separate stars, and that four moons rotate around Jupiter. So, Earth was not the single center of all celestial circular motions as the geocentric system claimed.

In September 1610 Galilei made a decision which was to become fatal. He left Padua and the Republic of Venice—at that time being the only independent state in Italy—and went to Florence, then under the influence of Rome. But the love for his toscan home and the prospects of becoming grand ducal mathematician and philosopher without any lecturing duties dispelled all his doubts.

During the first years in Florence, Galilei was occupied with the important discovery of the changing sun-spots and the full line of the phases of Venus. From the observation that all phases of Venus, from syringe to the full planet, were visible in the sky, one just had to deduce its motion around the sun. This fact was inconsistent with the geocentric system but fit perfectly in the Copernican theory instead.

In 1616 the Holy Officium declared the theory of sun being the world's center while earth rotated around sun and its own axis to be philosophically foolish and absurd as well as theologically heretical because it contradicted the Holy Writ. On February 26, 1616, Cardinal Roberto Bellarmino warned Galilei neither to defend the Copernican theory nor to believe in it, and Galilei promised to do so.

In his manor house in Bellosguardo near Florence, Galilei then wrote his astronomical main work, the *Dialogo* (Dialogue on the two main systems of the world, the ptolemaic and the copernican). After having made some changes, Rome and Florence permitted publication. The book came out in Florence in 1632. It is written as a conversation between three people, meeting in the Palazzo Sagredo in Venice. The language is Italian vernacular because the author wanted to influence the lay public concerned to educate themselves. Although Galilei can not be seen as original creator of Italian scientific prose, he did a lot for its development.

More than anything else, the work was meant to both do away with the prejudice against the rotation of Earth's axis and against its motion around the sun, and to show that the processes of motion on Earth are compatible with a moving earth.

The *Dialogo* expressed a new religiousness which at that time had something heretical about it. Man could but realize parts of truths, whereas the divine spirit realizes all. Those fragments man realized mathematically, however, "equal divine knowledge in its objective certainty" Galilei said. Here, divine and human comprehension of absolute truths are identical.

Soon after the *Dialogo* had been published, selling it was prohibited by the Pope. Galilei's opponents had succeeded in emphasizing several incriminating arguments against the book, e.g., that when dealing with the Copernican theory it would leave the hypothetical point of view and thus violate a law from 1616.

Galilei was ordered to Rome. The aged scholar, who in his last questioning was threatened with torture, was accused of heresy and defence of theories contrary to the Bible. He was sentenced to recant Copernican theory and, formally, to prison. Although his intellect secretly rebelled, Galilei recanted.

He spent his old age in a manor house near Florence, bothered by illness and under inquisitorial supervision. Here he wrote his main work in the field of physics, *Discorsi* (Conversation about and mathematical presentation of two new sciences), published in Leyden, 1638. The two new sciences are the science of the strength (which is a field of technical mechanics), and the beginnings of the science of moving bodies.

Like the *Dialogo*, *Discorsi* is a dialogue written in Italian vernacular. Only the systematical presentations are given in Latin. The first conversation begins in the Venice Arsenal where Galilei had got a lot of incentives for his work. The discussion of the third and fourth day is dedicated to falling and projectile motion. The extremely important and new findings in this field went back to his time in Padua.

The *Discorsi* mark the beginning of classical physics. At first Galilei's new physics is kinematics rather than dynamics though, due to a lack of the right notion of force. Neither did he manage to universally formulate the law of inertia.

[47] *Friedrich Wilhelm Herschel*, or William Herschel, b. Nov. 15, 1738, Hannover–d. Aug. 25, 1822, Slough (n. Windsor). At the age of 19 Herschel came to England. In 1766 the musics teacher became organist and concertmaster in Bath. Six years later he was joined by his sister Caroline, a singer. At about that time he began reading books on astronomy. However, he wanted to see for himself and built reflector telescopes. Until 1782 music and astronomy balanced each other. After having discovered Uranus in 1781, Herschel was engaged by the king. He had to answer orders for telescopes. With a gigantic telescope built in Slough Herschel discovered two new moons of Saturn in 1786, with a smaller one six moons of Uranus. His sister Caroline became his assistant. She discovered eight comets.

Herschel not only built gadgets of unequalled quality and observed methodically and persistently, but developed an understanding for the world's structure on an intellectual level as well. In 1783 he proved that the sun and the planets move through space. Stars visible in an area approached by an observer seem to make way by going sideways. About 100 nebula had been known until then; he discovered 2,500 others. In 1785 he realized that all stars visible with one's bare eye together with the Milky Way add up to a flat lens-shaped nebula. He also observed that binary stars are distant pairs of suns circling around each other. In 1800 the physicist discovered the infrared.

[48] *Francesco Maria Grimaldi*, b. Apr. 2, 1618, Bologna–d. Dec. 28, 1663, Bologna. The year 1665 saw the publication of his *Physicomathesis de lumine, coloribus et iride* which had quite an impact on development of physical optics because it reported on the discovery of diffraction of light. This phenomenon had been observed before, but G. was the first to interpret it as a new kind of propagation of light.

Grimaldi's book, a thick folio, is a *Summa* of his contemporaries' knowledge of chromatics. His own new thoughts are scattered over the whole book in a way that makes it difficult to find the few sentences proving he knew white light to be composed of spectral colors. He posessed this knowledge since carrying out Newton's experiment in which all spectral components, when merged by a concave lens, led to white light.

It is hardly known that Grimaldi can be seen as a cofounder of wave-theory of light. He described the difference between the individual spectral colors in terms analogous to wave length. Grimaldi knew that white light itself contains all spectral colors. Its separation into the individual spectral colors was called dispersion by him.

Besides, Grimaldi dealt with astronomy. Together with Giovanni Battista Riccioli, his brother monk, he compiled a lunar map the nomenclatura of which has been adopted.

[49] *Leonhard Euler*, b. Apr. 15, 1707, Basel–d. Sept. 18, 1783, St. Petersburg. From 1727 to 1741 Euler was member of the newly founded Petersburg Academy of Sciences and from 1741 to 1766 of the Preußische Akademie in Berlin. From there he went back to Petersburg. He wrote about 800 treatises, among them 20 thick monographs. About 60% of these works deal with pure mathematics, the others with applied mathematics, i.e., mathematical natural sciences in the broadest sense. Since 1911 the Euler-Kommission at the Schweizerische

Naturforschende Gesellschaft has been publishing the *opera omnia*. By now 62 volumes from the lot of about 80 have been edited.

Euler, leading mathematician of his time, was old-fashioned (Otto Spiess) as a physicist. His own opinions regarding this matter can be found in *Briefe an eine deutsche Prinzessin*. This work was originally written in French between 1760 and 1762, published between 1768 and 1772, and soon translated into all languages of the civilized world. It deals with those topics the enlightened 18th century subsumes to "education," namely theology, philosophy, and theory of music as well as findings in the field of physics.

Like René Descartes before him, Euler tried to reduce all phenomena in nature to pressure and collision of smallest material particles; still clinging to the *esprit de système* long overcome by 18th-century rationalism, Euler had an almost naive trust in theory and calculus. A paradoxical mathematical result (the free fall of a body in an imagined shaft going through the center of the earth towards the antipodeans with the body—according to Euler—suddenly changing direction at the center) was explained by him in his book on mechanics (1736): "This seems not to correspond to truth .... But anyway, one has to put more trust in calculus than in reason here." As a result the acknowledged *mathematicus acutissimus* had to put up with mockery and criticism from king Friedrich II, Voltaire, Benjamin Robins, and others believing in empiricism. The field of physics nevertheless owed a big deal to his mathematico-theoretical works. He consequently used the new language of infinitesimal calculus to formulate mechanics and thus became one of the founders of analytical mechanics. The minimum principle is named after Maupertuis, but in essence it was developed by Euler who had formulated it earlier and more clearly than Maupertuis. Its form still valid today, however, goes back to Louis Lagrange who modified Euler's version. Euler's hope (which he had published too rashly) to have solved the famous three-body problem did not come true, but the formulations enabled him to develop a theory of Moon's orbit. Since this problem had an impact in practical life (like tracing geographical length on the high sea), Euler's treatises were awarded several times.

Connected with the establishment of variational calculus was the calculation of elastic curves, i.e., determining the shape of elastic wires and sheets the edges of which are partly or totally clamped in. Euler did not stop at the equilibrium conditions but dealt with the oscillation of the strings and sheets, as well. He made progress with hydrodynamics, too. He formulated the so-called Euler equations of ideal incompressible liquids, and demonstrated that they could be transformed into the Bernoulli equation under certain mathematical conditions. Moreover, the Euler equations led to an advancement in acoustics. However, he did not yet succeed in solving the problem of velocity of sound. Here Newton had found a value too large.

Euler treated light as he did velocity of sound, i.e., different from Newton Euler supported a wave theory. He believed the ether to be the carrier of (longitudinal) luminous vibrations. He tried to cast doubt upon the emission theory by dealing with dispersion, double refraction, and colors of thin wafers with the means of wave theory. His opinion found no support due to the fact that Newton's authority still continued to have an effect. Besides, Euler just passed over the really physical arguments like interference.

Euler also disproved a Newtonian equation according to which the existence of achromats were impossible, and he tried to give the construction of an achromatic system of lenses of different axial refractions. This, in turn, led the optician John Dolland who had disapproved of Euler's thoughts initially, to construct an achromatic telescope in 1758.

[50] *John Dolland*, b. 1706, London–d. 1761. The manufacturer of optical instruments developed achromatic lenses. The son of a Huguenot silk weaver took up his father's trade. At the same time he studied Latin, Greek, mathematics, physics, and astronomy. In 1752 he gave up weaving and started building optical instruments instead. He quickly gained a good reputation in this field. In 1758 he published his findings about refraction of light which had led him to the construction of achromatic lenses.

[51] *Joseph Fraunhofer*, b. March 6, 1787, Straubing–d. June 7, 1826, Munich. Fraunhofer devoted his whole life to optics with admirable success. Being a glass grinder's son, he took up this trade and later on worked in an optical institute in Munich. Due to his achievements he became a joint-proprietor, and the company achieved world-wide reputation. The talented mechanic improved the processes of grinding and polishing optical lenses as well as the manufacturing and melting of glass. He constructed objectives for telescopes and microscopes, developed both test procedures for chromatic aberrations, and for the rectification of the deviation of lenses from an ideal sphere. Another of Fraunhofer's talents, the ability to calculate optical constructions mathematically, helped him to calculate the equation for achromatic telescope objectives.

Fraunhofer discovered the existence of lines in the solar spectrum, and he entered 576 different lines in it. He named the main lines with the letters *A* to *G*, and determined the wave length of six of them. Fraunhofer discovered the dark lines in the solar spectrum independent from William Byde Wollaston. He published his findings in memoranda of the Munich Academy.

Apart from a wire netting, Fraunhofer used a diffraction grating scratched on glass (300 pieces per mm) to examine the spectra. He also developed new methods to observe diffraction of light in parallel rays. In spite of several difficulties Fraunhofer's work was appreciated. His first work on the diffraction of light earned him an honorary doctorate at Erlangen university. He was elected corresponding member of Munich Academy in 1817 and six years later he got a professorship and a job as curator of the physics gallery. In 1824 he was ennobled by King Maximilian, and was awarded the civilian order of merit. Due to falling ill with tuberculosis, Fraunhofer died at a young age.

[52] *Olaf Römer*, b. Sept. 25, 1644, Aarhus–d. Sept. 19, 1710, Copenhagen. He studied under the natural scientist Erasmus Bartholinus in Copenhagen. In 1672 he accompanied Jean Picard on a trip to Paris. There he started working at the new observatory as a colleague of Giovanni Domenico Cassini. His field of work was the perfection of astronomical instruments. In 1672 he built a micrometer suitable to observe eclipses, and in 1675 a telescope through which one could take aim from both ends. A year later he constructed a Gitterrohr (lattice tube) with a crosshair to observe the surface of the moon. In 1677 he built a *Jovialium*, and in 1680 a planetarium. In 1675 Römer calculated the ideal form of cogwheels. His discovery of the finite speed of light became famous. Galilei had pointed out the possibility of using the eclipses of Jupiter's moons to determine degrees of longitude, provided that the moments of the moon's entry into or exit from Jupiter's shadow cone could be calculated in advance. Cassini and Römer worked on this problem. The moons' diving

into the shadow could be observed while Earth was approaching Jupiter, whereas their reappearance was visible while Earth left Jupiter. The average time needed for two dives was shorter than the time between the reappearances. Römer observed the phenomenon to fasten with Earth approaching while it slowed down when Earth left again. In addition to this, he predicted eclipses to come from the orbital period of the innermost moon as it was visible at the smallest distance from Earth to Jupiter. This, however, led to delays which were the longer, the farther away Earth was from Jupiter. From the diameter of Earth's orbit and the longest delay Römer in 1676 calculated the speed of light. After the edict of Nantes was repealed, Römer left Paris. In 1681 he went back to Copenhage and became professor for mathematics; besides, the king enlisted him for matters of the Mint, hydraulic engineering, and navigational lights. In 1683 the system of measures and weights worked out by Römer was officially introduced in Denmark.

Copenhagen had an observatory on a *round tower*. In 1690 R. gave it scientific importance by putting in two big instruments. For his house he had a *Meridianinstrument* built. This he installed centrically swivelling the meridian plane. To check if the axis was correct, Römer invented an auxiliary gadget. This enabled him to observe the times of corresponding hights of a star on either side of the meridian line. The *Meridianinstrument* itself served to check the transit time of stars through the meridian. The english passage instrument of John Flamsteed and Edmond Halley was modelled on this. In 1740 Römer built two more instruments, one of them being the first meridian circle. He also developed an alcohol thermometer with freezing point and boiling point as fixed points.

From 1690 onward he was determining right ascension and declination of more than a thousand stars. After his death all his instruments were gathered in the tower. When this burnt down in 1728, all the gadgets were destroyed together with 14 volumes of observational data. Luckily, Römer's assistant and successor Peter Horrobow had drafted the mechanical instruments and described them in his *Basis astronomiae* (publ. 1735).

[53]*Thomas Young*, b. June 13, 1773, Milverton (Somersetshire)–d. May 10, 1829, London. At the age of two he was able to read, two years later he had read the Bible twice, and memorized long poems. He easily learned classical and oriental languages, and was a talented painter and musician. Under the influence of his great-uncle he studied medicine and in 1800 he took over his uncle's surgery. Simultaneously he was professor of physics at the Royal Institution from 1801 to 1804. He published his lectures in two quartos in 1807. In 1811 he became physician at St. George's hospital. From 1818 onward he worked as a secretary for the office of longitudes and was in charge of the Nautical Almanac. His treatises deal with all sorts of topics. He was a philosopher and a scientist at the same time.

The contribution Young made to the field of optics was immense. It began in 1793 with a work on seeing. How does the eye focus on distance or nearness? It must have been by changes of the shape of the lens. Why do we see colors? Young imagined the retina to consist of three different nerve elements mediating the primary colors. This theory of trichromatics was taken up again by James Clerk Maxwell and Hermann Helmholtz.

Already in *Outlines of experiments and enquiries respecting sound and light* (Philosophical Transaction 1800) Young emphasized the advantages of Huygens' wave theory. While reflecting Newton's experiments, Young in May 1801 discovered the principle of interference. He imagined two groups of water waves floating from the open sea into a channel at the same time. The mountainous waves increased each other when meeting. "But if the elevations of one series are so situated as to correspond to the depression of the other, they must exactly fill up those depressions, and the surface of the water must remain smooth .... Now, I maintain that similar effects take place whenever two portions of light are thus mixed; and this I call the general law of the interference of light." Young knew about the properties of sound waves since having pursued studies in the field of acoustics. He had observed beatings quite often, and he interpreted the phenomena connected with light in the same way.

On November 12, 1801, he read his Bakerian lecture "On the theory of light and colors." He started out from remarks made by Newton which supported the wave theory. Waves the light path of which differ by half a wavelength extinguish each other. If, however, the difference is a whole wave length or a multiple of it, the waves intensify each other. Thus Young interpreted the regular Newton rings developing around the point of contact of a lens with a glass plate. With monochromatic light Newton had observed that for each color the layers of air progressed to a thickness which did not reflect light anymore. According to Young reflection did not stop there. More likely than not the rays which were reflected by the surfaces of the layers of air extinguished each other when meeting. Light added to light might result in darkness. Young proved interference to be active in the ultraviolet part of the spectrum as well by having the invisible reflex fall on a sheet of paper soaked with silver solution.

In another work Young extended this explanation to the colored streaks which line the shadows of narrow bodies. Grimaldi had started to analyze this phenomenon. The shadow of a narrow strip of cardboard put in behind a gap of light showed light and dark lines running parallel. Young explained the light beams passing the sides of the cardboard to be like wave trains coming from the edges. They either intensified or weakened each other, depending on their respective distances, and generated the lines when crossing each other. When blue light was used, the lines were closer together than with red light. From this fact the respective wave length could be deduced. Both the refraction patterns and the colors of thin wafers led to the same result. Newton had not seen the relation between these two phenomena. Young put both results down to interference. In a third treatise he proved that the lines on the cardboard were always caused by the meeting of two lights.

When Young presented his findings, nobody believed him. Newton's emission theory was universally acknowledged and nobody in England was inclined to accept this new wave theory. Especially Lord Henry Brougham and David Brewster disapproved of it. Young replied (in his lectures, edition 1845): "It was to be expected that objections would be raised against such a system. But I am sure that as soon as a sincere discussion about the pure facts will have taken place, the science of optics will reap the benefits of the facts, even if the theory turned out to be false." At the same time in France Augustin-Jean Fresnel, heavily criticized by Laplace and Biot, discovered the wave theory, too. This event helped to do justice to Young. In 1816 Francois Arago told Young about new discoveries concerning polarization. A year later Young wrote to him that to explain these phenomena it was necessary to imagine luminous vibrations vertically to the direction of propagation.

Young was a genius in other fields of science, too. In 1805 he calculated the size of molecules in a treatise on cohesion. He also pioneered the calculation of insurances. His contribution to the deciphering of hieroglyphs was quite important. In 1799 the stone of Rosette had been found. It was covered with three lines of writing, namely hieroglyphic, demotic, and greek. Young had the idea that those hieroglyphs which were surrounded by a setting did equal the proper names of the greek inscription, so that in this case they stood for letters rather than words. Some time later Jean-Francois Champollion managed to decode the alphabet and the grammar.

[54]*Etienne-Louis Malus*, b. June 23, 1775, Paris–d. Feb. 23, 1812, Paris. After having been one of the first pupils of the Ecole Polytechnique in Paris, he attended the school of engineering in Metz. Then he became an officer in the French engineer corps. He took part in the campaign to Egypt and survived the Plague in Jaffa. After his return he worked as an engineer in military services in Lille, Antwerpen, and Straßburg. He was promoted to lieutenant colonel and became examiner at the polytechnical school. In his leisure time Malus occupied himself with theoretical optics, especially with refraction and double refraction.

In 1808 he discovered polarization by reflection. In an Icelandic crystal spar a beam of light divides into two rays, an ordinary and an extraordinary one. In a second crystal (equal to the first one) each beam is divided but once. This light is called polarized. The phenomenon became known when Malus tested sunbeams which were reflected from the windows of a neighbouring building into his appartment. He had the sun beams fall through a calcareous spar and observed but one of two beams to come out again. The light had been polarized by the reflection of the window pane. An experiment showed that when the shine of a candle fell on the surface of water at an angle of 52 degrees 45' to the vertical axis, the reflected light was polarized like those beams leaving a calcareous spar. On that day Malus established a new field of optics.

[55]*Sir William Rowan Hamilton*, b. Aug. 4, 1805, Dublin–d. Sept. 2, 1865, Dunsik. He was a child prodigy with an incredible talent for languages and mathematics. Already as a child he had learned 13 languages, among them Arabic and Sanskrit. At the age of 22, he became Royal Astronomer of Ireland, head of the observatory, and professor for astronomy in Dublin.

As a scientist he dealt with geometrical optics. On a purely theoretical basis he predicted a conical refraction of light to be possible in symmetrical crystals, and he was right. He applied the methods developed in optics to analytical mechanics. In this field he introduced the so-called Hamilton principle which was important for celestial mechanics. He demonstrated that geometrical optics (which is valid for short wave lengths only) could be deduced from wave optics by a mathematical limiting process. He thought of geometrical optics as parallel to classical mechanics. A century later, this thought led Erwin Schrödinger to formulate a complete analogy between mechanics and optics, i.e., to establish a wave mechanics equivalent to wave optics.

From 1833 onward Hamilton was interested in nothing but pure mathematics. He studied the solution of algebraic equations of the fifth degree. More than anything else he tried to transfer the calculation with complex numbers from the second to the third dimension. Since quaternions were most useful for these calculations, Hamilton researched them from 1843 onward. He wrote two books on this problem.

[56]*Ernst Karl Abbe*, b. 1840, Eisenach–d. 1905, Jena. Since 1857 Abbe was studying mathematics, physics, astronomy, and philosophy under Bernhard Riemann and Wilhelm Weber. In 1861 he did a doctorate on "Empirical establishment of the theorem of the equivalence of heat and mechanical work." Two years later he qualified as a university lecturer under Karl Snell in Jena. At the end of the 1860s he started working for Carl Zeiss, a mechanic at Jena university. Together they improved a microscope Zeiss had built. This cooperation had a determining influence on Abbe. It both determined his main scientific field of interest, namely, optics, and became the basis for his life's work in the field of social reforms.

In 1868 he discovered the laws of sines of optical images with limited inclination of the principal ray. In 1869 and 1870 he introduced the terms entrance pupil, exit pupil, and numerical aperture into optics. He realized that it was impossible to improve microscopes any further. This was due to the fact that the objectives reduced the entering beams of light, what in turn led to diffraction. In 1872 Abbe gave the size of the smallest structure still visible (the so-called resolving power of microscopes): $d = \lambda/(n \sin \alpha)$ ($\lambda =$ wavelength of the light used, $n =$ refractive index of the immersion system, $\alpha =$ angular aperture of the objective). By making use of the physical theory of light in the field of practical optics he developed methods to precalculate all constructional parts of a microscope. It was thus unnecessary to do a lot of time-consuming tests. Moreover, Abbe tried to improve the properties of the other materials needed for constructing optical instruments. For this reason he himself, Carl Zeiss, and Otto Schott founded the Glastechnische Laboratorium Schott und Genossen in 1882. After Zeiss' death in 1889, Abbe started carrying out his social ideas. In 1891, having made millions by then, he established the Carl-Zeiss-Stiftung. He donated all his wealth to this foundation. He even gave up ownership of the optical laboratories and of the Schott glass factory. According to Abbe the motives leading to these decisions were of a social rather than caritative nature. The worker's son had known from his own experience the unhappiness of the former traders, brought about by progressive industrialization. The worker totally depended on the capitalist employer. The only protection provided by the trade regulations was the *Proletarierrecht* (rights of the proletarians). In Abbe's opinion, this was too little. He replaced it by a contract which granted an eight hour day, profit-sharing, and rights to a pension. These rights were unparalleled at the age of Abbe and had an enormous impact as an important example of public-spirited enterprise.

Index

δ-function, 13
μ-mesonic atoms, 36

aberration equation, 473
addition theorem for spherical harmonics, 83
admittance, 396
advanced potentials, 420
Ampère's second law, 192
analytic, 102
analytic continuation, 109
analytic properties of the index of refraction, 387
anomalous dispersion, 326
associated Legendre polynomials, 82

Bessel's differential equation, 372
Bessel's inequality, 73
betatron, 242
Biot-Savart law, 186, 468
Bohr magneton, 220
Bohr radius, 36
Boltzmann's distribution law, 153
boundary-value problem, 45
Brewster's angle, 348

capacitance, 19, 41
capacitive resistance, 297
capillary rise method, 168
Cauchy-Hadamard formula, 109
Cauchy-Riemann differential equations, 102, 103, 114
Cauchy's integral formula, 107
Cauchy's theorem, 105
causality, 393
center of inversion, 57

chain-of-circle procedure, 110
charge conservation, 185
charge density, 126
circularly polarized wave, 310
classical electron radius, 38
Clausius Mossotti formula, 151
coaxial cable, 360
coaxial line, 265
coil, 197, 244, 263
comets, 268
completeness relation, 73
complete set of functions, 70
complex conjugation, 101
complex differentiable, 102
complex functions, 102
complex numbers, 100
complex plane, 101
complex potential, 115
conductivity, 254
conjugate momenta, 492
conservation law of energy, 477
conservation law of momentum, 477
conservation laws, 477
constituive equations, 252
continuity equation, 185, 251, 462
    in differential form, 251
contraction, 461
convergence in square mean, 72
convolution theorem, 78, 395
Coulomb force, 5
Coulomb gauge, 205, 410
Coulomb law, 4
Coulomb potential, 5
covariance, 462
covariant Lagrangian formalism, 482
Curie's law, 225

Curie tempertature, 226
current density, 185
current four-vector, 462
cylindrical capacitor, 21

damping time, 322
de Broglie relation, 308
deelectrification, 144
deelectrification factor, 144
deformation polarization, 132
determination of nuclear radii, 36
diamagnetics, 225
diamagnetism, 227
dielectric displacement, 128
dielectric polarization, 151
dielectrics, 132
dielectric susceptibility, 136
dipole density, 21, 128
dipole layer, 21
dipole moment, 21, 25, 86
dipole vector, 24
Dirichlet boundary conditions, 47
dispersion relation for electromagnetic
      waves, 303
   in vacuum, 305
dispersion relation in a conducting
      medium, 318
dispersion relations, 398
dispersive media, 381
displacement current, 251
displacement polarization, 152
distributions, 13
Doppler shift formula, 473
Drude's formula, 326

effective current, 292
effective voltage, 292
Einstein's summation convention, 259
electret, 134
electric field intensity, 7
electric motors, 245
electric potential, 11
electric susceptibility, 130
electromagnetic waves in matter, 316
electromagnetic waves in vacuum, 302
electrostatic energy, 156
electrostatic unit, 6
electrostriction, 163

elliptically polarized wave, 310
energy density of the electric field, 30
energy-momentum tensor, 475
essential singularity, 112
Euler-Lagrange equations, 478
extinction coefficient, 321

Faraday's Law of Induction, 237
ferroelectric crystals, 134
ferroelectrics, 130
ferromagnetics, 225, 226
field equations, 462
field tensor, 462
first Ampère's law, 187
four-dimensional Laplacian operator, 461
four-divergence, 461
four-gradient, 303, 461
Fourier analysis, 306
Fourier integrals, 75
Fourier series, 73
Fourier transform, 77
four-potential, 463
four-tensors, 460
four-vectors, 460
frequency dependence of the conductivity,
      322
frequency dependence of the index of
      refraction, 327
Fresnel's formulas, 339

gauge invariant, 410
gauge transformation, 205, 410
Gauss theorem, 11, 27
Gaussian method, 185
Gaussian positions, 184
Gaussian system of units, 6
general integral theorems, 27
generalized Fourier coefficients, 71
generalized index of refraction, 320
generalized momenta, 478
generalized Ohm's law, 253, 294
generalized permittivity, 317
Green function, 48
   of the Poisson equation, 66
Green's Theorems, 45

group velocity, 363

Hamiltonian function, 481
Hamilton's principle, 478
Helmholtz coils, 196
Hertzian dipole, 429
Hessian normal form, 304
hexadecupole deformation, 94
hexadecupole moment, 95
holomorphic, 102
Huygen's principle, 425
hysteresis, 226, 232
  loop, 233

identity theorem for holomorphic
    functions, 110
identity theorem for power series, 109
impedance, 291
index of reflection, 333
index of refraction, 326, 333
induce charge, 51
inductance, 291
induction coefficients, 277
inductive resistance, 291
inessential singularity, 112
invariants, 470
inversion with respect to a sphere, 57
isolated singularity, 112

Jacobi identity, 470
Joulean heat, 256

Kirchhoff's integral representation, 422
Klein-Gordon equation, 392
Kramers-Kronig dispersion relations, 399

Laplace equation, 12, 47, 114
Laurent series, 110
Laurent's expansion, 112
laws of reflection, 339
laws of refraction, 339
Legendre differential equation, 84
Legendre polynomials, 79
Lenz's law, 238, 288
Liénard-Wiechert potentials, 421
light waves, 379
linear motor, 247
linearly polarized, 305

line-integrals, 104
Lorentz force, 191
Lorentz gauge, 410, 463
Lorentz transformation, 458

magnetic
  bottles, 195
  charge, 231
  dipoles, 182, 223
  energy, 278
  field, 181
  field in matter, 222
  field intensity, 223
  flux, 237
  induction, 187
  mirror, 195
  moment, 181, 213, 219
  monopoles, 193, 270
  susceptibility, 225
magnetization, 222
magnetizing current, 223
mass spectrometer, 200
Maxwell's equations, 250
  in a covariant form, 470
  in differential form, 251
  in vacuum, 302, 409
Maxwellian stress tensor, 257
mean free path, 452
method of electric images, 50
method of images, 50
Minkowski space, 458
mirror nuclei, 37
molecular polarizability, 150
monopole moment, 85
multipole expansion, 70, 84
  in Cartesian coordinates, 88
multipole moment, 85
multipoles constructed of point charges,
    96

natural width, 449
Neumann boundary condition, 47
normal dispersion, 326

oblate, 94
octupole moments, 94
Oerstedt's law, 193
ohmic power, 256

ohmic resistance, 254
Ohm's law, 254
optical theorem, 402
orientation polarization, 152, 153
orthogonal functions, 70
orthogonality relation, 70
orthonormalization, 78
orthonormal system, 70
oscillating circuit, 267
oscillator strengths, 326

parallel-plate capacitor, 19
paramagnetics, 225
paramagnetism, 225
Parseval's equation, 73
penetration depth, 320
permeability, 224
permittivity, 136
    of vacuum, 6
phase velocity, 363
photon mass, 6
photons, 308
piezoelectric effect, 134
piezoelectricity, 134
plane of incidence, 341
plasma frequency, 323
point dipole, 23
Poisson's equation, 12, 47
polar molecule, 133
polarization, 129
    charges, 124
    density, 126
    plane waves, 309
    plane, 309
    vector, 127
ponderomotive forces, 162
power series, 109
Poynting theorem, 255
Poynting vector, 255
    of electromagnetic waves, 308
principal value, 398
principle of superposition, 253
prolate, 94
pyroelectric effect, 134
pyroelectricity, 134

quadrupole deformation, 94
quadrupole moment, 87

quasi-stationary currents, 276

radiation impedance, 437
radiation pressure, 262, 268
radius of convergence, 109
Rayleigh scattering, 451
reflection coefficient, 336
reflection laws, 333
refraction laws, 333
regular, 102
relativistic Hamilton equations, 493
relativistic Lagrange equations, 485
remanence, 226
residue, 112
residue theorem, 112
resonant cavities, 354
retarded Green function, 419
retarded potential, 420
right-hand rule, 188
Rutherford atom model, 443

saturation magnetization, 225
scattering amplitude, 402
Schmidt's orthogonalization, 78
self-energy, 31
self-energy parts, 29
self-induction coefficients, 41, 278
skin effect, 328
Snellius' law of refraction, 341
spectral function, 77
speed of light, 313
spherical capacitor, 20
spherical harmonics, 80
Stokes theorem, 18, 28
stream function, 115
subtracted dispersion relations, 402
sum rules, 400
superposition, 4, 8
surface charge, 17
surface current, 228
susceptibility, 132
systems of orthonormal functions, 73
systems of units, 6

Taylor's theorem, 109
telegraph equations, 316
TEM waves, 358
tensor analysis, 461

tensor forces, 92
TE waves, 358, 370
Thomson formula, 296
three-body force, 4
time-dependent Green function, 413
TM waves, 358, 369
total reflection, 346
total scattering cross section, 404
transfer function, 396
transformer, 299
trochoid, 490
two-body forces, 4

vector diagram, 293
vector potential, 205

wattless current, 292
wave equation, 303, 410
wave guides, 354
wave vector, 304, 306
Weiss domains, 130
Weyl's eigendifferentials, 16
work function, 53

Yukawa potential, 5